Sixth Edition

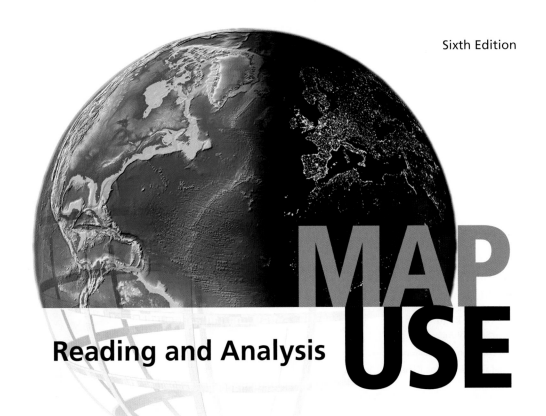

MAP USE

Reading and Analysis

A. Jon Kimerling

Aileen R. Buckley

Phillip C. Muehrcke

Juliana O. Muehrcke

ESRI PRESS ACADEMIC

REDLANDS, CALIFORNIA

ESRI Press, 380 New York Street, Redlands, California 92373-8100

Printed in the United States of America

Library of Congress Cataloging-in-Publication Data

Map use : reading and analysis / A. Jon Kimerling ... [et al.]. —6th ed.
     p. cm.
     Previous edition cataloged under Muehrcke, Phillip.
     Includes index.
     ISBN 978-1-58948-190-9 (hardcover : alk. paper)
     1. Maps. I. Kimerling, A. Jon. II. Muehrcke, Phillip. Map use. III. Title.
  GA105.3.M35   2009
  912—dc22   2008048032

Ask for ESRI Press titles at your local bookstore or order by calling 1-800-447-9778. You can also shop online at www.esri.com/esripress. Outside the United States, contact your local ESRI distributor.

ESRI Press titles are distributed to the trade by the following:

In North America:
Ingram Publisher Services
Toll-free telephone: (800) 648-3104
Toll-free fax: (800) 838-1149
E-mail: customerservice@ingrampublisherservices.com

In the United Kingdom, Europe, and the Middle East:
Transatlantic Publishers Group Ltd.
Telephone: 44 20 7373 2515
Fax: 44 20 7244 1018
E-mail: richard@tpgltd.co.uk

Cover design and production **Stefanie Tieman**
Interior design and production **Jennifer Hasselbeck**
Image editing **Jennifer Hasselbeck and Jay Loteria**
Editing **Michael Schwartz**
Copyediting and proofreading **Tiffany Wilkerson**
Media production **Jay Loteria**
Permissions **Kathleen Morgan**
Printing coordination **Cliff Crabbe and Lilia Arias**
Cover image courtesy of National Geophysical Data Center and NASA.

Contents

Acknowledgments

In previous editions of *Map Use,* we acknowledged the importance of contributions made by special teachers and colleagues, teaching assistants, and students at the University of Washington, University of Wisconsin, and Oregon State University. We continue to be grateful for the inspiration and assistance provided by these people. To these we add the valuable contributions made by ESRI.

We offer sincerest thanks to Jack Dangermond, president and founder of ESRI, who made the decision to purchase the rights to publish new editions of this work. Thanks also to Peter Adams, ESRI Press manager, and Judy Hawkins, ESRI acquisitions editor, for their continued support and encouragement and for their diligent effort to ensure the best publication possible, and to project editor Mike Schwartz who oversaw, with much attention, the editing of this book.

Others at ESRI Press worthy of mention include designer Jennifer Hasselbeck, who crafted a bright new look for the book while processing its many hundreds of images; copy editor Tiffany Wilkerson, who pored over the manuscript to catch even tiny grammatical errors or stylistic breaches; designer Jay Loteria for his hard work on the digital media; and editorial supervisor David Boyles, who patiently provided general team direction and thoroughly reviewed everyone's work prior to production.

ESRI Press contributors also included Kathleen Morgan, permissions and translations coordinator, who oversaw the process of gathering and citing sources for all illustrations; Kelley Heider, press projects coordinator and liaison to marketing and distribution; cartographer Michael Law; and editorial assistant Brian Harris.

We would particularly like to acknowledge the assistance of Fred Weston, an employee of Allan Cartography, who freely provided us a map illustration of the Willamette Valley shortly before his untimely death.

To our delight, hundreds of individuals have taken the time to comment on earlier editions of this book. Some of these people were merely lovers of maps; others were professors responsible for teaching introductory courses in map reading, analysis, and interpretation; and a number were students who had occasion to use the book in their studies. We were especially moved by letters from people who stumbled upon *Map Use* by accident at a friend's house or library and felt compelled to let us know how pleased they were with their discovery.

All these responses were gratefully received, and many were useful in crafting this improved sixth edition. We alone, of course, bear full responsibility for errors in the text or illustrations and for any controversial statements. This work reflects our deep love of maps and a desire to help others bring maps into their lives.

Acknowledgments would be incomplete without expressing gratitude to our families and many friends who helped us in so many ways. Particular thanks is given to Ann Kimerling, David Sandoval, and Ruth Buckley for their unending encouragement and patience during the creation of this sixth edition.

A. J. Kimerling
Corvallis, Oregon

A. R. Buckley
Redlands, California

P. C. Muehrcke and J. O. Muehrcke
Madison, Wisconsin

The essence of mapping is to let something "stand for" something else. In the special case of the geographical map, the representation is spatial and graphical, and the most common subject is the environment at or near the surface of the earth. These maps seem so intuitive and serve so many interests in modern society that it is easy to forget they are one of humanity's most sophisticated conceptual creations. Indeed, they may tell as much about how people think and communicate as they say about the nature of the environment.

To appreciate this last statement, consider what steps of abstraction are involved in making a map. First, we must be willing to think of ourselves as being separate from our surroundings. Second, we must be willing to deconstruct the environment into constituent parts that we then classify and label. Third, we must be able to gather meaningful and accurate raw data about the labeled features, attributes, and phenomena we have identified. Fourth, we must have the skill and capacity to process those data analytically to draw out essential characteristics. Fifth, we must have the capability and talent to manipulate and display the results graphically in a way that reveals something interesting or useful about the mapped environment.

Clearly, I have just outlined the essential components of a geographical information system (GIS). In fact, GIS technology has turned out to be a powerful mapping engine. It has greatly streamlined and facilitated the process of designing and producing maps. It has vastly expanded the range of topics mapped, as well as immensely increased the number of maps made each year. Most important, perhaps, it has greatly broadened the community of mappers.

For centuries, maps served primarily as a repository of geographical data. That role has been significantly diminished today by information-age technologies.

Foreword

Digital databases and digital storage media have largely superseded graphical map archives in our computer-oriented society. The plain reason is that digital information is far more convenient than maps to process, manipulate, and telecommunicate by electronic means.

Unburdened of this data storage role, increasingly maps are primarily designed to help solve problems. Maps are a conceptual aid found helpful by navigators, recreationalists, scientists, natural resource managers, civil engineers, business consultants, and policy makers. Almost uniformly, the need is for faster, more sensitive, and more robust problem solving and management. Fundamental to achieving these needs is greater integration of data collection, processing, and reporting that facilitates improved interaction with the data. Additionally, the detail and complexity of the tasks we face require more skilled problem solvers and managers than was true in the past. The challenges are likely only to get more difficult.

Naturally, this raises several questions: What does it take to use a map effectively? What special powers are needed to look at this greatly abstracted environmental representation and conjure up a useful image of the environment that was mapped? At the minimum it requires an ability to read the map. Mastering the graphical language of mapping requires a special set of skills. The process is far from transparent. Maps depicting raw data and concrete objects look similar to maps showing highly processed conceptual information. Thus, maps can easily mask the detail and complexity of the information mapped. The map legend is the key to unlocking information coded into these data-rich symbols. But most map legends fall far short of what the reader needs. The unfortunate result is that many people are not able to make sense of the highly processed data portrayals that are so characteristic of GIS mapping.

This book highlights these and other issues associated with using maps effectively. It is unique in the way it approaches the subject from a geographical-thought and spatial-communications perspective. The underlying theme is that maps do not merely show what is there, they are always more than what they seem. They are a window into how people think, adjust to their surroundings, and communicate spatially linked information with each other.

In pursuing these themes, the authors of this invaluable book place maps in the context of other representational systems, such as natural language, mathematics, music, art, and sculpture. They view mapping as a complex decision-making process where each decision has the potential to affect the final result in negative as well as positive ways. The authors stress that what you eventually view in map form is best seen as the answer to the following questions: What would the graphical display show if we made these mapping decisions? Would different mapping decisions produce a more insightful map? Can we learn to detect artifacts of the mapping process so we can avoid confusing them with environmental features? The authors also raise cautionary notes concerning the notion of accuracy when dealing with something as abstract as a map.

This book does an outstanding job of providing the keys needed to unlock the codes cartographers use to represent the environment. It is designed primarily for college-level students and instructors, as well as for professionals in a variety of environmental fields where maps are useful tools. It has also been written to appeal to casual map users who are looking for a comprehensive yet readable treatment of this important subject.

Jack Dangermond

Many books have been written on mapmaking, but since map use isn't simply the reverse of mapmaking, most are of limited value to you as a map user. By contrast, this book has been written for those who need to know how to employ maps as part of their spatial understanding of the world. Fully revised to better provide the context and demonstration of skills needed to read and properly analyze this complex form of communication, *Map Use: Reading and Analysis,* sixth edition, is a solid resource for an introductory cartography course.

Map Use is also a useful tool for people whose vocation or recreation requires knowing how to read and use maps because it takes the reader beyond visual representations and into the decision-making processes of cartographers.

Academics tend to treat maps as indoor objects, rarely including in their textbooks the fact that one of the most exciting ways to use maps is in the field. Conversely, military manuals and field guides to map and compass use have focused narrowly on wayfinding, virtually ignoring the role maps play in how we think about and communicate environmental information. In *Map Use,* sixth edition, we depart from this tradition by bridging the gap between these two extremes, pulling fragments of information from many fields into a coherent view of the environment. This book offers readers a comprehensive, philosophical, and practical treatment of map appreciation in three primary ways:

First, we define a map as a graphical representation of the environment that shows relations between geographical features. This encompassing definition lets us include a variety of important map forms that are otherwise awkward to categorize. Our definition should also accommodate any new cartographic forms developed in the future. Throughout this book, we have integrated discussions of standard planimetric maps, perspective diagrams, environmental photographs, and satellite images, rather than partitioning each into a separate category. This integrative approach, focused on commonality rather than uniqueness, showcases the enhanced environmental insight that these alternative mapping perspectives can inspire. It helps the reader understand that the definition of a map is very fluid. This fluidity lets us view our environment from countless perspectives to glean insights that would otherwise be lost to us.

Second, we have made a clear distinction between the tangible cartographic map and the mental or cognitive map of the environment that we hold in our heads. Ultimately, it is the map in our minds, not the map in our hands, with which we make decisions. Throughout the text, we stress that cartographic maps are valuable aids for developing better mental maps. We should strive to become familiar enough with the environment that we can move through it freely in both a physical and mental sense. Ideally, our cartographic and mental maps should merge into one so that our spatial understanding, communication, and behavior have the greatest chance of being tuned to the reality of our surroundings.

Third, where appropriate, we've made extensive reference to commercial products of special interest to the map user. A few years ago this would have seemed strange, since most mapping was done by large government agencies, but times have changed. The strong recent trend toward commercialization of all things cartographic now makes these products indispensable to efficient and effective map use. Computer software and digital data for mapping are being developed and sold by private industry. What you do with maps in the future will be strongly influenced by the nature of these commercial products.

Finally, we've worked to show that map use is relevant to daily life. Whenever possible, examples and illustrations have been taken from popular sources. Maps touch so many aspects of our daily lives that it is simple and natural to make points and reinforce ideas with illustrations from everyday communications. They are included in the text to demonstrate and reinforce basic mapping and map use principles.

The approach taken in *Map Use* leads to understanding the intentions or goals of a map, and ultimately to mapping choices—a depth of knowledge this book uniquely infuses in the reader. This contrasts with other books on the market that take a more simplistic mechanical view and treat the making and using of maps primarily as engineering and technical-illustration issues.

Learning to use a map in fact is a relatively easy and painless process, with an immense payoff: Maps let us see the environment from vantage points that are distant from us in time and space. Maps also let us visualize aspects of our surroundings that are intangible, imperceptible, or purely conceptual. Most important, maps let us focus our full attention on selected features by eliminating from the display environmental detail that would otherwise be distracting. They free us from all natural constraints, transcend our senses, and let us see our world anew.

its use for a wider audience. A natural way to do so is to improve GIS users' ability to think and communicate visually through the medium of maps.

Yet for all the technological advances, the philosophy underlying *Map Use* remains the same. As in earlier editions, we stress that a good map user must understand what goes into the making of a map. From mapmakers, we ask for little less than a miracle. We want the overwhelming detail, complexity, and size of our surroundings reduced to a simple representation that is convenient to carry around. We also want abstract maps to provide us with a meaningful basis for relating to the environment.

NEW TO THIS EDITION

The newest features of *Map Use: Reading and Analysis,* sixth edition, include nearly 500 four-color illustrations, an additional focus on online dynamic and animated maps, a comprehensive glossary of terms and text infused with references to GIS technology, whose maturation represents a notable transition.

When *Map Use* was first published, very little mapping was done in a computer environment. Today, not only is most mapping done with the aid of computers, but the map user is often the one who guides the mapping decision process. Especially with the aid of GIS software, the map user is the mapmaker. Even more significant, the map user can establish insightful dialogues with maps by manipulating the computerized (digital) data in various ways. To a large degree, GIS software is responsible for this exciting interchange between people, maps, and environmental data, and has changed the essential nature of map use to our immense benefit. At the same time, maps contribute greatly to GIS by providing a familiar visual interface through which GIS technology's powerful computing resources can be fully realized. Now that GIS technology is so widely available, it's appropriate to direct greater attention to enhancing

HOW THIS BOOK IS ORGANIZED

Map Use has been specifically designed and tested for use in a three-credit semester course of 15 weeks at the college freshman level. Material is presented at the upper high school to intermediate college level. It was designed for both the specialized and the general map user. It can be used equally as a basic reference work or as the textbook for a beginning map appreciation course. We have sought to cut through the confusing terms and details that characterize so many cartographic texts. To focus the reader's attention at the beginning of each chapter, we provide an overview of the most important concepts and how they fit together.

This edition focuses only on the topics of map reading and analysis. We haven't dropped the topic of map interpretation found in previous editions due to lack of relevance or interest. Far from it. It's through interpretation that we give meaning to what's mapped and, thereby, make sense of our surroundings and our place in the environment. For this reason, we've singled out the topic of map interpretation for special expanded treatment in future publishing initiatives.

We have structured the material in this book into two main sections under the headings "Map Reading" (part I)

and "Map Analysis" (part II). In most books, these terms have not had more than vague definitions and are often used interchangeably. Here they have been defined precisely, and the relationship of each to the other has been made clear.

The goal of part I, "Map Reading," is to develop an appreciation of how the mapmaker represents the environment in the reduced, abstract form of a map. In map reading one must mentally "undo" the mapping process. We discuss the geographical data that make up a map, the process required to transform that information from environment to mapping techniques, and map accuracy issues.

Once readers grasp the degree to which cartographic procedures can influence the appearance and form of a map, they're in a position to use maps to analyze spatial structures and relationships in the mapped environment. Part II, "Map Analysis," includes chapters on distance and direction finding. Here we explore position finding and navigation, including the use of GPS receivers, cartometrics, and spatial pattern analysis and comparison. With each of these topics, the concern is on estimating, counting, measuring, analyzing, and finding patterns in map features.

These two parts are followed by appendixes on digital cartographic databases, GPS terminology, and relevant mathematical tables. Each appendix is designed to complement material presented in the main body of the text. There also are resources on a companion Web site—www.esri.com/esripress/mapuse—that include Microsoft PowerPoint slides and lab exercises. Additionally, for instructors there will be a resource CD featuring lab keys and Microsoft PowerPoint slides.

Although a systematic development of subject matter is followed throughout this book, each section and chapter is autonomous from—and cross-referenced to—the rest of the material. This flexibility of design makes *Map Use* a useful text for instruction in the classroom, and a useful reference for practicing users. The book is organized to provide a logical development of concepts and a progressive building of skills from beginning to end. More experienced map users may wish to focus initially on sections or chapters of special interest and then refer to other parts to refresh their memories or clarify terms, concepts, and methods.

This book will have served its purpose if you finish it with a greater appreciation of maps than when you began. In even the simplest map, there is much to respect. Mapmakers have managed to shape the jumble of reality into a compact, usable form. They have done a commendable job. Now, as a map user, it is up to you.

AUTHORS' NOTE

This edition of *Map Use* marks several important transitions. First, it signals the end of an independent publishing enterprise by Phillip and Juliana Muehrcke. Back in the 1970s, the Muehrckes recognized that, although much had been written about how maps could be made, little was available to help people use these maps to enrich their personal and professional lives. They set out to fill this void by helping as many people as possible improve their ability to use maps in environmental thinking, communication, and navigation.

This called for a new style of publication, one that appealed equally to the casual and academic reader. To make the book accessible to the widest possible audience, Juliana and Phillip structured it around interesting examples and made it fun, easy to read, high quality, and relatively inexpensive.

Under the business name JP Publications, they published *Map Use: Reading, Analysis, and Interpretation* in 1978. Since then, that original volume has gone through five full editions and a number of intermediate revised editions. In later editions, A. Jon Kimerling joined the original authors.

It has been a long and successful journey, but things do change. So it is with a great sense of accomplishment and pride that JP Publications transfers publishing rights for *Map Use* to ESRI for this sixth and subsequent editions. ESRI's worldwide marketing and distribution capabilities will expose *Map Use* to a vast audience.

Also in this edition, Aileen R. Buckley has joined Juliana, Phillip, and Jon as a coauthor. Aileen studied with Jon Kimerling at Oregon State University as a graduate student. After receiving her PhD, she went on to several years of teaching and research at the University of Oregon, and then to a challenging career as a research cartographer with ESRI. There she has made outstanding contributions to GIS and cartography. She is recognized internationally as a top research scholar in private industry.

A. J. Kimerling
A. R. Buckley
P. C. Muehrcke
J. O. Muehrcke

INTRODUCTION

MENTAL MAPS

CARTOGRAPHIC MAPS

WHAT MAKES MAPS POPULAR?

FUNCTIONS OF MAPS

Reference maps

Thematic maps

Navigation maps

Persuasive maps

MAP USE

SELECTED READINGS

Introduction

It should be easier to read a map than to read this book. After all, we know that a picture is worth a thousand words. Everyone from poets to politicians works from the assumption that nothing could be easier to understand and follow than a map. The very term "map" is ingrained into our thinking. We use it to suggest clarification, as in "Map out your plan" or "Do I have to draw you a map?" How ironic, then, to write a book using language that is, supposedly, more complicated than the thing we're trying to explain!

The problem is that maps aren't nearly as simple and straightforward as they seem. Using a map to represent our detailed and complexly interrelated surroundings can be quite deceptive. This isn't to say that maps themselves are unclear. But it's the environment, not the map, that you want to understand. A map lets you view the environment as if it were less complicated. There are advantages to such a simplified picture, but there's also the danger that you'll end up with an unrealistic view of your surroundings. People who manage critical natural and human resources all too often make decisions based on maps that inherently are oversimplified views of the environment.

In this book, we'll define a **map** as a spatial representation of the environment that is presented graphically. By **representation,** we mean something that stands for the environment, portrays it, and is both a likeness and a simplified model of the environment. This definition encompasses such diverse maps as those on walls, those that appear ephemerally on a computer screen and then are gone, and those held solely in the mind's eye, known as **cognitive** or **mental maps.** You may envision the environment by using **cartographic maps,** which are what most people think of as traditional maps drawn on paper or nowadays displayed on computer screens. Or you can use mental maps, which often are made light of by many people although they are really the ultimate maps that you use to make decisions about the environment. Let's look more closely at mental and cartographic maps.

MENTAL MAPS

As a child, your mental map was probably based on *direct experience,* connected pathways such as the routes from your home to school or playground. You had a self-centered view of the world in which you related everything to your own position. The cartoon in figure I.1 graphically portrays this type of mental map. As an adult, you can appreciate this cartoon because you see how inefficient the child's mental map is. But the truth is that you will often resort to this way of visualizing the environment when thrown into unfamiliar surroundings. If you go for a walk in a strange city, you will remember how to get back to your hotel by visualizing a pathway like that in the cartoon. Landmarks will be strung like beads along the mental path. Even if you might be able to guess at a more direct route back, you may feel more comfortable following the string of landmarks to assure that you do not get lost.

Most of your mental maps are more detailed than this, however. For one thing, you take advantage of *indirect* as well as direct experience. You acquire information through TV, photographs, books and magazines, the Internet, and other secondary sources. You can transcend your physical surroundings and visualize distant environments, even those on the other side of the planet at different historical periods. Your mental maps become incredibly complex as they expand to encompass places and times you have never seen and may never be able to visit.

As you grow older, your self-centered view of the world is replaced by a **geocentric view.** Rather than relating everything to your own location, you learn to mentally orient yourself with respect to the external environment. Once you learn to separate yourself from your environment, you don't have to structure your mental map in terms of connected pathways. You can visualize how to get from one place to another "as the crow flies"—the way you would go if you weren't restricted to roads and other connected routes. It's your adult ability to visualize the "big picture" that makes the cartoon amusing.

Sharing your mental map with others, either in conversation or on maps that you draw, is much easier when you use a **geometrical reference framework,** or a framework on which you can easily describe and determine locations, distances, directions, and other geographic relationships. The system of **cardinal directions** (north, south, east, and west) is such a framework. You can pinpoint the location of something by stating its cardinal direction and distance from a starting location. You can say, for example, that the store is two miles north of a particular road intersection or the police station is 200 meters west of the courthouse.

This visualization of space is based on **Euclidian geometry,** the geometry you learned in elementary and high school. It's the geometry that says that parallel lines never cross, that the shortest distance is a straight line, that space is three-dimensional, and so on. The ability to visualize the environment in terms of Euclidian geometry is an essential part of developing a geocentric mental map.

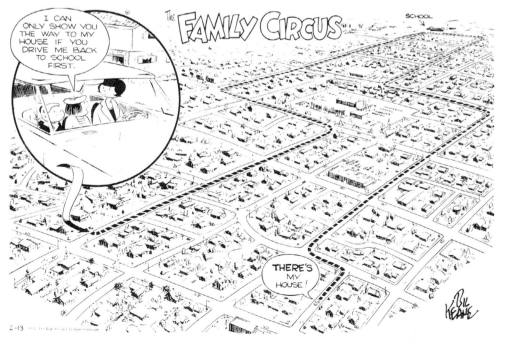

Figure I.1 **The geometry of a child's mental map is based on direct experience and connected pathways.**

Figure I.2   **The United States as seen through the eyes of a resident of Michigan's Upper Peninsula.**

But even if you develop mental maps based on Euclidian geometry, they will only be correct over small areas. That's because the earth is spherical, and the spherical geometry of the earth's surface is inherently non-Euclidian. (We'll see in chapter 1, for example, that north–south lines on the earth, called **meridians,** aren't parallel but, rather, converge to a common point at the poles.) Very few of us have well-developed spherical mental maps.

Even if you're able to visualize the world geocentrically in terms of Euclidian or spherical geometry, it's hard for most people to transform their mental map to a cartographic map in a geometrically accurate manner. Try drawing, from memory, a map of the area in which you live. The hand drawn map will tell you a great deal about the geometrical accuracy of your mental map. Not only will you probably draw the places you know best with the greatest detail and spatial accuracy, you'll probably draw things important to your life and leave off those that you don't care about.

Few people's mental maps correspond with cartographic maps. Figure I.2 shows the distorted visual image that a person from Michigan's Upper Peninsula might have of the country. Tongue in cheek as this map may be, it captures the fact that people visualize their own region as far more important than the rest of the world. In the same way, your mental maps emphasize your own neighborhood, with distant places less well visualized.

It's important to recognize these biases in your mental maps. The quality of your mental maps is crucial, because your behavior in the environment largely depends on them. You relate to your surroundings as you visualize them, not necessarily as they really are. If the discrepancy between your mental maps and the real world is great, you may act in self-defeating or even disastrous ways. Luckily, you don't have to rely solely on mental maps, since cartographic maps have been created for a multitude of places and features in the environment.

CARTOGRAPHIC MAPS

A **cartographic map** is a graphic representation of the environment. By graphic, we mean that a cartographic map is something that you can see or touch. Cartographic databases or digital image files are not in themselves maps, but are essential to current methods for the creation of maps. In a similar vein, an exposed piece of photographic film or paper doesn't become a photograph until it has been developed into a slide or paper print.

Cartographic maps come in many forms. Globes, physical landscape models, and Braille maps for the blind are truly three-dimensional objects, but most maps are two-dimensional line drawings or images of the earth taken from aircraft or orbiting satellites that have been cartographically enhanced. Cartographic maps have been carved, painted, or drawn on a variety of media for thousands of years, and printed maps have been produced for the last five centuries. Today you are just as likely to see maps displayed electronically on a computer monitor or other screen device.

What gives a graphic representation of the environment its "mapness?" Many mapmakers say that cartographic maps have certain characteristics, the five most important being the following:

1. Maps are *vertical or oblique views* of the environment, not profile views like a photograph of a side of your home taken from the street.

2. Maps are created at a reduced *scale,* meaning that there has been a systematic reduction from ground distance to map distance, as we will see in chapter 2.

3. Except for globes and landscape models faithfully representing the earth's curvature, maps are made on a *map projection* surface. A map projection is a mathematically defined transformation of locations on the spherical earth to a flat map surface, as we explain in chapter 3.

4. Maps are *generalized* representations of the environment. Mapmakers select a very limited number of features from the environment to display on the map, then display these features in a simplified manner. Insignificant features won't be shown, the sinuosity of linear features and area boundaries will be reduced, and several small ground features may be aggregated into a single feature on the map, as explained in chapter 10.

5. Maps are *symbolized* representations of the environment. The generalized features are then shown graphically with different map symbols. The mapmaker will use different point symbols, line widths, gray tones, colors, and patterns to symbolize the features, as we describe further in chapters 7 and 8. Names, labels, and numbers that annotate the map are also important map symbols.

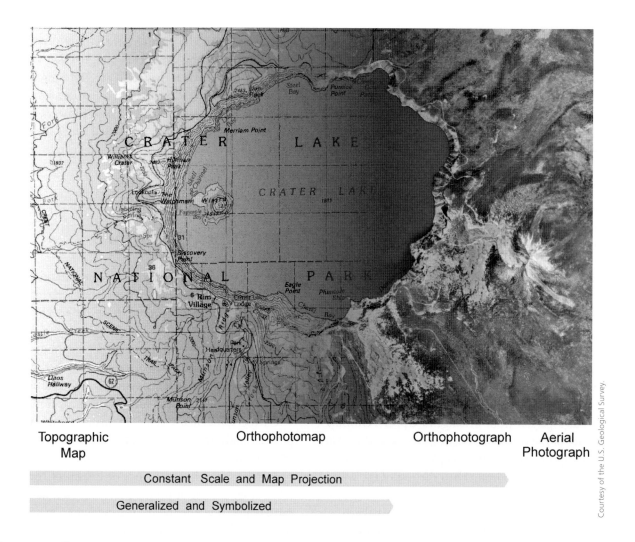

Courtesy of the U.S. Geological Survey.

Topographic Map   Orthophotomap   Orthophotograph   Aerial Photograph

Constant   Scale   and   Map   Projection

Generalized   and   Symbolized

Figure I.3   Different types of maps lie along a "mapness" continuum. Their position on the continuum is defined by how many characteristics of maps they possess. In the examples above, all are vertical views.

A cartographic map need not have all five characteristics of maps, but it should have at least one. You can think of different types of maps as being at different places on a "mapness" continuum defined by the degree to which they exhibit these four characteristics. This continuum is illustrated in figure I.3 for a gradation of map types depicting part of Crater Lake National Park in Oregon.

The **topographic map,** which shows elevations or landforms as well as a limited set of other features, in the left quarter of the illustration strongly reflects all five characteristics and is a good example of what most people think of as a map. The **orthophotomap** in the center of the illustration also has all five characteristics, since topographic map symbols have been printed over a geometrically corrected aerial photo called an **orthophotograph** (**orthophoto** for short). An orthophoto is corrected to a constant scale on a map projection surface, and hence has four of the five characteristics. Finally, the **aerial photograph** from which the orthophoto was made is not on a map projection surface and varies in scale with elevation differences on the ground. The aerial photo only has the characteristic of being a graphically represented vertical view of the environment, but is still a form of cartographic map. All the other maps are also vertical views.

As you can see from this illustration, there are multitudes of cartographic maps, each somewhere along the "mapness" continuum. The variety is so great that from now on we'll shorten the term "cartographic map" to simply "map," in accord with what you're used to hearing these products called.

WHAT MAKES MAPS POPULAR?

In scrutinizing the nature of maps, the obvious question is, "What accounts for their widespread popularity?" There are four main factors:

1. *Maps are convenient to use.* They are usually small and flat for ease of storage and handling. Thus, they bring reality into less unwieldy proportion for study.

2. *Maps simplify our surroundings.* Without them, our world often seems a chaos of unrelated phenomena that we need to organize in order to understand our environment. The selection of information found on a map, on the other hand, is clear at a glance. The world becomes intelligible.

3. *Maps are credible.* They claim to show how things really are. The coordination between symbol and reality seems so straightforward that we're comfortable letting maps "stand for" the environment. When we manipulate maps, we expect the results to apply in our surroundings. Maps, even more than the printed word, impress people as authentic. We tend to accept the information on maps without question.

4. *Maps have strong visual impact.* Maps create a direct, dramatic, and lasting impression of the environment. Their graphic form appeals to our visual sense. It's axiomatic that "seeing is believing" and "a picture is worth a thousand words."

These factors combine to make a map appealing and useful. Yet these same four factors, when viewed from a different perspective, can be seen as limitations.

Take convenience. It's what makes fast food popular. When we buy processed foods, we trade quality for easy preparation. Few would argue that the result tastes like the real thing made from fresh ingredients. The same is true of maps. We gain ease of handling and storage by creating a prepared image of the environment. This representation of reality is bound to make maps imperfect in many ways.

Simplicity, too, can be seen as a liability as well as an asset. Simplification of the environment through mapping appeals to our limited information processing ability, while at the same time it reduces the complexity and—potentially—the intricacies we need to understand. By using maps, you can reduce the overwhelming and confusing natural state of reality. But the environment remains unchanged. It's just your view of it that lacks detail and complexity.

You should also question the credibility of maps. The mapmaker's invisible (to you) hand isn't always reliable or rational. Some map features are distortions; others are errors; still others have been omitted through oversight or design. So many perversions of reality are inherent in mapping that the result is best viewed as an intricate, controlled representation. Maps are like statistics—people can use them to show whatever they want. And once a map is made, it may last hundreds of years, although the world keeps changing. For all these reasons, a map's credibility is open to debate.

Also be careful not to confuse maps' visual impact with proof or explanation. Just because a map leaves a powerful visual impression doesn't make it meaningful or insightful. A map is a snapshot of the portion of the environment at a point in time. From this single view, it is sometimes difficult or impossible to understand the processes that caused the patterns we can see on maps. For explanations you must look beyond maps and confront the real world.

FUNCTIONS OF MAPS

Maps function as media for the communication of geographic information, and it is instructive to draw parallels between maps and other communication media. You can first think of maps as a reference library of geographic information. Maps serving this function, called **reference maps,** are more efficient geographic references of the locations of different features rather than maps with a certain theme. Reference maps let you instantly see the position of features and estimate directions and distances between them. Explaining these spatial relationships among features in writing would take hundreds of pages.

Maps can also function like an essay on a particular topic. Like a well-written theme, a map can focus on a specific subject and be organized so that the subject stands out above the geographical setting. We call maps that function as geographic essays **thematic maps.**

Maps are *tools for navigation,* equal in utility to a compass or GPS receiver. When you get into your car and drive across your city, you are navigating the land. When you find your route on a subway or bus system, you are navigating a transit system. In the first case, you will use a **road map,** and in the second case you will use a **transit map.**

When you step into an airplane and fly to a distant city, you must do air navigation (assuming you are the pilot). And when you motor or sail between two destinations on a body of water, you are marine navigating. In these two cases, you will use **navigation charts** to plan your route in advance and to guide you on your trip.

Maps are also *instruments of persuasion.* Like a written advertisement or television endorsement, some maps are made to persuade you to buy a particular product, to make a certain business decision, or to take a targeted political action. These maps often are more sales hype or propaganda than a graphic representation of the environment, and you should view such maps with suspicion.

Let's take a closer look at maps with each of these different functions.

Reference maps

The earliest known maps, dating back several thousand years, are of the reference type. On reference maps, symbols are used to locate and identify prominent landmarks and other pertinent features. An attempt is made to be as detailed and spatially truthful as possible so that the information on the map can be used with confidence. These maps have a basic "Here is found..." characteristic and are useful for looking up the location of specific geographic features. On reference maps, no particular

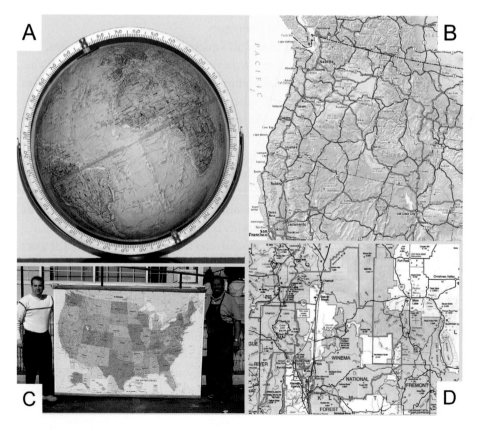

Figure I.4   Reference map examples include (A) a world globe, (B) the northwest section of a United States road and recreation atlas map, (C) the Terra Grande United States wall map, and (D) a section of the Oregon State Highway Map.

B. Courtesy of Benchmark Maps; C. Created by The Map Shop, ESRI business partner since 1997; D. Courtesy of Oregon Department of Transportation.

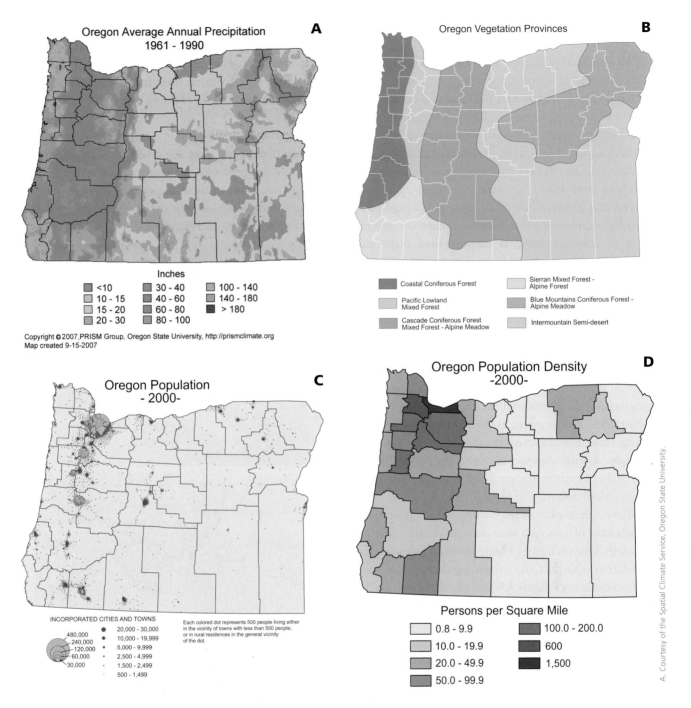

Figure I.5 Thematic maps for Oregon show **(A)** annual precipitation as colored regions of a continuous surface, **(B)** vegetation provinces as uniform areas, **(C)** rural and urban population by dots and graduated circles, and **(D)** population density as uniform colors within counties.

feature is emphasized over the others. As much as possible, all features are given equal visual prominence.

The topographic map and remote sensor images (orthophotomap, orthophotograph, and aerial photograph) in figure I.3 are excellent examples of reference maps, because they show a variety of phenomena with about the same emphasis given to each. Reference maps are often produced in national mapping series, such as

the U.S. Geological Survey (USGS) **topographic map series.** The topographic map segment in figure I.3 is from such a series.

Topographic maps show and label **natural features,** including mountains, valleys, plains, lakes, rivers, and vegetation. They also show **cultural features,** such as roads, boundaries, transmission lines, and buildings. One thing that distinguishes topographic maps from other map types is the use of **contour lines** to portray the shape and elevation of the land.

The geographic reference information on topographic maps and remote sensor images makes them useful to professional and recreational map users alike. They are used in engineering, energy exploration, natural resource conservation, environmental management, public works design, commercial and residential planning, and outdoor activities like hiking, camping, and fishing.

Globes and atlases are reference maps that show natural and cultural features in more generalized form than topographic maps. School wall maps are another form of reference map, as are the road maps and recreation guides produced for each state (figure I.4).

Thematic maps

Unlike reference maps, which show many types of features but emphasize no particular one over the others, thematic maps show a single type of feature that is the theme of the map. While reference maps focus on the location of different features, thematic maps stress the geographical distribution of the theme. A climate map (figure I.5A) showing how average annual precipitation changes continuously across the state of Oregon is a good example. A map showing the areal extent of Oregon vegetation provinces (figure I.5B) likewise shows the geographic distribution of physical features.

Many thematic maps show the geographic distribution of concepts that don't physically exist on the earth. One example is a map showing Oregon's rural population with dots and urban population with variations in circle sizes (figure I.5C). Another example is the Oregon population density map in figure I.5D. Although you can't actually see population density in the physical environment, maps showing the spatial distribution of such statistical themes are very useful to experts in demographics and other fields.

Thematic maps ask, "What if we wanted to look at the spatial distribution of some aspect of the world in this particular way?" Figure I.5D, for example, asks, "What if we wanted to look at population density by taking the 2000 population census totals for Oregon counties as truth, dividing each total by the area of the county to get a population density, generalizing the densities into seven defined categories, and representing each density category with a special symbol?"

Take another look at each map in figure I.5 and note how each theme is superimposed on a background of county outlines. Most thematic maps have similar background information to give a geographical context to the theme. Be careful not to use the geographical reference

information to find specific locations or to make precise measurements. Remember, that's not the intent of thematic maps; it's what reference maps are for. When using thematic maps, focus on their function of showing the geographic distribution of the theme.

Navigation maps

As you'll see in chapter 13, several types of maps are specially designed to assist you in land, water, and air navigation. Many of these maps are called *charts*—maps created specifically to help the navigator plan voyages and follow the planned travel route.

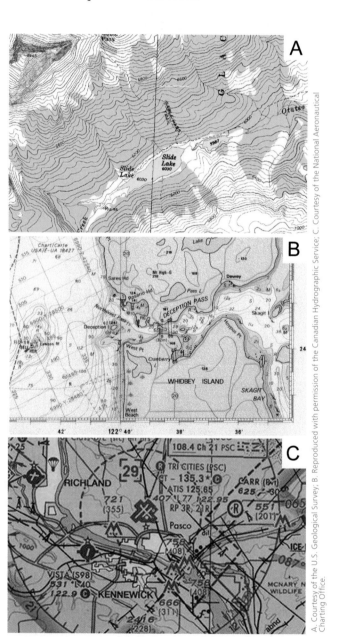

Figure I.6 (A) Topographic maps, (B) nautical charts, and (C) aeronautical charts are important tools of land, water, and air navigation, respectively.

A topographic map, such as the segment in figure I.6A, not only is a valuable reference map, but it also is one of the most important land navigation tools. Hikers, off-road vehicle enthusiasts, and land management professionals use topographic maps to find their way across the land. The topographic map shows ground features such as roads, trails, lakes, and streams that are both landmarks and obstacles. Contour lines on the maps are equally important, as they allow you to determine elevation changes and estimate the slopes you'll encounter along a route. This information will help you estimate the time and physical effort it will take to complete your trip. In addition, topographic maps are drawn on map projections that allow you to measure distances and directions

between locations along your route (in chapters 11 and 12, you'll learn how to make these measurements).

Nautical charts, such as the southeastern corner of the San Juan Islands, Washington, chart in figure I.6B, are maps created specifically for water navigation. Recreational and commercial boat navigators use the detailed shoreline, navigational hazard, and water depth information on the chart to plan the "tracks" that they will follow between ports or anchorages. As you'll see in chapter 13, each chart is made on a special map projection that allows you to quickly and easily measure the distance and direction of each track. Another type of nautical chart found in a *Current Atlas* (see chapter 13 for an example) gives you information about the currents you must deal with

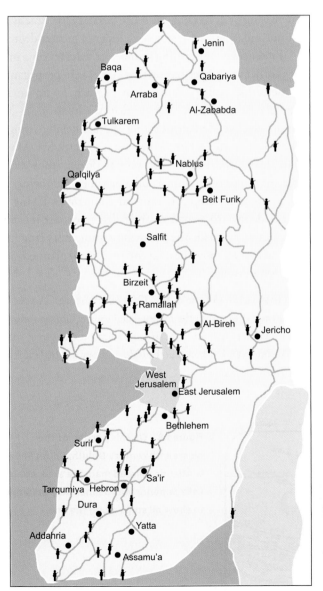
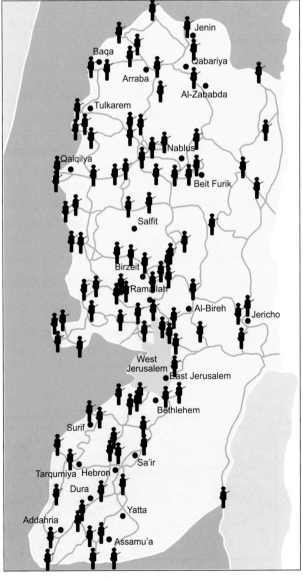

Figure I.7 Disproportionately large or small map symbols may create the impression that the West Bank was more or less crowded with Israeli military checkpoints than it actually was at the time when the information for the map was collected (2002).

on a particular day and hour. This information is of critical importance in planning your time of departure and estimating the time of arrival at your destination.

Aeronautical charts are maps designed for the air navigator. Figure I.6C is part of an aeronautical chart covering the state of Washington. Air navigation involves planning and following safe routes between airfields, and the chart is filled with information important to safe flying. Notice the detailed information shown on the chart for the Tri-cities airport. In chapter 13, you'll learn what these map symbols mean, and you will see how pilots use the information on the chart to find distances, directions, and travel times between destinations. Air navigation also involves maintaining a safe altitude above the ground, and you'll see that aeronautical charts show the heights of towers and other obstructions to navigation, as well as contours and special ground elevation symbols that help navigators quickly determine the minimum safe in-flight altitude.

Persuasive maps

Maps have always played a role in decision making, and mapmakers can deliberately try to persuade you to choose a particular product or support a certain position. Some of these persuasive maps distort or misrepresent to such an extent that they become **propaganda maps.** Such propaganda is common, especially on advertising, political, and religious maps. Since all maps distort reality, what could be easier than to make this distortion serve a special company, organization, or point of view? Unless we know enough to question every map, how would we suspect anything was wrong? Let's look at several examples of persuasive maps that either are or border on propaganda.

One type of propaganda involves disproportionate symbols as a means of persuasion. Mapmakers must make symbols overly large in relation to the size of the feature on the ground; otherwise, the symbols wouldn't show up at reduced map scales. In propaganda mapping, mapmakers carry this normal aspect of cartographic symbolization to extremes.

Take two maps of Israeli military checkpoints in the West Bank that we made from information gathered in 2002 (figure I.7) Notice on the map on the right the large soldier symbol for each checkpoint. By using such large symbols, we have made the West Bank appear crowded with military checkpoints. The map on the left has the same number and placement of soldier symbols one-quarter the size of those on the right map. Now the West Bank appears far less crowded with checkpoints. This is an interesting example because the two maps of the same information have opposite propagandizing effects of increasing or decreasing the sense of safety for Israelis, or of intimidating Palestinians to a lesser or greater degree.

Presenting a misleading number of features on a map is another tool of persuasion that grades into propaganda. Nineteenth century railroad maps such as figure I.8 are classic examples. Notice that the map scale has been selectively enlarged along the artificially straight main line from Duluth, Minnesota, to Sault Ste. Marie, Michigan. The map shows several dozen stops along the main and feeder lines, some of which are towns. By mapping every stop, the railroad company is trying to persuade investors, settlers, and riders to choose it over a competing railroad. The following commentary from the *Inland Printer* shows how far this practice of planned map distortion went:

"This won't do," said the General Passenger Agent, in annoyed tones, to the mapmaker. "I want Chicago moved down here half an inch, so as to come on our direct route to New York. Then take Buffalo and put it a little farther from the lake.

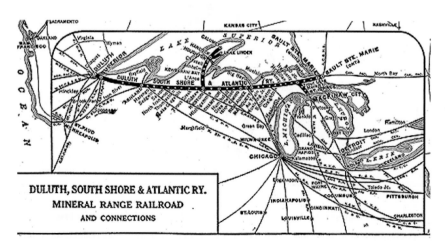

Figure I.8 On this portion of the nineteenth century Duluth, South Shore & Atlantic railroad map, the area along Lake Superior was deliberately enlarged to show all stops along the line.

"You've got Detroit and New York on different latitudes, and the impression that that is correct won't help our road.

"And, man, take those two lines that compete with us and make 'em twice as crooked as that. Why, you've got one of 'em almost straight.

"Yank Boston over a little to the west and put New York a little to the west, so as to show passengers that our Buffalo division is the shortest route to Boston.

"When you've done all these things I've said, you may print 10,000 copies—but say, how long have you been in the railroad business, anyway?"

(*New York Herald,* "Some Railway Map-Making," Inland Printer, vol. 15, 1895, p. 500)

Map simplification can also be used for persuasion purposes. Figure I.9 shows two maps of illegal West Bank settlements in the mid-1980s. Both maps come from the same weekly news magazine. One map depicts 16 settlements, while the other shows 30 settlements. Which is correct? The truth, revealed deep in the magazine's text, is that there were 45 illegal Israeli settlements in the West Bank region when these maps were made. The legend, at the least, should have provided this information. The makers of these maps, whether intentionally or not, presented a picture that favored the Israeli cause. You should never overlook the possibility for such political bias in mapping.

It pays to be especially cautious of novel or artful advertising maps, because they are invariably more eye-catching than factual. Consider figure I.10, made to promote Bend, Oregon, as the center of Oregon in terms of travel distance. Distance rings in 50-mile increments centered on Bend have been drawn on the map to show the travel distance to other locales. The idea of drawing equally spaced distance rings outward from Bend is absurd, since travel to the city is by road, not air. The

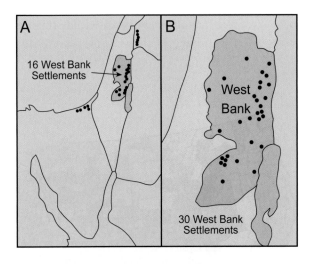

Figure I.9   The need for map simplification can easily be used to create a map that borders on propaganda. Here West Bank settlements on maps of two scales are shown.

Figure I.10   On this deceptive map, distances from Bend, Oregon, are shown by concentric distance circles spaced 50 miles apart. The travel distances the map shows are "as the crow flies," not by road.

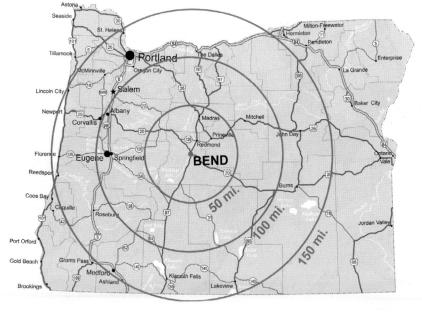

shortest road distance from Bend to Corvallis, for example, is 130 miles, not the 100 miles shown by the distance rings.

Whether you realize it or not, mapmakers are constantly molding your attitudes. Of course, they aren't the only people guilty of persuasion and propaganda. But the effect of map propaganda is especially insidious because so many people believe that maps are neutral and unbiased. The consequences are often dramatic: A year's vacation is ruined or a retirement nest egg is spent on a land parcel in a swamp.

MAP USE

Map use is the process of obtaining useful information from one or more maps to help you understand the environment and improve your mental map. Map use consists of three main activities: reading, analysis, and interpretation.

In **map reading,** you determine what the mapmakers have depicted and how they've gone about it. If you carefully read the maps in figure I.11, for instance, you can describe the maps as showing mortality from all types of cancer for white males and females from 1980 to 1990 within data collection units called Health Service Areas. Reading the main map legends, you learn that the mortality rate is the number of deaths per one hundred

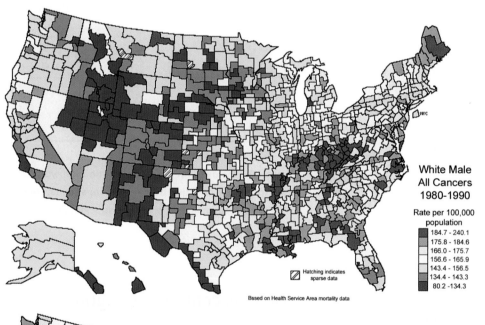

**White Male
All Cancers
1980-1990**

Rate per 100,000
population

	184.7 - 240.1
	175.8 - 184.6
	166.0 - 175.7
	156.6 - 165.9
	143.4 - 156.5
	134.4 - 143.3
	80.2 -134.3

Hatching indicates sparse data

Bssed on Health Service Area mortality data

Figure I.11 White male and female 1980–1990 cancer mortality rates in the United States by Health Service Area.

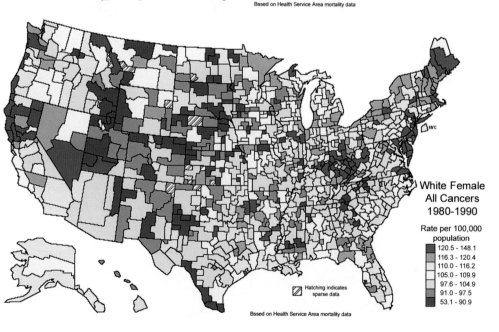

**White Female
All Cancers
1980-1990**

Rate per 100,000
population

	120.5 - 148.1
	116.3 - 120.4
	110.0 - 116.2
	105.0 - 109.9
	97.6 - 104.9
	91.0 - 97.5
	53.1 - 90.9

Hatching indicates sparse data

Bssed on Health Service Area mortality data

thousand people, and that mortality rates have been generalized into seven categories shown by a two-color sequence with lighter to darker tones for each color. Light to dark blue and brown represent low to high mortality on both maps, but the range for each category differs on the two maps. You need to discover this to understand that males have higher overall deaths from cancer. There is no explanation of how the category limits were selected, but the maps appear to have about the same number of Health Service Areas in each category. Finally, the smaller legends tell you that areas with sparse mortality data are shown with a diagonal line pattern, and you can see a few patterned areas on both maps.

Part I of this book examines these and many other facets of map reading. You will become familiar with map scale, map projections and coordinate systems, land partitioning methods, different ways of portraying landforms, maps that show qualitative and quantitative information, and ways of expressing map accuracy.

Learning to read the information on maps is only the first step. Your curiosity or a work assignment may lead you to go further and analyze the information on one or more maps. Part II of this book is devoted to **map analysis.** In this stage of map use, you make measurements and look for spatial patterns. We have seen that topographic maps and navigational charts are tools for the measurement of distances, directions, and surface areas.

Analysis of the spatial patterns on the maps in figure I.11 is particularly thought-provoking. While it's comforting to believe that cancer is unpredictable, analysis of these maps shows that's not true. If you focus on the patterns on these maps, you find that cancer mortality rates are far from random across the country. Regional clustering of high and low rates occurs for both males and females. You see a cluster of high mortality in Kentucky and western West Virginia (the brown counties to the right of the center) and low mortality in Utah and southeastern Idaho (the blue-green counties toward the upper left), for example. It's unlikely that this high and low clustering of deaths occurred by chance.

If you next focus on the spatial correspondence of mortality rates on the two maps, you find that eastern Kentucky and western West Virginia appear to have the highest cancer mortality for both men and women. The upper Midwest and Northeast, along with Northern California (the brown counties along the western coast), have the next highest overall male and female mortality. Some areas have high mortality rates for one sex but not

for the other. For instance, the Mississippi Delta region along the south-central coast has high mortality rates for males but not for females.

You may see clusters of high and low mortality on the two maps, but someone else may not see things the same way. Quantitative measures of spatial patterns on a map and spatial association among patterns on two or more maps add rigor and repeatability to your map analysis, and we have devoted two chapters (chapters 17 and 18) to these important aspects of map analysis.

After analyzing the maps in figure I.11 and finding spatial patterns of high and low cancer mortality, your curiosity may now be aroused still further. You may wonder how to explain the patterns and the spatial correspondences between the two maps. Finding such explanations takes you into the realm of **map interpretation.** To understand why things are related spatially, you have to search beyond the map. To do so, you may draw on your personal knowledge, fieldwork, written documents, interviews with experts, or other maps and images.

In your search, you'll find that cancer deaths are associated with many factors, including industrialized working environments, mining activities, chemical plants, urban areas, ethnic backgrounds, genetics, and personal habits of eating, drinking, and smoking. You'll find that people die from cancer because of contaminants in their air, water, food, clothing, and building materials. Some local concentrations of high or low cancer mortality, however, don't seem to fit this pattern, suggesting that there must be other causes or that people have migrated in or out of the area. In this book, we will focus on the first two of the map use processes—reading and analysis.

Since maps reflect a variety of aspects of environmental knowledge, map use is intertwined with many disciplines. It is impossible to appreciate them in isolation. The more different fields you study, the better you will be at using maps. Map interpretation grows naturally out of an appreciation of a variety of subjects. The reverse is also true. An appreciation of maps leads to a better understanding of the world around you; for the subject of maps, after all, is the world itself.

This brings us to a final, important point. As you gain an understanding of map use, be careful not to confuse the mapped world with the real world. Remember, the reason you're using maps is to understand the physical and human environment. The ultimate aim of map use is to stimulate you to interact with your environment and to experience more while you do.

SELECTED READINGS

Ames, G. P. 2003. Forgetting St. Louis and other map mischief. *Railroad History* 188 (Spring–Summer): 28–41.

Arnheim, R. 1969. *Visual thinking.* Berkeley, Calif.: University of California Press.

Balchin, W. G. V. 1976. Graphicacy. *The American Cartographer* 3 (1) (April): 33–38.

Castner, H. W. 1990. *Seeking new horizons: A perceptual approach to geographic education.* Montreal: McGill-Queen's University Press.

Dent, B. D. 1998. *Cartography: Thematic map design.* 5th ed. New York: McGraw-Hill.

Downs, R. M., and D. Stea 1977. *Maps in mind: Reflections on cognitive mapping.* New York: Harper & Row Publishers.

Gershmehl, P. J., and S. K. Andrews. 1986. Teaching the language of maps. *Journal of Geography* 85 (6) (November–December): 267–70.

Head, C. G. 1984. The map as natural language: A paradigm for understanding. *Cartographica* 21: 1–32.

Keates, J. S. 1996. *Understanding maps.* 2nd ed. Essex: Addison Wesley Longman Ltd.

Kitchin, R. M. 1994. Cognitive maps: What they are and why study them? *Journal of Environmental Psychology* 14: 1–19.

Lloyd, R. 1997. *Spatial cognition: geographical environments.* Boston: Kluwer Academic Publishers.

MacEachren, A. M. 1995. *How maps work: Representation, visualization and design.* New York: The Guilford Press.

Monmonier, M. 1991. *How to lie with maps.* Chicago: University of Chicago Press.

Monmonier, M., and G. A. Schnell. 1988. *Map appreciation.* Englewood Cliffs, N.J.: Prentice Hall.

Pickle, L. W., et al. 1996. *Atlas of United States mortality.* Hyattsville, Md.: U.S. Department of Health and Human Services.

Robinson, A. H., and B. Bartz-Petchenik. 1976. *The nature of maps: Essays toward understanding maps and mapping.* Chicago: University of Chicago Press.

Tufte, E. R. 1997. *Visual explanations: Images and quantities, evidence and narrative.* Cheshire, Conn.: Graphics Press.

Art, said Picasso, is a lie that makes us realize the truth. So is a map. We don't usually associate the precise craft of the mapmaker with the fanciful realm of art. Yet a map has many ingredients of a painting or a poem. It is truth compressed in a symbolic way, holding meanings it doesn't express on the surface. And, like any work of art, it requires imaginative reading.

To read a map, you translate its features into a mental image of the environment. The first step is to identify map symbols. The process is usually quite intuitive, especially if the symbols are self-evident and if the map is well designed. Obvious as this step might seem, however, you should look first at the map and its surrounds or marginalia, both to confirm the meaning of familiar symbols and to make sure of the logic underlying unfamiliar or poorly designed ones. This is usually found in the legend, but sometimes explanatory labels and text boxes are found on the map itself. Too many people look for the symbol explanation only after becoming confused. Such a habit is not only inefficient but potentially dangerous.

In addition to clarifying symbols, the map marginalia contain other information, such as scale, orientation, and data sources important to map reading, and sometimes includes unexpectedly revealing facts. But the marginalia are still only a starting point. The map reader must make a creative effort to translate the world as represented on the map into an image of the real world, for often there is a large gap between the two. Much of what exists in the environment has been left off the map, while many things on the map do not

Part One
Map Reading

occur in reality but are instead interpretations of characteristics of the environment, like population density or average streamflow.

Thus, map and reality are not—and cannot be—identical. No aspect of map use is so obvious yet so often overlooked. Most map reading mistakes occur because the user forgets this vital fact and expects a one-to-one correspondence between map and reality.

Since the exact duplication of a geographical area is impossible, a map is actually a metaphor. The mapmaker asks the map reader to believe that an arrangement of points, lines, and areas on a flat sheet of paper or a computer screen is equivalent to some facet of the real world in space and time. To gain a fuller understanding, the map reader must go beyond the graphic representation to what the symbols refer to in the real world.

A map, like a painting, is just one special version of reality. To understand a painting, you must have some idea of the medium used by the artist. You wouldn't expect a watercolor to look anything like an acrylic painting or a charcoal drawing, even if the subject matter of all three were identical. In the same way, the techniques used to create maps will greatly influence the final portrayal. As a map reader, you should always be aware of the mapmaker's invisible hand. Never use a map without asking yourself how it has been biased by the methods used to make it.

If the mapping process operates at its full potential, communication of environmental information takes place between the cartographer and the map reader. The mapmaker translates reality into the clearest possible picture that a map can give, and the map reader converts this picture back into a mental image of the environment. For such communication to occur, the map reader must know something about how maps are created.

The complexities of mapping are easier to study if we break them up into simpler parts. Thus, we have divided "Map Reading" (part I) into 10 chapters, each dealing with a different aspect of mapping. Chapter 1 examines geographical coordinate systems for the earth as a sphere, an oblate ellipsoid, and a geoid. Chapter 2 looks at ways of expressing and determining map scale. Chapter 3 introduces different map projections and the types of geometric distortions that occur with each projection. Chapter 4 focuses on different grid coordinate systems used on maps. Chapter 5 discusses land partitioning systems used in the United States and other countries and how they are mapped. Chapter 6 is devoted to the different methods of relief portrayal found on maps. Chapter 7 examines various methods for mapping qualitative information, and chapter 8 does the same for quantitative information. Chapter 9 is an overview of image maps. Finally, chapter 10 explores various aspects of map accuracy and uncertainty.

These 10 chapters should give you an appreciation of all that goes into mapping and the ways that different aspects of the environment are shown on maps. As a result, you'll better understand the large and varied amount of geographic information you can glean from a map. In addition, once you realize how intricate the mapping process is, you can't help but view even the crudest map with more respect, and your map reading skill will naturally grow.

chapter
one   **THE EARTH AND EARTH COORDINATES**

1

The earth and earth coordinates

Of all the jobs maps do for you, one stands out. They tell you where things are and let you communicate this information efficiently to others. This, more than any other factor, accounts for the widespread use of maps. Maps give you a superb **locational reference system**—a way of pinpointing the position of things in space.

There are many ways to determine the location of a feature shown on a map. All of these begin with defining a geometrical figure that approximates the true shape and size of the earth. This figure is either a **sphere** or an **oblate ellipsoid** (slightly flattened sphere) of precisely known dimensions. Once the dimensions of the sphere or ellipsoid are defined, a **graticule** of east–west lines called **parallels** and north–south lines called **meridians** is draped over the sphere or ellipsoid. The angular distance of a parallel from the equator and a meridian from what we call the **prime meridian** gives us the **latitude** and **longitude** coordinates of a feature. The locations of elevations measured relative to an average gravity or sea-level surface called the **geoid** can then be defined by three-dimensional coordinates.

THE EARTH AS A SPHERE

We have known for over 2,000 years that the earth is spherical in shape. We owe this idea to several ancient Greek philosophers, particularly **Aristotle** (fourth century BC), who believed that the earth's sphericity could be proven by careful visual observation. Aristotle noticed that as he moved north or south the stars were not stationary—new stars appeared in the northern horizon while familiar stars disappeared to the south. He reasoned that this could occur only if the earth were curved north to south. He also observed that departing sailing ships, regardless of their direction of travel, always disappeared from view hull first. If the earth were flat, the ships would simply get smaller as they sailed away. Only on a sphere would hulls always disappear first. His third observation was that a circular shadow is always cast by the earth on the moon during a lunar eclipse, something that would occur only if the earth were spherical. These arguments entered the Greek literature and persuaded scholars over the succeeding centuries that the earth must be spherical in shape.

Determining the size of our spherical earth was a daunting task for our ancestors. The Greek scholar **Eratosthenes,** head of the then-famous library and museum in Alexandria, Egypt, around 250 BC, made the first scientifically based estimate of the earth's circumference. The story that has come down to us is of Eratosthenes reading an account of a deep well at Syene near modern Aswan about 500 miles (800 kilometers) south of Alexandria. The well's bottom was illuminated by the sun only on June 21, the day of the summer solstice. He concluded that the sun must be directly overhead on this day, with rays perpendicular to the level ground (figure 1.1). Then he reasoned brilliantly that if the sun's rays were parallel and the earth was spherical, a vertical column like an obelisk should cast a shadow in Alexandria on the same day. Knowing the angle of the shadow would allow the earth's circumference to be measured if the north–south distance to Syene could be determined. The simple geometry involved here is "if two parallel lines are intersected by a third line, the alternate interior angles are equal." From this he reasoned that the shadow angle at Alexandria equaled the angular difference at the earth's center between the two places.

The story continues that on the next summer solstice Eratosthenes measured the shadow angle from an obelisk in Alexandria, finding it to be 7°12′, or $1/50$ th of a circle. Hence, the distance between Alexandria and Syene is $1/50$ th of the earth's circumference. He was told that Syene must be about 5,000 stadia south of Alexandria, since camel caravans traveling at 100 stadia per day took 50 days to make the trip between the two cities. From this distance estimate, he computed the earth's circumference as 50 × 5,000 stadia, or 250,000 stadia. In Greek times a **stadion** varied from 200 to 210 modern yards (182 to 192 meters), so his computed circumference was somewhere between 28,400 and 29,800 modern statute miles (45,700 and 47,960 kilometers), 14 to 19 percent greater than the current value of 24,874 statute miles (40,030 kilometers).

We now know that the error was caused by overestimation of the distance between Alexandria and Syene, and to the fact that they are not exactly north–south of each other. However, his method is sound mathematically and was the best circumference measurement until the 1600s. Equally important, Eratosthenes had the idea that careful observations of the sun would allow him to determine angular differences between places on earth, an idea that we shall see was expanded to other stars and recently to the Global Positioning System (GPS), a "constellation" of 24 earth-orbiting satellites that make it possible for people to pinpoint geographic location and elevation with a high degree of accuracy using ground receivers (see chapter 14 for more on GPS).

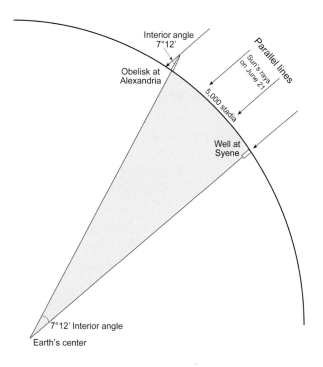

Figure 1.1 Eratosthenes's method for measuring the earth's circumference.

THE GRATICULE

Once the shape and size of the earth were known, mapmakers required some system for defining locations on the surface. We are again indebted to ancient Greek scholars for devising a systematic way of placing reference lines on the spherical earth.

Parallels and meridians

Astronomers before Eratosthenes had placed on maps horizontal lines marking the **equator** (forming the circle around the earth that is equidistant from the north and south poles) and the **tropics of Cancer and Capricorn** (marking the northernmost and southernmost positions where the sun is directly overhead on the summer and winter solstices, such as Syene). Later the astronomer and mathematician **Hipparchus** (190–125 BC) proposed that a set of equally spaced east–west lines called **parallels** be drawn on maps (figure 1.2). To these he added a set of north–south lines called **meridians** that are equally spaced at the equator and converge at the north and south poles. We now call this arrangement of parallels and meridians the **graticule**. Hipparchus's numbering system for parallels and meridians was and still is called **latitude** and **longitude.**

Latitude and longitude

Latitude on the spherical earth is the north–south angular distance from the equator to the place of interest (figure 1.3). The numerical range of latitude is from 0° at the equator to 90° at the poles. The letters N and S, such as 45°N for Fossil, Oregon, are used to indicate north and south latitude. **Longitude** is the angle, measured along the equator, between the intersection of the reference meridian, called the **prime meridian,** and the point where the meridian for the feature of interest intersects the equator. The numerical range of longitude is from 0° to 180° east and west of the prime meridian, twice as long as parallels. East and west longitudes are labeled E and W, so Fossil, Oregon, has a longitude of 120°W.

Putting latitude and longitude together into what is called a **geographic coordinate** (45°N, 120°W) pinpoints a place on the earth's surface. There are several ways to write latitude and longitude values. The oldest is the **Babylonian sexagesimal system** of degrees (°), minutes ('), and seconds ("), where there are 60 minutes in a degree and 60 seconds in a minute. The latitude and longitude of the capitol dome in Salem, Oregon, is 44°56'18"N, 123°01'47"W, for example.

Sometimes you will see longitude west of the prime meridian and latitude south of the equator designated with a negative sign instead of the letters W and S. Latitude and longitude locations can also be expressed in **decimal degrees** through the following equation:

Decimal degrees = dd + mm / 60 + ss / 3600

where *dd* is the number of whole degrees, *mm* is the number of minutes, and *ss* is the number of seconds. For example:

44°56'18" = 44 + 56 / 60 + 18 / 3600 = 44.9381°

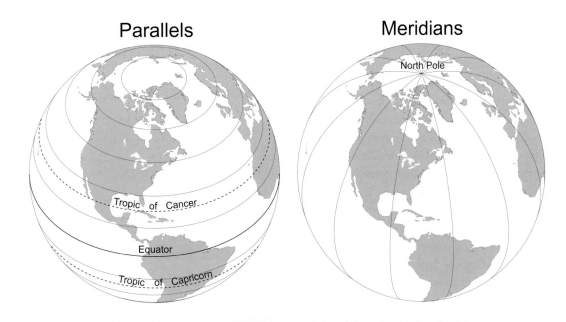

Figure 1.2 Parallels and meridians.

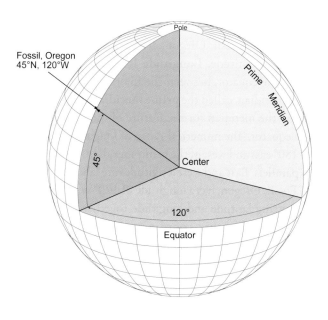

Figure 1.3 Latitude and longitude on the sphere allow the positions of features to be explicitly identified.

in Greenwich, a suburb of London. This is called the **Greenwich meridian.**

You may occasionally come across a historical map using one of the prime meridians in table 1.1, at which time knowing the angular difference between the prime meridian used on the map and the Greenwich meridian becomes very useful. As an example, you might see in an old Turkish atlas that the longitude of Seattle, Washington, is 151°16′ W (based on the Istanbul meridian), and you know that the Greenwich longitude of Seattle is 122°17′ W. You can determine the Greenwich longitude of Istanbul through subtraction:

$$151°16′ - 122°17′ = 28°59′E$$

The computation is done more easily in decimal degrees as described earlier:

$$151.27 - 122.28 = 28.99 \text{ degrees}$$

Decimal degrees are often rounded to two decimal places, so that the location of the Oregon state capitol dome would be written in decimal degrees as 44.94, –123.03. If we can accurately define a location to the nearest 1 second of latitude and longitude, we have specified its location to within 100 feet (30 meters) of its true location on the earth.

Prime meridians

The choice of **prime meridian** (the meridian at 0° used as the reference from which longitude east and west is measured) is entirely arbitrary because there is no physically definable starting point like the equator. In the fourth century BC Eratosthenes selected Alexandria as the starting meridian for longitude, and in medieval times the Canary Islands off the coast of Africa were used since they were then the westernmost outpost of western civilization. In the eighteenth and nineteenth centuries, many countries used their capital city as the prime meridian for the nation's maps, including the meridian through the center of the White House in Washington, D.C., for early nineteenth-century maps (see table 1.1 for a listing of historical prime meridians). You can imagine the confusion that must have existed when trying to locate places on maps from several countries. The problem was eliminated in 1884 when the **International Meridian Conference** selected as the international standard the British prime meridian—defined by the north–south optical axis of a telescope at the Royal Observatory

Amsterdam, Netherlands	4°53′01″E
Athens, Greece	23°42′59″E
Beijing, China	116°28′10″E
Berlin, Germany	13°23′55″E
Bern, Switzerland	7°26′22″E
Brussels, Belgium	4°22′06″E
Copenhagen, Denmark	12°34′40″E
Ferro, Canary Islands	17°40′00″W
Helsinki, Finland	24°57′17″E
Istanbul, Turkey	28°58′50″E
Jakarta, Indonesia	106°48′28″E
Lisbon, Portugal	9°07′55″W
Madrid, Spain	3°41′15″W
Moscow, Russia	37°34′15″E
Oslo, Norway	10°43′23″E
Paris, France	2°20′14″E
Rio de Janeiro, Brazil	43°10′21″W
Rome (Monte Mario), Italy	12°27′08″E
St. Petersburg, Russia	30°18′59″E
Stockholm, Sweden	18°03′30″E
Tokyo, Japan	139°44′41″E
Washington, D.C., USA	77°02′14″W

Table 1.1 Prime meridians used previously on foreign maps, along with longitudinal distances from the Greenwich meridian.

THE EARTH AS AN OBLATE ELLIPSOID

Scholars assumed that the earth was a perfect sphere until the 1660s when Sir Isaac Newton developed the theory of gravity. Newton thought that mutual gravitation should produce a perfectly spherical earth if it were not rotating about its polar axis. The earth's 24-hour rotation, however, introduces outward centrifugal forces perpendicular to the axis of rotation. The amount of force varies from zero at each pole to a maximum at the equator, obeying the following equation:

$$centrifugal\ force = mass \times velocity^2 \times distance\ from\ the\ axis\ of\ rotation$$

To understand this, imagine a very small circular disk the diameter of a dinner plate (about 10 inches or 25 centimeters) at the pole and two very thin horizontal cylindrical columns, one from the center of the earth to the equator (along the **equatorial axis**) and the second perpendicular from the **polar axis** (the axis from the center of the earth to the poles) to the 60th parallel (figure 1.4). The disk at the pole has a tiny mass, but both the velocity and distance from the axis are very small, so the **centrifugal force** (the apparent force caused by the inertia of the body that draws a rotating body away from the center of rotation) is nearly 0. The column to the equator is the earth's radius in length, and the velocity increases

from 0 at the center of the earth to a maximum of around 1,040 miles per hour (1,650 kilometers per hour) at the equator. This means that the total centrifugal force on the column is quite large. From the pole to the equator there will be a steady increase in centrifugal force. We see this at the 60th parallel where the column would be half the earth's radius in length and the velocity would be half that at the equator.

Newton noted that these outward centrifugal forces counteract the inward pull of gravity, so the net inward force decreases progressively from the pole to the equator. The column from the center of the earth to the equator extended outward slightly because of the increased centrifugal force and decreased inward gravitational force. A similar column from the center of the earth to the pole experiences zero centrifugal force and hence does not have this slight extension. Slicing the earth in half from pole to pole would then reveal an ellipse with a slightly shorter polar radius and slightly longer equatorial radius; we call these radii the **semiminor** and **semimajor axes,** respectively (figure 1.5). If we rotate this ellipse 180 degrees about its polar axis, we obtain a three-dimensional solid that we call an **oblate ellipsoid.** In the 1730s, scientific expeditions to Ecuador and Finland measured the length of a degree of latitude at the equator and near the Arctic

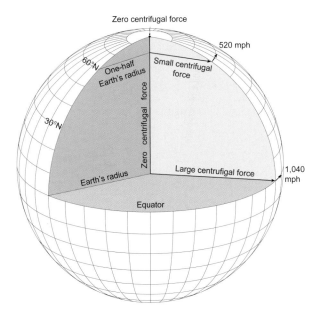

Figure 1.4   A systematic increase in centrifugal force from the pole to equator causes the earth to be an oblate ellipsoid.

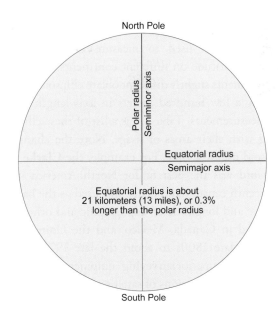

Figure 1.5   The form of the oblate ellipsoid was determined by measurements of degrees at different latitudes beginning in the 1730s. Its equatorial radius was about 13 miles (21 kilometers) longer than its polar radius. The north–south slice through the earth's center in this figure is true to scale, but our eye cannot see the deformation because it is so minimal.

Name	Date	Equatorial radius (km)	Polar radius (km)	Areas of use
WGS 84	1984	6,378.137	6,356.75231	Worldwide
GRS 80	1980	6,378.137	6,356.7523	Worldwide (NAD 83)
Australian	1965	6,378.160	6,356.7747	Australia
Krasovsky	1940	6,378.245	6,356.863	Soviet Union
International	1924	6,378.388	6,356.9119	Remainder of world not covered by older ellipsoids (European Datum 1950)
Clarke	1880	6,378.2491	6,356.5149	France; most of Africa
Clarke	1866	6,378.2064	6,356.5838	North America (NAD 27)
Bessel	1841	6,377.3972	6,356.079	Central Europe; Sweden; Chile; Switzerland; Indonesia
Airy	1830	6,377.5634	6,356.2569	Great Britain; Ireland
Everest	1830	6,377.2763	6,356.0754	India and the rest of South Asia

Table 1.2   Historical and current oblate ellipsoids.

Circle, proving Newton correct. These and additional meridian-length measurements in following decades for other parts of the world allowed the semimajor and semiminor axes of the oblate ellipsoid to be computed by the early 1800s, giving about a 13-mile (21-kilometer) difference between the two, only one-third of one percent.

Different ellipsoids

During the nineteenth century better surveying equipment was used to measure the length of a degree of latitude on different continents. From these measurements slightly different oblate ellipsoids varying by only a few hundred meters in axis length best fit the measurements. Table 1.2 is a list of these ellipsoids, along with their areas of usage. Note the changes in ellipsoid use over time. For example, the **Clarke 1866 ellipsoid** was the best fit for North America in the nineteenth century and hence was used as the basis for latitude and longitude on topographic and other maps produced in Canada, Mexico, and the United States from the late 1800s to about the late 1970s. By the 1980s, vastly superior surveying equipment coupled with millions of observations of satellite orbits allowed us to determine oblate ellipsoids that are excellent average fits for the entire earth. Satellite data are important because the elliptical shape of each orbit monitored at ground receiving stations mirrors the earth's shape. The most recent of these, called the **World Geodetic System of 1984 (WGS 84)**, replaced the Clarke 1866 ellipsoid in North America and is used as the basis for latitude and longitude on maps throughout the world. You'll see in table 1.2 that the WGS 84 ellipsoid has an equatorial

radius of 6,378.137 kilometers (3,963.191 miles) and a polar radius of 6,356.752 kilometers (3,949.903 miles).

The oblate ellipsoid is important to us because parallels are not spaced equally as on a sphere, but vary slightly in spacing from the pole to the equator. This is shown in figure 1.6, a cross section of a greatly flattened oblate ellipsoid. Notice that near the pole the ellipse curves less than near the equator. We say that on an oblate ellipsoid the **radius of curvature** (the measure of how curved the surface is) is greatest at the pole

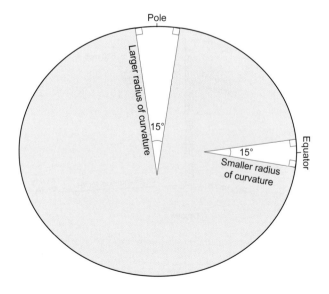

Figure 1.6   This north–south cross section through the center of a greatly flattened oblate ellipsoid shows that a larger radius of curvature at the pole results in a larger ground distance per degree of latitude relative to the equator.

and smallest at the equator. The north–south distance between two points on the surface equals the radius of curvature times the angular difference between them. For example, the distance between two points one degree apart in latitude between 0° and 1° at the equator is 68.703 miles (110.567 kilometers), shorter than the 69.407 miles (111.699 kilometers) between two points at 89° and 90° north latitude. This example shows that the spacing of parallels decreases slightly from the pole to the equator.

Geodetic latitude

Geodetic latitude is defined as the angle made by the horizontal equator line and a line perpendicular to the ellipsoidal surface at the parallel of interest (figure 1.7). Geodetic latitude differs from latitude on a sphere because of the unequal spacing of parallels on the ellipsoid. Lines perpendicular to the ellipsoidal surface only pass through the center of the earth at the poles and equator, but all lines perpendicular to the surface of a sphere will pass through its center. This is why the latitude defined by these lines on a sphere is called **geocentric latitude.**

Defining geodetic latitude in this way means that geocentric and geodetic latitude are identical only at 0° and 90°. Everywhere else geocentric latitude is slightly smaller than the corresponding geodetic latitude, as shown in table 1.3. Notice that the difference between geodetic and geocentric latitude increases in a symmetrical fashion from zero at the poles and equator to a maximum of just under one-eighth of a degree at 45°.

Geocentric	Geodetic	Geocentric	Geodetic
0°	0.000°	50°	50.126°
5	5.022	55	55.120
10	10.044	60	60.111
15	15.064	65	65.098
20	20.083	70	70.082
25	25.098	75	75.064
30	30.111	80	80.044
35	35.121	85	85.022
40	40.126	90°	90.000°
45°	45.128°		

Table 1.3   Geocentric and corresponding geodetic latitudes (WGS 84) at 5° increments.

Geodetic longitude

It turns out that there is no need to make any distinction between geodetic longitude and geocentric longitude. While the definition of geocentric longitude is mathematically different from geodetic longitude, the end result is essentially the same. The angle between the line from a point on the surface of the earth to the center of the earth and then to the prime meridian determines the longitude. Because of the nature of the ellipsoidal model, this turns out to be the same as the angle for geocentric longitude, which was described above.

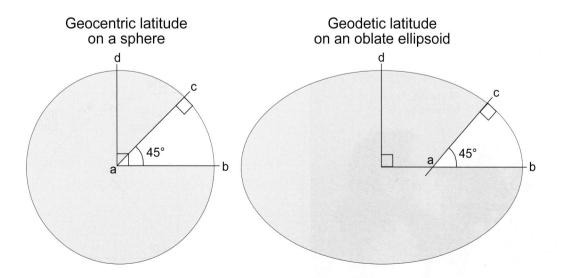

Figure 1.7   Geocentric and geodetic latitudes of 45°. On a sphere, circular arc distance b–c is the same as circular arc distance c–d. On the greatly flattened oblate ellipsoid, elliptical arc distance b–c is less than elliptical arc distance c–d. On the WGS 84 oblate ellipsoid, arc distance b–c is 3,097.50 miles (4,984.94 kilometers) and arc distance c–d is 3,117.43 miles (5,017.02 kilometers), a difference of about 20 miles (32 kilometers).

DETERMINING GEODETIC LATITUDE AND LONGITUDE

The oldest way to determine geodetic latitude and longitude is with instruments for observing the positions of celestial bodies. The essence of the technique is to establish celestial lines of position (east–west, north–south) by comparing the predicted positions of celestial bodies with their observed positions. A handheld instrument, called a **sextant,** historically was the tool used to measure the angle (or altitude) of a celestial body above the earth's horizon (figure 1.8). Before GPS, it was the tool that nautical navigators used to find their way using the moon, planets, and stars, including our sun.

Astronomers study and tabulate information on the actual motion of celestial bodies that helps to pinpoint latitude and longitude. Since the earth rotates on an axis defined by the north and south poles, stars in the northern hemisphere's night sky appear to move slowly in a circle centered on **Polaris** (the **North Star**). The navigator needs only to locate Polaris to find north. In addition, because the star is so far away from the earth, the angle from the horizon to Polaris is the same as the latitude (figure 1.9).

In the southern hemisphere, latitude is harder to determine because four stars are used to interpolate due south. Because there is no equivalent to Polaris over the south pole, navigators instead use a small constellation called **Crux Australis** (the **Southern Cross**) to serve the same function (figure 1.10). Finding south is more complicated because the Southern Cross is a collection of five stars that are part of the constellation Centaurus. The four outer stars form a cross, while the fifth much dimmer star (Epsilon) is offset about 30 degrees below the center of the cross.

Geodetic longitude can be determined in a straightforward manner. As we saw earlier, the prime meridian at 0° longitude passes through Greenwich, England. Therefore, each hour difference between your time and that at Greenwich, called **Greenwich mean time** or **GMT,** is roughly equivalent to 15° of longitude from Greenwich. To determine geodetic longitude, compare your local time with Greenwich mean time and multiply by 15° of longitude for each hour of difference. The difficulty is that time is conventionally defined by broad time zones, not by the exact local time at your longitude. Your exact local time must be determined by celestial observations.

In previous centuries, accurately determining longitude was a major problem in both sea navigation and mapmaking. It was not until 1762 that a clock portable enough to take aboard ship and accurate enough for longitude finding was invented by the Englishman John Harrison. This clock, called a **chronometer,** was set to Greenwich mean time before departing on a long voyage. The longitude of a distant locale was found by noting the Greenwich mean time at local noon (the highest point of the sun in the sky, found with a sextant). The time difference was simply multiplied by 15 to find the longitude.

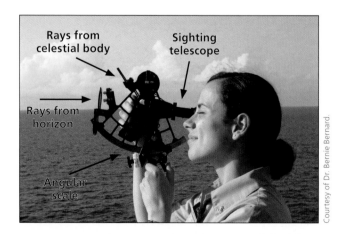

Courtesy of Dr. Bernie Bernard.

Figure 1.8   A sextant is used at sea to find latitude from the vertical angle between the horizon and a celestial body such as the sun.

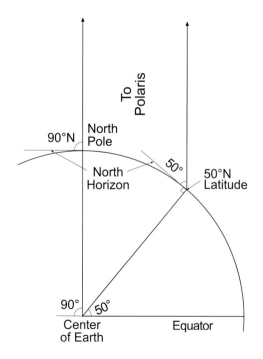

Figure 1.9   In the northern hemisphere, it is easy to determine your latitude by observing the height of Polaris above your northern horizon.

Figure 1.10   The Southern Cross is used for navigation in the southern hemisphere.

PROPERTIES OF THE GRATICULE

Circumference of the authalic and other spheres

When determining latitude and longitude we sometimes use a different approximation to the earth than the oblate ellipsoid. Using a sphere leads to simpler calculations, especially when working with small-scale maps of countries, continents, or the entire earth (see the next chapter for more on small-scale maps). On these maps differences between locations on the sphere and the ellipsoid are negligible. The value of the earth's spherical circumference used in this book is for what cartographer's call an **authalic sphere.** The authalic (meaning "area-preserving") sphere is a sphere with the same surface area as a reference ellipsoid we are using. The equatorial and polar radii of the WGS 84 ellipsoid are what we used to calculate the radius and circumference of the authalic sphere that is equal to the surface area of the WGS 84 ellipsoid. The computations involved are moderately complex and best left to a short computer program, but the result is a sphere with a radius of 3,958.76 miles (6,371.017 kilometers) and circumference 24,873.62 miles (40,030.22 kilometers).

There are other properties of an oblate ellipsoid that we can use to define the circumference of a spherical earth. A **rectifying sphere,** for example, is one where the length of meridians from equator to pole on the ellipsoid equals one-quarter of the spherical circumference. For the WGS 84 ellipsoid, the rectifying sphere is of radius 3,956.55 miles (6,367.449 kilometers) and circumference 24,859.73 miles (40,007.86 kilometers). We can also use the 3,964.038 mile (6378.137 kilometer) equatorial radius of the WGS 84 ellipsoid, giving a sphere of circumference 24,901.46 miles (40,075.017 kilometers).

This sphere is a little over three-tenths of a percent larger in surface area than the WGS 84 ellipsoid. These circumference values are all very close to each other, differing by less than two-tenths of a percent.

Spacing of parallels

As we saw earlier, on a spherical earth the north–south ground distance between equal increments of latitude does not vary. However, it is important to know how you want to define the distance. As we will see below, there are different definitions for terms you may take for granted, such as a "mile."

Using the WGS 84 authalic sphere circumference, latitude spacing is always 24,874 miles ÷ 360° or 69.09 statute miles per degree. Expressed in metric and nautical units, it is 111.20 kilometers and 60.04 nautical miles per degree of latitude (see table C.1 in appendix C for the metric and English distance equivalents used to arrive at these values.)

You can see that there is about a 15 percent difference in the number of statute and nautical miles per degree. **Statute miles** are what we use for land distances in the United States, while **nautical miles** are used around the world for maritime and aviation purposes. A statute mile is about 1,609 meters, while the international standard for a nautical mile is 1,852 meters exactly (about 1.15 statute miles). The original nautical mile was defined as 1 minute of latitude measured north–south along a meridian. Kilometers are also closely tied to distances along meridians, since the meter was initially defined as one ten-millionth of the distance along a meridian from the equator to the north or south pole.

We have seen that parallels on an oblate ellipsoid are not spaced equally, as they are on a sphere, but decrease slightly from the pole to the equator. You can see in table C.2 (in appendix C) the variation in the length of a degree of latitude for the WGS 84 ellipsoid, measured along a meridian at one-degree increments from the equator to the pole. You can see that the distance per degree of latitude ranges from 69.407 statute miles (111.699 kilometers) at the pole to 68.703 statute miles (110.567 kilometers) at the equator. The graph in figure 1.11 shows how these distances differ from the constant values of 69.09 and 69.07 statute miles (111.20 and 111.16 kilometers) per degree for the authalic and rectifying spheres, respectively. For both spheres, the WGS 84 ellipsoid distances per degree are about 0.3 miles (0.48 kilometers) greater than the sphere at the pole and 0.4 miles (0.64 kilometers) less at the equator. The ellipsoidal and spherical distances are almost the same in the mid-latitudes, somewhere between 45 and 50 degrees.

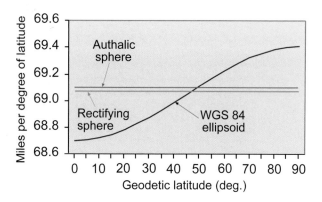

Figure 1.11   Distances along the meridian for one-degree increments of latitude from the equator to the pole on the WGS 84 ellipsoid, authalic sphere, and rectifying sphere.

Converging meridians

A quick glance at any world globe (or figure 1.2) shows that the length of a degree of longitude, measured east–west along parallels, decreases from the equator to the pole. The precise spacing of meridians at a given latitude is found by using the following equation:

69.09 miles (or 111.20 kilometers) / deg. × cosine (latitude). *

At 45° north or south of the equator, for example, cosine (45°) = 0.7071. Therefore, the length of a degree of longitude is

   69.09 × 0.7071, or 48.85 statute miles
   (111.20 × 0.7071 or 78.63 kilometers)

This is roughly 20 miles (32 kilometers) shorter than the 69.09-mile (111.20-kilometer) spacing at the equator.

Great and small circles

A **great circle** is the largest possible circle that could be drawn on the surface of the spherical earth. Its circumference is that of the sphere, and its center is the center of the earth so that all great circles divide the earth into halves. Notice in figure 1.3 that the equator is a great circle dividing the earth into northern and southern hemispheres. Similarly, the prime meridian and the 180° meridian at the opposite side of the earth (called the **antipodal meridian**) form a great circle dividing the earth into eastern and western hemispheres. All other pairs of meridians and their antipodal meridians are also great circles. A great circle is the shortest route between any two points on the earth and hence great circle routes are fundamental to long-distance navigation.

Any circle on the earth's surface that intersects the interior of the sphere at any location other than the center is called a **small circle,** and its circumference is smaller than a great circle. You can see in figure 1.3 that all parallels other than the equator are small circles. The circumference of a particular parallel is given by the following equation:

24,874 miles (or 40,030 kilometers) × cosine (latitude)

For example, the circumference of the 45th parallel is

24,874 × 0.7071 or 17,588 statute miles
(40,030 × 0.7071 or 28,305 kilometers)

Quadrilaterals

Quadrilaterals are areas bounded by equal increments of latitude and longitude, 10° by 10° in figure 1.3, for example. Since meridians converge toward the poles, the shapes of quadrilaterals vary from a square on the sphere at the equator to a very narrow spherical triangle at the pole. The equation *cosine(latitude)* gives the **aspect ratio** (width/height) of any quadrilateral. A quadrilateral centered at 45°N will have an aspect ratio of 0.7071, whereas a quadrilateral at 60°N will have an aspect ratio of 0.5. A map covering 10° by 10° will look very long and narrow at this latitude.

---

* Inexpensive engineering calculators can be used to compute trigonometric functions such as cosine. You can also find trig function calculators on the Internet.

GRATICULE APPEARANCE ON MAPS

Small-scale maps

World or continental maps such as globes and world atlas sheets normally use coordinates based on an authalic sphere. There are several reasons for this. Prior to using digital computers to make these types of maps numerically, it was much easier to construct them from spherical coordinates. Equally important, the differences in the plotted positions of spherical and corresponding geodetic parallels become negligible on maps that cover so much area.

You can see this by looking again at table 1.3. Earlier we saw that the maximum difference between spherical and geodetic latitude is 0.128° at the 45th parallel. Imagine drawing parallels at 45° and 45.128° on a map scaled at one inch per degree of latitude. The two parallels will be drawn a very noticeable 0.128 inches (0.325 centimeters) apart. Now imagine drawing the parallels on a map scaled at one inch per 10 degrees of latitude. The two parallels will now be drawn 0.013 inches (0.033 centimeters) apart, a difference that would not even be noticeable given the width of a line on a piece of paper. This scale corresponds to a world wall map approximately 18 inches (46 centimeters) high and 36 inches (92 centimeters) wide.

Large-scale maps

Parallels and meridians are shown in different ways on different types of maps. Topographic maps in the United States and other countries have tick marks showing the location of the graticule. All U.S. Geological Survey 7.5-minute topographic maps, for example, have graticule ticks at 2.5-minute intervals of latitude and longitude (figure 1.12). The full latitude and longitude is printed in each corner, but only the minutes and seconds of the intermediate edge ticks are shown. Note the four "+" symbols used for the interior 2.5-minute graticule ticks.

The graticule is shown in a different way on nautical charts (figure 1.13). Alternating white and dark bars spaced at the same increment of latitude and longitude line the edge of the chart. Because of the convergence of meridians, the vertical bars on the left and right edges of the chart showing equal increments of latitude are longer than the horizontal bars at the top and bottom. Notice the more closely spaced ticks beside each bar, placed every tenth of a minute on the chart in figure 1.13. These ticks are used to more precisely find the latitude and longitude of mapped features.

Aeronautical charts display the graticule another way. The chart segment for a portion of the Aleutian Islands in Alaska (figure 1.14) shows that parallels and meridians are drawn at 30-minute latitude and longitude intervals. Ticks are placed at 1-minute increments along each graticule line, allowing features to be located easily to within a fraction of a minute.

GEODETIC LATITUDE AND LONGITUDE ON LARGE-SCALE MAPS

You will always find parallels and meridians of geodetic latitude and longitude on detailed maps of smaller extents. This is done to make the map a very close approximation to the size and shape of the piece of the ellipsoidal earth that it represents. To see the perils of not doing this, you only need to examine one-degree quadrilaterals at the equator and pole, one ranging from 0° to 1° and the second from 89° to 90° in latitude.

You can see in table C.2 (in appendix C) that the ground distance between these pairs of parallels on the WGS 84 ellipsoid is 68.703 and 69.407 miles (110.567 and 111.699 kilometers), respectively. If the equatorial quadrilateral is mapped at a scale such that it is 100 inches (254 centimeters) high, the polar quadrilateral mapped

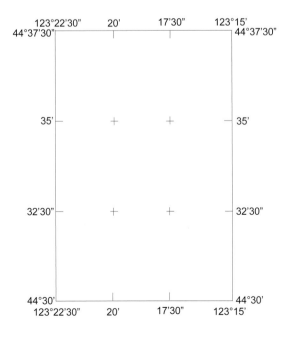

Figure 1.12 Graticule ticks on the Corvallis, Oregon, 7.5-minute topographic map.

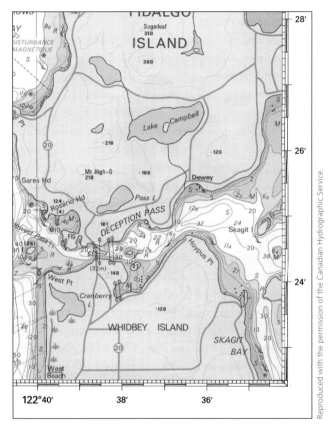

Figure 1.13 Graticule bars on the edges of a nautical chart segment.

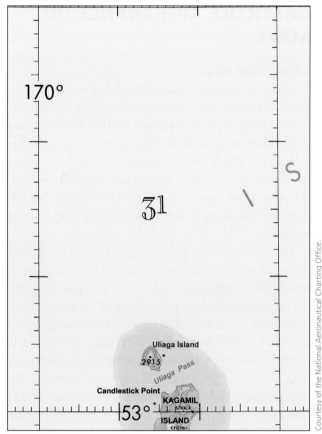

Figure 1.14 Graticule ticks on a small segment of an aeronautical chart.

at the same scale will be 101 inches (256.5 centimeters) long. If we mapped both quadrilaterals using the authalic sphere having 69.09 miles (111.20 kilometers) per degree, both maps would be 100.6 inches (255.5 centimeters) long. Having both maps several tenths of an inch (or around a centimeter) longer or shorter than they should really be is an unacceptably large error for maps used to make accurate measurements of distance, direction, or area.

Horizontal reference datums

To further understand the use of different types of coordinates on detailed maps of smaller extents, we must first look at datums—the collection of very accurate **control points** (points of known accuracy) surveyors use to georeference all other map data (see chapter 5 for more on control points and georeferencing). Surveyors determine the precise geodetic latitude and longitude of horizontal control points spread across the landscape. You may have seen a horizontal control point **monument** (a fixed object established

by surveyors when they determined the exact location of a point) like figure 1.15 on the ground on top of a hill or other prominent feature. From the 1920s to the early 1980s these control points were surveyed relative to the surface of the Clarke 1866 ellipsoid, together forming what was called the **North American Datum of 1927 (NAD 27)**. Topographic maps, nautical and aeronautical charts, and many other large-scale maps of this time period had graticule lines or ticks based on this datum. For example, the southeast corner of the Corvallis, Oregon, topographic map first published in 1969 (figure 1.16) has an NAD 27 latitude and longitude of 44°30′N, 123°15′W.

By the early 1980s, better knowledge of the earth's shape and size and far better surveying methods led to the creation of a new horizontal reference datum, the **North American Datum of 1983 (NAD 83)**. The NAD 27 control points were corrected for surveying errors where required, then these were added to thousands of more recently acquired points. The geodetic latitudes and longitudes of all these points were

Courtesy of the National Oceanic and Atmospheric Administration.

Figure 1.15 Horizontal control point marker cemented in the ground.

determined relative to the **Geodetic Reference System of 1980 (GRS 80)** ellipsoid, which is essentially identical to the WGS 84 ellipsoid.

The change of horizontal reference datum meant that the geodetic coordinates for control points across the continent changed slightly in 1983, and this change had to be shown on large-scale maps published earlier but still in use. On topographic maps the new position of the map corner is shown by a dashed "plus" sign, as in figure 1.16. Many times the shift is in the 100 meter range and must be taken into account when plotting on older maps the geodetic latitudes and longitudes obtained from GPS receivers and other modern position finding devices.

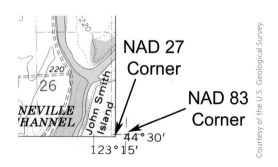

Courtesy of the U.S. Geological Survey.

Figure 1.16 Southeast corner of the Corvallis, Oregon, topographic map showing the difference between its NAD 27 and NAD 83 position.

Europe in the early 1900s faced another problem—separate datums for different countries that did not mesh into a single system for the continent. Military map users in World War II found different latitudes and longitudes for the same ground locations on topographic maps along the borders of France, Belgium, the Netherlands, Spain, and other countries where major battles were fought. The **European Datum of 1950 (ED 50)** was created after World War II as a consistent reference datum for most of western Europe, although Belgium, France, Great Britain, Ireland, Sweden, Switzerland, and the Netherlands continue to retain and use their own national datums. Latitudes and longitudes for ED 50 were based on the **International Ellipsoid of 1924.** Users of GPS receivers will find that, moving westward through Europe from northwestern Russia, the newer **European Terrestrial Reference System 1989 (ETRS 89)** longitude coordinates based on the WGS 84 ellipsoid gradually shift to the west of those based on the International Ellipsoid of 1924. In Portugal and western Spain the WGS 84 longitudes are approximately 100 meters to the west of those found on topographic maps based on ED 50. Moving southward, WGS 84 latitudes gradually shift northward from those based on ED 50, reaching a maximum difference of around 100 meters in the Mediterranean Sea.

Great Britain and Ireland are examples of countries that continue to use ellipsoids defined in the nineteenth century to best fit their region. Topographic maps in both nations use the **Airy 1830 ellipsoid** as the basis for the **Ordnance Survey Great Britain 1936 (OSGB 36)** datum for geodetic latitude and longitude coordinates. Along the south coast of England, WGS 84 latitudes are about 70 meters to the south of those based on OSGB 36. This southward shift gradually diminishes to zero near the Scottish border and then becomes a northerly difference that reaches a maximum value of around 50 meters at the northern extremes of Scotland. In Ireland, WGS 84 longitudes are around 50 meters to the east of their OSGB 36 equivalents and gradually increase to a maximum difference of around 120 meters along the southeast coast of England.

THE EARTH AS A GEOID

When we treat the earth as a smooth authalic sphere or oblate ellipsoid, we neglect mountain ranges, ocean trenches, and other surface features that have vertical relief. There is justification for doing this, as the earth's surface is truly smooth when we compare the surface undulations to the 7,918-mile (12,742-kilometer) diameter of the earth based on the authalic sphere. The greatest relief variation is the approximately 12.3-mile (19.8-kilometer) difference between the summit of Mt. Everest (29,035 feet or 8,852 meters) and the deepest point in the Mariana Trench (36,192 feet or 11,034 meters). This vertical difference is immense on our human scale, but it is only $1/640$th of the earth's diameter. If we look at the difference between the earth's average land height (2,755 feet or 840 meters) and ocean depth (12,450 feet or 3,795 meters), the average roughness is only 4,635m/12,742km, or $1/2,750$th of the diameter. It has been said that if the earth were reduced to the diameter of a bowling ball, it would be smoother than the bowling ball!

The earth's global-scale smoothness aside, knowing the elevations and depths of features is very important to us. Defining locations by their geodetic latitude, longitude, and elevation gives you a simple way to collect elevation data and display this information on maps. The top of Mt. Everest, for example, is located at 27°59′N, 86°56′E, 29,035 feet (8,852 meters), but what is this elevation relative to? This leads us to another approximation of the earth called the **geoid,** which is a surface of equal gravity used as the reference for elevations.

Vertical reference datums

Elevations and depths are measured relative to what is called a **vertical reference datum,** an arbitrary surface with an elevation of zero. The traditional datum used for land elevations is **mean sea level (MSL)** (see chapter 6 for more on mean sea level). Surveyors define MSL as the average of all low and high tides at a particular starting location over a **metonic cycle** (the 19-year cycle of the lunar phases and days of the year). Early surveyors chose this datum because of the measurement technology of the day. Surveyors first used the method of **leveling,** where elevations are determined relative to the point where mean sea level is defined, using horizontally aligned telescopes and vertically aligned leveling rods (see chapter 6 for more on leveling). A small circular monument was placed in the ground at each surveyed benchmark elevation point. A **benchmark** is a permanent monument that establishes the exact elevation of a place.

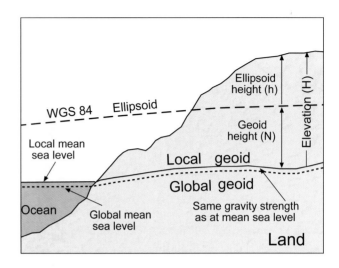

Figure 1.17 The geoid is the surface where gravity is the same as at mean sea level. Elevations traditionally have been measured relative to the geoid, but modern GPS-determined heights are relative to the WGS 84 ellipsoid.

Later, surveyors could determine elevation by making gravity measurements at different locations on the landform and relating them to the strength of gravity at the point used to define MSL. Gravity differences translate into elevation differences.

Mean sea level is easy to determine along coastlines, but what about inland locations? What is needed is to extend mean sea level across the land. Imagine that the mean sea level is extended under the continental land masses, which is the same thing as extending a surface having the same strength of gravity as mean sea level (figure 1.17). This imaginary equal gravity surface doesn't form a perfect ellipsoid, however, because differences in topography and earth density affect gravity's pull at different locations. The slightly undulating nearly ellipsoidal surface that best fits mean sea level for all the earth's oceans is called a **global geoid.** The global geoid rises and falls approximately 100 meters above and below the oblate ellipsoid surface in an irregular fashion. World maps showing land topography and ocean bathymetry use land heights and water depths relative to the global geoid surface.

The mean sea level datum based on the geoid is so convenient that it is used to determine elevations around the world and is the base for the elevation data found on nearly all topographic maps and nautical charts. But be aware that the **local geoid** used in your area is probably slightly above or below (usually within two meters) the global geoid elevations used on world maps. This difference is caused by mean sea level at one or more locations being used as the vertical reference datum for your nation or continent, not the average sea level for all the oceans.

In the United States, for example, you may see elevations relative to the **National Geodetic Vertical Datum of 1929 (NGVD 29)** on older topographic maps. This datum was defined by the observed heights of mean sea level at 26 tide gauges, 21 in the United States and 5 in Canada. It also was defined by the set of elevations of all benchmarks resulting from over 60,000 miles (96,560 kilometers) of leveling across the continent, totaling over 500,000 vertical control points. In the late 1980s, surveyors adjusted the 1929 datum with new data to create the **North American Vertical Datum of 1988 (NAVD 88)**. Topographic maps, nautical charts, and other cartographic products made from this time forward have used elevations based on NAVD 88. Mean sea level for the continent was defined at one tidal station on the St. Lawrence River at Rimouski, Quebec, Canada. NAVD 88 was a necessary update of the 1929 vertical datum since about 400,000 miles (650,000 kilometers) of leveling had been added to the NGVD since 1929. Additionally, numerous benchmarks had been lost over the decades—and the elevations at others had been affected—by vertical changes caused by rising of land elevations since the retreat of glaciers at the end of the last ice age, or subsidence from sedimentation and the extraction of natural resources like oil and water.

GPS has created a second option for measuring elevation (see chapter 14 for more on GPS). GPS receivers calculate what is called the **ellipsoidal height** (h), the distance above or below the surface of the WGS 84 ellipsoid along a line from the surface to the center of the earth (figure 1.17). An ellipsoidal height is not an elevation, since it is not measured relative to the mean sea level datum for your local geoid. Therefore, you must convert GPS ellipsoidal height values to mean sea level datum elevations (H) before you can use them with existing maps. You do this by subtracting the geoid height (N) at each point from the ellipsoid height (h) measured by the GPS receiver using the equation $H = h - N$. The look-up table needed to make this conversion usually is stored in your GPS receiver's computer. In the conterminous United States, geoid heights range from a low of –51.6 meters in the Atlantic Ocean to a high of –7.2 meters in the Rocky Mountains (figure 1.18). Worldwide, geoid heights vary from –105 meters just south of Sri Lanka to 85 meters in Indonesia.

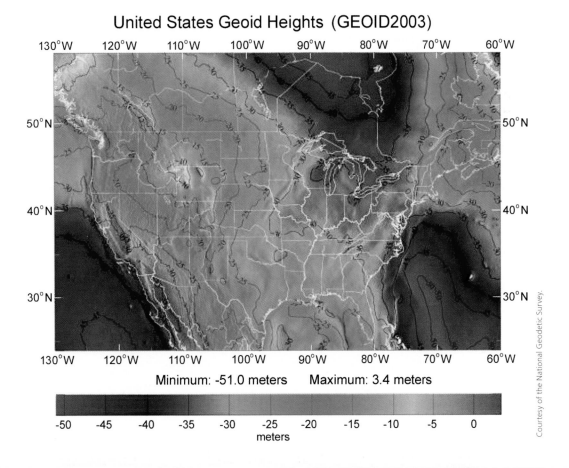

Minimum: -51.0 meters   Maximum: 3.4 meters

Courtesy of the National Geodetic Survey.

Figure 1.18   Geoid heights in the United States and vicinity (from the National Geodetic Survey GEOID2003 model).

SELECTED READINGS

Greenhood, D. 1964. *Mapping*. Chicago: University of Chicago Press.

Iliffe, J. C. 2000. *Datums and map projections for remote sensing, GIS and surveying*. Caithness, Scotland: Whittles Publishing.

La Condamine, C. M. de. 1747. A succinct abridgement of a voyage made within the inland parts of South America as it was read to the Academy of Science, Paris, April 28, 1745. London.

Maling, D. H. 1992. *Coordinate systems and map projections*. 2d ed. New York: Pergamon Press.

Maupertius, P. L. M. de. 1738. *The figure of the earth determined from observations made by order of the French king, at the polar circle* (translation). London.

Meade, B. K. 1983. Latitude, longitude, and ellipsoidal height changes NAD-27 to oredicted NAD-83. *Surveying and Mapping* 43: 65–71.

Robinson, A. H., et al. 1995. Basic geodesy. In *Elements of cartography*. 6th ed. New York: John Wiley & Sons, Inc.

Smith, J. R. 1988. *Basic geodesy. An introduction to the history and concepts of modern geodesy without mathematics*. Rancho Cordova, Calif.: Landmark Enterprises.

Snyder, J. P. 1987. *Map projections–a working manual*. U.S. Geological Survey Professional Paper 1395. Washington, D.C.: U.S. Government Printing Office.

Sobel, D. 1995. *Longitude: The true story of a lone genius who solved the greatest scientific problem of his time*. New York: Walker & Co.

Wallis, H. M., and A. H. Robinson, eds. 1987. *Cartographical innovations*. London: Map Collector Publications.

Wilford, J. N. 1981. *The mapmakers*. New York: Alfred A. Knopf.

U.S. Department of the Army. 2001. Grids. In *Map reading*. FM 3–25.26 Washington, D.C.: Department of the Army.

chapter
two **MAP SCALE**

2

Map scale

Maps are usually smaller in size than the environment they represent. The amount of size reduction is known as the **map scale,** which tells you the relationship between distances on the map and their corresponding ground distances. To use maps effectively, you'll need to convert measurements from map units to ground units. As you might expect, an understanding of map scale is central to performing this task. In this chapter, we'll explore what you need to know about map scale to become a skilled map user.

EXPRESSING SCALE

Map scale is always given in the following form: "This distance on the map represents this distance on the earth's surface." The relationship between map and ground distance can be expressed in a number of ways—most commonly as a representative fraction, a verbal scale, or a scale bar.

Representative fraction

A common way to describe scale is with a **representative fraction** (RF), which is the ratio between map and ground distance. The RF simplifies calculations involving scale. An RF is written either as *¹/x* or *1:x*. The numerator is always 1 and represents map distance, while the denominator (x) indicates distance on the ground in the same units of measurement as the distance on the map:

¹/x = map distance/ground distance

and

scale denominator x = ground distance/map distance

The advantage of having identical units on the top and bottom of the fraction is that map measurements may be made in centimeters, inches, or whatever distance unit you choose. For example, an RF of 1/24,000 or 1:24,000 means one inch on the map represents 24,000 inches on the ground, one centimeter on the map represents 24,000 centimeters on the ground, and so on. We can also say that the scale reduction is 24,000 to 1.

Verbal scale

Another familiar way to express scale is to use a descriptive **verbal scale.** The verbal scale expresses large RF denominator values in more familiar (larger) units of measurement. We express the scale verbally as so many "inches to 1 mile" or "centimeters to 1 kilometer" but also

as so many "miles to 1 inch" or "kilometers to 1 centimeter." An RF of 1:24,000 can be expressed with a verbal scale as "1 inch to 2,000 feet," while an RF of 1:100,000 can be expressed as "1 centimeter to 1 kilometer." These are also sometimes seen as "inches to the mile" or "centimeters to the kilometer" statements. Verbal scales for commonly used map scales are given in table 2.1, along with an example of a type of map published at this scale.

Many verbal scales are only close approximations to the RF and should not be used in mathematical calculations. An RF of 1:250,000 is really "3.9457 miles to 1 inch," but the verbal scale is much easier to remember if rounded to 4 miles. The reason the values can be rounded is because of how they are used. Verbal scales are often used with rulers to determine the length of a feature. Since the divisions of the rulers are fairly large, to state the map scale with multiple decimal places is no more useful than stating the rounded values. However, if you want to indicate more precision, you can use more decimal places in the verbal scale.

At first it may be confusing to find that one map indicates scale as "1 centimeter to 1 kilometer" and another as "4 miles to 1 inch." This lack of consistency should cause little trouble, however, since the smaller unit of measurement (inches or centimeters) always refers to the map while the larger measurement unit refers to the ground.

Scale bar

A third way to show map scale is to use a **scale bar** (also called a bar scale) placed on the map to simplify map distance measurements. These look like a small ruler printed on the map. You'll usually read the scale bar from left to right, beginning at 0. Sometimes the scale bar is extended to the left of the 0 point, using smaller markings (figure 2.1). A scale bar extension allows you to determine distance not only in whole units but also in fractions of units such as tenths of a mile or kilometer.

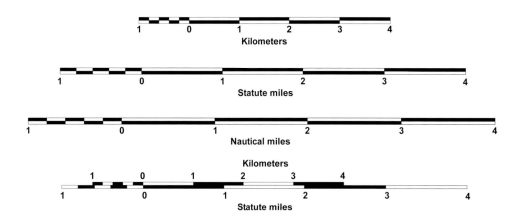

Figure 2.1 Scale bars in kilometers, statute miles, and nautical miles taken from maps identical in scale. Note the scale bar extensions to the left of the 0 point that are used when making more precise distance measurements.

RF	Verbal Scale	Example Maps	Scale Class
1:1,250	50 inches to 1 mile; 1 cm to 12.5 meters	OSGB Landplan Map	large scale
1:10,000	10 cm to 1 kilometer; 6¼ inches to 1 mile	OSGB Landplan Map	
1:24,000	1 inch to 2,000 feet	USGS Topographic Map	
1:25,000	2½ inches to 1 mile; 4 cm to 1 kilometer	OSGB Explorer Series Map	
1:50,000	1¼ inches to 1 mile; 2 cm to 1 kilometer	OSGB Landranger Map	
1:63,360	1 inch to 1 mile	USGS Topographic Map	
1:100,000	1 inch to 1½ miles; 1 cm to 1 kilometer	USGS Topographic Map; OSGB Travel Tour Map	
1:250,000	1 inch to 4 miles; 1 cm to 2.5 kilometers	USGS Topographic Map; OSGB Travel Road Map	
1:500,000	1 cm to 5 kilometers	Aeronautical Chart	medium scale
1:625,000	1 inch to 10 miles; 1 cm to 6.25 kilometers	OSGB Great Britain Route Map	
1:1,000,000	1 inch to ~16 miles; 1 cm to 10 kilometers	Aeronautical Chart; various OSGB wall maps	small scale
1:10,000,000	1 cm to 100 kilometers	Continental Maps	
1:100,000,000	1 inch to ~1,600 miles; 1 cm to 1,000 kilometers	World Maps	

Table 2.1 Different ways of expressing map scale commonly used by two national mapping agencies: the U.S. Geological Survey (USGS) and Ordnance Survey Great Britain (OSGB).

The marks on the scale bar are arranged so as to provide whole numbers or fractions of miles or kilometers of ground distance. This means that the marks won't necessarily represent whole numbers of centimeters or inches—there will almost always be some fraction left over. In other words, although the scale bar looks like a ruler, its markings will not coincide with those on your ruler. Rare exceptions would be a scale of 1:100,000, since at this scale one centimeter on the map would be exactly one kilometer on the map, and 1:63,360, since one inch on the map is exactly one mile on the ground.

The scale bar has three features that make it especially useful. First, if the map is enlarged or reduced using some method of photocopying or screen display, the scale bar changes size in direct proportion to the physical size of the map. The verbal scale and representative fraction, on the other hand, are incorrect if you reduce or enlarge the map. Second, both kilometers and miles can be shown conveniently on the same scale bar (see bottom of figure 2.1). This is called a **stacked scale bar** because the two scale bars using different units are stacked one on top of the other. And finally, the scale bar is easy to use when figuring distances on a map, as we'll show you in chapter 11.

On one special map, the mapmaker sometimes replaces the standard scale bar with a **variable scale bar.** An example of this type of scale bar is shown in figure 2.2. The only type of map that it is appropriate to use a variable scale bar on is a map on which scale varies systematically in the north–south direction (along the parallels), and the projection is conformal so distortion in any direction is the same at a given latitude (see chapter 3 for more on map projections). The only map that meets these qualifications is a map on the Mercator projection. If you have seen this used on any other type of map, it was used incorrectly.

To use such a scale bar, first decide at what latitude you want to use the map scale, and then find the scale bar for this latitude. In effect, you are working with a scale bar that is stretched to match the local map scale. If your latitude falls between two of the scale bars, you can add in a scale bar for your latitude as a horizontal line that you place in the correct vertical position on the variable scale bar. You would draw the scale bar for 45° as a horizontal line half way between the 40° and 50° scale bars, for instance.

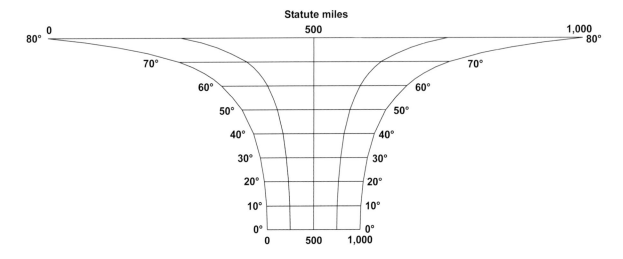

Figure 2.2 Variable scale bar such as those seen on a Mercator world map projection.

LARGE- AND SMALL-SCALE MAPS

Since map distance is always stated in the numerator of the RF as 1, it follows that the smaller the denominator, the closer to a 1 to 1 ratio, and consequently the larger the scale will be. Thus, a map scale of 1:20,000 is twice as large as a scale of 1:40,000. If that sounds backward, remember that the terms **large-scale** and **small-scale** maps come from the numerical value of the representative fraction $1/x$. The number $1/1,200$ is much larger than $1/100,000,000$. Thus the OSGB Landplan Map at a scale of 1:1,250 is a large-scale map, and a world map at a scale of 1:100,000,000 is a small-scale map.

Common as it is to classify maps by their scales, there is no general agreement as to where the class limits should be set. If we sort maps into two groups—large- and small-scale—then 1:500,000 would be a likely dividing point between the two. World atlas and wall maps of continental coverage would then fall into the small-scale group, while topographic, city street guides, and other detailed maps of small areas would be in the large-scale class. One reason for dividing maps into large and small scale is that you can use large-scale maps to make accurate measurements but you cannot do so on small-scale maps. If a more detailed three-way grouping is used, maps with scales of 1:1,000,000 and smaller ("16 or more miles to 1 inch") would probably be classed as small scale, and those with scales of 1:250,000 and larger ("4 or fewer miles to 1 inch") would be large scale. Maps ranging in scale between these extremes would then be referred to as **medium scale**. The last two columns in table 2.1 show the progression of scales from large through medium to small along with the types of maps associated with these scales. Any such classification, of course, is arbitrary and

shouldn't be given meaning beyond the organizational convenience it provides.

When you choose a map, be sure to note whether it is a large-, medium-, or small-scale product. Check that the features of interest to you are displayed at the correct scale for your purposes. If you need to study a small ground area in great detail with little generalization of features, you need a large-scale map. When you need to make accurate distance, direction, and area measurements, you should only use large-scale maps. This is because the change in scale across the surface of a large-scale map is negligible, so you can trust the map as a geometrically exact representation of the small piece of the earth it covers. However, if you are more interested in a generalized presentation of a large area like a state, country, continent, or the entire globe, you will choose a small-scale map. The scale is changing continuously across small-scale maps, so the RF or scale bar printed on these maps gives the scale at a particular point or along a given line but not for the entire map.

Multiscale maps

You may have used commercial software such as ArcGIS Explorer or ArcGlobe, or visited Web sites such as Google Maps or MapQuest, to see a series of maps at increasingly larger or smaller scales. The maps you view are what we refer to as multiscale maps. Multiscale maps are really a series of maps at varying scales each with an amount and type of information appropriate to the scale shown. As you zoom in on the map, more geometric detail and often additional themes of information are shown. These maps are seamless, allowing you to pan across the entire map extent. The six maps in figure 2.3, for example, show different levels of detail for a hydrologic map of an area in the state of Texas.

Often, cartographers use software such as ESRI's ArcMap to create the series of maps at set scales. The maps are carefully designed so that when you zoom in or out, the content remains legible and clear. The maps are then "cached" as a collection of prerendered map tiles used for display. The map tiles display quickly because the map does not have to be drawn on the fly for each scale you want to see. You see a progressive increase in detail at larger scales because the cartographer has designed the maps to change the degree of generalization at each map scale. For example, different types of roads are displayed at different scales, from freeways only at the smallest scale to freeways, highways, arterials, and the full street network at the largest scale. And streams that are lines at smaller scales become more detailed polygons at larger scales. Notice that the labeling of the streams in the maps

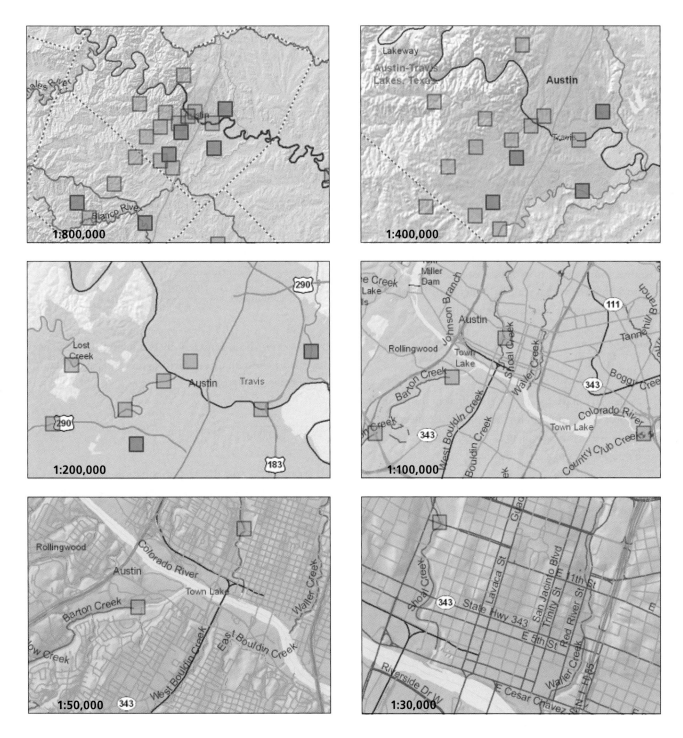

Figure 2.3   A multiscale hydrologic map at a series of scales as indicated by the representative fraction. Notice the increasing detail shown as the map scale increases.

in figure 2.3 increases as the map scale increases, until all streams are named at the largest scale. In this map series, you will also see that roads become more geometrically precise as the map scale increases, although major roads and streets are artificially widened on the map to emphasize their importance in the road network.

CONVERTING SCALE

The scale depicted on the map may be in the wrong form to best serve your purpose. Therefore, you may need to make conversions between a verbal scale, RF, and scale bar. **Scale conversions** are based on applying common distance equivalents such as 1 foot = 12 inches, 1 statute mile = 5,280 feet, 1 statute mile = 63,360 inches, or 1 kilometer = 100,000 centimeters. Table C.1 in appendix C is a more complete list of English and metric distance equivalents.

You may, for instance, have a map with a verbal scale and wish to know the RF. The first thing to remember with any scale conversion is that the ratio $^1/x$ is always map distance (numerator) to ground distance (denominator). Suppose that the verbal scale is "3 inches to 10 miles." In converting to an RF, the ratio is the following:

$$\frac{1}{x} = \frac{3\text{ inches}}{10\text{ miles}}$$

But you can't have an RF with different units in its numerator and denominator. So you must convert miles to inches—this is no problem if you remember that there are 63,360 inches in a statute mile:

$$\frac{1}{x} = \frac{3\text{ inches}}{10\text{ miles}}$$

$$\frac{1}{x} = \frac{3\text{ inches}}{10\text{ miles} \times \dfrac{63,360\text{ inches}}{1\text{ mile}}}$$

The miles in the denominator cancel out, as do the remaining inches in the numerator and denominator:

$$\frac{1}{x} = \frac{3}{633,600}$$

Remember that the numerator of an RF is always 1. So in this case you will also have to reduce the numerator from 3 to 1 by dividing the numerator and denominator by 3. To solve for x, you invert the equations to get the correct map scale of 1:211,200:

$$x = \frac{633,600}{3}$$
$$x = 211,200$$

The computations are even easier with metric units. If the verbal scale is "4 centimeters to 1 kilometer," then the equation is the following:

$$\frac{1}{x} = \frac{4\text{ centimeters}}{1\text{ kilometer}}$$

To convert kilometers to centimeters, you first convert the denominator knowing that there are 100,000 centimeters in a kilometer:

$$\frac{1}{x} = \frac{4\text{ centimeters}}{1\text{ kilometer}}$$

$$\frac{1}{x} = \frac{4\text{ centimeters}}{1\text{ kilometer} \times \dfrac{100,000\text{ centimeters}}{1\text{ kilometer}}}$$

In this case, the kilometers in the denominator cancel out, as do the remaining centimeters in the numerator and the denominator:

$$\frac{1}{x} = \frac{4}{100,000}$$

To solve for x, you invert the equation and divide the 100,000 by 4 to get the correct map scale of 1:25,000:

$$x = \frac{100,000}{4}$$
$$x = 25,000$$

Sometimes you may find yourself in the opposite situation. You know the RF but want to know how many "inches to 1 mile" (or "miles to 1 inch") or "centimeters to 1 kilometer" (or "kilometers to 1 centimeter") the map scale represents. If it is a "inches to 1 mile" verbal scale that you want, divide 63,360 by the denominator of the RF. To convert an RF of 1:200,000 to an "inches to 1 mile" verbal scale, the conversion would be as follows:

$$\frac{63,360_{\text{ inches / mile}}}{200,000}$$

$$\frac{63,360}{200,000}\text{ inches} = 1\text{ mile}$$

$$0.32\text{ inches} = 1\text{ mile}$$

If it is a "miles to 1 inch" verbal scale that you need, then you merely divide the denominator of the RF by the number of inches in a mile, or 63,360. For example, an RF of 1:200,000 is converted to a "miles to 1 inch" verbal scale as follows:

$$\frac{200,000}{63,360_{\text{ inches / mile}}}$$

$$1\text{ inch} = \frac{200,000}{63,360}\text{ miles}$$

$$1\text{ inch} = 3.16\text{ miles}$$

As we saw earlier, we often round this number off to state that the map scale is "3 miles to 1 inch."

If you want "centimeters to 1 kilometer," divide the denominator of the RF. To convert an RF of 1:200,000 to a verbal scale of "centimeters to 1 kilometer," the conversion is as follows:

$$\frac{100,000_{\text{ centimeters/ kilometer}}}{200,000}$$

$$\frac{100,000\text{ centimeters}}{200,000} = 1\text{ kilometer}$$

$$0.50\text{ centimeters} = 1\text{ kilometer}$$

This gives us a verbal scale of "0.5 centimeters to 1 kilometer."

If you want "kilometers to 1 centimeter," divide the denominator of the RF by 100,000. To convert an RF of 1:200,000 to a verbal scale of "kilometers to 1 centimeter," the conversion is as follows:

$$\frac{200,000}{100,000_{\text{ centimeters/ kilometer}}}$$

$$\frac{200,000}{100,000}\text{ kilometers} = 1\text{ centimeter}$$

$$2\text{ kilometers} = 1\text{ centimeter}$$

The verbal scale is then "2 kilometers to 1 centimeter."

Sometimes you may want to create a scale bar from a verbal scale or RF. Imagine that you have a map at a scale of ¾ mile to an inch, or 1:47,520. This would be considered a large-scale map, so your scale bar should have mile (or kilometer) increments, because this will facilitate ground distance measurements. Since an inch represents ¾ mile, ⅓ inch represents ¼ mile, and each mile covers $\frac{4}{3}$ inches on the map. So you would mark off $\frac{4}{3}$-inch intervals on your scale bar to indicate the mile increments. To make it more useful universally, you could also show kilometers.

DETERMINING MAP SCALE

It's good to know that, no matter what sort of map scale (RF, verbal scale, or scale bar) you encounter, you can change it to the type of scale you want. But what if you come across a map with no scale depicted at all, or an RF or verbal scale that seems incorrect because the map appears to have been enlarged or reduced after it was designed? This happens more often than you might expect. You may want to know the scale of a photocopied portion of a map on which no scale has been shown, for instance. Digital maps that you copy from the Internet are another good example, since they often are scans of original maps that do not include the RF or scale bar. Even if the RF is visible, it may not be correct because the map is being displayed at a different pixel density than when it was created. For example, a map created at 150 pixels per inch, but displayed on your computer monitor at a default density of 72 pixels per inch, will be slightly more than twice as large in scale as the RF indicates.

You can figure out the map scale if you know the ground distance between any two points on the map. You just measure the distance between these same two points on the map. The ratio of map to ground distance, in the same units of measurement, will be the map's scale. How, though, do you find the ground distance between the two points? One method is to use some terrestrial feature whose length is known, as described below. Keep in mind that scale will vary across the extent of the map, so for small-scale maps in particular, the scale you come up with will only be accurate at a particular point or along a given line on the map.

Determining map scale from a known terrestrial feature

Some features have standard lengths. If you can identify one of these on your map, you can easily figure out the map scale. A regulation U.S. football field, for example, is 100 yards long. If the map distance of the field is 0.5 inch, then 0.5 inch on the map represents 100 yards on the ground. To determine the map scale, you merely convert yards to inches and reduce the numerator to 1 by solving the RF ratio:

$$1/x = \frac{0.5 \text{ inches}}{100 \text{ yards} \times \dfrac{3 \text{ feet}}{1 \text{ yard}} \times \dfrac{12 \text{ inches}}{1 \text{ foot}}}$$

$$1/x = \frac{0.5 \text{ inches}}{3,600 \text{ inches}}$$

$$x = \frac{3,600 \text{ inches}}{0.5 \text{ inches}}$$

$$x = 7,200.00$$

This gives an RF of 1:7,200. In Canada, a football field is 110 yards, so you would have to rework the calculations for the different length. This method works best for large-scale maps, since most features of standard lengths are short and their lengths and widths are accurately shown only on large-scale maps.

Determining map scale from reference material

If you can't find a feature of standard length on your map, you can still determine scale if you turn to other reference material, such as road maps, topographic sheets, atlases, or other maps of similar scale to yours. From these sources, you should be able to find out the distance of something on your map—for instance, the distance between two prominent road intersections. With this information in hand, you can compute the map scale as described above, for determining map scale from a known terrestrial feature.

You can follow the same procedure to determine the RF of a small-scale map. For a world map, you can make use of the fact that the earth's circumference is approximately 25,000 statute miles (we saw in chapter 1 that the actual figure is 24,901.46 statute miles or 40,075.017 kilometers for a major authalic sphere using the semimajor axis of the WGS 84 ellipsoid). If, for example, the equator is shown in its entirety and is

measured on the map as 8 inches long, you can find the map's RF as follows:

$$1/x = \frac{8 \text{ inches}}{24,901.46 \text{ miles} \times \dfrac{63,360 \text{ inches}}{1 \text{ mile}}}$$

$$x = 24,901.46 \times \frac{63,360}{8}$$

$$x = 197,219,563$$

Rounding this number gives a RF of 1:197,219,600 to seven significant figures. The verbal scale for "miles to 1 inch" is calculated as

$$1 \text{ inch} = \frac{197,219,600}{63,360} \text{ miles}$$

$$1 \text{ inch} = 3,112.68 \text{ miles}$$

which can be rounded to "3113 miles to 1 inch." In metric units, 8 inches is 20.32 centimeters, and the earth's authalic sphere circumference is 40,075.017 kilometers, so the calculation would be as follows:

$$1/x = \frac{20.32 \text{ centimeters}}{40,075.017 \text{ kilometers} \times \dfrac{100,000 \text{ centimeters}}{1 \text{ kilometer}}}$$

$$x = 40,075.017 \times \frac{100,000}{20.32}$$

$$x = 197,219,572$$

As you can see, this gives the same RF—1:197,219,600 to seven significant figures. To determine the verbal scale in metric units, the calculation would be

$$1 \text{ centimeter} = \frac{197,219,566.9}{100,000} \text{ kilometers}$$

$$1 \text{ centimeter} = 1,972.20 \text{ kilometers}$$

or "1,972 kilometers to 1 centimeter." Remember that this is only the scale at the equator, since the scale may be much larger or smaller at other places on a small-scale map.

Another method you can use for both large- and small-scale maps is to determine the RF from the length of a feature measured on a reference map of known scale. For example, on a map with a scale of 1:100,000 you might measure the distance between two road intersections as 0.7 centimeters, and as 0.5 centimeters on your map.

Since the ratio of measured distances equals the ratio of the two RFs, you can use the following proportion:

$$\frac{1/100,000}{1/x} = \frac{0.7 \text{ centimeters}}{0.5 \text{ centimeters}}$$

This equation reduces to

$$x/100,000 = \frac{0.7 \text{ centimeters}}{0.5 \text{ centimeters}}$$

$$x = \frac{0.7 \text{ centimeters}}{0.5 \text{ centimeters}} \times 100,000$$

$$x = 140,000.00$$

giving an RF of 1:140,000.

Determining map scale from the spacing of parallels and meridians

It isn't always convenient or even possible to find a feature such as a football field or a lake of known length on your map or to use other reference materials. But on many maps, especially those of small scale, parallels and meridians are shown. Thus, you can determine scale by finding the ground distance between these lines. Finding the ground distance between parallels is quite simple, since a degree of latitude varies only slightly from pole to equator. Small-scale world maps usually use the authalic sphere as the approximation to the earth, so parallels are equally spaced at 69.093 miles (111.195 kilometers) per degree of latitude, as we saw in chapter 1. For large-scale maps, an ellipsoidal approximation to the earth's shape has been used. You saw in chapter 1 that on the WGS 84 ellipsoid, the first degree of latitude is 68.703 statute miles (110.567 kilometers) north or south of the equator, while the distance from the 89th parallel to the adjacent pole is 69.407 miles (111.699 kilometers). Thus, the variation in a degree of latitude from equator to pole is only 0.704 miles or 1.132 kilometers. This difference is so small that for many small-scale maps it can be ignored.

To find the length of a degree of latitude on the WGS 84 ellipsoid, you can refer to table C.2 in appendix C. In order to use this information in determining map scale, you must be able to find at least two parallels on your map. Let's say that on a small-scale world map you find two parallels separated by two degrees of latitude and there is a map distance of 5 inches between the parallels. To find the ground distance between these parallels,

you multiply the number of degrees separating them by the length of a degree:

$$2 \text{ degrees} \times \frac{69.093 \text{ miles}}{1 \text{ degree}} = 138.18 \text{ miles}$$

Now you can find the ratio between map distance and ground distance:

$$\frac{1}{x} = \frac{5 \text{ inches}}{138.18 \text{ miles}}$$

$$\frac{1}{x} = \frac{5 \text{ inches}}{138.18 \text{ miles} \times \dfrac{63,360 \text{ inches}}{1 \text{ mile}}}$$

$$\frac{1}{x} = \frac{5 \text{ inches}}{8,755,100 \text{ inches}}$$

$$x = \frac{8,755,100}{5}$$

$$x = 1,751,020.00$$

This gives an RF of 1:1,751,020 or a verbal scale of "27.64 miles to 1 inch."

In metric, 5 inches is 12.7 centimeters, so this would be as follows:

$$2 \text{ degrees} \times \frac{111.195 \text{ kilometers}}{1 \text{ degree}} = 222.40 \text{ kilometers}$$

Again, you can find the ratio between map distance and ground distance:

$$\frac{1}{x} = \frac{12.7 \text{ centimeters}}{222.40 \text{ kilometers}}$$

$$\frac{1}{x} = \frac{12.7 \text{ centimeters}}{222.40 \text{ kilometers} \times \dfrac{100,000 \text{ centimeters}}{1 \text{ kilometer}}}$$

$$\frac{1}{x} = \frac{12.7 \text{ centimeters}}{22,240,000 \text{ centimeters}}$$

$$x = \frac{22,240,000}{12.7}$$

$$x = 1,751,180$$

This gives an RF of nearly the same value— the difference lies in the rounding of numbers. The verbal scale for an RF of 1:1,751,180 is "17.5 kilometers to 1 centimeter."

Unfortunately, two parallels are not always shown on a map. You can still compute the map scale, however, if two meridians are shown. The complication is that, because meridians converge at the poles, the distance between a degree of longitude varies from 69.09 miles (111.20 kilometers) along the equator to 0 kilometers at either pole (see table C.3 in appendix C). This means that the distance between meridians, called the **longitudinal distance,** depends on the mapped region's latitude. Luckily, there's a simple functional relationship between latitude and longitudinal distance if you again assume a sphere for the earth's approximation to make a small-scale map: longitudinal distance decreases by the cosine of the latitude. To find the longitudinal distance for a degree of longitude at a given latitude, multiply the length of a degree of latitude (69.093 miles or 111.195 kilometers) by the cosine of the latitude. This relationship is written mathematically as follows:

Longitudinal distance = cos(latitude) × 69.093 miles

or

Longitudinal distance = cos(latitude) × 111.195 kilometers

Now how can you use this equation to determine map scale? The procedure is actually straightforward if viewed as a series of simple steps. You begin by measuring the map distance between two meridians. For example, on the Aleutian Islands aeronautical chart segment in figure 2.4 you find that at 53°N, meridians spanning one-half degree of longitude on the map are 2.66 inches (6.75 centimeters) apart.

Next you determine ground distance between these two meridians. To do that, use your calculator to find the cosine of 53°N. In this case, for miles, longitudinal distance is calculated as follows:

cos(53°) × 69.093 miles =
0.6018 × 69.093 =
41.58 miles

For kilometers, it is as follows:

cos(53°) × 111.195 kilometers =
0.6018 × 111.195 =
66.92 kilometers

Now you know that one degree of longitude at 53°N covers a ground distance of 41.58 miles or 66.92 kilometers. But since the measured map distance spanned one-half rather than one degree of longitude, the ground distance in this problem would be half the computed value, or 0.5 × 41.58 = 20.79 miles (33.46 kilometers).

You now have all the information necessary to find the map scale: the distance between two meridians on the map and the corresponding ground distance. Simply compute the ratio between map and ground distance:

$$\frac{1}{x} = \frac{2.66 \text{ inches}}{20.79 \text{ miles}}$$

$$\frac{1}{x} = \frac{2.66 \text{ inches}}{20.79 \text{ miles} \times \dfrac{63,360 \text{ inches}}{1 \text{ mile}}}$$

$$\frac{1}{x} = \frac{2.66 \text{ inches}}{1,317,254 \text{ inches}}$$

$$x = \frac{1,317,254}{2.66}$$

$$x = 495,208.4$$

This gives an RF of 1:495,208, which would be rounded to 1:495,000 or a verbal scale of "7.82 miles to 1 inch." For metric, 2.66 inches equals 6.756 centimeters:

$$\frac{1}{x} = \frac{6.756 \text{ centimeters}}{33.46 \text{ kilometers}}$$

$$\frac{1}{x} = \frac{6.756 \text{ centimeters}}{33.46 \text{ kilometers} \times \dfrac{100,000 \text{ centimeters}}{1 \text{ kilometer}}}$$

$$\frac{1}{x} = \frac{6.75 \text{ centimeters}}{3,346,000 \text{ centimeters}}$$

$$x = \frac{3,346,000}{6.756}$$

$$x = 495,234$$

This gives an RF of 1:495,234, which would also be rounded to 1:495,000 or a verbal scale of "4.95 kilometers to 1 centimeter."

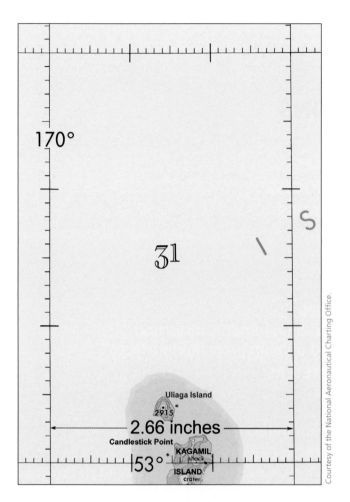

Courtesy of the National Aeronautical Charting Office.

Figure 2.4   You can begin to compute the RF of this Aleutian Islands aeronautical chart segment measuring one-half degree of longitude as 2.66 inches long on the chart (latitude and longitude ticks are spaced one minute apart on the chart).

SELECTED READINGS

ASCE Committee on Cartographic Surveying. 1983. *Map uses, scales, and accuracies for engineering and associated purposes.* New York: American Society of Civil Engineers.

Dickinson, G. C. 1969. *Maps and air photographs.* London: Edward Arnold Publishers, Ltd. 99–107, 148.

Espenshade, E. B. 1951. Mathematical scale problems. *Journal of Geography* 50 (3): 107–13.

Greenhood, D. 1964. *Mapping.* Chicago: University of Chicago Press, 39–53, 180–84.

Hodgkiss, A. G. 1970. *Maps for books and theses.* New York: Pica Press, 37–43, 169–70, 239–42.

Quattrochi, D. A., and M. F. Goodchild. 1997. *Scale in remote sensing and GIS.* Boca Raton, Fla.: Lewis Publishers.

Robinson, A. H., et al. 1995. *Elements of cartography.* 6th ed. New York: John Wiley & Sons, Inc.

U.S. Army. 1969. Scale and distance. In *Map reading,* FM 21–26, 4–1 to 4–4. Department of the Army Field Manual.

chapter
three **MAP PROJECTIONS**

3

Map projections

A map projection is a geometrical transformation of the earth's spherical or ellipsoidal surface onto a flat map surface. Much has been written about map projections, yet people still find this subject to be one of the most bewildering aspects of map use. Many people readily admit that they don't understand map projections. This shortcoming can have unfortunate consequences. For one thing, it hinders their ability to understand international relations in our global society. It also makes them easy prey for politicians, special interest groups, advertisers, and others who, through lack of understanding or by design, use map projections in potentially deceptive ways.

Potentially, there are an infinite number of map projections, each of which is better suited for some uses than for others. How, then, do we go about distinguishing one projection from another and choosing among them? One way is to organize the wide variety of projections into a limited number of groups or families on the basis of shared attributes. Two approaches commonly used include classifying the projections into families based on their geometrical distortion properties (relating to distance, shape, direction, and area) and examining the nature of the surface used to construct the projection (a plane, a cone, or a cylinder), which helps us understand the pattern of spatial distortion over the map surface. The two approaches go hand in hand because the map user is concerned, first, with what spatial properties are preserved (or lost) and, second, with the pattern and extent of distortion. In this chapter, we will look closely at the distortion properties of map projections and the surfaces used in the creation of projections.

Before we discuss map projection properties and families, we can help to clarify the issue of map projections with a discussion of the logic behind them. Why are projections necessary? We'll begin our discussion with globes, a form of map you have probably looked at since childhood. Your familiarity with globes makes it easy for us to compare them with flat maps.

GLOBES VERSUS FLAT MAPS

Of all maps, **globes** give us the most realistic picture of the earth as a whole. Basic geometric properties such as distance, direction, shape, and area are preserved because the globe is the same scale everywhere (figure 3.1). Globes have a number of disadvantages, however. They don't let you view all parts of the earth's surface simultaneously—the most you can see is a **hemisphere** (half of the earth). Nor are globes useful to see the kind of detail you might find on the road map in your glove compartment or the topographic map you carry when hiking.

Globes also are bulky and don't lend themselves to convenient handling and storage. You wouldn't have these handling and storage problems if you used a baseball-sized globe. But such a tiny globe would be of little practical value for map reading and analysis, since it would have a scale reduction of approximately 1:125,000,000. Even a globe 2 feet (60 centimeters) in diameter (the size of a large desk model) still represents a 1:20,000,000 scale reduction. It would take a globe about 40–50 feet (12–15 meters) in diameter—the height of a four-story building!—to provide a map of the scale used for state highway maps. A globe nearly 1,800 feet (about 550 meters) in diameter—the length of six U.S. football fields!—would be required to provide a map of the same scale as the standard 1:24,000-scale topographic map series in the United States.

Another problem with globes is that the instruments and techniques that are suited for measuring distance, direction, and area on spherical surfaces are relatively difficult to use. Computations on a sphere are far more complex than those on a plane surface. (For a demonstration of the relative difference in difficulty between making distance computations from plane and spherical coordinates, see chapter 11.)

Finally, globe construction is laborious and costly. High-speed printing presses have kept the cost of flat map reproduction to manageable levels but have not yet been developed to work with curved media. Therefore, globe construction is not suited to the volume of map production required for modern mapping needs.

It would be ideal if the earth's surface could be mapped undistorted onto a flat medium, such as a sheet of paper or computer screen. Unfortunately, the spherical earth is not what is known as a **developable surface,** defined by mathematicians as a surface that can be flattened onto a plane without geometrical distortion. The only developable surfaces in our three-dimensional world are cylinders, cones, and planes, so all flat maps necessarily distort the earth's surface geometrically. To transform a spherical surface that curves away in every direction from every point into a plane surface that doesn't exhibit curvature in any direction from any point means that the earth's surface must be distorted on the flat map. Map projection affects basic properties of the representation of the sphere on a flat surface, such as scale, continuity, and completeness, as well as geometrical properties relating to direction, distance, area, and shape. What we would like to do is minimize these distortions or preserve a particular geometrical property at the expense of others. This is the map projection problem.

THE MAP PROJECTION PROCESS

The concept of map projections is somewhat more involved than is implied in the previous discussion. Not one but a series of geometrical transformations is required. The irregular topography of the earth's surface is first defined relative to a much simpler three-dimensional surface, and then the three-dimensional surface is projected onto a plane. This progressive flattening of the earth's surface is illustrated in figure 3.2.

The first step is to define the earth's irregular surface topography as elevations above or sea depths below a more regular surface known as the geoid, as discussed in chapter 1. This is the surface that would result if the average level of the world's oceans (mean sea level) were extended under the continents. It serves as the datum, or starting reference surface, for elevation data on our maps (see chapter 1 for more on datums).

Courtesy of Replogle Globes, Inc.

Figure 3.1   A typical world globe.

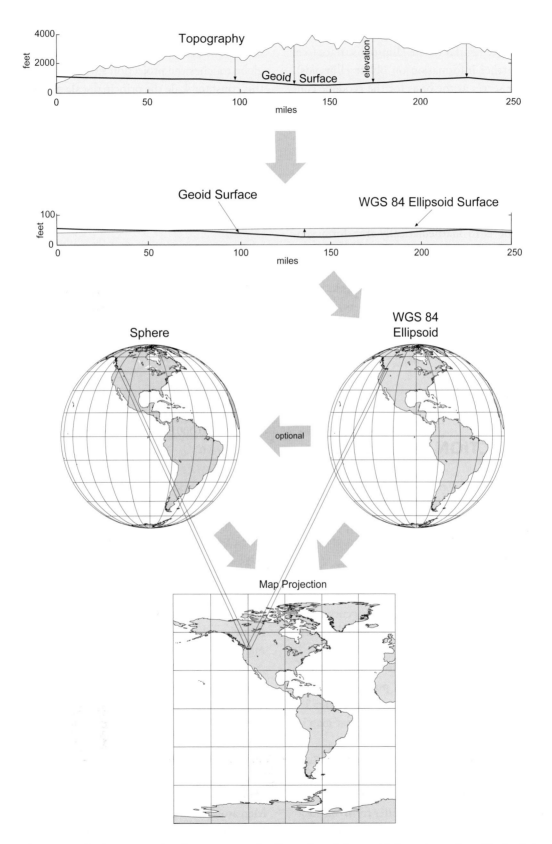

Figure 3.2   The map projection process involves a progressive flattening of the earth's irregular surface. The land elevation or sea depth of every point on the earth's surface is defined relative to the geoid surface. The vertical differences between the geoid and ellipsoid are so small that horizontal positions on the geoid surface are the same as on an oblate ellipsoid surface such as WGS 84. The geodetic latitude and longitude of the point is now defined. Geodetic latitudes and longitudes may be converted into spherical coordinates for small-scale maps. Finally, the geodetic or spherical coordinates are transformed into planar (x,y) map projection surface coordinates by manipulating the data on a computer.

The second step is to project the slightly undulating geoid onto the more regular oblate ellipsoid surface (see chapter 1 for further information on oblate ellipsoids). This new surface serves as the basis for the geodetic control points determined by surveyors and as the datum for the geodetic latitude and longitude coordinates found on maps. Fortunately, the vertical differences between the geoid and oblate ellipsoid surface are so small (less than 100 meters on land) that latitude and longitude on the ellipsoid is used as the horizontal position on the geoid as well. An additional step taken in making a small-scale flat map or globe is to mathematically transform geodetic coordinates into geocentric coordinates on a sphere, usually equal in surface area to the ellipsoid.

The third step involves projecting the ellipsoidal or spherical surface onto a plane through the use of map projection equations that transform geographic or spherical coordinates into planar (x,y) map coordinates. The greatest distortion of the earth's surface geometry occurs in this step.

MAP PROJECTION PROPERTIES

Now that you understand the map projection process, you can see that it will lead to some distortion on the flat map. Let's look now at how the properties of scale, completeness, correspondence relations, and continuity will be affected by the map projection process.

Scale
Because of the stretching and shrinking that occurs in the process of transforming the spherical or ellipsoidal earth surface to a plane, the stated map scale (see chapter 2 for more information on scale) is true only at selected points or along particular lines called **points and lines of tangency**—we will talk about these more later in this chapter. Everywhere else the scale of the flat map is actually smaller or larger than the stated scale.

To grasp the idea of scale variation on map projections, you first must realize that there are in fact two map scales. One is the **actual scale**—this is the scale that you measure at any point on the map; it will differ from one location to another. Variation in actual scale is a consequence of the geometrical distortion that results from flattening the earth.

The second is the **principal scale** of the map. This is the scale of the **generating globe**—a globe reduced to the scale of the desired flat map. This globe is then transformed into a flat map (figure 3.3). The constant scale of the generating globe is the scale stated on the flat map, which is correct only at the points or lines where the globe touches the projection surface. Cartographers call this constant scale the principal scale of the map.

To understand the relation between the actual and principal scale at different places on the map, we compute a ratio called the **scale factor (SF),** which is defined as the following:

$$SF = \frac{Actual\ Scale}{Principal\ Scale}$$

We use the representative fractions of the actual and principal scales to compute the SF. An actual scale of 1:50,000,000 and a principal scale of 1:100,000,000 would thus give an SF of the following:

$$SF = \frac{\dfrac{1}{50,000,000}}{\dfrac{1}{100,000,000}}$$

$$SF = \frac{100,000,000}{50,000,000}$$

$$SF = 2.0$$

If the actual and principal scales are identical, then the SF is 1.0. But, as we saw earlier, because the actual scale varies from place to place, so does the SF. An SF of 2.0 on a small-scale map means that the actual scale is twice as large as the principal scale (figure 3.3). An SF of 1.15 on a small-scale map means that the actual scale is 15 percent larger than the principal scale. On large-scale maps, the SF should vary only slightly from 1.0 (also called **unity**), following the general rule that the smaller the area being mapped, the less the scale distortion.

Completeness
Completeness refers to the ability of map projections to show the entire earth. You'll find the most obvious distortion of the globe on "world maps" that don't actually show the whole world. Such incomplete maps occur when the equations used for a map projection can't be applied to the entire range of latitude and longitude. The Mercator world map (figure 3.19) is a classic example. Here the y-coordinate for the north pole is infinity,[1] so the map usually extends to only the 80th parallel north and south. Omitting these high latitudes may be acceptable for maps showing political boundaries, cities, roads, and other cultural features, but maps showing physical phenomena such as average temperatures, ocean currents, or landforms normally are made on globally complete

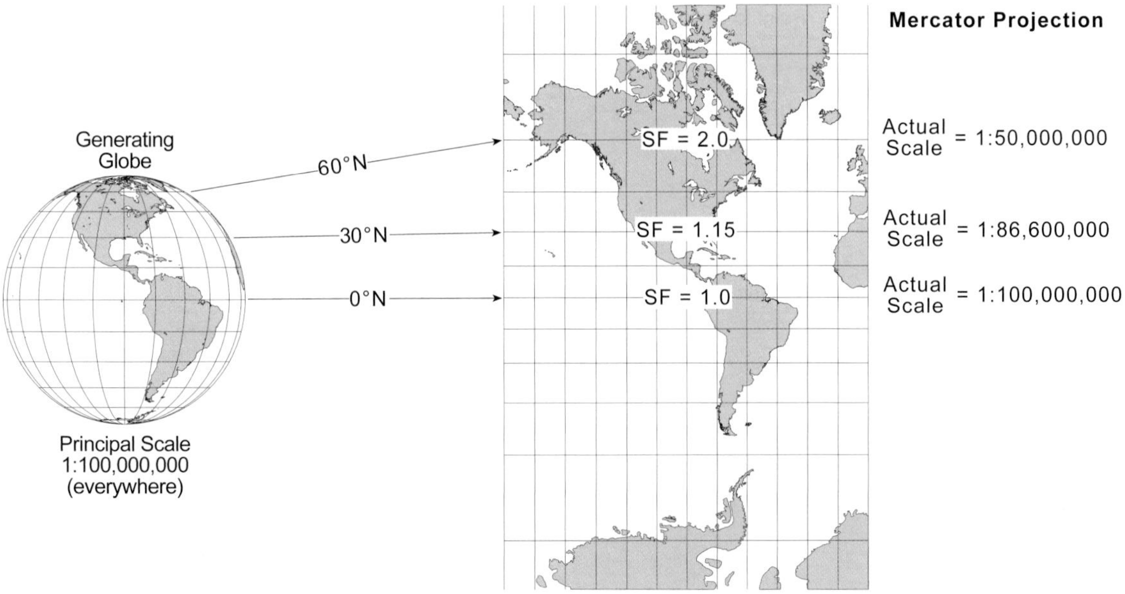

Figure 3.3   Scale factors (SFs) greater than 1.0 indicate that the actual scales are larger than the principal scale of the generating globe.

projections. The gnomonic projection (figure 3.15) is a more extreme example, since it is limited mathematically to covering less than a hemisphere (the SF is infinitely large 90 degrees away from the projection center point).

Correspondence relations

You might expect that each point on the earth would be transformed to a corresponding point on the map projection. Such a **point-to-point correspondence** would let you shift attention with equal facility from a feature on the earth to the same feature on the map, and vice versa. Unfortunately, this desirable property can't be maintained for all points on many world map projections. As figure 3.4 shows, one or more points on the earth may be transformed into straight lines or circles on the boundary of the map projection, most often at the north and south poles. You may notice that the SF in the east–west direction must be infinitely large at the poles on this projection, yet it is 1.0 in the north–south direction.

Continuity

To represent an entire spherical surface on a plane, the continuous spherical surface must be interrupted at some point or along some line. These breaks in **continuity** form the map border on a world projection. Where the map-maker places the discontinuity is a matter of choice. On some maps, for example, opposite edges of the map are the same meridian (figure 3.4). Since this means features next to each other on the ground are found at opposite sides of the map, this is a blatant violation of proximity

relations and a source of confusion for map users. Similarly, a map may show the north and south poles with lines as long as the equator. This means features adjacent to each other, but on opposite sides of the meridian used to define the edge of the map, will be far apart along the top or bottom edge of the map, while on the earth's surface they are at almost exactly the same location. Maps of individual continents (except Antarctica) or nations almost always show these areas without breaks in continuity. To do otherwise would needlessly complicate reading, analyzing, and interpreting the map.

MAP PROJECTION FAMILIES

As we mentioned previously, it is sometimes convenient to group map projections into families based on common properties so that we can distinguish and choose among them. The two most common approaches to grouping map projections into families are based on geometric distortion and the projection surface.

Projection families based on geometric distortion

You can gain an idea of the types of geometric distortions that occur on map projections by comparing the graticule (latitude and longitude lines) on the projected surface with the same lines on a globe (see chapter 1 for more on the graticule). Figure 3.4 shows how you might make

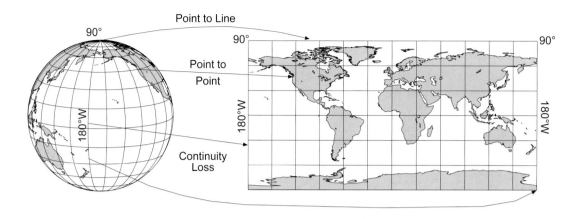

Figure 3.4 The point-to-point correspondence between the globe and map projection may become point-to-line at some locations. There also may be a loss of continuity, in which the same line on the globe forms two edges of the map.

this comparison (in figure 3.4, the drawing of a globe on the left is actually a map projection of a hemisphere.) Ask yourself several questions: To what degree do meridians converge? Do meridians and parallels intersect at right angles? Do parallels shorten with increasing latitude? Are the areas of quadrilaterals on the projection the same as on the globe?

We can use these observations to understand better how cartographers assign the types of geometric distortion you see on the map projection into the categories of distance, shape, direction, and area. Let's look at each type of distortion, beginning with variations in distance.

Distance

The preservation of spherical great circle distance on a map projection, called **equidistance,** is at best a partial achievement (see chapter 1 for more on great circles). For a map projection to be truly distance-preserving (**equidistant,**) the scale would have to be equal in all directions from every point. This is impossible on a flat map. Since the scale varies continuously from location to location on a map projection, the great circle distance between the two locations on the globe must be distorted on the projection. For some equidistant projections such as the azimuthal equidistant projection that we will learn more about later in this chapter, cartographers make the SF a constant 1.0 radially outward from a single point like the north pole (figure 3.14). Great circle distances are correct along the lines that radiate outward from that point. Long-distance route planning is based upon knowing the great circle distance between beginning and ending points, and equidistant maps are excellent tools for determining these distances on the earth.

Equidistant map projections are used to show the correct distance between a selected location and any other point on the projection. While the correct distance will be shown between the select point and all others, distances between all other points will be distorted. Note that no flat map can preserve both distance and area.

Shape

When *angles* on the globe are preserved on the map (thus preserving *shape*), the projection is called **conformal,** meaning "correct form or shape." Unlike the property of equidistance, conformality can be achieved at all points on conformal projections. To attain the property of conformality, the map scale must be the same in all directions from a point. A circle on the globe will thus appear as a circle on the map. But to achieve conformality, it is necessary either to enlarge or reduce the scale by a different amount at each location on the map (figure 3.5). This means, of course, that the map area around each location must also vary. Tiny circles on the earth will always map as circles, but their sizes will differ on the map. Since tiny circles on the earth are projected as circles, all directions from the center of the circles are correct on the projected circles. The circles must be tiny because the constant change in scale across the map means that they will be projected as ovals if they were hundreds of miles in diameter on the earth. You can also see the distortion in the appearance of the graticule—while the parallels and meridians will intersect at right angles, the distance between parallels will vary. Typically there will be a smooth increase or decrease in scale across the map. Conformality applies only to directions or angles at—or in the immediate vicinity of—points, thus shape is preserved only in small areas. It does not apply to areas of any great extent; the shape of large regions can be greatly distorted.

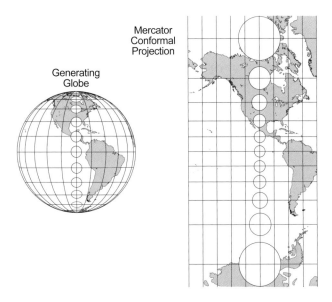

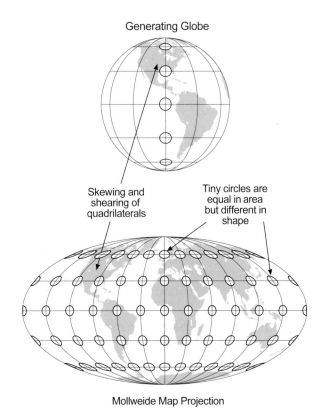

Figure 3.5 On a map using a conformal projection, identical tiny circles (greatly magnified here) on the generating globe are projected to the flat map as different sized circles according to the local SF. The appearance of the graticule in all the projected circles is the same as on the globe.

Figure 3.6 Tiny circles on the generating globe (greatly enlarged here) are projected as ellipses of the same area but different shape. This causes the directions of the major and minor axes for each ellipse to be skewed. Similarly, quadrilaterals are skewed and sheared on an equal area projection.

Conformal maps are best suited for tasks that involve plotting, guiding, or analyzing the motion of objects over the earth's surface. Thus, conformal projections are used for aeronautical and nautical charts, topographic quadrangles, and meteorological maps, as well as when the shape of environmental features is a matter of concern.

Direction

While conformal projections preserve angles locally, it's sometimes important to preserve directions globally. Projections that preserve global *directions* are called **true direction projections.** You will also sometimes see these referred to as **azimuthal projections.** As you will see in the next section, this term is also sometimes used instead of "planar" to describe projections. To avoid confusion, we will use the terms "true direction" and "planar" in this book. Unfortunately, no projection can represent correctly all directions from all points on the earth as straight lines on a flat map. But scale across the map can be arranged so that certain types of direction lines are straight, such as in figure 3.13 where all meridians radiating outward from the pole are correct directionally. All true direction projections correctly show the **azimuth** or direction from a reference point, usually the center of the map (see chapter 12 for more on azimuths). Hence, great circle directions on the ground from the reference point can be measured on the map.

True direction projections are used to create maps that show the great circle routes from a selected point to a desired destination. Special true direction projections for maps like this are used for long-distance route planning in air and sea navigation.

Area

When the relative *size* of regions on the earth is preserved everywhere on the map, the projection is said to be **equal area** and to have the property of **equivalence.** The demands of achieving equivalence are such that the SF can only be the same in all directions along one or two lines, or from at most two points. Since the SF and hence angles around all other points will be deformed, the scale requirements for equivalence and conformality are mutually exclusive. No projection can be both conformal and equal area—only a globe can be.

Adjusting the scale along meridians and parallels so that shrinkage in one direction from a point is compensated for by exaggeration in another direction creates an equal area map projection. For example, a small circle on the globe with an SF of 1.0 in all directions may be projected as an ellipse with a north–south SF of 2.0 and east–west SF of 0.5 so that the area of the ellipse is the same as the circle. Equal area world maps, therefore, compact, elongate, shear, or skew circles on the globe as well as the quadrilaterals of the graticule (figure 3.6). The distortion of shape, distance, and direction is usually most pronounced toward the map's margins.

Despite the distortion of shapes inherent in equal area projections, they are the best choice for tasks that call for area or density comparisons from region to region. Examples of geographic phenomena best shown on equal area map projections include world maps of population density, per capita income, literacy, poverty, and other human-oriented statistical data.

Projection families based on map projection surface

So far we've categorized map projections into families based on the types of geometric distortions that occur on them, realizing that particular geometric and other properties can be preserved under special circumstances. We can also categorize map projections based on the different surfaces used in constructing the projections.

As a child, you probably played the game of casting hand shadows on the wall. The surface your shadow was projected onto was your **projection surface.** You discovered that the distance and direction of the light source relative to the position of your hand influenced the shape of the shadow you created. But the surface you projected your shadow onto also had a great deal to do with it. A shadow cast on a corner of the room or on a curved surface like a beach ball or lampshade was quite different from one thrown upon the flat wall.

You can think of the globe as your hand and the map projection as the shadow on the wall. To visualize this, imagine a transparent generating globe with the graticule and continent outlines drawn on it in black. Then suppose that you place this globe at various positions relative to a source of light and a surface that you project its shadow onto (figure 3.7). In the case of maps, the flat surface that the earth's features are projected onto is called the **developable surface.** There are three basic projection families based on the developable surface—**planar, conic,** and **cylindrical.** Depending on which type of projection surface you use and where your light source is placed, you will end up with different map projections cast by the globe. All projections that you can create in this light-casting way are called **true perspective.**

Strictly speaking, very few projections actually involve casting light onto planes, cones, or cylinders in a physical sense. Most are not purely geometric and all use a set of mathematical equations in map projection software that transform latitude and longitude coordinates on the earth into x,y coordinates on the projection surface. The x,y coordinates for a coastline, for example, are the endpoints for a sequence of short straight lines drawn by a plotter or displayed on a computer monitor. You see the sequence of short lines as a coastline. Projections that don't use developable surfaces are more common because they can be designed to serve any desired purpose, can be made conformal, equal area, or equidistant, and can be readily produced with the aid of computers. Yet even nondevelopable surface projections usually can be thought of as variations of one of the three basic projection surfaces.

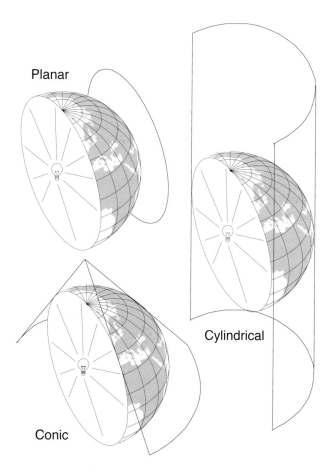

Figure 3.7 Planes, cones, and cylinders are used as true perspective map projection surfaces.

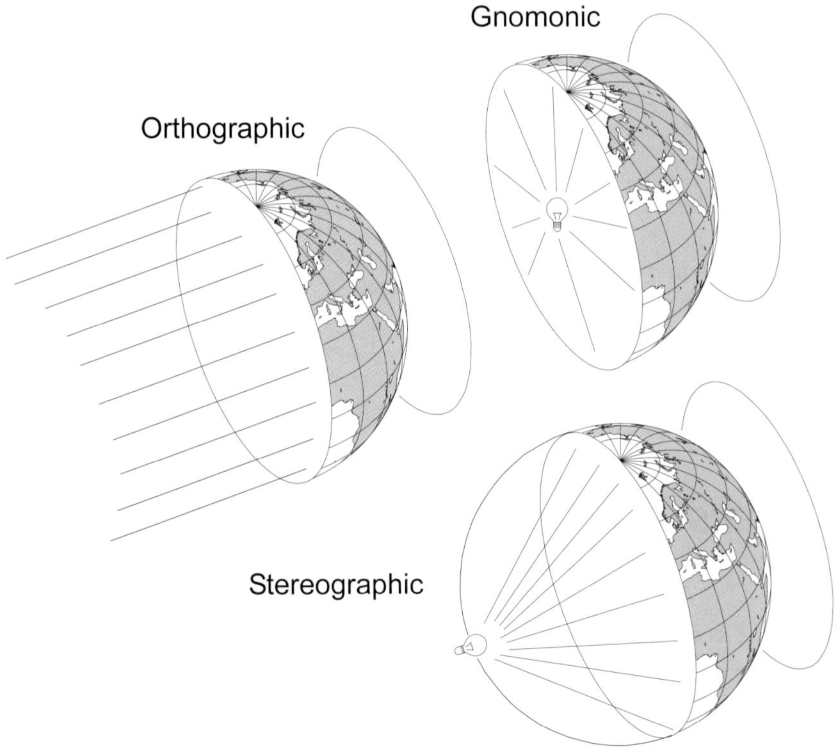

Gnomonic

Orthographic

Stereographic

Figure 3.8 **The three true perspective planar projections (orthographic, gnomonic, and stereographic) can be thought of as being constructed by changing the light-source location relative to the generating globe (they are actually constructed by mathematical equations in mapping software).**

Planar

Planar projections (also called **azimuthal projections**) can be thought of as being made by projecting onto a flat plane that touches the generating globe at a point or slices through the generating globe. If you again consider your childhood shadow-casting game, you will recall that the projection surface was only one of the factors influencing the shape of the shadow cast upon the wall. The other influence was the distance of the light source from the wall. Thus, there will be basic differences within the **planar projection family,** depending on where the imaginary light source is located.

This family includes three commonly used **true perspective planar projections,** which can be conceived of as projecting onto a plane tangent to the generating globe at a point of tangency (figure 3.8): **orthographic** (light source at infinity), **stereographic** (light source on the surface of the generating globe opposite the point of tangency), and **gnomonic** (light source at the center of the generating globe). Other projections, such as the **azimuthal equidistant** and **Lambert azimuthal equal area** (described later in this chapter), are mathematical constructs that cannot be created geometrically and must be built from a set of mathematical equations.

Cylindrical

Cylindrical projections can be thought of as being made by projecting onto a cylinder that touches the generating globe along any small circle or slices through the generating globe. As with planar projections, cylindrical projections can also be distinguished by the location of the light source relative to the projection surface. **True perspective cylindrical projections** can be thought of as projecting onto a cylinder tangent to the generating globe along a great circle line of tangency (usually the equator). This family includes two commonly used true perspective projections (figure 3.9): **central cylindrical** (light source at the center of the generating globe) and **cylindrical equal area** (linear light source akin to a fluorescent light with parallel rays along the polar axis). These are described in more detail later in this chapter. The whole world can't be projected onto the central cylindrical projection because the polar rays will never intersect the cylinder. For this reason, it is often cut off at some specified latitude north and south.

Conic

The last family of map projections based on the surface used to construct the projection is **conic projections,** which can be thought of as projecting onto a cone with the line of tangency on the generating globe along any small circle or a cone that slices through the generating globe. We usually select a mid-latitude parallel as a line of tangency. This projection family includes one **true perspective conic projection,** which can be conceived of as projecting onto a cone tangent to the generating globe along a small circle line of tangency, usually in the mid-latitudes. The one commonly used true perspective projection in this family is the **central conic projection** (light source at the center of the generating globe).

To fully understand map projection families based on the projection surface, it is also necessary to understand various projection parameters, including case and aspect, among others.

MAP PROJECTION PARAMETERS

Tangent and secant case

The projection surface may have either a tangent or secant relationship to the globe. A **tangent case** projection surface will touch the generating globe at either a point (called a **point of tangency**) for planar projections or along a line (called a **line of tangency**) for conic or cylindrical projections (figure 3.10). The scale factor (SF) is 1.0 at the point of tangency for planar projections or along the line of tangency for cylindrical and conic projections. The SF increases outward from the point of tangency or perpendicularly away from the line of tangency.

A **secant case** planar projection surface intersects the generating globe along a small circle line of tangency (figure 3.10). Secant case conic projections have two small circle lines of tangency, or standard parallels, usually at the mid-latitudes. Secant case cylindrical projections have two small circle lines of tangency that are equidistant from the parallel where the projection is centered. For example, a secant case cylindrical projection centered at the equator might have the 10°N and 10°S parallels as lines of tangency. All secant case projections have an SF of 1.0 along the lines of tangency. Between the lines of tangency, the SF decreases from 1.0 to a minimum value half way between the two lines, while outside the lines, the SF increases from 1.0 to a maximum value at the edge of the map. In other words, the SF is slightly smaller in the middle part of the map and slightly larger than the stated scale at the edges. For planar projections, the SF increases outward from the circle of tangency and decreases inward to a minimum value at the center of the circle of tangency.

The advantage of the secant case is that it minimizes the overall scale distortion on the map. This is because the SF is 1.0 along a circle instead of at a single point (for a secant case planar projection) or two lines instead of one (for secant case conic or cylindrical projections). As

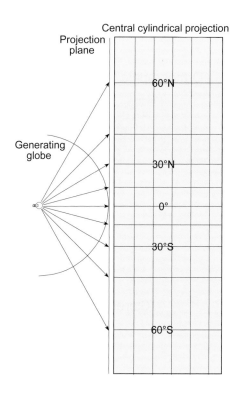

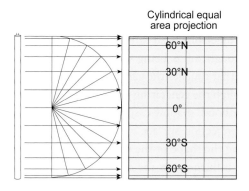

Figure 3.9   True perspective central cylindrical and cylindrical equal area projections.

a result, the central part of the projection has a slightly smaller SF than the stated scale and the edges don't have SFs as large as with the tangent case, thus providing a greater area of minimum distortion.

Obviously, there is minimal distortion around the point or line of tangency. This explains why earth curvature may often be ignored without serious consequence when using flat maps for a local area, but only if the line—or lines—of tangency are set to fall within the area being mapped. But as the distance from the point or line of tangency increases, so does the degree of scale distortion. By the time a projection has been extended to include the entire earth, scale distortion may have greatly affected the earth's appearance on the map.

PLANAR PROJECTIONS
TANGENT CASE

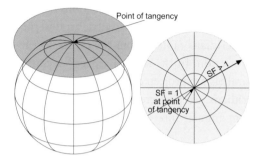

CYLINDRICAL PROJECTIONS
TANGENT CASE

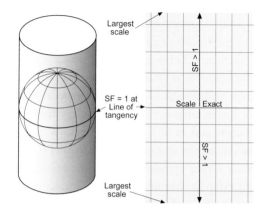

SECANT CASE

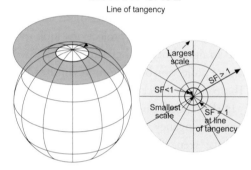

SECANT CASE

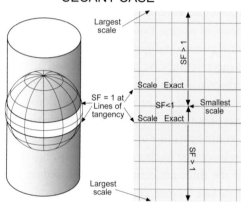

CONIC PROJECTIONS

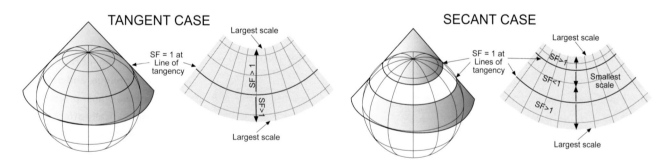

Figure 3.10 Tangent and secant cases of the three basic projection surfaces.

Secant case projections are common for all three projections families—planar, conic, and cylindrical. You will also find tangent case planar and cylindrical, but tangent case conic projections are little used in mapping, as their relatively small area of minimal scale distortion limits their practical value.

Although secant case planar projections have less overall distortion, the tangent case is often used instead for equidistant projections. Secant case cylindrical projections are best suited for world maps since they have less overall distortion than tangent case projection—this distortion is minimized in the mid-latitudes, which is also where the majority of the earth's population is found. Secant case conic projections are best suited for maps of mid-latitude regions, especially those elongated in an east–west direction. The United States and Australia, for example, meet these qualifications and are frequently mapped using secant case conic projections.

Aspect

Map projection **aspect** refers to the location of the point or line(s) of tangency on the projection surface (figure 3.11). A projection's point or line(s) of tangency can in theory touch or intersect anywhere on the developable surface. When a tangent case projection point or line of tangency is at or along the equator, the resulting projection is said to be in **equatorial aspect.** When the point or line of tangency is at or encircles either pole, the projection is said to be in **polar aspect.** With cylindrical projections, the term **transverse aspect** is also used. Transverse aspect occurs when the line of tangency for the projection is shifted 90 degrees so that it follows a pair of meridians (figure 3.12). Any other alignment of the point or line(s) of tangency to the globe is a projection in **oblique aspect.**

Aspects of secant case projections are defined in a manner similar to tangent case aspects. For example, a polar aspect, secant case planar projection has the north or south pole as the center of its circle of tangency. A transverse aspect, secant case cylindrical projection has two lines of tangency—equally spaced from the meridian —that would be its line of tangency in its tangent case.

The aspect that has been used the most historically for each of the three projection families based on the developable surface (planar, conic, and cylindrical) is referred to as the **normal aspect.** The normal aspect of planar projections is polar; the normal aspect of conic projections is

the apex of the cone above a pole; and the normal aspect of cylindrical projections is equatorial.

The normal aspect of planar projections is polar—the parallels of the graticule are concentric around the center and the meridians radiate from the center to the edges of the projection. In the normal aspect for conic projections, parallels are projected as concentric arcs of circles, and meridians are projected as straight lines radiating at uniform angular intervals from the apex of the cone (figure 3.10).

The graticule appears entirely different on normal aspect and transverse aspect cylindrical projections. You can recognize the normal aspect of cylindrical projections by horizontal parallels of equal length, vertical meridians of equal length that are also equally spaced, and right angle intersections of meridians and parallels. Transverse aspect cylindrical projections look quite different. The straight line parallels and meridians on the normal aspect projection become curves in the transverse aspect. These curves are centered on the vertical line of tangency in the tangent case or halfway between the two vertical lines of tangency in the secant case (figure 3.12).

As we have seen, the choice of projection surface aspect leads to quite different appearances of the earth's land masses and the graticule. Yet the distortion properties of a given projection surface remain unaltered when the aspect is changed from polar to oblique or equatorial. The map on the left in figure 3.12 is a good example.

Planar Projection Aspects

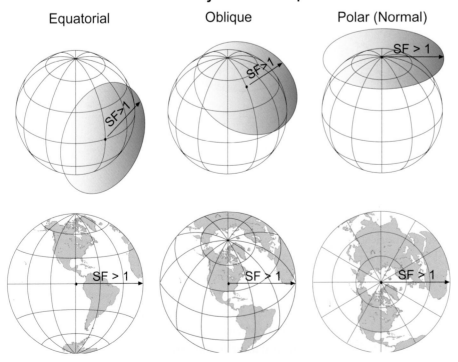

Equatorial Oblique Polar (Normal)

Figure 3.11 Tangent case equatorial, oblique, and polar aspects of the planar projection surface dramatically affect the appearance of the graticule and land areas. Scale distortion increases radially away from the point of tangency, no matter where the point is located on the globe.

In its normal aspect tangent case, the SF is 1.0 at the equator and increases north and south at right angles to the equator. In its transverse aspect tangent case, the SF is 1.0 along the pair of meridians that form the line of tangency (that is, a selected meridian and its antipodal meridian). The SF again increases perpendicularly to the line of tangency, which is due east and west at the intersection of each parallel and the pair of meridians. The SF will be exactly the same distance above and below the equator on the normal aspect and exactly the same distance again to the left and right of the meridian pair on the transverse aspect projection.

Other map projection parameters

There are a few other map projection parameters that are useful to know about as they can have an impact on the appearance and appropriate use of a map projection. The **central meridian** (also called the **longitude of origin** or less commonly the *longitude of center*) defines the origin of the longitudinal x-coordinates. This is usually modified to so that it is in the center of the area being mapped. The **central parallel** (also called the **latitude of origin** or less commonly the *latitude of center*) defines the origin of the y-coordinates—an appropriately defined map projection will be modified so that this parameter is defined relative to the mapped area, although this parameter may not be located at the center of the projection. In particular, conic projections use this parameter to set the origin of the y-coordinates below the area of interest so that all y-coordinates are positive.

CYLINDRICAL PROJECTION ASPECTS

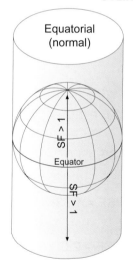

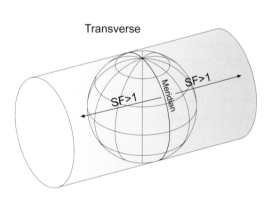

Figure 3.12 **Tangent case equatorial and transverse aspects of the cylindrical projection surface dramatically affect the appearance of the graticule and land areas. Scale distortion increases perpendicularly away from the line of tangency, no matter where the line is located on the globe.**

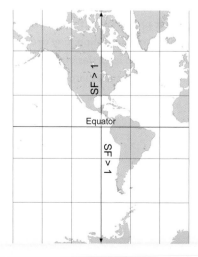

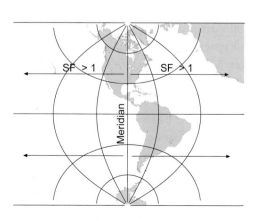

COMMONLY USED MAP PROJECTIONS

As stated above, one way that map projections are commonly grouped into families is based on the projection surface—a plane, cone, or cylinder. We'll use these surfaces to structure our discussion of commonly used map projections.

Planar projections

Orthographic

The **orthographic projection** is how the earth would appear if viewed from a distant planet (figure 3.13). Since the light source is at an infinite distance from the generating globe, all rays are parallel. This projection appears to have been first used by astronomers in ancient Egypt, but it came into widespread use during World War II with the advent of the global perspective provided by the air age. It is even more popular in today's space age, often used to show land-cover and topography data obtained from remote sensing devices. The generating globe and half-globe illustrations in this book are orthographic projections, as is the map on the front cover of the book. The main drawback of the orthographic projection is that only a single hemisphere can be projected. Showing the entire earth requires two hemispherical maps. Northern and southern hemisphere maps are commonly made, but you may also see western and eastern hemisphere maps.

Stereographic

Projecting a light source from the antipodal point on the generating globe to the point of tangency creates the **stereographic projection** (figure 3.14). This is a conformal projection, so shape is preserved in small areas. The Greek scholar **Hipparchus** is credited with inventing this projection in the second century BC. It is now most commonly used in its polar aspect and secant case for maps of polar areas. It is the projection surface used for the Universal Polar Stereographic grid system for polar areas, as we will see in the next chapter. A disadvantage of the stereographic conformal projection is that it is generally restricted to one hemisphere. If it is not restricted to one hemisphere, then the distortion near the edges increases to such a degree that the geographic features in these areas are basically unrecognizable. In past centuries, it was used for atlas maps of the western or eastern hemisphere.

Figure 3.13 The orthographic projection best shows the spherical shape of the earth.

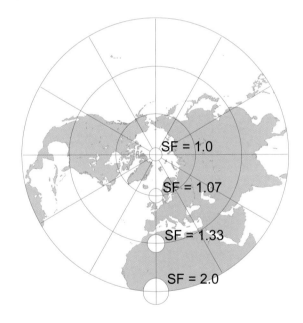

Figure 3.14 Polar stereographic projection of the northern hemisphere. Since this is a conformal projection, tiny circles on the generating globe are projected as circles of the same size at the point of tangency to four times as large at the equator.

Gnomonic

Projecting with a light source at the center of the generating globe to a tangent plane produces the **gnomonic projection.** One of the earliest map projections, the gnomonic projection was first used by the Greek scholar **Thales of Miletus** in the sixth century BC for showing different constellations on **star charts,** which are used to plot planetary positions throughout the year. The position of constellations in the sky over the year was used as a calendar, telling farmers when to plant and harvest crops, and when floods would occur. Horoscopes and astrology also began with the ancient Greeks over two thousand years ago. Many believed that the position of the sun and the planets had an effect on a person's life and that future events in their lives could be predicted based upon the location of celestial bodies in the sky.

The gnomonic projection is the only projection with the useful property that all great circles on the globe are shown as straight lines on the map (figure 3.15). Since a great circle route is the shortest distance between two points on the earth's surface, the gnomonic projection is especially valuable as an aid to navigation (see great circle directions in chapter 12). The gnomonic projection is also used for plotting the global dispersal of seismic and radio waves. Its major disadvantages are increasing distortion of shape and area outward from the center point and the inability to project a complete hemisphere.

Azimuthal equidistant

The **azimuthal equidistant projection** in its polar aspect has the distinctive appearance of a dart board—equally spaced parallels and straight-line meridians radiating outward from the pole (figure 3.16). This arrangement of parallels and meridians results in all straight lines drawn from the point of tangency being great circle routes. Equally spaced parallels mean that great circle distances are correct along these straight lines. The ancient Egyptians apparently first used this projection for star charts, but during the air age it also became popular for use by pilots planning long-distance air routes. In the days before electronic navigation, the flight planning room in major airports had a wall map of the world that used an oblique aspect azimuthal equidistant projection centered on the airport. You will also find them in the public areas of some airports. All straight lines drawn from the airport are correctly scaled great circle routes. This is one of the few planar projections that can show the entire surface of the earth.

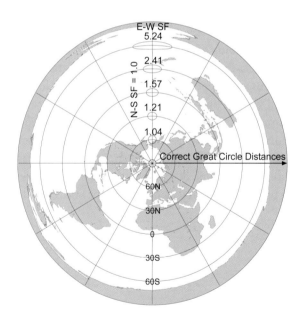

Figure 3.16 **Polar aspect azimuthal equidistant world map projection tangent at the pole. Great circle distances are correct along straight lines outward from the point of tangency at the north pole since the north–south SF is always 1. The east–west SF increases to a maximum of infinity at the south pole, where there is a point-to-line correspondence.**

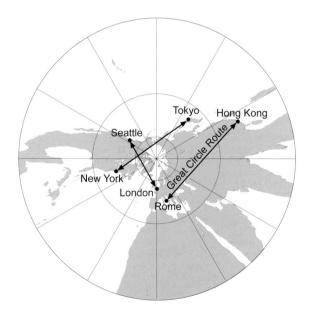

Figure 3.15 **Polar gnomonic projection of the northern hemisphere from 15°N to the pole. All straight lines on the projection surface are great circle routes. Note the severe shape distortion of this projection compared to the polar stereographic projection shown in figure 3.13.**

Lambert azimuthal equal area

In 1772 the mathematician and cartographer **Johann Heinrich Lambert** published equations for the tangent case planar **Lambert azimuthal equal area projection,** which, along with other projections he devised, carries his name. This planar equal area projection is usually restricted to a hemisphere, with polar and equatorial aspects used most often in commercial atlases (figure 3.17). More recently, this projection has been used for statistical maps of continents and countries that are basically circular in overall extent, such as Australia, North America, and Africa. You will also see the oceans shown on maps that use the equatorial or oblique aspects of this projection. The Lambert azimuthal equal area projection is particularly well suited for maps of the Pacific Ocean, which is almost hemispheric in extent.

Cylindrical projections

Equirectangular

The **equirectangular projection** is also called the **equidistant cylindrical** or **geographic projection.** This simple map projection, nearly 2,000 years old, is attributed to **Marinus of Tyre,** who is thought to have constructed the projection about in 100 AD. Parallels and meridians are mapped as a grid of equally spaced horizontal and vertical lines twice as wide as high (figure 3.18). The equal spacing of parallels means that the projection is equidistant in the north–south direction with a constant SF of 1.0. In the east–west direction the SF increases steadily from a value of 1.0 at the equator to infinity at each pole, which is projected as a straight line.

You may see world maps showing elevation data or satellite imagery on this projection. This choice of projection is not based on any geometrical advantages, but rather on the simplicity of creating flat world maps when they had to be done by hand.

Figure 3.17 The Lambert azimuthal equal area projection is often used for maps of continents that have approximately equal east–west and north–south extents. This map of North America in the box is part of an oblique aspect of the projection centered at 45°N, 100°W.

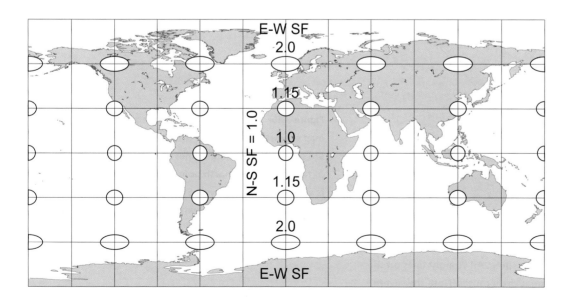

Figure 3.18 Equirectangular world map projection.

Mercator

Constructed by **Gerhardus Mercator** in 1569, the Mercator projection is a tangent case cylindrical conformal projection. As with all conformal projections, shape is preserved in small areas. This projection offers a classic example of how a single projection can be used both poorly and well. Looking at the projection (figure 3.19), we can imagine Mercator started the construction of his projection with a horizontal line to represent the equator and then added equally spaced vertical lines to represent the meridians. Mercator knew that meridians on the globe converge toward the poles, so the meridians he had drawn as parallel vertical lines must become progressively more widely spaced toward the poles than they would be on the generating globe. He progressively increased the spacing of parallels away from the equator so that the increase matched the increased spacing between the meridians. As a result of this extreme distortion toward the poles, he cut his projection off at 80°N and S. This not only produced a conformal map projection, but also the only projection on which all lines of constant compass direction, called **rhumb lines,** are straight lines on the map.

Navigators who used a magnetic compass immediately saw the advantage of plotting courses on the Mercator projection, since any straight line they drew would be a line of constant bearing. This meant they could plot a course on the map and simply maintain the associated bearing during passage to arrive at the plotted location. You can see how navigators would prefer a map on which compass bearings would appear as straight rhumb lines (see chapter 13 for more on rhumb line plotting). The Mercator projection has been used ever since for nautical charts, such as small-scale piloting charts of the oceans. The large-scale nautical charts used for coastal navigation can be thought of as small rectangles cut out of a world map that uses the Mercator projection.

Of course, using these maps for navigation is not really as simple as this. Recall that the gnomonic projection is the only projection on which all great circles on the generating globe are shown as straight lines on the flat map. Lines drawn on a Mercator projection show constant compass bearing, but they are not the same as the great circle route, which is the shortest distance between two points on a globe. Therefore, the Mercator projection is often used in conjunction with the gnomonic projection to plot navigational routes. The gnomonic projection is used first to determine the great circle route between two points, and then the route is projected to transform the great circle to a curve on a Mercator projection. Finally, the curve is translated into a series of shorter straight line segments representing portions of the route, each with constant compass direction (see figure 12.28 in chapter 12 for an illustration of this process).

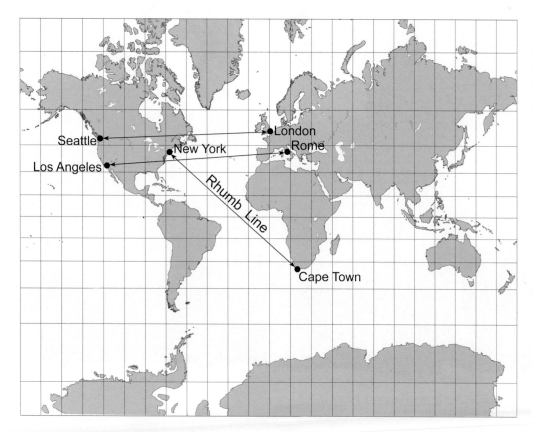

Figure 3.19 **Mercator projection shown with rhumb lines between selected major world cities.**

The use of the Mercator projection in navigation is an example of a projection used for its best purpose. A poor use of the Mercator projection is for wall maps of the world. We saw earlier that this projection cannot cover the entire earth, and is often cut off at 80°N and S. Cutting off part of the world does create a rectangular projection surface with a height-to-width ratio that fits walls very well. The problem, of course, is the extreme scale enlargement and consequent area distortion at higher latitudes. The area exaggeration of North America, Europe, and Russia gives many people an erroneous impression of the size of the land masses.

Gall-Peters

The **Gall-Peters projection** is a variation of the cylindrical equal area projection. Its equations were published in 1885 by Scottish clergyman James Gall as a secant case of the cylindrical equal area projection that lessens shape distortion in higher latitudes by placing lines of tangency at 45°N and 45°S. **Arno Peters,** a German historian and journalist, devised a map based on Gall's projection in 1967 and presented it in 1973 as a "new invention" superior to the Mercator world projection (figure 3.20).

The projection generated intense debate because of Peters's assertion that this was the only "nonracist" world map. Peters claimed that his map showed developing countries more fairly than the Mercator projection, which distorts and dramatically enlarges the size of Eurasian and North American countries. His assertion was a bit of a straw dog, since the Mercator projection was designed and admirably suited for navigation and never intended for comparing country sizes.

The first English version of the Gall-Peters projection map was published in 1983, and it continues to have passionate fans as well as staunch critics.

Although the relative areas of land masses are maintained, their shapes are distorted. According to prominent cartographer Arthur Robinson, the Gall-Peters map is "somewhat reminiscent of wet, ragged long winter underwear hung out to dry on the Arctic Circle." Although several international organizations have adopted the Gall-Peters projection, there are other equal area world projections, such as the Mollweide projection (figure 3.24), that distort the shapes or land masses far less. Maps based on the Gall-Peters projection continue to be published and are readily available, though few major map publishers use the projection today.

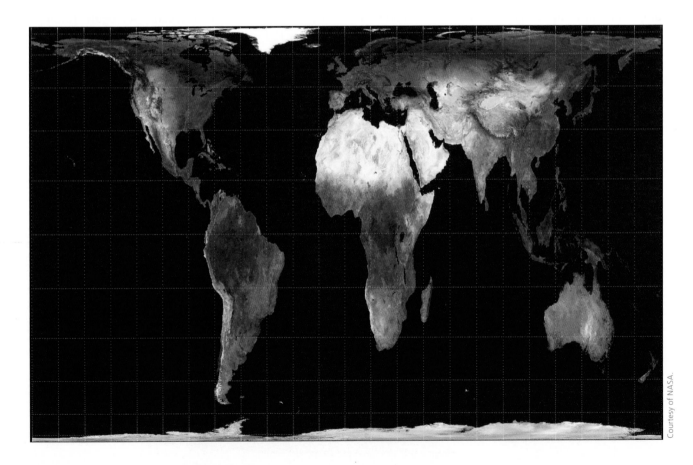

Figure 3.20 Gall-Peters projection world map.

Transverse Mercator

We saw earlier that Lambert constructed his azimuthal equal area projection in 1772. That same year, he also constructed the **transverse Mercator projection,** along with the Lambert azimuthal equal area projection and another described later in this chapter. In Europe this projection is called the Gauss-Krüger, in honor of the mathematicians **Carl Gauss** and **Johann Krüger** who later worked out formulas describing its geometric distortion and equations for making it on the ellipsoid. Lambert's idea for the transverse Mercator projection was to rotate the Mercator projection by 90° so that the line of tangency became a pair of meridians—that is, any selected meridian and its antipodal meridian (figure 3.12, bottom right). The resulting projection is conformal, as is the Mercator projection, but rhumb lines no longer are straight lines. Along the central meridian of the projection (the vertical meridian that defines the y-axis of the projection), the SF is 1.0, and the scale increases perpendicularly away from the central meridian. Thus, narrow north–south strips of the earth are projected with no local shape distortion and little distortion of area.

You're likely to see the transverse Mercator projection used to map north–south strips of the earth called **gores** (figure 3.21), which are used in the construction of globes. Because printing the earth's surface directly onto a round surface is very difficult, instead, a map of the earth is printed in flat elongated sections and then attached to a spherical object. The narrow, 6°-wide zones of the universal transverse Mercator grid system (described in chapter 4) are based on a secant case transverse Mercator projection. North–south trending zones of the U.S. state plane coordinate system (also explained in chapter 4) are also based on secant cases of the projection. Most 1:24,000-scale U.S. Geological Survey (USGS) topographic maps are projected on these state plane coordinate system zones.

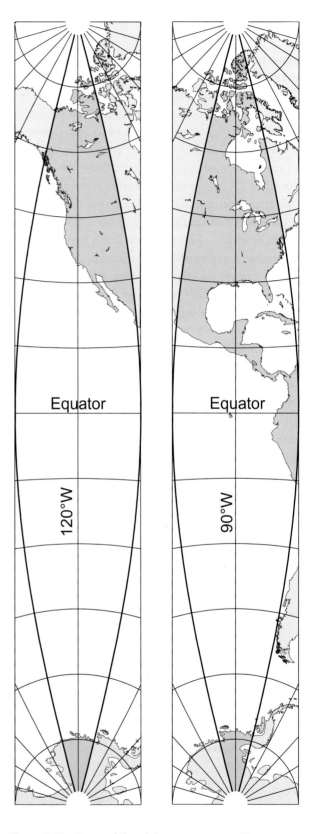

Figure 3.21 Gores of the globe on transverse Mercator projections 30° wide at the equator centered at 90°W and 120°W. To make a world globe, the highlighted portion of each map would be cut out and pasted onto the globe and other gores would be used to cover the remaining area.

Conic map projections

Lambert conformal conic

The **Lambert conformal conic projection** is another of the widely used map projections constructed by Lambert in 1772. It is a secant case normal aspect conic projection with its two standard parallels placed so as to minimize the map's overall scale distortion. The standard parallels for maps of the conterminous United States are placed at 33°N and 45°N to keep scale distortion at the map's edges to less than 3 percent (figure 3.22). Just as the transverse Mercator projection is used as the basis for the state plane coordinate system zones in north–south trending states, the Lambert conformal conic projection is used as the basis for system zones in east–west trending states in the United States like Oregon and Wisconsin (see chapter 4 for more on the state plane coordinate system using examples from these two states). As noted earlier, these projections are in turn used for the 1:24,000-scale USGS topographic maps within the state.

One major use of the Lambert conformal conic projection is for aeronautical charts. All U.S. 1:500,000-scale sectional charts can be thought of as smaller rectangles of the national map described in the above paragraph. Recall that navigators (like aviators) prefer navigational charts that use conformal projections, which preserve shapes and directions locally. Equally important is the fact that straight lines drawn on the 1:500,000-scale charts are almost great circle routes on the earth's surface (figure 3.22).

Albers equal area conic

Mathematically devised in 1805 by the German mathematician **Heinrich C. Albers,** the **Albers equal area conic projection** was first used in 1817 for a map of Europe. This is probably the projection used most often for statistical maps of the conterminous United States and other mid-latitude east–west trending regions. For example, you will see this projection used for U.S. statistical maps created by the Census Bureau and other federal agencies. The secant case version that has been used for nearly 100 years for the conterminous United States has standard parallels placed at 29.5°N and 45.5°N (figure 3.23). This placement reduces the scale distortion to less than 1 percent at the 37th parallel in the middle of the map, and to 1.25 percent at the northern and southern edges of the country. USGS products, such as the national tectonic and geologic maps also use the Albers projection, as do recent reference maps and satellite image mosaics of the country at a scale of around 1:3,000,000.

The reason for the widespread use of the Albers projection is simple—people looking at maps using this projection can assume that the areas of states and counties on the map are true to their areas on the earth.

Pseudocylindrical and other projections

We have seen that map projections can be classed into families based on the nature of the surface used to construct the projection. This classification results in planar, conic, and cylindrical families. But these three families constitute only a small portion of the vast array of

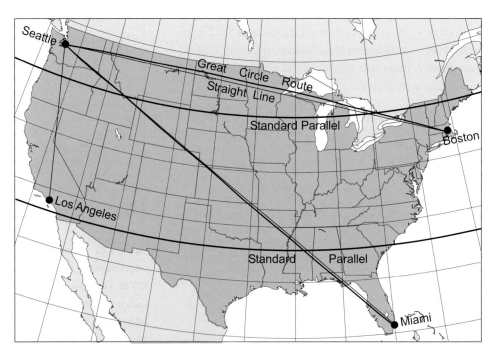

Figure 3.22 Lambert conformal conic projection used for a map of the United States. Straight lines are very close to great circle routes, particularly on north–south paths.

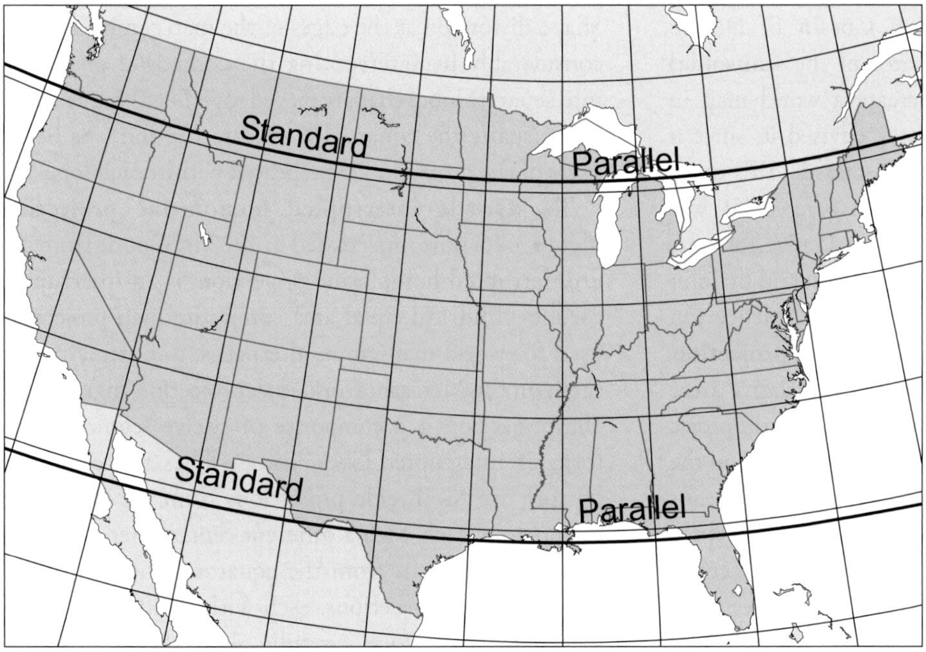

Figure 3.23 Albers equal area conic projection of the United States.

projections that have been constructed by cartographers. We also noted that map projections could be classified based on their geometric distortion properties. Using this classification, we have shown that some preserve areas or shapes, while others have no special property aside from holding overall scale distortion to a minimum or presenting a pleasing visual image. For many purposes, a projection that "looks right" is more important than a projection that rigidly provides area, distance, shape, or direction fidelity.

Let's look at a last set of projections that can be described using one or both of the classifications above. These include pseudocylindrical and other projections that are of special interest. **Pseudocylindrical** map projections can be conceived of as juxtaposing a number of partial cylindrical maps. They are similar to cylindrical projections in that parallels are horizontal lines and meridians are equally spaced. The difference is that all meridians except the vertical-line central meridian are curved instead of straight.

Mollweide

You've probably seen world maps in the shape of an ellipse twice as wide as it is high. Most likely you were looking at the **Mollweide projection,** constructed in 1805 by the German mathematician **Carl B. Mollweide.** This elliptical equal area projection most commonly uses the equator as the standard parallel and the prime meridian as the central meridian of the projection (figure 3.24). Parallels are horizontal lines, but they are not equally spaced as on

the generating globe. Instead, they reduce in distance as the poles are approached. The elliptical shape of this projection makes it look more "earth-like," and the overall distortion in shape is less than on other equal area world projections such as the Gall-Peters.

You'll find the Mollweide projection used for maps that show a wide range of global phenomena, from population to land cover to major diseases. Cartographers have devised other orientations of the projection by adjusting the central meridian to better show the oceans or to center attention on a particular continent.

Figure 3.24 Mollweide projection used for a world map.

Sinusoidal

According to some sources, **Jean Cossin** of Dieppe, France, appears to be the originator of the **sinusoidal projection,** which he used to create a world map in 1570. Others suggest that Mercator devised it, since it was included in later editions of his atlases. This easily constructed equal area projection (figure 3.25) was used by Nicholas Sanson (ca. 1650) of France for atlas maps of the world and separate continents, and by John Flamsteed (1729) of England for star maps. Hence you may also see it called the **Sanson-Flamsteed projection.**

In addition to correctly portraying the relative areas of continents and countries, the sinusoidal world projection has an SF of 1.0 along the central meridian, and the east–west SF also is 1.0 anywhere on the map. This means that the projection is equidistant in the east–west direction and in the north–south direction along the central meridian, but only in these directions. Note the severe shape distortion at the edges of the map (figure 3.25).

Homolosine

You may have seen world map projections made by compositing different projections along certain parallels or meridians. The **uninterrupted homolosine projection** (figure 3.26, top), constructed in 1923 by American geography professor **J. Paul Goode,** is a composite of two pseudocylindrical projections. The sinusoidal projection is used for the area from 40°N to 40°S latitude, while Mollweide projections cover the area from 40°N to 40°S to the respective poles. Since the component projections are equal area, the homolosine projection is as well. Notice in figures 3.25 and 3.26 that the shapes of the continents look less distorted on the homolosine projection than on the sinusoidal projection, particularly in polar areas.

An **interrupted projection** is one in which the generating globe is segmented in order to minimize the distortion within any **lobe** (section) of the projection. Shape distortion at the edges of the map can be lessened considerably by interrupting the composite projection into separate lobes that are pieced together along a central line, usually the equator. With interruption, the better parts of the projection are repeated within each lobe.

The **Goode interrupted homolosine projection** (figure 3.26, bottom), created in 1923 by Goode from the uninterrupted homolosine projection, is an interrupted pseudocylindrical equal area composite map projection used for world maps as an alternative to portraying the earth on the Mercator world map projection (figure 3.19). The projection is a composite of twelve segments that form six interrupted lobes. The six lobes at the top and bottom are Mollweide projections from 40°N or 40°S to the pole, each with a different central meridian. The six interior regions from the equator to 40°N or 40°S are sinusoidal projections, each with a different central meridian. If you look carefully along the edges of the lobes, you can see a subtle discontinuity at the 40th parallels. The two northern sections are usually shown with some land areas repeated in both regions to show the Greenland land mass without interruption.

You will find the Goode homolosine projection used for maps in commercial world atlases to show a variety of global information. It is a popular projection for showing physical information about the entire earth, such as land elevations and ocean depths, land-cover and vegetation types, and satellite image composites. When viewing these maps, remember that the continuity of the earth's surface has been lost completely by interrupting the projection. This loss of continuity may not be apparent if the graticule is left off the projection, as may be the case with some global distribution maps found in world atlases.

Robinson

In 1963, the American academic cartographer **Arthur H. Robinson** constructed a pseudocylindrical projection that is neither equal area nor conformal but makes the continents "look right." A "right-appearing" projection is called **orthophanic.** For the **Robinson projection** (figure 3.27), Robinson visually adjusted the horizontal line parallels and curving meridians until they appeared suitable for a world map projection to be used in atlases and for wall maps. To do this, Robinson represented the poles as horizontal lines a little over half the length of the equator. You may have seen this projection used for world maps created by the National Geographic Society. It has also been used for wall maps of the world that show the shape and area of continents with far less distortion than wall maps that use the Mercator projection.

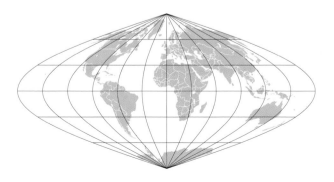

Figure 3.25   Sinusoidal equal area world map projection.

Figure 3.26 Uninterrupted
homolosine and Goode
interrupted homolosine equal
area world projections.

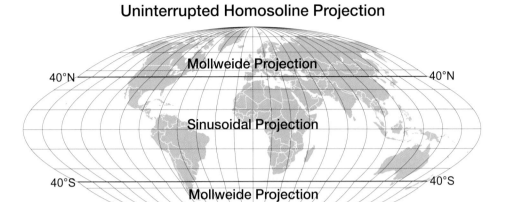

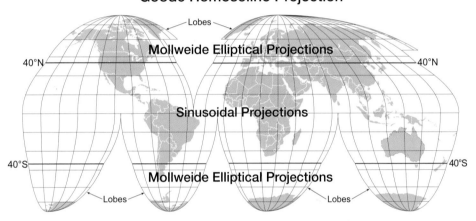

Figure 3.26 Uninterrupted
homolosine and Goode
interrupted homolosine equal
area world projections.

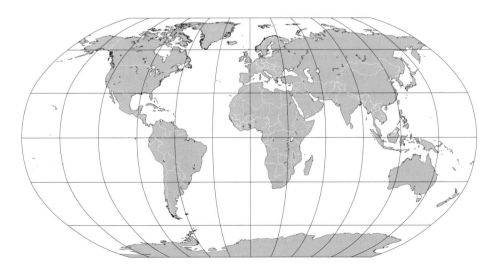

Figure 3.27 Robinson
pseudocylindrical world map
projection.

Aitoff

Other map projections can be constructed by mathematically modifying widely used projections developed centuries earlier. In 1889, the Russian cartographer **David Aitoff** published a modification of the equatorial aspect azimuthal equidistant projection that today carries his name (figure 3.28). For the **Aitoff projection,** Aitoff simply doubled the horizontal scale of the azimuthal equidistant projection, creating an elliptical projection with the same two-to-one width-to-height ratio as the Mollweide projection (figure 3.24). Unlike the Mollweide projection, parallels are not straight horizontal lines, and the map is neither equal area nor equidistant. The Aitoff projection is an interesting compromise between shape and area distortion, suggesting the earth's shape with less polar shearing than on maps that use the Mollweide projection.

The Winkel tripel projection was not used widely until 1998 when the National Geographic Society announced that it was adopting the Winkel tripel projection as its standard for maps of the entire world. As a result, use of the Winkel tripel projection has increased dramatically over the last few years.

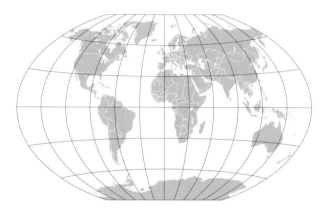

Figure 3.29   Winkel tripel projection used to create a world map.

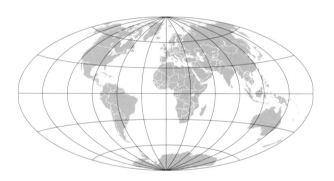

Figure 3.28   Aitoff projection used to create a world map.

Winkel tripel

New map projections may also be constructed as mathematical combinations of two existing projections. Perhaps the best known example is the **Winkel tripel projection** constructed in 1921 by the German cartographer Oswald Winkel. The term **tripel** is not someone's name, but rather a German word meaning a combination of three elements. Winkel used the term to emphasize that he had constructed a compromise projection that was neither equal area, conformal, nor equidistant, but rather minimized all three forms of geometric distortion. He accomplished this by averaging the x- and y-coordinates computed for the equirectangular and Aitoff world projections at the same map scale. The resulting projection is similar to the Robinson projection, but if you look closely you will see that parallels are not the straight horizontal lines characteristic of pseudocylindrical projections. Rather, they are slightly curving, nonparallel lines (figure 3.29).

PROJECTIONS ON THE SPHERE AND ELLIPSOID

Mapmakers have a general rule that small-scale maps can be projected from a sphere, but large-scale maps always must be projected from an ellipsoidal surface such as the WGS 84 ellipsoid. We saw in chapter 1 that small-scale world or continental maps such as globes and world atlas sheets normally use coordinates based on an authalic or other auxiliary sphere. This was because prior to using digital computers to make these types of maps numerically, it was much easier to construct them from spherical geocentric coordinates. Equally important, the differences in the plotted positions of spherical and corresponding geodetic coordinates are negligible on small-scale maps.

Large-scale maps must be projected from an ellipsoidal surface because, as we saw in chapter 1, the spacing of parallels decreases slightly but significantly from the pole to the equator. We noted that on the WGS 84 ellipsoid the distance between two points one degree apart in latitude (between 0° and 1°) at the equator is 68.703 miles (110.567 kilometers), shorter than the 69.407 mile (111.699 kilometer) distance between two points at 89°N and 90°N.

Let's see the differences in length when we project one degree of geodetic latitude along a meridian at the pole, at the 45th parallel, and at the equator using a transverse Mercator projection at a scale of 1:1,000,000 (figure 3.30). At this scale, the length of a degree of latitude at the pole is projected as slightly over 1 millimeter longer than a degree of latitude at the equator. This difference on these two projections may seem minimal, but on the ground it represents a distance of nearly a kilometer! At larger map scales the difference in length becomes more noticeable—for instance, slightly over 1 centimeter on polar and equatorial maps at a scale of 1:100,000. Topographic and other maps at this scale and larger are projected from an ellipsoid so that accurate distance and area measurements can be made on them. The same holds true for large-scale equidistant projections such as the polar aspect azimuthal equidistant projection (figure 3.16), since the spacing of parallels on the projection must be slightly lessened from the pole outward to reflect their actual spacing on the earth.

We saw in chapter 1 and in table C.2 in appendix C that the only place on the earth where the 69.05 mile (111.12 kilometer) per degree spacing of parallels on the authalic or rectifying sphere is essentially the same as on the WGS 84 ellipsoid is in the mid-latitudes close to the 45th parallel. You will find that the transverse Mercator or another map projection based on the sphere or the WGS 84 ellipsoid will have the same spacing of parallels in locations that straddle the 45th parallel.

Every map projection has its own virtues and limitations. You can evaluate a projection only in light of the purpose for which a map is to be used. You shouldn't expect that the best projection for one situation will be the most appropriate for another. Fortunately, the map projection problem effectively vanishes if the cartographer has done a good job of considering projection properties and if you are careful to take projection distortion into consideration in the course of map use.

At the same time, however, there is the very real possibility that the mapmaker through ignorance or lack of attention can choose an unsuitable projection or projection parameters. Therefore is it of utmost importance to map users to understand the projection concepts discussed in this chapter to be assured that the maps we use have been made with careful consideration of all the decisions required to make an appropriate map projection.

Since most map use takes place at the local level, where earth curvature isn't a big problem, global map projections aren't a great concern for many users. With regions as small as those covered by topographic map quadrangles, your main projection-related problem is that, while

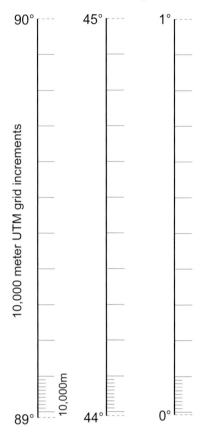

UTM Projections of One Degree of Geodetic Latitude along a Meridian

Figure 3.30 Transverse Mercator map projections of one degree of geodetic latitude on the WGS 84 ellipsoid along a meridian at the pole, midway between pole and equator, and at the equator.

the individual sheets match in a north–south direction, they don't fit together in an east-west direction. Yet even this difficulty won't be a serious handicap unless you try to create a large map mosaic.

The age of computers has changed how we think about map projections. You no longer need to "make do" with inappropriate projections. Computer-generated projections are available for almost any use. Furthermore, it is now practical for you to sit down at a computer and construct your own projections. Most important, perhaps, you can manipulate projections on the computer in search of the ideal projection base for a given application. All these benefits can only be realized, of course, if you know enough about projections to take advantage of these opportunities that computers provide.

Being map-savvy also helps you evaluate projections published in magazines and newspapers. World maps in the popular media often lack latitude-longitude grids.

This masks the extreme spatial distortion inherent in such maps and gives the impression that the continents are truly represented. Thus, the map user is well advised to reconstruct the latitude-longitude grid mentally as a first step in map reading.

Understanding the concepts above will go a long way toward helping you make the right decisions, but there are also books and Web sites dedicated to map projections. National mapping agencies such as the USGS and swisstopo, the Swiss national mapping agency, also offer useful map projection information on their Web sites and in their publications.

NOTES

1. The equation for the sphere is $y = ln(tan(\pi/4 + \varphi/2))$, where φ is the latitude in **radians** (there are 2π radians in a 360° circle, so 1 radian is approximately 57.295 degrees). The value of this equation at 90° $(\pi/2\ radians)$ is infinity.

SELECTED READINGS

American Cartographic Association. 1986. *Which map is best? Projections for world maps.* Bethesda, Md.: American Congress on Surveying and Mapping.

——. 1988. *Choosing a world map—Attributes, distortions, classes, aspects.* Bethesda, Md.: American Congress on Surveying and Mapping.

——. 1991. *Matching the map projection to the need.* Bethesda, Md.: American Congress on Surveying and Mapping.

Bugayevskiy, L. M., and J. P. Snyder. 1995. *Map projections: A reference manual.* London: Taylor & Francis.

Canters, F., and H. Decleir. 1989. *The world in perspective: A directory of world map projections.* New York: John Wiley & Sons, Inc.

Chamberlin, W. 1947. *The round earth on flat paper: A description of the map projections used by cartographers.* Washington, D.C: National Geographic Society, 39–126.

Dent, B. D. 1996. *Cartography: Thematic map design.* 4th ed. Dubuque, Iowa: Wm. C. Brown Publishers, 24–48.

Hsu, M. L. 1981. The role of projections in modern map design. *Cartographica* 18 (2): 151–86.

Maling, D. H. 1992. *Coordinate systems and map projections.* 2d ed. New York: Pergamon Press.

Pearson, F. 1990. *Map projections: Theory and applications.* Boca Raton, Fla.: CRC Press.

Richardus, P., and R. K. Adler. 1972. *Map projections: For geodesists, cartographers and geographers.* New York: American Elsevier Publishing Co.

Robinson, A. H., et al. 1995. *Elements of cartography.* 6th ed. New York: John Wiley & Sons, Inc, 59–90.

Snyder, J. P. 1987. *Map projections–A working manual.* Professional paper 1395, Washington, D.C.: U.S. Geological Survey.

——. 1993. *Flattening the Earth: A thousand years of map projections.* Chicago: University of Chicago Press.

Snyder, J. P., and H. Steward, eds. 1988. *Bibliography of map projections.* Bulletin 1856. Washington, D.C.: U.S. Geological Survey.

Snyder, J. P, and P. M. Voxland. 1989. *An album of map projections.* Professional Paper 1453. Washington, D.C.: U.S. Geological Survey.

Tobler, W. R. 1962. A classification of map projections. *Annals of the Association of American Geographers* 52: 167–75.

chapter
four **GRID COORDINATE SYSTEMS**

4

Grid coordinate systems

There are several ways to pinpoint locations on maps. We have seen in chapter 1 that the latitude–longitude graticule has been used for over 2,000 years as the worldwide locational reference system. Geocentric latitude and longitude coordinates on the sphere or geodetic latitudes and longitudes on the oblate ellipsoid, still key to modern position finding, are not as well suited for making measurements of length, direction, and area on the earth's surface. The basic difficulty is the fact that latitude-longitude is a coordinate system giving positions on a rounded surface. It would be much simpler if we could designate location on a flat surface using horizontal and vertical lines spaced at regular intervals to form a square grid. We could then simply read coordinates from the square grid of intersecting straight lines.

In chapter 3, we explained that most maps are created by projecting the earth's surface onto a flat surface, such as a sheet of paper or a computer screen. The advantage of the flat map projection surface is that we can locate something by using a simple two-axis coordinate reference system. This coordinate system is the basis for the square grid of horizontal and vertical lines on the map.

We call a plane-rectangular coordinate system based upon and mathematically placed on a map projection a **grid coordinate system.** To devise such a system for large areas, we have to deal somehow with the earth's curvature. From chapter 3, we know that transferring something spherical to something flat always introduces geometrical distortion. But we also know that map projection distortion caused by the earth's ellipsoidal shape is minimal for fairly small regions. If we superimpose a square grid onto flat maps of small areas, we can achieve positional accuracy good enough for many map uses. All grid coordinate systems are based on Cartesian coordinates, invented in 1637 by the famous French philosopher and mathematician **René Descartes.**

CARTESIAN COORDINATES

If you superimpose a square grid on the map, with divisions on a horizontal **x-axis** and a vertical **y-axis** where the axes cross at the system's **origin,** you have established the familiar **Cartesian coordinate system** (figure 4.1). You can now pinpoint any location on the map precisely and objectively by giving its two coordinates (x,y). The Cartesian coordinate system is divided into four **quadrants** (I–IV) based on whether the values along the x- and y-axes are positive or negative. Mapmakers use only quadrant I for grid coordinate systems so that all coordinates will be positive numbers relative to the (0,0) grid origin (figure 4.1).

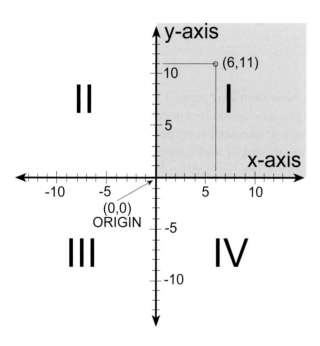

Figure 4.1   The structure of the Cartesian coordinate system. The basic notation for this system includes appropriate use of a negative sign for the x- or y-coordinates in quadrants II, III and, IV. We use only quadrant I (shaded) on maps because it is desirable to have all coordinates as positive numbers.

GRID COORDINATES

The simple way of defining map positions based on Cartesian coordinates has definite advantages over using the spherical graticule to define positions. Measuring x- and y-coordinates from horizontal and vertical axes with equally spaced distance increments greatly simplifies locating environmental features because you do not have to deal with the decreasing separation between meridians converging toward the pole.

Grid coordinate systems based on the Cartesian coordinate system are especially handy for such map analysis procedures as finding the distance or direction between locations or determining the area of a mapped feature like a lake. We will examine in depth one type of grid coordinate system—universal transverse Mercator (UTM)—that is found on many topographic maps, military maps, and navigational charts, among other types of maps. We will also look in detail at the state plane coordinate system, which is found on many large-scale maps in the United States.

Universal transverse Mercator system

A grid coordinate system can be used worldwide if enough zones are defined to ensure reasonable geometric accuracy. The best known grid coordinate system of international scope is the **universal transverse Mercator (UTM) system.** The UTM grid extends around the world from 84°N to 80°S. Sixty north–south zones are used, each six degrees in longitude (figure 4.2). Each zone has its own central meridian and uses a secant case transverse Mercator projection centered on the zone's central meridian for each of the 60 zones (see chapter 3 for more on this map projection). This projection makes it possible to achieve a geometrical accuracy of one part in 2,500 maximum scale error (scale factors ranging from 0.9996 to 1.0004) within each zone.

Each zone is individually numbered from west to east, beginning with Zone 1 from 180°W to 174°W. Zones 10 through 19 cover the conterminous United States, Zones 49 through 56 cover Australia, and so on. Each zone has separate origins for the northern and southern hemispheres. To understand how the origins are specified, it is useful to first understand a few terms and concepts.

An **easting** is simply the east–west x-coordinate in a grid coordinate system—that is, it is the distance east from the origin. In both the northern and southern hemispheres, an easting value of 500,000 meters (written 500,000mE) is assigned to the central meridian of each UTM zone. This value, called the **false easting,** is added

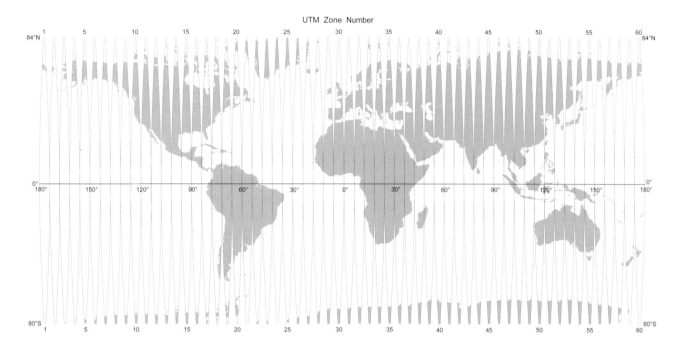

Figure 4.2   The 60 zones of the universal transverse Mercator (UTM) grid coordinate system.

to all x-coordinates so that there are no negative eastings in the zone.

Similarly, a **northing** is the north–south, y-coordinate in a grid coordinate system. In the northern hemisphere, a northing value of 0mN is assigned to the equator so that all northings are positive numbers. Since no **false northing** value is added, a UTM northing is simply the distance in meters north of the equator. In the southern hemisphere, the equator is given a false northing of 10,000,000mN. There are no negative y-values in the southern UTM zone because this large false northing value places the origin of the zone very close to the south pole.

Taking Zone 10, which covers much of the western seaboard of the United States, as an example (figure 4.3), we can examine the origin for the northern hemisphere. The x-axis follows the equator. The central meridian for this zone is 123°W and the longitude range is 120°W to 126°W. The origin lies on the equator 500,000 meters west of the central meridian at 123°W. In the southern hemisphere, the zone's origin lies 500,000 meters west and 10,000,000 meters south of the intersection of the equator and central meridian at 123°W. These origins were selected to guarantee that all UTM eastings and northings are positive numbers.

Since the equator has different northings for the northern and southern hemisphere, every location that lies exactly on the equator has two UTM grid coordinate pairs. For example, the coordinates for the intersection

of the equator central meridian of Zone 10 are the following:

500,000mE, 0mN in the northern hemisphere (labeled Zone 10 North)

and

500,000mE, 10,000,000mN in the southern hemisphere (labeled Zone 10 South).

You read and record x-coordinates first to the east and then y-coordinates to the north of the zone's origin, giving rise to the use of the terms "eastings" and "northings" for the x- and y-coordinates. Since north is conventionally at the top of the map, it may be helpful to remember a simple rule: always read coordinates right-up. You give the easting in meters, the northing in meters, the zone number, and the zone hemisphere (north or south). Thus, you would designate the location of the state capitol dome in Madison, Wisconsin, like this:

305,900mE, 4,771,650mN, Zone 16 North.

The near-global extent of the UTM grid makes it a valuable worldwide referencing system. The UTM grid is indicated on many foreign maps and on all recent U.S. Geological Survey (USGS) quadrangles in the topographic, orthophotoquad, and orthophotomap series. All GPS vendors program the UTM specifications into their receivers.

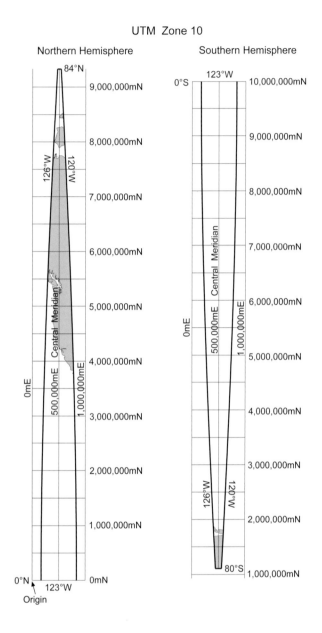

Figure 4.3 **The complete universal transverse Mercator grid for Zone 10.**

Universal polar stereographic system

As mentioned above, UTM grid zones extend from 84°N to 80°S. To have complete global coverage, a complementary rectangular coordinate system called the **universal polar stereographic (UPS) system** was created. The UPS grid consists of a north zone and a south zone. Each zone is superimposed upon a secant case polar stereographic projection that covers a circular region over each pole. The north zone (figure 4.4) extends from 84°N to the pole, where the UPS coordinate at the grid center is 2,000,000mE, 2,000,000mN. These large numbers were selected so that all eastings and northings would be positive numbers. The south zone extends from the pole to 80°30'S (80 degrees plus an additional 30 minutes of latitude extending into UTM grid). As with the north zone, the grid center is 2,000,000mE, 2,000,000mN to assure that all coordinates are positive.

Virtually all large-scale maps of these high latitudes, such as topographic sheets for Antarctica, are based on the UPS grid. The UTM grid system was not extended to each pole because the 60 zones converge at the poles, meaning that a new zone would be encountered every few miles.

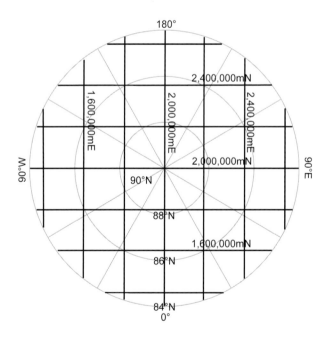

Figure 4.4 **North zone of the UPS grid system.**

Note that UTM coordinates will differ when different datums are used (see chapter 1 for more on datums), so you should check the datum information in the GPS receiver to assure that the coordinates are being recorded in the correct system.

Because meridians and not state boundaries delimit UTM zones, it usually takes more than one UTM zone to cover a state completely. For example, Oregon falls into Zones 10 and 11, and Wisconsin falls into Zones 15 and 16 (figure 4.2).

State plane coordinate system

The **state plane coordinate (SPC) system** was created in the 1930s by the land surveying profession in the United States as a way to define property boundaries that would simplify computation of land parcel perimeters and areas. The idea was to completely cover the United States and its territories with grids laid over map projection surfaces so that the maximum scale distortion error would not exceed one part in 10,000. Thus, a distance measured over a 10,000-foot course would be accurate to within a foot of the true measure. This level of accuracy could not be achieved if only one grid covered the whole country, because the area is too large. The solution was to divide each state into one or more zones and make a separate grid for each zone.

The United States was divided originally into 125 zones, each having its own projection surface based on the Clarke 1866 ellipsoid and NAD 27 geodetic latitudes and longitudes (see chapter 1 for further details). Most states have several zones, as figure 4.5 shows. Secant case Lambert conformal conic projections are used for states of predominantly east–west extent, and secant case transverse Mercator projections are used

for states of greater north–south extent. This explains the orientation of the individual zones. For states with more than one zone, the names North, South, East, West, and Central are used to identify zones, except in California where Roman numerals are used. In recent years, Nebraska, South Carolina, and Montana have combined their two or three zones into a single zone covering the entire state.

The logic of the SPC system zones is quite simple. Zone boundaries follow state and county boundaries because surveyors have to register land surveys in a particular county. Each zone has its own central meridian that defines the vertical axis for the zone. An origin is established to the west and south of the zone, usually 2,000,000 feet west of the central meridian for Lambert conformal conic zones and 500,000 feet west of the central meridian for transverse Mercator zones (figure 4.5). This means that the central meridians will usually have an x-coordinate of either 500,000 feet for the transverse Mercator or 2,000,000 feet for the Lambert conformal conic zones. These large numbers for zone centers were selected so that all x-coordinates will be positive numbers. Although different for each zone, the origin is

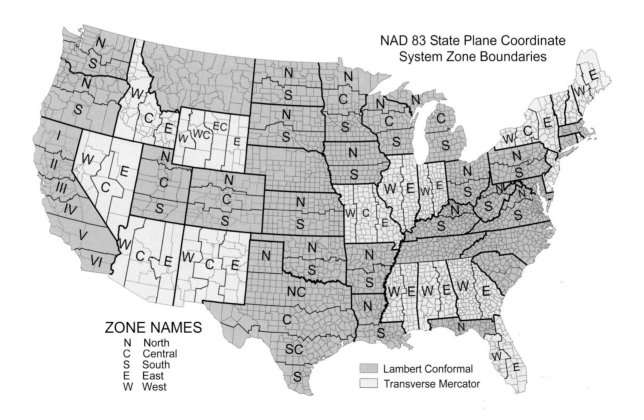

Figure 4.5 Zones of the state plane coordinate system for the contiguous United States. Notice that the transverse Mercator projection is used for zones that are oriented north–south, and the Lambert conformal conic projection is used for zones oriented east–west. Multizone states have zones identified by North, South, Central, East, and West, except for the Roman numerals used in California. States without letters have only one grid zone.

always at a parallel to the south of the zone to ensure that all y-coordinates will be positive numbers.

You read and record SPC coordinates in the same manner as UTM coordinates—first to the east and then to the north of the zone's origin. Specifically, the correct form of SPC notation is to give the easting in feet, the northing in feet, the state, and the zone name. For example, you would give the location of the state capitol dome in Madison, Wisconsin, in abbreviated form like this:

2,164,600 ft E, 392,280 ft N, Wisconsin, south zone

To illustrate this, we will look at the case of Oregon. The Lambert conformal conic projection is the base surface for the north and south SPC zones (figure 4.6). The central meridian is the same for both zones, with an easting of 2,000,000 feet in the original system. The origin for northings in each zone is a parallel just south of the counties in the zone, called the latitude of origin (see chapter 3 for more on this and other projection parameters). The intersection of this parallel and central meridian has a northing of 0 feet.

In 1983 the SPC system was modernized by switching to NAD 83 and the GRS 80 ellipsoid. Zones were redefined in metric units, so the Oregon central meridian now has an easting of 2,500,000 meters, and the intersection of the parallel of origin and central meridian has a northing of 0 meters. These changes in values and the switch from English to metric units were done in part to make it easy to tell if the map is based on NAD 27 or NAD 83. You will need to check the map legend to find out if old or new SPC system coordinates are used on the map.

The SPC system served the needs of the states when it was created, and state plane coordinates have been widely used for public works and land surveys. However, the SPC system is now largely obsolete as far as surveyors and other professional map users are concerned. One reason is that accuracy of 1 part in 10,000 in locating points is now easily exceeded using modern surveying methods. Also, each SPC zone is a separate entity with its own grid definition—a fact that frustrates and discourages uses across zone boundaries. Nevertheless, the SPC grid is useful for some map analysis applications, such as distance and direction finding (see chapters 11 and 12 for more on these). Additionally, you can enter SPC zone parameters into your GPS receiver as a user-defined grid (see chapter 14 for more information on GPS and maps).

U.S. state grids

The widespread use of computer mapping and geographic information systems (GIS) in state and local government has kindled the desire for grids tailored to each state's specific needs. States that fall into two UTM zones, for example, often create a special state coordinate grid by shifting the central meridian of a UTM zone to the center of the state. Wisconsin, for one, routinely records and reports data formatted in the **Wisconsin transverse Mercator (WTM) system** (figure 4.7). The UTM and WTM grids have the same geometric accuracy, but WTM avoids the problem of two UTM zones covering the state.

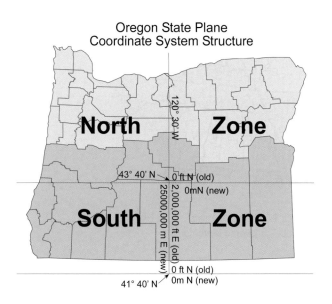

Figure 4.6   North and south zones of the Oregon state plane coordinate system.

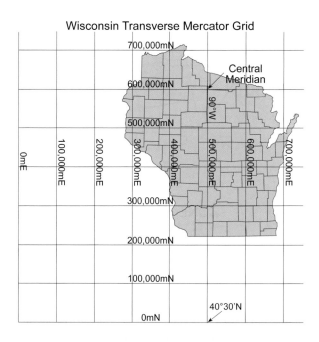

Figure 4.7   The Wisconsin transverse Mercator grid shifts the UTM zone to the center of the state.

The same approach can be taken when a state has two state plane zones. The **Oregon Lambert system** is a good example (figure 4.8). A single grid has replaced Oregon's two SPC zones based on a conformal conic projection. The new grid has the same central meridian but a different parallel of origin. The easting value at the central meridian (400,000mE) is entirely different from the SPC values, and the standard parallels are different. Georgia, Illinois, Kentucky, and Pennsylvania also have developed single-zone grid coordinate systems for use by their state agencies dealing with land and resource management.

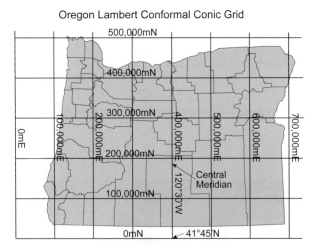

Figure 4.8   The Oregon Lambert system combines two SPC grids into a single grid covering the entire state.

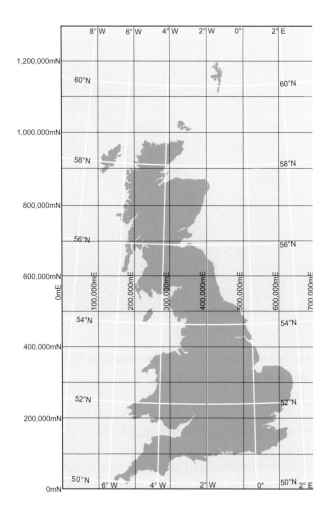

Figure 4.9   The British National Grid coordinate system covers all of England, Scotland, and Wales.

European grid coordinate systems

Many nations have a national grid system covering their territory, the most notable being Australia, Belgium, Finland, Great Britain, Ireland, Italy, the Netherlands, New Zealand, Sweden, and Switzerland. Many systems are based on a single transverse Mercator map projection that covers their country. Let's look in detail at the national grid systems used in two European countries, Great Britain and Switzerland.

British National Grid coordinate system

Great Britain's national mapping agency, the **Ordnance Survey of Great Britain (OSGB),** devised the **British National Grid coordinate system**, which is heavily used by land surveyors, as well as for maps based on surveys made by the Ordnance Survey or commercial map firms. Grid coordinates covering Great Britain, including its outlying islands, are also used in publications such as Admiralty nautical charts, guide books, or government planning documents.

The grid is based on the OSGB 36 datum (Ordnance Survey Great Britain 1936), defined in 1936 and based on the Airy 1830 ellipsoid. This datum is still the best fit to the earth for Great Britain (see chapter 1 for more on the Airy ellipsoid). A transverse Mercator projection of the Airy ellipsoid is used with its origin set at 49°N, 2°W (figure 4.9) so that the 2°W central meridian is a vertical line. The National Grid is placed over the map projection surface so that the central meridian has an easting of 400,000mE and the northing at 49°N is given a value of −100,000mN. Since all eastings and northings must be positive numbers, the grid actually starts a little south of the 50°N parallel where the northing is 0mN.

Swiss coordinate system

The **Swiss coordinate system** is the grid coordinate system used in Switzerland for land surveying and topographic mapping by the **Swiss Federal Office of Topography (swisstopo),** the Swiss national mapping agency. You can see by the curved parallels in figure 4.10

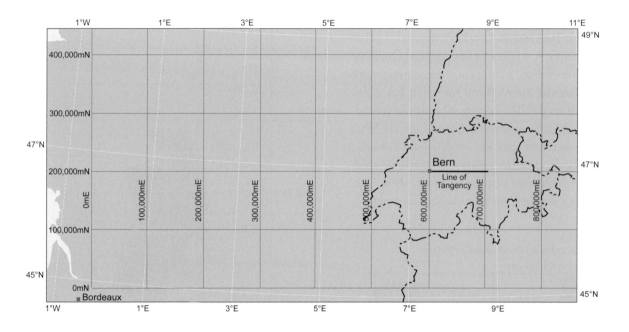

Figure 4.10 The Swiss coordinate system has its origin near Bordeaux, France, so that eastings and northings have distinct ranges of numerical values.

that the geometry of the Swiss grid differs from a UTM zone. The curved parallels are the result of using an oblique Mercator map projection as the surface upon which the grid is placed. An oblique aspect Mercator projection differs from the normal and transverse aspects (see chapter 3) in that the line of tangency is a great circle other than the equator or a meridian. The great circle used as the line of tangency for the oblique Mercator projection begins as a straight line perpendicular to the meridian running through Bern, Switzerland's capital.

You may wonder why the origin of the metric grid is near Bordeaux, France, five hundred kilometers to the west of Switzerland. This origin point puts Switzerland in Cartesian coordinate quadrant I where all eastings and northings are positive numbers. But why are the eastings that cover the country relative to a starting point near the Atlantic Ocean? The answer lies in reducing confusion when stating coordinates. Notice that the range of northings is from a little less than 100,000 meters to 300,000 meters, whereas the range of eastings is from a little less than 500,000 meters to more than 800,000 meters. You can see that within Switzerland no location can have the same easting and northing number because eastings must always be greater than northings.

You can see the Swiss grid on topographic maps such as the segment in figure 4.11. Notice that the numbers 645 and 145 for the 1,000-meter grid are the full eastings and northings (645,000mE, 145,000mN) with the three trailing digits (000) dropped. This notation is commonly

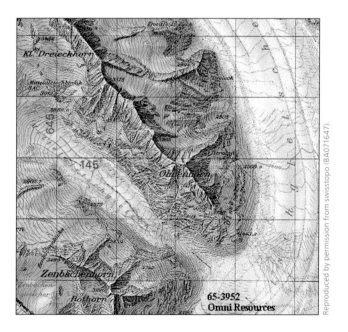

Figure 4.11 Swiss topographic map segment with Swiss coordinate system grid lines.

found on maps using grid coordinates, as we will learn more about in the next section. You should be aware that beginning in 1995 the eastings were increased by 2,000,000 meters and the northings by 1,000,000 meters. This increase means that the grid values 645 and 145 are printed as 2645 and 1145 on post-1995 editions of this and other Swiss topographic maps.

GRID COORDINATE DETERMINATION ON MAPS

Grid coordinate appearance on maps

We have seen that UTM grid coordinates appear on U.S. and foreign topographic maps and that state plane coordinates are printed on USGS quadrangles of the topographic series covering our states. The grid appearance varies among map series, so let's focus first on United States large-scale topographic maps. Military topographic maps have UTM grid lines superimposed at one kilometer intervals, whereas civilian maps show both UTM and SPC grids by grid ticks along the edges of each map. Figure 4.12 shows how state plane and UTM grid ticks appear on two sections of the Madison West 1:24,000-scale topographic map.

Black ticks along the outer margin of USGS topographic maps indicate SPC 10,000-foot grid lines. The northings or eastings of these ticks are given by the value at one tick on each edge of the map (a 370,000 foot northing at the tick on the left edge of the Madison West quadrangle in figure 4.12, for example.) Other ticks are spaced every 5 inches on the 1:24,000-scale (1 inch to 2,000 feet) topographic maps. UTM grid ticks are spaced at 1,000- or 10,000-meter intervals, depending on the map scale. On USGS quadrangles of the 1:24,000-scale topographic series, 1,000-meter grid ticks are printed in blue. These ticks are labeled (in black) with their easting and northing values along the map margin. The principal digits—those showing the eastings and northings in kilometers—are printed in larger type with the three trailing digits (000) dropped from all but one label along each edge of the map.

UTM grid coordinates are shown in different ways on foreign topographic maps, as you can see in figure 4.13. The use of single digits to identify the 10,000-meter easting and northing increment on the Vancouver, Canada, 1:250,000-scale map (left) is similar to how easting and northing ticks are labeled on USGS topographic maps of the same scale. Notice, however, that light blue grid lines, not grid ticks, are drawn on the map. Grid lines are shown prominently in black on the Sonneberg, Germany, 1:50,000-scale topographic map from the 1960s. The full UTM easting and northing is given for each grid line, with the three trailing digits (000) dropped on all numbers except those at each map corner. This style of showing the UTM grid is identical to that used for United States military topographic maps, which is not surprising when you realize that the German map was

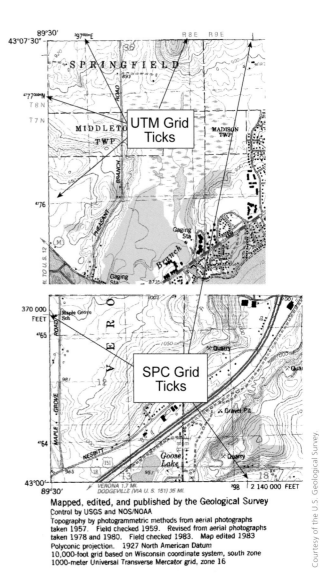

Figure 4.12   Appearance of 1,000-meter and 10,000-foot grid ticks on a USGS 1:24,000-scale topographic map.

reprinted by the U.S. Army with the grid likely added to the original map.

Grid orientation

Grid lines or ticks are rarely oriented parallel to the edges of topographic and other quadrilateral-formatted maps. We can see the reason for this by looking at UTM Zone 10 in figure 4.3. Notice that only at the equator do horizontal UTM grid lines intersect the zone edges perpendicularly. Moving toward either pole, the converging edge meridians intersect the horizontal grid lines at increasingly acute angles to a maximum of slightly less than three degrees from perpendicular at the top and bottom of the zone. Let's now look at the southwest and southeast corners of USGS 1:100,000-scale topographic

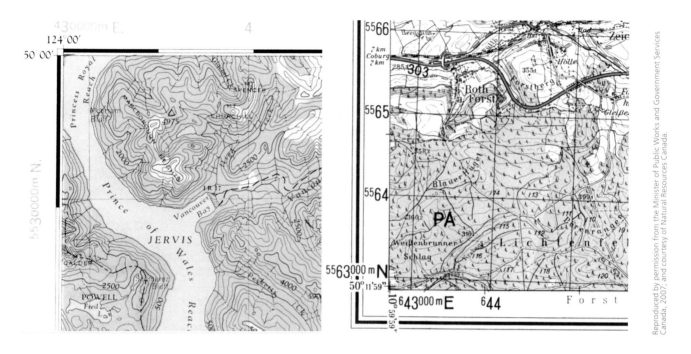

Figure 4.13 UTM grid coordinates are shown differently on these corner pieces of Canadian 1:250,000-scale (left) and German 1:50,000-scale (right) topographic maps.

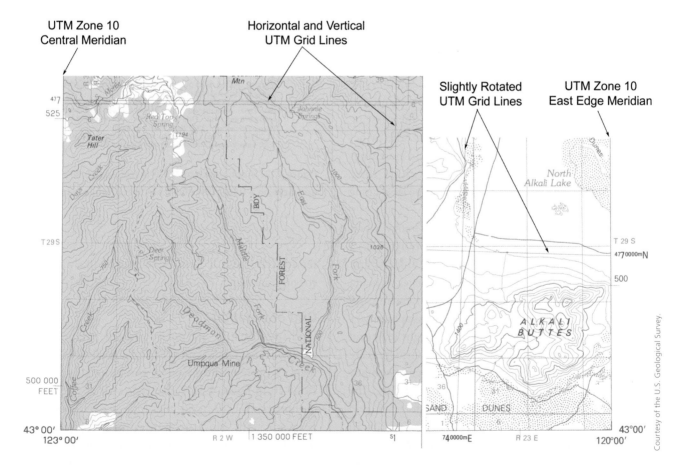

Figure 4.14 UTM grid lines at the zone center (left) are vertical and horizontal, but are slightly rotated clockwise to the east edge meridian of the zone (right) when it is drawn as a vertical line on the map (segments of Diamond Lake (left) and Christmas Valley (right), Oregon, USGS 1:100,000-scale topographic maps in UTM Zone 10).

maps that are at the same 43°N latitude but at the center and east edge of UTM Zone 10 (figure 4.14). On the left map at the 123°W central meridian for Zone 10, notice that the UTM grid lines are vertical and horizontal (the vertical 500,000mE line at the central meridian is not shown on the map because it is the same line as the meridian). The slight clockwise rotation (2 degrees at this latitude) of the east UTM edge meridian to vertical on the map to the right means that the grid lines are slightly rotated clockwise. This slight rotation of grid lines from horizontal and vertical at zone edges is termed **grid convergence.**

Grid convergence at zone boundaries, then, is zero at the equator and maximum at the top and bottom of the UTM zone. You may wonder how to deal with grid coordinates on a map like figure 4.15 that spans UTM grid Zones 15 and 16. Measuring half of your coordinates in each zone would be very confusing and difficult to work with if you wanted to calculate lengths, directions, and areas from the coordinates. The solution is to extend the zones outward to cover the entire map. This allows you to choose one of the two zones for your map work. UTM zones, for example, can be overlapped with acceptable distortion up to 30 minutes of longitude with their neighboring zones. For the same reason, state plane coordinate system zones also extend above and below—or to each side of—the counties they cover.

Grid coordinate determination

Although the structure of grid coordinate systems is relatively easy to understand, it may take practice to gain skill in using coordinates on maps. Sometimes you'll want to determine a feature's SPC or UTM coordinates. At other times you'll need to find the position on the map of a feature whose coordinates are given.

For instance, you want to determine the UTM coordinates of the gravel pit on the 1:24,000-scale topographic map segment in figure 4.16. If UTM grid lines aren't printed on the map, use the marginal grid ticks and a straightedge to construct the grid lines lying immediately to the south and north, and east and west, of the gravel pit. Note the coordinate values of the grid lines lying to the west and south of the gravel pit (497,000mE and 4,764,000mN). Next, measure the map distance from these lines to the gravel pit (3.73 and 1.00 cm), and form ratios between these values and the grid interval distance in map units (4.17 cm per kilometer for a 1:24,000-scale map). Multiply these proportions (0.894 and 0.240) by the grid interval distance in ground units (1,000m), and add the results to the west and south grid line values. Thus, the UTM coordinate of the gravel pit is the following:

Easting = 497,000m + 894m = 497,894mE

and

Northing = 4,764,000m + 240m = 4,764,240mN

Now imagine that you want to plot the location of a feature for which you know the grid coordinates. This problem is essentially the reverse of determining the UTM coordinates of a mapped feature such as the gravel

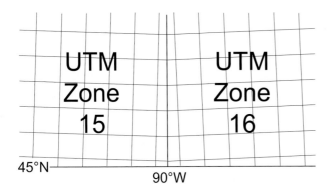

Figure 4.15 UTM grid lines usually are not parallel with graticule lines, so they will intersect at a slight angle. Maps that span two grid zones will have grid convergence, a problem solved by extending one or both grids across the entire map.

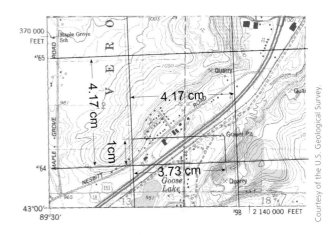

Courtesy of the U.S. Geological Survey.

Figure 4.16 To determine the UTM coordinates of the gravel pit or to plot the gravel pit's location from UTM coordinates, follow the steps outlined in the text.

pit in figure 4.16, which we know is at 497,894mE, 4,764,240mN.

First, determine the UTM northings and eastings for the grid lines falling immediately below and to the left of the coordinate (4,764,000mN and 497,000mE). If these grid lines aren't drawn on the map, use the grid ticks and a straightedge to draw them. Next, subtract the grid line value immediately west of the easting from the easting (497,894m – 497,000m = 894m), and subtract the grid line value immediately south of the northing from the northing (4,764,240m – 4,764,000m = 240m). Form proportions between these differences and the 1,000m grid interval distance (0.894 and 0.240). Now multiply these proportions by the grid interval in map units (4.17cm per kilometer) to obtain the easterly and northerly differences in map units (3.73cm and 1.0cm). Finally, plot these distances from the grid lines to the west and south. The two plotted lines will intersect at the gravel pit.

If you'll be working with a single coordinate system over and over on maps of a certain scale, it may pay to use a simple measurement aid called a roamer. Commercially available roamers are clear, plastic devices that have calibrated rulers etched into their surface (figure 4.17). The rulers match standard topographic map scales. Check your local map or outdoor recreation store to get one of these handy ready-made products. You can also easily construct your own roamer by marking off the grid interval distance along each edge of a piece of paper, starting at the corner. Then divide these distances into units fine enough for the precision of your measurements. Millimeters or tenths of inches normally suffice.

By aligning the roamer with the north–south and east–west grid lines, you can determine map references and plot coordinate locations quickly and accurately (figure 4.18). You'll need a separate roamer for each map scale and grid system, of course. Therefore, you may want to put UTM and SPC roamers for standard USGS quadrangles on opposite corners of the same sheet of paper.

GRID CELL LOCATION SYSTEMS

The grid coordinate systems that we've described so far all use point coordinates. Several other reference grids that are found on maps provide grid cell locations rather than point coordinates. These grid cell locations consist of an alphanumeric code that locates cells by their column and row. In this section, we'll explore several grid cell location systems.

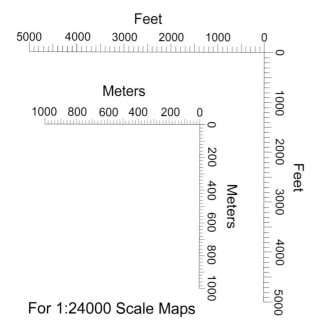

Figure 4.17 **Roamers for determining state plane and UTM coordinates on 1:24,000-scale topographic maps. You can copy this illustration onto a transparency and use it in your map reading work, but be sure it is scaled correctly first.**

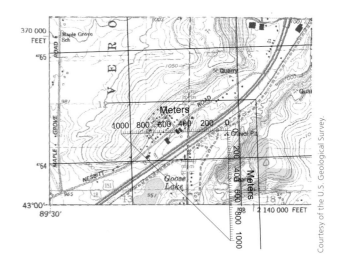

Figure 4.18 **You can use a roamer to determine rectangular coordinates for a feature or to plot a feature's position from known coordinates, such as the gravel pit on this topographic map.**

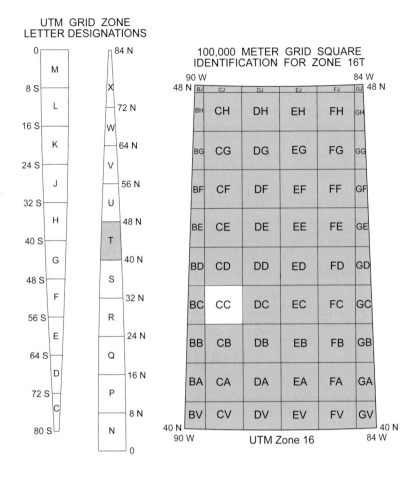

UTM GRID ZONE
LETTER DESIGNATIONS

100,000 METER GRID SQUARE
IDENTIFICATION FOR ZONE 16T

UTM Zone 16

Figure 4.19 With the MGRS, each of the 60 UTM zones is first divided into 8-degree quadrilaterals of latitude and lettered from south to north. Each of these quadrilaterals is divided into 100,000 meter squares that are given a two-letter code.

Military Grid Reference System

The U.S. **Military Grid Reference System (MGRS)** is a grid cell location scheme used by North Atlantic Treaty Organization (NATO) nations. When devising this system, the U.S. Army aimed to minimize the confusion of using long numerical coordinates (up to 15 digits may be required for very small features) and numerical grid zone specifications. This was achieved by substituting single letters for several numerals.

The MGRS is used with UTM and UPS grids, although we will discuss only the UTM version of the MGRS in detail, as most map users have little call for locating features in polar regions. In the MGRS, each of the 60 UTM zones is divided into 19 quadrilaterals covering 8 degrees and one (the northernmost) covering 12 degrees of latitude. Quadrilaterals are assigned the letters C through X consecutively, beginning at 80°S latitude (figure 4.19). The letters I and O are omitted to avoid possible confusion with similar appearing numerals. Each grid cell is designated by the appropriate alphanumeric code, referring first to the zone number (column) and next to the row letter. Madison, Wisconsin, for example, is located in the quadrilateral designated 16T (UTM Zone 16 and quadrilateral T).

Next, each quadrilateral is divided into 100,000-meter-square cells, and each cell is identified by a two-letter code. The first letter is the column designation; the second letter is the row designation. The letters I and O are again omitted to avoid possible confusion with numerals. The 100,000-meter-square cell containing Madison, Wisconsin, is designated 16TCC, for example. To assist the map user, the 100,000-meter-square cell identification letters for each map sheet are generally shown in the sheet miniature, which is part of the grid reference box found in the lower margin of the map (figure 4.20).

For more precise designation of grid cells, the MGRS uses the standard UTM numerals. Thus, the regularly spaced lines that make up the UTM grid on any large-scale map are divisions of the 100,000-meter-square cell. These lines are used to locate a point with the desired precision within the cell. Dividing the cell by 10 adds a pair of single-digit numerals that designate a cell of 10,000 meters on a side (figure 4.21). For example, Wisconsin's state capitol dome is located in the 10,000-meter cell designated 16TCC07. Further division by 10 requires a pair of two-digit numbers, yielding the designation 16TCC0751 for the 1,000-meter cell containing the capitol. The process can be continued until the

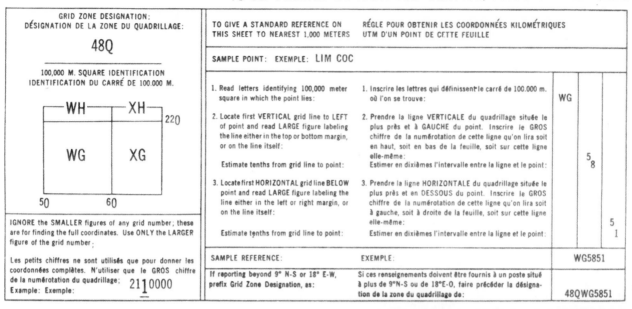

Figure 4.20 A map sheet miniature showing the 100,000-meter-square cell identification letters can be found in the marginal grid reference box on maps that use the MGRS.

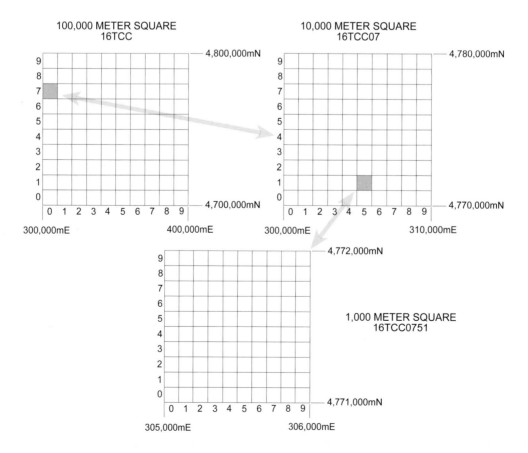

Figure 4.21 MGRS coordinate designations within 100,000-meter-square cells involve progressive subdivisions by 10 and the use of standard UTM numerals.

desired level of precision is achieved. In each instance, the coordinate pair designates the southwest corner of the grid cell at the specified level of precision.

MGRS coordinate system cells are printed on 1:25,000-, 1:50,000-, and 1:250,000-scale military topographic maps. The 10,000- or 1,000-meter cell row and column numbers are printed over the UTM grid lines for easy identification, along with the two-letter code for the 100,000-meter square. This allows grid cells like the two 10,000-meter squares in the corner of the McDermitt 1:250,000-scale military topographic map (figure 4.22) to be easily identified as 11TMS23 and 11TMS24.

United States National Grid

In 2005, the U.S. **Department of Homeland Security (DHS),** a department of the federal government charged with protecting the United States from terrorist attacks and responding to natural disasters, proposed that the **United States National Grid (USNG)** be used for all geographical referencing and mapping. This proposal was put forward in response to the ever-increasing use of portable

Global Positioning System (GPS) devices, GPS-enhanced cell phones, and automated vehicle location (AVL) technology. Users of these devices require a nationally uniform grid reference system for accurate and consistent identification, communication, and mapping of ground coordinates. A single national grid reference system is important because people cannot easily convert between different grid systems like UTM and state plane coordinates without using calculators or conversion tables.

The USNG (officially known as the United States National Grid for Spatial Addressing) is a grid cell location system that is based on the UTM coordinate system and the alphanumeric grid square referencing system used in the MGRS. As we previously mentioned, 10 UTM zones, numbered 10 through 19, span the conterminous United States from 126° to 66° west longitude. The MGRS divides each UTM zone into quadrilaterals 8 degrees in latitude, and the letters R, S, T, and U identify the four quadrilaterals from 24° to 56° north latitude. The land area of the conterminous United States falls within 32 of the 40 quadrilaterals shown in figure 4.23.

Let's take a look at a particular quadrilateral in a UTM grid zone, such as quadrilateral 17R that covers the peninsula of Florida. As with the MGRS, this quadrilateral is divided into 100,000-by-100,000-meter-square cells, and each cell is identified by a two-letter code. The first letter is the column designation and the second is the row designation. For example, Saint Lucie Inlet, approximately 30 miles (50 kilometers) north of West Palm Beach, Florida, on the southeast coast lies in 100,000-meter cell NL, or 17RNL to be more exact.

You can subdivide cell NL into 10,000-by-10,000-meter and then 1,000-by-1,000-meter cells in the manner shown for the MGRS in figure 4.21. Here's how to give the MGRS and USNG coordinates for finer subdivisions of 1,000-by-1,000-meter cell 17RNL7513 in the northwest corner of the Saint Lucie Inlet, Florida topographic map (figure 4.24). Reading right-up, the six-digit coordinate for the 100-by-100-meter square at the upper left corner of the cell is 17RNL750139, the eight-digit coordinate for the 10-by-10-meter square in the lower left corner is 17RNL75001300, and the ten-digit code for the 1-by-1-meter square at the upper right corner is 17RNL7599913999. Identifying ground locations to 10- or 1-meter resolution is important in high-accuracy navigation, as well as in scientific studies and land-management activities in urban areas. Consequently, GPS receivers—from recreational to survey-grade instruments—calculate and display geographic positions in USNG format to this level of precision (see chapter 14 for more on GPS and maps).

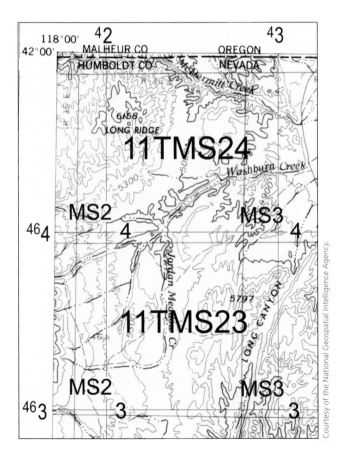

Courtesy of the National Geospatial Intelligence Agency.

Figure 4.22   MGRS cell boundaries are printed on military topographic maps, along with column and row identifiers. The full MGRS identifiers have been added to two 10,000-meter squares.

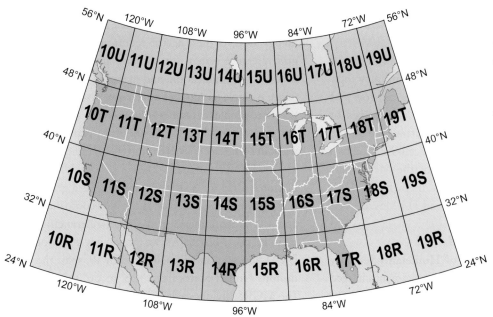

Figure 4.23 Two-letter codes for the 40 6°-by-8° quadrilaterals from the United States Military Grid Reference System that form the USNG.

British National Grid reference system

The **British National Grid (BNG) reference system** is commonly used in England, Scotland, and Wales. A similar grid cell location system used in Ireland (including Northern Ireland) is the **Irish National Grid reference system.** The OSGB devised the BNG system to overlay the British National Grid coordinate system described earlier in this chapter. BNG references are commonly given for geographic features and locations cited in publications such as guide books or government planning documents for the British Isles.

Figure 4.25 shows that the BNG system divides the land area covered by England, Scotland, and Wales into 56 100,000-by-100,000-meter (100-by-100-kilometer) squares identified by a two-letter code. For the first letter the grid is divided into 500,000-by-500,000-meter (500-by-500-kilometer) squares. Four of these contain significant land area within Great Britain: S, T, N, and H. (The "O" square contains a tiny area of Yorkshire coastline, mostly below the mean high tide line.)

Each 500-by-500-kilometer square is divided into 25 100-kilometer-by-100-kilometer squares that have a second letter code from A to Z (omitting I) starting with A in the northwest corner to Z in the southeast corner. For example, square NA is in the northwest corner of large square N, while square NZ is in the southeast corner.

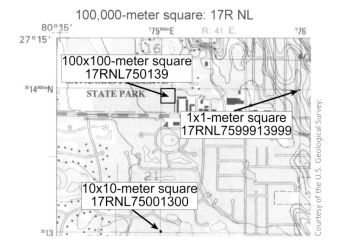

Figure 4.24 Northwest corner of the Saint Lucie Inlet, Florida, 1:24,000-scale topographic map, annotated with coordinates for 100-by-100-meter, 10-by-10-meter, and 1-by-1-meter squares in the USNG.

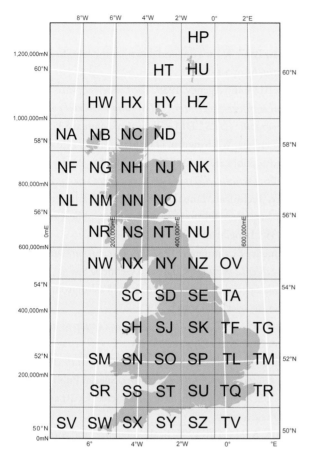

Figure 4.25 The BNG system divides the area covered by England, Scotland, and Wales into 56 100,000-by-100,000-meter squares identified by a two-letter code.

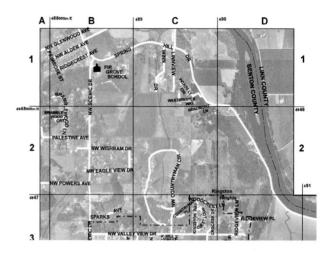

Figure 4.26 An arbitrary grid cell locator system identifies streets by column letter and row number. In this North Albany, Oregon, example, the location of Palestine Avenue is given as B-2.

Proprietary grids

In recent years, mapping has become more and more commercialized. Commercial vendors have developed and marketed several zone reference systems for use with their products. These **proprietary grids,** developed by private companies, may be map related, or they may be related to the use of maps in conjunction with specialized instruments. Let's look at a few examples.

Map publishers

Publishers of maps and atlases often develop their own version of a grid cell location system. You have probably seen **arbitrary grid cell location systems** superimposed over city, state highway, recreational, and atlas maps (figure 4.26). It is common for these to have grid columns that are lettered and rows that are numbered (or vise versa), so that a cell might have a column, row identification such as "C-2." The grid cells usually are keyed to a place name index. If you are looking for a particular street, you simply look it up in the index, where its grid cell identification can be found.

Finding a particular feature within a grid cell can be difficult if the density of names is great or the cell is large relative to the extent of the feature. It can also be difficult to find a feature that spans many cells such as an interstate, river, or railroad. In these cases, the location is sometimes identified by the cell locations where the feature starts and where it ends on the map. The problem of searching for a difficult-to-spot feature is even greater when only margin ticks indicate cell boundaries. Remember, too, that each grid cell is specific to the map for which it was drawn. If people ask you where a street is and you tell them "C-2," the information won't be useful unless they have the same map you do. Keep this in mind when you use maps that include insets. Because each inset may have its own grid cell locator system, the same place may be represented in several grid cells.

The arbitrary grid cell system is the basis for locating features with some brands of GPS equipment (see chapter 14 for more on GPS and maps). For example, the well-known map publisher, **Thomas Brothers Maps**, has created the **Page and Grid system** for use with its three scales of maps covering the continental United States. Through special arrangement with **Trimble Navigation,** a vendor of GPS receivers, Thomas Brothers Maps supplies a grid-related product called **Thomas Guides.** Trimble, in turn, markets a coordinate extension called **Trimble Atlas** (supplied by Thomas Brothers Maps), which is designed to increase the locational precision of the Page and Grid system. This proprietary grid reference system assigns a unique page number to each geographical area. Map grids are half-mile squares, ranging from

Grid A1 through Grid J7, and the grids always appear at the same location on each page, so you can also quickly and easily estimate distances and locate cities, streets, addresses, and points of interest.

The more specific the zone reference, the larger the scale of the map referenced. The result is a tailored package. A GPS receiver locates the user on a map, providing the ability to zoom in or out in scale to see the desired level of detail. As more vendors market electronic navigation systems that display maps on a portable computer screen, such linking of zone coordinates to map features will likely become more common.

Amateur radio operators

Another proprietary grid is used by amateur (ham) radio operators. This terrestrial reference scheme is based on the **Maidenhead grid system,** which partitions the earth into progressively smaller quadrilaterals of latitude and longitude. The first two letters in the reference divide the earth into 20°-by-10° fields. Pairs of numbers designate 2°-by-1° squares within these fields. Two more letters are used to define 5'-by-2.5' sub-squares within each square. Thus a six-character code can locate any place on earth within a rectangular zone of up to 5½-by-3 miles.

As with the Page and Grid scheme, Trimble has extended the Maidenhead grid so that it provides more precise spatial referencing. This extension makes the grid more suitable for use with its GPS receivers. The extension, called the Trimble Grid Locator, adds a pair of numbers and a pair of letters to the six-character Maidenhead code.

SELECTED READINGS

Atwill, L. 1997. What's up (and down) at the USGS. *Field & Stream,* May, 54–5.

Defense Mapping Agency. 1989. *The Universal grids: Universal transverse mercator (UTM) and universal polar stereographic (UPS).* DMA Technical Manual 8358.2 (September). Washington, D.C. http://earth-info.nga.mil/GandG/publications/tm8358.2/TM8358_2.pdf.

———. 2006. *Datums, ellipsoids, grids, and grid reference systems.* DMA Technical Manual 8358.1 (June). Washington, D.C. http://earth-info.nga.mil/GandG/publications/tm8358.1/toc.html.

Department of Commerce. 1953. *Plane coordinate intersection tables (2 minute).* Coast and Geodetic Survey, Wisconsin, Special Publication No. 308, Washington, D.C. (series to cover each state).

Maling, D. H. 1992. *Coordinate systems and map projections.* 2d ed. New York: Pergamon Press.

Mitchel, H. C., and L. G. Simmons. 1945. *The state coordinate systems: A manual for surveyors.* Special Publication No. 235, U.S. Department of Commerce, Coast and Geodetic Survey.

Robinson, A. H., et al. 1995. Scale, reference, and coordinate systems. In *Elements of cartography,* 6th ed., 92–111. New York: John Wiley & Sons, Inc.

Stem, James E. 1989. *State plane coordinate system of 1983.* NOAA Manual NOSNGS5 (January). Rockville, Md. http://www.ngs.noaa.gov/PUBS_LIB/Manual-NOSNGS5.pdf.

chapter
five **LAND PARTITIONING**

5

Land partitioning

In the last chapter, we looked at grid coordinate systems, which allow you to use numeric values to identify the location of a point or cell on the surface of the earth. Many people confuse land partitioning systems with grid coordinate systems, often with less than satisfactory results because of their differences. In this chapter, we will clarify these differences through an exploration of a variety of land partitioning systems.

Since the beginning of recorded history people have created spatial referencing systems convenient for land partitioning. **Land partitioning** is the dividing of property into **parcels,** also called **tracts,** which are areas with some implication for landownership or land use. You may also see the term **lot,** which is a special type of parcel within a subdivision that is recorded on a map.

One of the first steps in the management of an area of land is to divide it into tracts that are then recorded on **plats,** or maps drawn to scale showing the lots into which the area has been divided. A **platted subdivision** is therefore a subdivision that has been mapped to show the subdivided lots; the map is then placed on public record. Landownership, zoning, taxation, and resource management are just a few purposes served by such a system. The system has to be conceptually simple enough that it can be generally understood and technically simple enough that it can be readily implemented in the field at the time of settlement or development.

Land partitioning systems are used throughout the world. For example, when European settlers first arrived in the United States, they brought with them a host of land partitioning methods. Both irregular and regular land partitioning systems were soon in vogue in the colonies. Geometrically irregular schemes, like the metes and bounds system used in Great Britain and the rest of Europe, prevailed since the late seventeenth century in the colonial states, Texas, and portions of Louisiana. Other irregular systems, such as French long lots, Spanish and Mexican land grants and donation land claims, have also been used. "Regular" systems of laying out the land were also developed, such as the geometrically systematic town plans that are common in New England. We'll start with irregular and then look at the more regular systems that are still used today.

IRREGULAR SYSTEMS

Metes and bounds

Much land in the United States colonized before 1800 was characterized by widely scattered settlements. This was particularly true in the South, where the climate was moderate and agriculture was widely practiced. These conditions encouraged people to settle far apart rather than to cluster together as they did in the North. In the South, a family was granted land of a certain size, say 400 acres (about 162 hectares), through a gift or purchase. The family was then permitted to select their 400 acre parcel from anywhere within the remaining unsettled area.

English settlements in the relatively humid Northeast and Southeast were located mainly with respect to soil and timber resources. Consequently, the shape of parcels was decided mainly by the geography of the local setting; people naturally picked the best land they could, regardless of its shape. The convenience of the land parcel description was only an afterthought. An irregular parcel, such as the one shown in figure 5.1, is harder to describe than a simple geometrical figure such as a square. The problem was to define these asymmetrical parcels well enough to make clear whose land was where. The solution was to start from an established point, then describe a connected path around a parcel's boundaries, noting landmarks along the way. A parcel might be described using only natural features:

> "That parcel of land enclosed by a boundary beginning at the falls on Green River, thence downstream to the confluence of the North Fork, thence up the North Fork to the first falls, thence along Black Rock Escarpment to the point of beginning."

This form of land description is referred to as **metes and bounds.** With this method, minimal surveying skill (measurement of angles and distances on the ground so that they can be accurately plotted on a map) was required to delineate a property boundary. The legal property description was tied to geographic features and remained useful as long as neighbors agreed with the place names and accepted the boundaries, and as long as the landmarks still existed in the landscape. Some landmarks, like trees and rocks, were missing when the parcel description was later retraced.

Parcels didn't have to be described using only natural features, of course. The parcel shown in figure 5.1(A), for instance, could be given the following metes and bounds description, based on anthropogenic as well as natural features:

> "Parcel beginning at the point where Pine Road crosses Beaver Creek, thence due east to the big rock pile, thence northeasterly to the confluence of Beaver Creek and Pine Creek, thence northwesterly to the S. corner of Tom Smith's fence, thence back to the point of beginning."

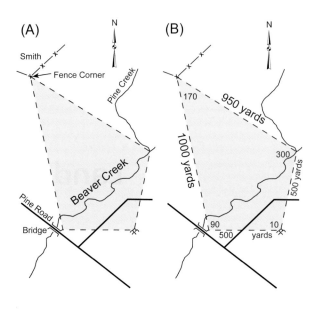

Figure 5.1 **Metes and bounds descriptions may consist of (A) landmarks and approximate directions or (B) precise distance and direction readings.**

After land survey methods were developed, it became less common to use geographic features in land parcel descriptions. Descriptions became more mathematical, based on surveyed distances and directions from an established point. The land survey form of the feature-based parcel description shown in figure 5.1(B) would read:

> "Parcel beginning at the N.E. corner of the Pine Road bridge over Beaver Creek, thence along a compass sighting of 90° for 500 yards, thence along a compass sighting of 10° for 500 yards, thence along a compass sighting of 300° for 950 yards, and thence along a compass sighting of 170° for 1,000 yards back to the point of beginning."

The description also would likely include the parcel's size in common areal units, such as acres or hectares.

Areas that have been surveyed by metes and bounds are easily identified on topographic maps and aerial photographs (figure 5.2). The characteristic pattern of irregular fields and winding roads intersecting at oblique angles tells you that this is a metes and bounds survey area.

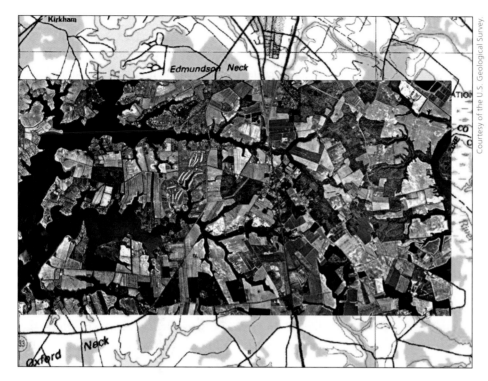

Figure 5.2 Topographic map segment and air photo overlay of an area just south of Easton, Maryland, showing the appearance of metes and bounds surveys. Irregular field boundaries and roads reflect the original settlement pattern described in the metes and bounds of the land survey.

French long lots

French settlements in Quebec and Louisiana from the 1630s until around 1760 have a unique land parcel arrangement resulting from the French feudal system of giving land grants called **seigneuries** to soldiers and other elite citizens. People given a seigneury usually divided the block of land into a number of narrow long lots used for farming. Settlement usually took place along rivers or lakes, which provided the chief source of transportation and communication for the French. Boundaries ran back from the waterfront as roughly parallel lines, creating narrow ribbon farms or **long lots,** also called *arpent sections* or *French arpent land grants* (an arpent is a French measurement that equals approximately 192 feet or 58 meters in length or 0.84 acres or 0.34 hectares in area). This allowed the settler to have a dock on the river, a home on the natural levee formed by the river, and a narrow strip of farmland that often ended at the edge of a marsh or swamp. Through subsequent subdivision, the parcels often became so narrow that they were no longer practical to farm, but their boundaries still exist legally and are plotted on maps.

In the United States, long lot boundaries are often shown on U.S. Geological Survey (USGS) topographic maps (figure 5.3). Long lots are particularly apparent along the Mississippi River in Louisiana, but you will also see them in other areas of early French settlement.

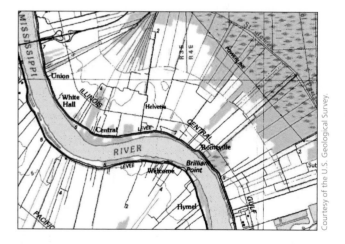

Figure 5.3 French long lot boundaries and lot numbers like these along the Mississippi River in Louisiana are shown on 1:24,000-scale USGS topographic maps.

Spanish and Mexican land grants

A **land grant** is an area of land whose title was given to its owner before the territory was part of the United States. After the territory was acquired by the U.S. government, the title was confirmed officially. During the period from the late 1600s to around 1850, when much of the southwestern United States was part of Spain and later the Mexican Republic, three types of land grants were issued by these governments:

1. **Pueblo grants** issued to communities of Native Americans were among the earliest, and today Indian reservations in New Mexico and other states are often based on these grants.

2. **Private grants** of property by the government that could be sold by the owner were made to individuals as a reward for service to the government.

3. **Community grants** were made to groups of settlers. Individuals in the group were given small tracts to settle and cultivate, but most of the grant was held in common for grazing, timber, and other purposes.

To receive a land grant, you had to physically step on the land, run your fingers through the soil, and make a commitment to live on the land, cultivate it, and defend it with your life if necessary.

Land grant boundaries were made easy to recognize. Physical features like hilltops, rivers, and arroyos always defined boundaries. Settlers were very concerned with water resources, so many grants straddle rivers and lakes. Descendents of the original settlers still live on these lands, and the boundaries of large grants appear on 1:24,000-scale USGS topographic maps, as in figure 5.4.

Donation land claims

The **Donation Land Claim (DLC) Act of 1850** granted 320 acres (almost 130 hectares) of federal land to any qualified settler who had resided on public lands before 1851 in the Oregon Territory (Idaho, Oregon, Washington, and western Montana). Settlers claiming public lands between 1851 and 1854 were awarded a smaller **donation land claim,** a 160 acre (about 65 hectare) parcel that had to be surveyed with north–south and east–west boundaries. These boundaries had to conform to the Public Land Survey (described later in this chapter) if the survey had already been made. DLCs were numbered and shown on U.S. Government Land Office maps. DLC boundaries and parcel numbers appear on large-scale topographic maps of land in the former Oregon Territory, as in figure 5.5.

All the land partitioning systems we've discussed so far are unsystematic settlement schemes, in which the land was settled before surveys were made. This free-for-all system encouraged inefficient partitioning of area—at least from the government's point of view. The first people moving into a region had a virtual monopoly over the choicest lands. The only land available to later

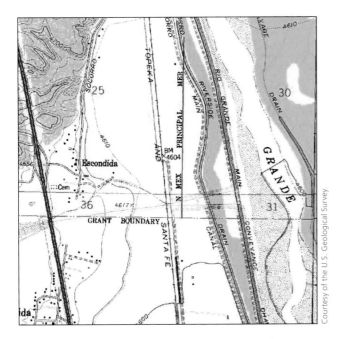

Courtesy of the U.S. Geological Survey.

Figure 5.4   Spanish and Mexican land grant boundaries are shown on 1:24,000-scale USGS topographic maps as dashed red lines with the words "GRANT BOUNDARY" inside the grant. This Mexican land grant is near Socorro, New Mexico.

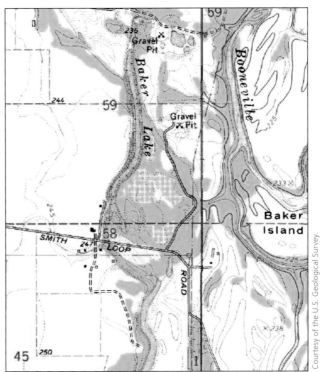

Courtesy of the U.S. Geological Survey.

Figure 5.5   Donation land claims south of Corvallis, Oregon, are shown with dashed red boundary lines and claim numbers, always greater than 36. Notice that claims 58 and 59 have north–south and east–west boundaries except for the line for claim 59 at the right edge.

settlers had swamps, steep slopes, or poor soils. In many areas, fragments of poor land remained unowned and unwanted long after a region was "fully" settled.

Furthermore, land claims often overlapped, and boundary descriptions were often in error. In homogeneous environments, it was hard to establish accurate borders in the first place. In any setting, boundary mistakes naturally occurred as the land passed through various owners and environmental changes. Good fences, as Robert Frost pointed out, did make good neighbors, because they clarified the location of the irregular boundaries. It used to be the custom for neighboring landowners to walk together around their lots each spring. The excursion was far more than a social outing; it made sure that all owners agreed on the borders between their lands. Problems arose, however, when later generations ignored the boundary-walking tradition. Many markers were shifted or destroyed through time: Fences fell away; trees died or were cut down; lakes dried up; rock piles were moved. The result has been a host of legal battles over property boundaries of unsystematically settled land.

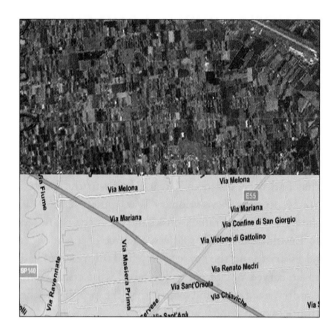

Figure 5.6 The road pattern seen on this satellite image and road map segment from ArcGlobe of an agricultural area near Ravenna, Italy, is a remnant of Roman centuriation land subdivision.

REGULAR SYSTEMS

The alternative to the scattered settlement of irregular parcels was to survey and divide the land systematically before settlers arrived. Surveying land into geometrically regular parcels in a systematic manner is an ancient practice. The pattern of agricultural land subdivision in both ancient Egypt and China was a grid, and cities in these—and succeeding—cultures often were planned as a rectangular grid of streets and lots.

Centuriation system

In Roman times, agricultural land in the Po River valley in northern Italy and elsewhere throughout the empire was subdivided into a grid of square parcels using the **Centuriation system** of land surveying. Field surveyors called **agrimensores** divided large tracts of land into square **centuria,** in modern terms about 132 acres (53.5 hectares) in area. Each centuria was further divided into 100 square **heredia,** each of which was allotted to a family. Agrimensores used a rudimentary surveying instrument called a **cross-staff** (which measured the angle between the north star and the horizon) to lay out the vertical and horizontal boundaries of centuria as a grid of lines that began in the middle of the area with

a north–south line called the **cardo maximus** and an east–west line called the **decumanus maximus.** Main roads were constructed along these two initial lines and narrower public lanes followed centuria boundary lines. You can see on topographic maps and satellite images remnants of centuriation in certain parts of Italy and other Mediterranean locales that were Roman colonies. The right-angled grid of roads in the agricultural land near Ravenna, Italy, shown on the satellite image and road map segments in figure 5.6, outline centuria surveyed during Roman imperial times.

In the United States, rectangular grids of streets and lots were used for planning towns in many of the colonies, occasionally in the South but mostly in the North. Town sites were laid out and surveyed, and plat maps of the parcels prepared and recorded, all prior to settlement. These were part of the foundation for the U.S. Public Land Survey System (PLSS), the dominant public land partitioning system in the United States, which, in turn, served as the model for the Dominion Land Survey in central and western Canada, both of which are described in more detail below. It is possible that the Roman centuriation system influenced Thomas Jefferson's design of the PLSS, although there is no direct historical evidence to support this claim.

NATIONAL PUBLIC LAND SURVEY SYSTEMS

United States Public Land Survey System

In 1783, the United States Congress of the brand new confederation of 13 states was faced with an urgent need for a national land policy. They had to devise some way to manage the vast lands east of the Mississippi River that had been ceded by Great Britain to the United States after the Revolutionary War. Quick action was important for several reasons: Land had been promised to Revolutionary War soldiers; a source of income was necessary to run the new country; and future states had to be carved out of the wilderness. Most crucial, the country needed a land policy that the people on the frontier could understand.

By the time the lands west and north of the Ohio River (the Northwest Territories, covering northwest Ohio and all of modern day Indiana, Illinois, Michigan, and Wisconsin, as well as the northeastern part of Minnesota) were opened to settlement in the late 1700s, the newly formed U.S. government had come up with what seemed to be an orderly way to transfer land to settlers. The solution was called the **Land Ordinance of 1785,** which established the **USPLSS,** or more commonly **PLSS.** Otherwise known as the **Township and**

Range System, this plan called for regular, systematic partitioning of land into easily understood parcels prior to settlement. Settlers were able to select surveyed parcels to their liking, but the legal description of each parcel had to be carefully recorded.

The PLSS was first implemented in the Northwest Territories and subsequently in the even vaster territories acquired by the United States through the Louisiana Purchase, Red River Cession, Florida Cession, Oregon Country Cession, Mexican Cession, Gadsden Purchase, and Alaska Purchase. To establish the land partitioning system in these areas, the first step was to select arbitrarily an **initial point** (figure 5.7) for the public land survey. Government land surveyors determined the parallel and meridian that intersected at the initial point. The parallel was called the **baseline** or **geographer's line** (surveyors were called geographers in those days), and the meridian was called the **principal meridian.**

Thirty-five principal meridians and baselines were established within the conterminous United States (figure 5.8), and five were established within Alaska. The testing ground was Ohio, where seven methods were tried before the final structure was developed. Each principal meridian was given a number or name (though the baselines were not), which was used to identify surveyed parcels within the region. The final system structure was developed in Ohio using the 1st principal meridian and baseline in the northwest corner of the state. The system was then used to survey the rest of the nation to the west. Notice that at first small areas were surveyed from the initial point (east of the Mississippi River), but that later surveys in the West covered large areas, often a territory that was subsequently divided among one or more states (for example, see Oregon and Washington in figure 5.8). The small surveys in the West were for Indian reservation lands.

In figure 5.8, it appears that the entire area covered by a principal meridian and baseline has been surveyed completely, but this is not the case. Some land has never been surveyed under the PLSS, mainly because it was reserved for national forests, reservations, or other government use. Set-aside areas such as these did not need to be surveyed for later partitioning for subsequent land grants or purchases.

Townships and sections

Once an initial point was established, **range lines** were surveyed along meridians at 6-mile intervals east and west of the principal meridian. **Township lines** were surveyed along parallels at 6-mile intervals north and south of the baseline (figure 5.9). The 6-by-6 mile quadrilaterals

Courtesy of Richard Garland.

Figure 5.7   Initial points were surveyed and defined physically with survey markers. This is the current marker at the intersection of the Willamette principal meridian and baseline in Portland, Oregon.

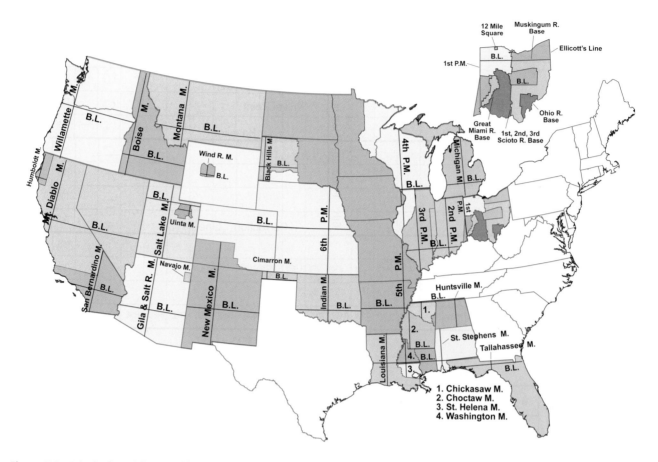

Figure 5.8 **Principal meridians and baselines of the Public Land Survey System for the conterminous United States. An initial point marks the intersection of these two lines. Areas surveyed from each initial point are shown with different brown tones.**

bounded by these intersecting township and range lines were called **survey or congressional townships,** which are identified by the number of columns and rows they are away from the initial point. Rows are called **townships** and columns **ranges,** corresponding to the surveyed township and range lines. The notation is to indicate township or range, then to note the number, and finally to add a suffix indicating the direction away from the principal meridian or baseline. So T3N, R5W identifies the township at the intersection of the third row (township) north and fifth column (range) west of the initial point.

The surveyors encountered a problem, however. Because meridians converge toward the north pole, the east–west distance between the meridians decreases (see chapter 1 for more on this concept). Thus, the township grid could not be extended indefinitely north or south from the initial survey point without townships becoming distorted in shape and area. To reduce the problem of unequal township dimensions, **correction lines** were established at every fourth township (24 miles or 38.62 kilometers), along which new range lines were surveyed to the east and west of the principal meridian

at 6-mile intervals. The effect of this pattern of surveying is seen in figure 5.9. The length of township lines north of the baseline decreases progressively until the correction line is reached, at which time they are resurveyed and hence appear progressively offset to the east or west. This means that township boundaries are not truly 6 miles on a side. Nor are they exactly 36 square miles in area, since their areas progressively decrease northward from the baseline and correction lines.

Each township was further subdivided into 36 square mile (93 square kilometer) parcels—ideally of 640 acres (2.59 square kilometers), called **sections.** (Recall, though, that the area distortion caused by the convergence of meridians toward the poles will affect the actual section area.) Every section was then given a number from 1 to 36, depending on its position in the township (figure 5.9). A zigzag method of numbering the sections was adopted, beginning with section 1 in the upper right corner of the township and ending with section 36 in the lower right corner. This was done so that every section would share a common side with the section with the number before and after it.

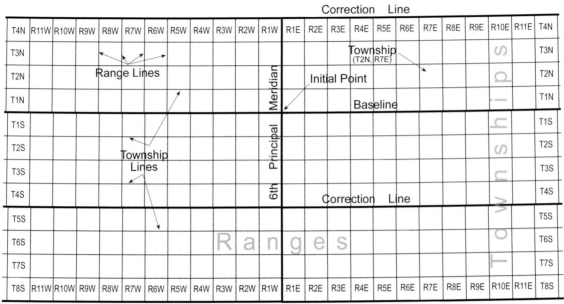

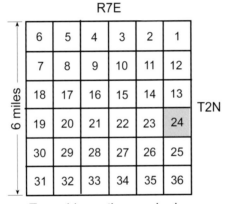

R7E

Township section numbering

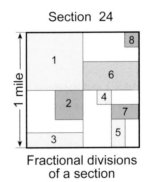

Section 24

Fractional divisions
of a section

TYPICAL FRACTIONAL LAND DIVISIONS

(1) NW¼, Sec. 24, T2N, R7E, 6th. P.M.
(2) NE¼, SW¼, Sec. 24, T2N, R7E, 6th. P.M.
(3) S½, S½, SW¼, Sec. 24, T2N, R7E, 6th. P.M.
(4) NE¼, NW¼, SE¼, Sec. 24, T2N, R7E, 6th. P.M.
(5) W½, SE¼, SE¼, Sec. 24, T2N, R7E, 6th. P.M.
(6) S½, NE¼, Sec. 24, T2N, R7E, 6th. P.M.
(7) S½, NE¼, SE¼, Sec. 24, T2N, R7E, 6th. P.M.
(8) NE¼, NE¼, NE¼, Sec. 24, T2N, R7E, 6th. P.M.

Figure 5.9 The structure and notation of the PLSS provides a systematic means of describing land parcels as small as 10 acres (4 hectares).

Fractional divisions

A section could be divided successively into smaller units such as **half sections** (one half of one section) and **quarter sections** (one half of one half of a section.) These can be further subdivided to produce even smaller units. For example, a **quarter-quarter section** is one-sixteenth of a section (one quarter of one quarter of a section). An easy way to correctly identify quarter sections is to think of a compass placed at the center of the section. The compass direction from the center of the section to the center of each quarter section is northeast, southeast, southwest, or northwest. With this system, each land parcel's legal description is unique and unambiguous. In figure 5.9, for example, parcel 4 (the 10-acre or 0.04-square-kilometer piece of land in section 24) would be described in abbreviated form as the following:

"NE¼, NW¼, SE¼, Sec. 24, T.2N, R.7E, 6th P. M."

In expanded form, the description reads as follows:

"The northeast quarter of the northwest quarter of the southeast quarter of section 24 of township 2 north, range 7 east, 6th principal meridian."

To locate a parcel from its legal description, the trick is to read backward, beginning with the principal meridian and working back through the township and range, the section, and the fractional section. To give the parcel's legal description, you simply reverse the above procedure and work up from the smallest division to the principal meridian. It is useful to understand the structure and notation of the system, because the PLSS is the basis for abstracts, deeds, and other landownership documents in much of the United States.

Government lots and platted lots

There are some exceptions to the use of the PLSS fractional land division in areas covered by the system. One exception occurs when most of a small parcel (less than 10 acres or 4 hectares) borders a body of water. In this case, a subdivision of the section called a **government lot** will be designated. These lots may be regular or irregular in shape, and its acreage may vary from that of regular fractional divisions of a section. A government lot is designated by its lot number, such as lot 3 (figure 5.10). Parcels in a platted subdivision within a section, called **platted lots,** are also specified by lot numbers. Other exceptions are parcels often—but not necessarily—small in size, which are given descriptions in metes and bounds rather than PLSS because of their irregular shapes.

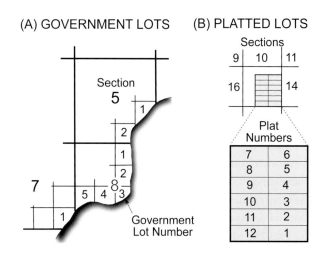

Figure 5.10   Land divisions for small parcels, such as (A) government lots and (B) platted lots, are designated with separate numbers in the PLSS.

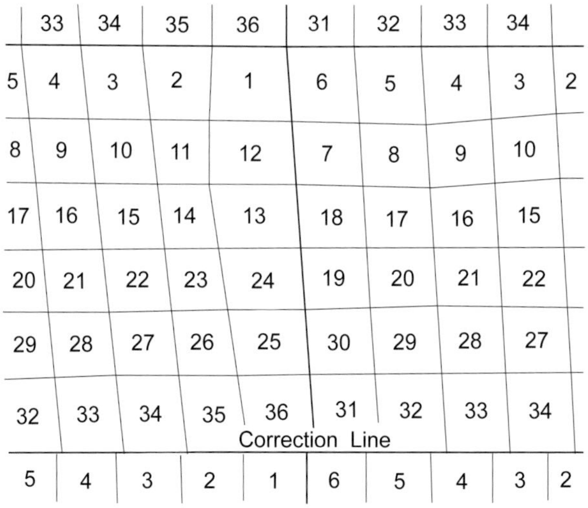

33	34	35	36	31	32	33	34	
5	4	3	2	1	6	5	4	3	2
8	9	10	11	12	7	8	9	10
17	16	15	14	13	18	17	16	15
20	21	22	23	24	19	20	21	22
29	28	27	26	25	30	29	28	27
32	33	34	35	36	31	32	33	34

Correction Line

| 5 | 4 | 3 | 2 | 1 | 6 | 5 | 4 | 3 | 2 |

Figure 5.11   Survey inaccuracies are evident in this actual segment of the PLSS section grid.

Survey irregularities

In practice, the idealized structure of the PLSS frequently broke down. Inaccuracies in the original survey occurred because of the relatively crude instrumentation of the period, rugged terrain, or in some cases, simply sloppy work by surveyors, who were paid on the basis of total miles surveyed (figure 5.11). These inaccuracies have persisted, primarily because historical boundaries as originally surveyed hold legal precedence over newer surveys of those boundaries.

Variations in the shape and size of sections resulting from survey inaccuracies weren't the only obstacles to partitioning the land into square-mile sections. The pinching effect of convergence of the meridians toward the poles on the ellipsoidal earth complicated the surveyor's job of laying out a regular grid of sections within a township. To systematize the distribution of shape and area distortions, surveyors determined section corners in a standardized order so that errors accumulated along the western and northern tiers of sections within each township (figure 5.9).

Another source of irregularity in the PLSS grids is where surveys starting from different initial points meet. The section lines from one survey will likely not mesh with those of another, particularly if the boundary between the two lies along an irregular feature such as a river. These junctures result in a mismatch of the two grids, as well as in the partial sections along their borders.

An irregular pattern of township and section lines also occurs at the boundary between the areas surveyed under different principal meridians, when the boundary is an irregular feature like a river. For example, the boundary between areas surveyed under the Michigan principal meridian and 4th principal meridian follows the Michigan–Wisconsin state line, and you will find offset and partial townships all along the boundary between the two states (figure 5.12).

Another source of deviation from the ideal pattern of the PLSS grid is that in some areas townships were not surveyed systematically outward from the initial point. A good example is the surveying of townships in Oregon along the baseline for the Willamette principal meridian. The earliest townships were surveyed outward from the initial point in Portland, but independent surveys also were completed at the coast and from three separate starting points along the baseline in eastern Oregon. These other starting points were calculated so that in theory they would mesh perfectly with the surveys from the initial point. But inherent inaccuracies in surveys of these early periods, as discussed previously in this chapter, made this impossible to achieve. You can see on topographic maps the shortened and elongated townships that mark the places where these independent surveys come together. The result of the many independently surveyed parts of PLSS grids is a mismatch where they come together. Along the boundary where these independent surveys converge, some confusing PLSS descriptions can occur. This is especially true for two land parcels straddling the boundary, as two deeds are required.

A similar problem may occur when the PLSS grid converges with earlier land partitioning systems; this can happen in regions not surveyed at the time of settlement or where boundaries were defined in an older system like DLCs or Spanish land grants. The prior Spanish land grant boundary in figure 5.4 is a classic example—both section lines and the principal meridian stop at the land grant boundary.

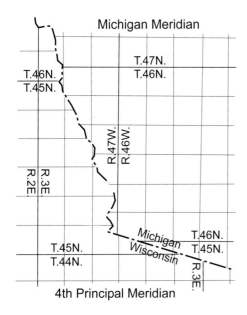

Figure 5.12　**Irregularities in PLSS township boundaries may be attributed to the convergence of surveys from different principal meridians.**

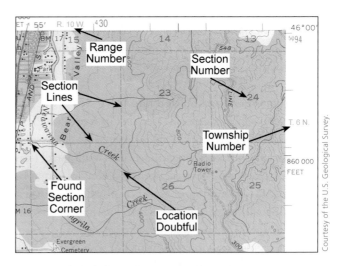

Figure 5.13　**Appearance of PLSS township and section information on a typical USGS topographic map (northeast corner of Cannon Beach, Oregon, 1:62,500 quadrangle).**

PLSS appearance on maps

The PLSS grid is found on many different types of maps. On maps made by government agencies, for example, townships and sections are often included because they are so closely associated with the boundaries of civil townships, counties, and other political units. The PLSS grid probably finds its fullest expression on USGS quadrangle maps in the topographic series. On these maps, township and section lines, section corners, section numbers, and marginal township-range notations are all printed in red (figure 5.13). Dashed red lines indicate section lines with doubtful locations.

Township and section lines also are found on a host of governmentally produced maps dealing with land resources, such as U.S. Forest Service and Bureau of Land Management ownership maps. Privately produced road and recreation atlases, some available in digital as well as paper form, often include the PLSS grid. Surveys of city and rural lots in PLSS areas also are tied to section or fractional section corners. These are commonly shown on land subdivision plat maps and tax assessor cadastral maps (described in more detail later in this chapter) both of which are used as land records.

As we've seen, the PLSS grid shown on these topographic and land resource management maps lets us partition a region into easily defined area parcels. By slightly modifying the PLSS, you also can use it as a grid cell location referencing system (see chapter 4 for more on grid cell location systems). To do so, you state the location of environmental features relative to the corner or center of a standard PLSS parcel. You can pinpoint many features this way, since the cultural landscape has been aligned to the PLSS grid in many parts of the country. For instance, Wisconsin's capitol dome is centered on the northeast corner of section 23, T7N, R9E, 4th P.M.

Canada's Dominion Land Survey

The **Dominion Land Survey (DLS)** was the public land survey system used to divide most of Canada's prairie and western provinces into townships and sections for agricultural and other purposes. In 1869, two years after the Dominion of Canada was created from British North America, Canada purchased what was called Rupert's Land from the Hudson's Bay Company. The government wanted this massive territory, including much of what is now the Northwest Territories, Nunavut, Alberta,

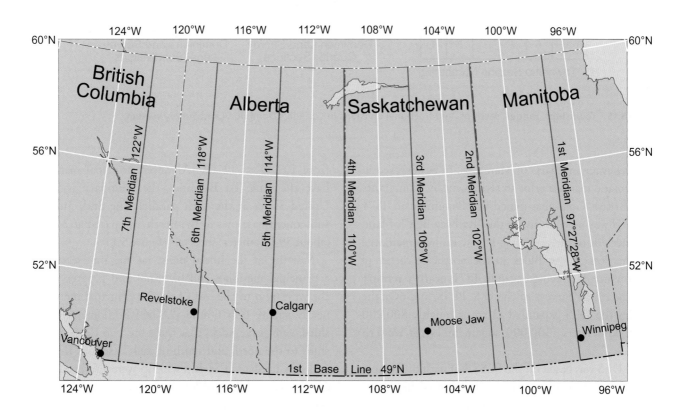

Figure 5.14   Principal meridians and 1st Baseline of the Canadian Dominion Land Survey.

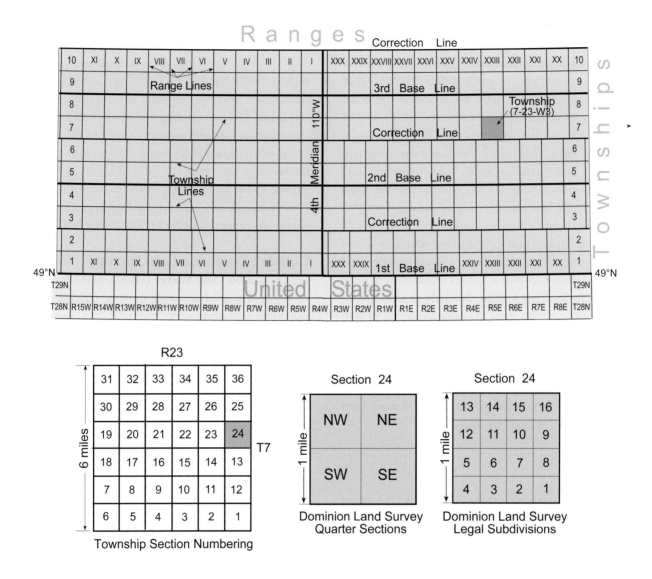

Figure 5.15 **Townships, ranges, sections, and fractional divisions in Canada's Dominion Land Survey system.**

Saskatchewan and parts of Manitoba, to be surveyed in a systematic manner prior to settlement by immigrants. After British Columbia joined the Dominion in 1871, the DLS was used to divide the "Railway Belt" of land on either side of the Canadian Pacific Railway main line and the Peace River Block in the northeast corner of the province. Surveying began in 1871, nearly a century after the initial surveys of the U.S. PLSS, and over the decades the DLS expanded to cover roughly 800,000 square kilometers (300,000 square miles) of western Canada.

The DLS was based on the PLSS, but with differences in how townships and sections were surveyed and identified numerically. By the late 1870s, seven principal meridians and baselines had been surveyed (figure 5.14), the first just west of Winnipeg, Manitoba, and the second through seventh at increments of 4 degrees of longitude

from 102°W to 122°W. All seven principal meridians have the same 1st Baseline—the boundary with the United States at the 49th parallel. The 2nd, 3rd, and subsequent baselines to the north are spaced at 24-mile (about 39-kilometer) intervals (figure 5.15).

Starting at the initial point (at the intersection of a principal meridian and baseline) and progressing to the west (also to the east of the 1st principal meridian), nearly square 6-by-6-mile (9.8-by-9.8-kilometer) townships were surveyed so that there are two tiers of townships to the north and south of each baseline. The west and east borders of each township were surveyed every 6 miles as meridians of longitude that converge toward the north pole as in the PLSS. Therefore, the northern border of every township is slightly shorter than the southern border, and the southern border is 6 miles in width only along each baseline.

Halfway between each pair of baselines are correction lines where the wider townships to the south of the second baseline meet the narrower townships to the north of the first baseline. This is where the townships surveyed north and south from each baseline converge. The boundaries of the townships are offset to the east or west of each other along the correction lines, and on maps you can see these offsets by east–west jogs every 24 miles (the length of four 6-mile townships between correction lines). On the ground, you can see them in roads, property boundaries, and other anthropogenic features that follow these survey lines.

Townships are identified by township and range numbers. Township 1 is in the first row north of the 1st baseline, and range 1 is the first column west of the principal meridian. Township numbers increase to the north. Range numbers begin in the first column west of each principal meridian (with range 1 at each principal meridian) and increase with ranges (except that east of the 1st principal meridian they also are numbered eastward). On maps, township numbers are given in Arabic numerals, and Roman numerals are often used for range numbers. However, in written descriptions of land parcels Arabic numerals are used for both townships and ranges. For example, the notation for the township highlighted in figure 5.15 is

"Township 7, range 23 west of the 3rd principal meridian," abbreviated as "7-23-W3."

As with the U.S. PLSS, each township is divided into 36 sections, each approximately 1 square mile in area. Notice in figure 5.15 that sections are numbered within townships in reverse order to PLSS sections, beginning in the lower right corner of the township and progressing northward row-by-row in zigzag fashion. For example, the notation for the township highlighted in figure 5.15 is

"Section 24, township 7, range 23 west of the 3rd principal meridian," abbreviated as "24-7-23-W3."

Each section can be divided into other fractional divisions such as half sections and quarter sections. Division into four northwest, northeast, southeast, and southwest quarter sections was used primarily for agricultural lands. The written description of the upper left quarter of section 24 shown in figure 5.15 is

"The northwest quarter of section 24, township 7, range 23 west of the 3rd principal meridian," abbreviated "NW-24-7-23-W3."

A section may also be divided into as many as 16 **legal subdivisions** (LSDs) which are other fractions of a section. Although LSDs may be square, rectangular, and sometimes even triangular, a half-quarter section of roughly 80 acres or 32 hectares is common. LSDs of 40-acre quarter-quarter sections numbered from south to north in a zigzag manner are shown in figure 5.15 for section 24. The oil and natural gas industry uses LSDs for describing the locations of wells and pipelines.

LAND RECORDS

Today, everyone uses and relies on **land records,** which are a publicly owned and managed system, defined by statewide or provincial standards, that record real estate ownership, transfers, taxation, and development. Traditionally, the basic spatial unit for this record keeping is the land parcel. As mentioned earlier, a **land parcel** results from the division of property into areas with some implication for landownership or land use. The parcel is the smallest unit of ownership or, as in the case of a farm field, a unit of uniform use. Physical features, resource reserves, market value, ownership, improvements, accessibility, and restrictions on land use are but a few of the attributes often found in these records.

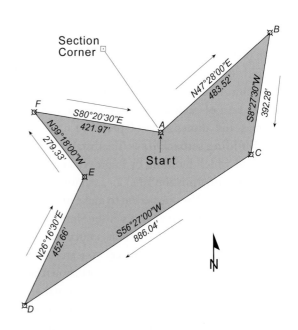

Figure 5.16 A typical closed traverse: the starting point is determined relative to a known control point, in this case a section corner. The bearings and distances for each property line are shown.

Types of land records

Subdivision plats

Land surveyors carry out the subdivision of U.S. PLSS fractional lots and other land parcels. The boundaries of most lots were probably determined using plane surveying methods similar to the metes and bounds survey illustrated in figure 5.1(B). Property boundary corners are defined by the distance and bearing from the previous corner, with the first corner defined by the distance and direction from a previously surveyed point of known integrity, such as a PLSS section corner, called a **reference** or **control point.** The corner-to-corner survey of the property lines for the entire parcel is called a **traverse.**

A traverse is generally classified as either an open traverse or a closed traverse (figure 5.16). An **open traverse** ends at a survey station whose relative position was not known previously. A **closed traverse** forms a closed loop with the first point surveyed being also the last; these types of traverses provide checks against errors in the distances and bearings, since they must ultimately lead back to the original starting point. The requirement of ending at the starting point lets surveyors check the accuracy of their work and make adjustments to all the points and curves along the traverse to force an exact fit. This accuracy check makes the extra effort of closing the traverse worthwhile.

Land subdivision surveys are recorded on a **subdivision plat,** which is a map that shows subdivided lots; these must be legally recorded in the city or county surveyor's office or an equivalent bureau. A typical subdivision plat map (figure 5.17) shows each property line along with its distance and bearing from the previous survey point. In addition to these property line dimensions, the area of each lot must be shown, along with the names and dimensions of proposed and existing streets. **Building setback lines** (the distance from a lot line beyond which building or improvements may not extend without permission from an authority) and **easements** (a right held by one person to make limited use of another person's property) also will be shown, making the subdivision plat map the best source of detailed information about any land parcel you are interested in purchasing. The geometry of the lots shown on the subdivision plat map is translated to the actual ground location of those lot boundaries by the placement of survey marks or **monuments** at corners of the subdivision, at lot corners, and street intersections during the original survey of the subdivision. Specific monumentation requirements are, however, usually a function of the state or local requirements.

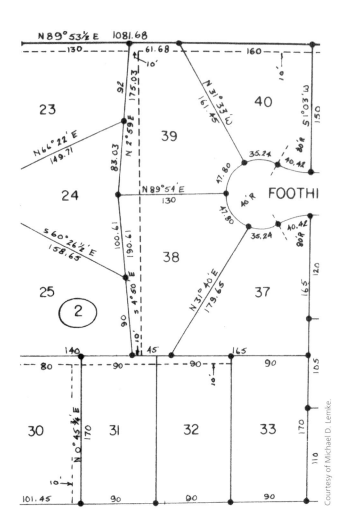

Figure 5.17 Subdivision plats show detailed land survey information and monuments for individual lots.

Courtesy of Michael D. Lemke.

The cadastre and cadastral maps

We call the written records kept on land parcels the **cadastre.** There are a number of cadastres of different types. **Fiscal cadastres** of property valuation and land taxation are among the oldest, not surprising, since governments have taxed property since the beginning of recorded history. Fiscal cadastres include the owner's name and address, parcel description and size, the parcel's assessed value and any improvements, and the current tax levy. In the United States, the fiscal cadastre for rural areas is the responsibility of the county government and is housed in the county treasurer's office in the county seat (often the courthouse). The fiscal cadastre for city property is the responsibility of the municipal government.

Another type of cadastre is the **legal cadastre,** which consists of records concerning proprietary interests in land parcels. These records contain the current owner's name and address, legal description of the property, deed, title, abstract, and legal encumbrances (such as easements, mineral rights, and transfer restrictions). In the United States, some of this information is held by the property taxing authority, but the rest may be scattered among several governmental agencies and nongovernmental entities, such as property abstract and title companies in business to help people track down these elusive records to serve a variety of purposes such as providing title insurance.

In practice, there is a great deal of overlap and duplication between fiscal and legal cadastres, and they are of limited value for land administrators, managers, and planners who make decisions involving natural resources, land use, or infrastructure considerations (fire, police, ambulance, disaster relief services) in the course of their work. These land information specialists need extensive, reliable information about land parcel features and their attributes, such as land slope and aspect, soil characteristics, drainage, vegetation cover, number of residents, building construction, access road width and surface material, utility service, and zoning restrictions. The body of land records containing this information is called the **multipurpose cadastre,** the third type of cadastre, which is currently attracting the most attention for reliable and accessible land information.

Cadastres are generally made up of two complementary parts. One part contains the written record, or **register,** which provides information concerning landownership. It may include all manner of documents, forms, official seals, and stamps of approval that characterize bureaucratic activities. These written records have traditionally been widely scattered among government offices, each with a different mission and authority. Assembling the register material can therefore be frustrating, time-consuming, and costly. The second part of most cadastres, thoroughly cross-referenced with the first part, contains detailed geographic descriptions of each parcel. These descriptions may be in the form of the original subdivision plat or in the form of cadastral maps made from the subdivision plats (figure 5.18). In either case, you can determine the relative location and areal extent of each parcel from these records.

A key function of the cadastral map is to provide the foundation for a **system of land rights transfer.** It is necessary to know a parcel's boundaries before it can be conveyed without ambiguity from one owner to the next. **Land conveyance,** or the transfer of the title to land by one or more persons to another person or other people, involves more than a geographic description of a parcel, of course, because such a description says nothing about possible restrictions or encumbrances on the property. It is for this reason that the cadastral map is cross-referenced with the register, and it is important to consult both land records when transferring property rights.

Courtesy of the County of Fairfax, Virginia.

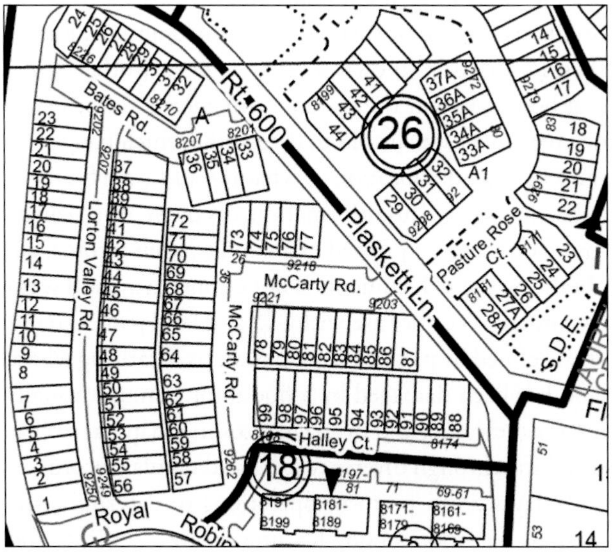

Figure 5.18 A cadastral map shows property boundaries and tax lot numbers that are tied to the fiscal and legal cadastre, and increasingly the multipurpose cadastre.

Courtesy of the City of Corvallis, Oregon, Public Works Department.

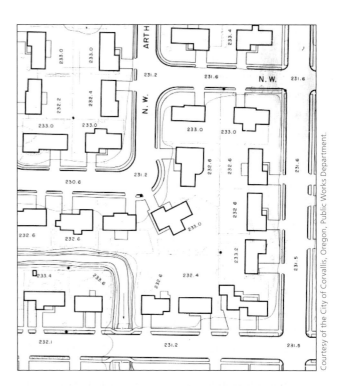

Figure 5.19 An engineering plan, such as this 1:1,200-scale segment for Corvallis, Oregon, shows property boundaries and facilities critical to public works management.

Engineering plans

In more affluent or technologically progressive cities, city engineering or public works departments usually maintain a series of very large-scale maps called **engineering plans.** A city might be covered by 1:1,200-scale map sheets (1 inch equals 100 feet), each covering a 2,000-foot-by-3,000-foot (609-meter-by-914-meter) rectangle defined by state plane coordinates. Each sheet is often composed of a set of map overlays for features such as property boundaries, streets and sidewalks, building footprints and street addresses, and detailed elevation contours. Additionally, the municipality that manages these plans also may be responsible for the management of utilities such as water and electrical. In these cases, there will also be information about the related facilities, such as water and sewer lines, and telephone and power poles. A typical engineering plan (figure 5.19) will include several, but usually not all, of these features.

Nowadays, these map overlays are most desirably stored as separate data layers in a geographic information system (GIS). All layers in a GIS database are georeferenced, which allows them to be used in overlay operations. **Georeferencing** is the procedure used to bring data layers into alignment via known ground location control points or, alternatively, the procedure of bringing a map or data layers into alignment with the earth's surface via a common coordinate system. The result is that for all georeferenced layers every location on one layer is precisely matched to its corresponding locations on all the other layers. Once the layers are georeferenced, they can be used in overlay operations. **Overlay** is the process of superimposing two or more maps or map themes to better understand the relationships between the geographic features of interest and their attributes.

The information for engineering plans comes from subdivision plats, engineering records, and interpretation of large-scale aerial photography covering the city. These map overlays are part of a city's multipurpose cadastre, with each overlay stored as a data layer in a land information system. Today, some city engineering and public works departments have Web sites where engineering plan overlays can be viewed and downloaded to home computers.

Land information systems

The cadastre represents an important part of the geographic information database used in modern **land information systems,** also referred to as **LIS.** Computerized LIS integrate the various cadastres into a useful whole. Such systems stress data compatibility, sharing, and cooperation. They also incorporate powerful analytical and graphic tools to help users analyze data, generate and manage land records, produce maps, and make better land management decisions. LIS must meet two requirements:

1. All entries should be spatially referenced, using some form of coordinate referencing system to a known level of control. In other words, there must be some provision for moving from the descriptive or graphic data records to the appropriate ground positions.

2. There must be compatibility among the spatial referencing entries. It is preferable that all spatial reference data be identified with some terrestrial system. Latitude–longitude coordinates are ideal because they are universal, but grid coordinates, such as state plane and universal transverse Mercator coordinates, also are acceptable if we are willing to accept the zone boundary problems they introduce. If several spatial reference systems are used in recording data, they should at least be transformable from one to the other.

A major problem for the cadastre, which subsequently affects LIS, is that land partitioning systems developed to accommodate early settlers aren't best suited for today's primary concern: land records management. The PLSS is a case in point. It was an excellent system when the government's main concern was selling or giving away land parcels as quickly and efficiently as possible. It also served land transfer needs fairly well in subsequent years. But for purposes of land management, the PLSS has been an administrative nightmare.

The problem is that because of technological barriers faced at the time of the original PLSS surveys (which were mostly performed during the nineteenth century) the locations of township and section corners were not originally defined by their latitude and longitude. Rather, these locations are known only in reference to other points and boundaries in the system. This means that there is no convenient way in LIS to determine which parcel contains a particular location or what resources are found in a specific parcel. This information, of course, often can be found on archived printed maps and written records, but obtaining these documents is not a convenient and seamless process as is the case with LIS and land parcel boundaries defined by parallels and meridians. Learning whose land has been damaged by a flood, for example, usually involves rescaling and overlaying several maps, as well as searching through diverse textual and tabular records.

Today we can perform operations to determine the spatial association among features far more quickly and accurately with computers using GIS software than we can manually (see chapter 18 for more on spatial association). For this reason, many thousands of PLSS section corners have been located on the ground and given latitude–longitude, UTM, or state plane coordinates. This, however, is an exceptionally expensive and time-consuming process given the need to first determine the location of those corners originally set in the nineteenth century, which, in many cases, have rarely or never been revisited since their original establishment. Although the old system will undoubtedly persist for years to come, we can expect land parcels in the future to be described by geographic or grid coordinates.

SELECTED READINGS

Crossfield, J. K. 1984. Evolution of the United States public land survey system. *Surveying and Mapping* 44 (3): 259–65.

Estopinal, S. V. 1989. *A guide to understanding land surveys.* Eau Claire, Wis.: Professional Education Systems.

Hart, J. F. 1975. Land division in America. In *The look of the land.* Englewood Cliffs, N.J.: Prentice-Hall, 45–66.

Johnson, H. B. 1976. *Order upon the land: The U.S. rectangular land survey and the upper Mississippi country.* London: Oxford University Press.

National Research Council. 1980. *Need for a multipurpose cadastre.* Washington, D.C.: National Academy Press.

——. 1982. *Modernization of the public land survey system.* Washington, D.C.: National Academy Press.

——. 1982. *Procedures and standards for a multipurpose cadastre.* Washington, D.C.: National Academy Press.

Robillard, Walter G., et al. 2006. *Evidence and procedures for boundary location.* 5th ed. Hoboken, N.J.: John Wiley & Sons, Inc.

Thrower, N. J. W. 1972. Cadastral survey and county atlases of the United States. *The Cartographic Journal* 9 (1) (June): 43–51.

Trewartha, G. T. 1946. Types of rural settlement in colonial America. *Geographical Review* 36 (4): 568–96.

U.S. National Atlas. 2008. The Public Land Survey System (PLSS). http://www-atlas.usgs.gov/articles/boundaries/a_plss.html.

Ventura, S. J. 1991. *Implementation of land information systems in local government—Steps toward land records modernization in Wisconsin.* Madison: Wisconsin State Cartographer's Office.

Vonderohe, A. P., et al. 1991. *Introduction to local land information systems for Wisconsin's future.* Madison: Wisconsin State Cartographer's Office.

von Meyer, Nancy. 2004. *GIS and land records.* Redlands, Calif.: ESRI Press.

Wilson, Donald A. 2006. *Interpreting land records.* Hoboken, N.J.: John Wiley & Sons, Inc.

chapter
six   RELIEF PORTRAYAL

ABSOLUTE RELIEF MAPPING METHODS
Spot elevations, benchmarks, and soundings
Contours
Isobaths
Hypsometric tinting

RELATIVE RELIEF MAPPING METHODS
Planimetric perspective maps
Oblique perspective maps

COMBINING METHODS ON MAPS
Relief shading and contours
Relief shading and hypsometric tinting

STEREOSCOPIC VIEWS
Stereopairs
Stereoscopic polarization
Anaglyphs
Chromadepth maps

DYNAMIC RELIEF PORTRAYAL
Animated methods
Interactive methods

DIGITAL ELEVATION MODEL DATA
ETOPO5, ETOPO2, and GLOBE
Shuttle radar topography mission
National Elevation Dataset
Coastal Relief Model
Lidar

SELECTED READINGS

6

Relief portrayal

The **terrain surface** provides the foundation upon which we play out our lives. Nothing in the environment is immune from the vertical differences on the earth's surface that we call **relief.** Our mobility, orientation, and environmental understanding are all affected by relief. Yet it's easy to forget the significance of relief in our environment, because many of us live in an essentially flat world. The floor of our house is flat; most yards and streets are flat. Thus, we tend to think of geographical position in purely horizontal terms. This is often a perfectly good way to simplify our world, but ignoring or misunderstanding the terrain surface can have tragic consequences. Ships run aground, airplanes crash into mountainsides, and lost hikers die of exposure—all because the terrain wasn't understood properly.

In mapping, a terrain surface is a three-dimensional representation of data about the elevations of our physical environment, and cartographers use **relief portrayal** techniques to map those data. Maps treat relief in several ways. Some maps ignore it and give only horizontal information. These **planimetric maps** are useful when the mapped area is essentially flat or when facts about an area's relief aren't important to your needs. In such situations, relief information would merely clutter the map with unnecessary detail.

At other times, understanding the terrain surface is crucial to establishing your position and studying spatial associations with other things like vegetation and patterns of rainfall. When relief information is important, it's best to turn to **topographic** and other maps that show the three-dimensional nature of the terrain surface. There are two types of relief portrayal—**absolute relief mapping methods** for showing precise elevation information and **relative relief mapping methods** for showing different landform features and giving a general impression of their relative heights.

ABSOLUTE RELIEF MAPPING METHODS

Engineers, scientists, surveyors, and other map users who work analytically with the terrain surface require maps that show more than the general form of terrain features. In these cases, the map users need information about the **absolute relief,** or the actual elevation values at locations in the landscape. Absolute relief methods provide the numerical elevation and water depth information they need. The most accurate elevation and depth data shown on maps are surveyed elevations at individual points on the earth.

Spot elevations, benchmarks, and soundings

On aeronautical charts, topographic maps, engineering plans, and other large-scale maps, the elevation of the surface is given numerically at individual survey points. These elevation values, relative to the **mean sea level (MSL)** datum (see chapter 1 for more on mean sea level), are called **spot elevations.** Spot elevations are often located at positions or on features that are positively identifiable and easily recoverable (that is, they can be easily found again). Searching for survey marks, or **mark recovery** (also sometimes improperly called **benchmarking**), at road intersections or forks is much easier than at changeable locations such as stream confluences or vegetation patch corners. On topographic maps, as shown in figure 6.1, spot elevations are symbolized with an "×" followed by the elevation value above MSL. Where spot elevations are located at a road intersection, the symbol is omitted since the location of the reference elevation is known.

At some of these locations, the elevation has been determined by precise leveling methods (see chapter 1 for more on leveling), and a permanently fixed brass plate, called a **benchmark,** has been installed in the ground (figure 6.2). Benchmarks are symbolized on topographic quadrangles made by the U.S. Geological Survey (USGS) and other national mapping agencies by a small "×" next to the elevation value (figure 6.1), with the identifier "BM" (for benchmark) preceding the elevation value.

Water depth readings are called **soundings.** For thousands of years, mariners have measured shallow water depths using a calibrated **lead line** (line with a lead weight tied to one end that is lowered into the water to determine depth by markings on the line) or **sounding pole** (pole marked with water depth values). Today soundings are obtained by **electronic depth measuring instruments** that determine the amount of time that acoustic pulses take to reach the bottom and return to the instrument. This travel time is then converted to distance above the bottom since the velocity of acoustic waves is a known quantity. Our modern electronic instruments collect soundings more rapidly, accurately, and at greater depths than poles or lead lines allowed. Global Positioning System (see chapter 14 for more on GPS) receivers integrated into the instruments also give the exact latitude, longitude, or grid coordinates at the instant each sounding is obtained.

Sounding values are not relative to MSL, but rather to specific definitions of "low water." The two datums used in North America to define low water are the arithmetic average of all the low tide levels recorded over a 19-year period, called **mean low water (MLW),** and the arithmetic average of the lower of the two daily low tides recorded

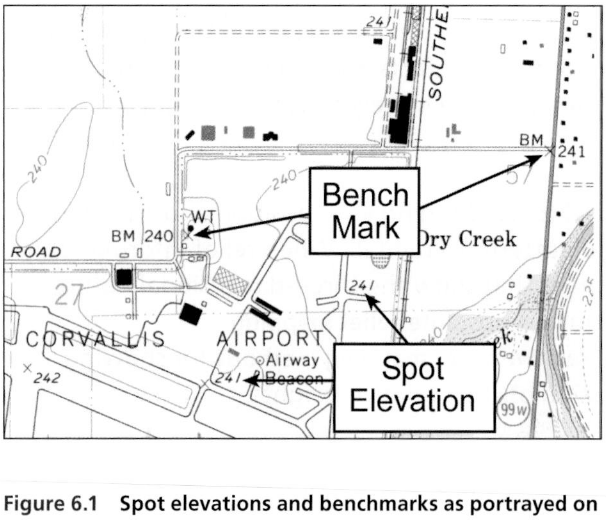

Figure 6.1 Spot elevations and benchmarks as portrayed on a 1:24,000-scale topographic map.

Figure 6.2 Precisely surveyed elevation points are identified on the ground by circular brass benchmarks.

Courtesy of the National Ocean Service.

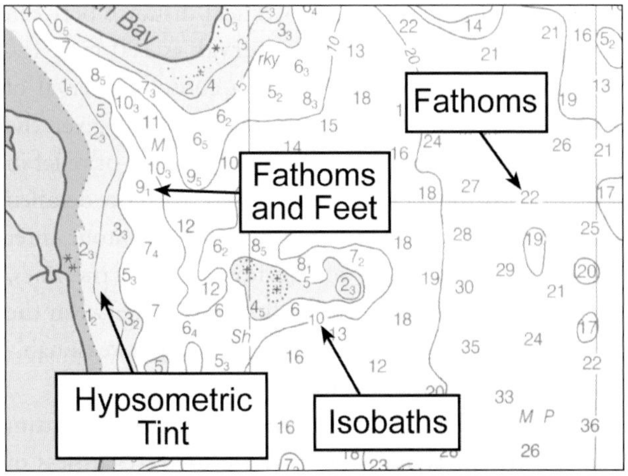

Figure 6.3 Soundings on United States nautical charts usually are shown in fathoms for depths greater than 11 fathoms. Fathoms and feet are used together for shallower areas.

over the same 19-year period, called **mean lower low water (MLLW).** Canadian nautical charts use MLW, whereas in the United States MLLW is the official U.S. National Ocean Service (NOS) nautical chart datum. The datum used to define high water to support harbor and river navigation is **mean high water (MHW).**

The MLLW datum is used on nautical charts for water depths because ship captains need to decide whether there is enough clearance for their vessels between the changing water surface and a fixed submerged obstacle, such as the harbor bottom or a wreck. To do so, they must determine the minimum depth likely to be encountered at a given position. If overhead clearance is the concern, as when moving under a bridge, the situation is reversed and the MHW datum is more useful.

Sounding values have been printed on nautical charts since the early 1600s. Charts covering the coastline of the United States (figure 6.3) traditionally have used the **fathom** (6 feet) as the unit of measurement for depth. In areas shallower than 11 fathoms where more exact depths are required for safe navigation, fathoms and feet are used together. The value 1_5, for instance, is read as one fathom and five feet, or 11 feet. Nautical charts made in Canada and most other nations show depths in meters, and recent U.S. charts are using meters as well.

Contours

Contours are lines of equal elevation above a datum. If a contour were actually drawn on the earth, it would trace a horizontal path constant in elevation. It is common to use contours on topographic maps to show variations in relief and landform features like hills and valleys. The portion of the 1:24,000-scale USGS topographic quadrangle shown in figure 6.4 is representative of this type of map.

Courtesy of the U.S. Geological Survey.

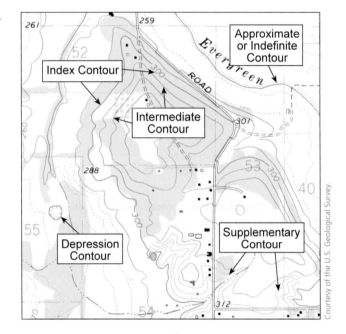

Figure 6.4 A portion of a 1:24,000-scale USGS topographic quadrangle for Corvallis, Oregon, with a 20-foot contour interval. Index, intermediate, approximate or indefinite, supplementary, and depression contours are used to portray special landform features and to make the map easier to read and analyze.

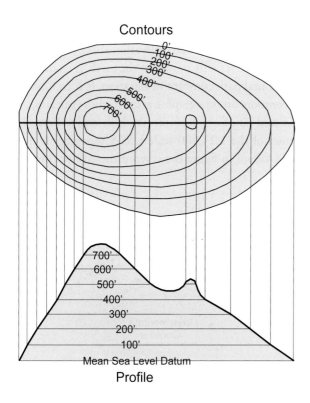

Contours

Profile

Mean Sea Level Datum

Figure 6.5 **The logic of the contour method of relief portrayal is illustrated here by converting the contours from a horizontal transect line (top) into a profile (bottom).**

To understand the logic behind contours, imagine that you are on a small sandy island in the ocean. It has a hill just over 700 feet (213 meters) high and another just over 500 feet (152 meters) high (figure 6.5). If you walked around the island at the shoreline when the tide was at mean sea level, you would trace the zero or **datum contour** and would return precisely to your starting point. It is important to remember that all contours eventually close, although the closed curve won't always be seen on a single map sheet.

Now suppose that you start climbing up the highest hill until you reach an elevation of 100 feet above mean sea level. If you again follow the contour around the island, all points on your path will be at the same elevation, and again you will return to your starting point. The effect is the same as if you had walked along the shoreline after the ocean level was raised 100 feet. If you did the same thing for elevations of 200 feet, 300 feet, and so forth, and the paths you walked were projected vertically onto a flat map, the result would be something like figure 6.5. If the island were viewed in profile, it would look as though it had been sliced into layers by imaginary horizontal planes. (The procedure used to create a profile from a contour map is explained under "Profiles" in chapter 16.)

In our island example, the vertical distance between contours, called the **contour interval,** is 100 feet. The mapmaker, of course, may select a contour interval of 20 feet, 50 feet, or whatever seems appropriate given the extent of area being mapped and the amount of relief in that area. The smaller the interval used, the more detailed the relief portrayal. Once you know the contour interval, it's a simple matter of "reading between the lines." If the spot you're interested in lies halfway between the 500 foot and 600 foot contours, you might reasonably guess that it's around 550 feet high.

On most maps showing contours, the MSL datum and a constant contour interval are used, regardless of the nature of the terrain on the map. You are usually safe in treating these factors as constants. This leaves only the size of the interval as a variable factor. In general, you will find that the greater the relative relief on a map, the larger the contour interval. Mapmakers use this relationship between relief and contour interval to keep contours from becoming too dense in areas of high relief and too sparse in areas of low relief.

A special problem arises when the terrain changes markedly from one part of the map to another. When this occurs, it is quite possible that several different but constant intervals may have been used on the same map. This might happen, for example, with map sheets spanning high to low relief. This is called **variable contour line density.** Without variable density, closely spaced contours would clutter the map and become difficult to differentiate, and widely spaced contours would make it hard to determine slopes and predict elevations at points falling between contours.

Types of contours

On USGS and other topographic maps, you are likely to see a variety of contour line types, including index, intermediate, approximate or indefinite, supplementary, and depression contours (figure 6.6). Less commonly, you will see carrying, feathering, cut and fill, glacial, and submerged or underwater contours. Let's look at each of these in more detail.

On most maps, you'll find that not every contour is marked with its elevation value; instead, every fourth, fifth, or tenth line is labeled, depending on the scale of the map. These labeled contours, called **index contours,** are usually drawn with a thicker line than the **intermediate contours,** which are drawn between the index contours (figure 6.4) so that they are easily seen. Index contour labels normally are spaced widely, meaning that you may have to trace the index contour for some distance before you see its value. On U.S. maps, the labels

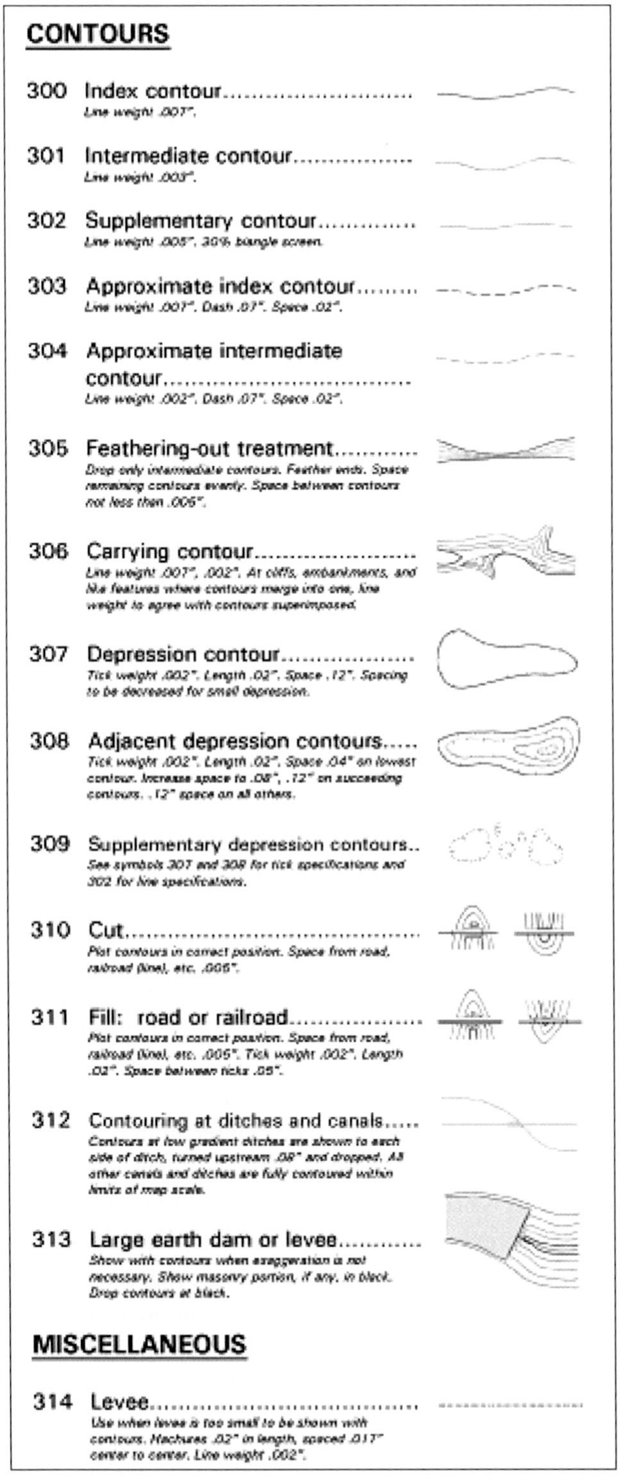

Figure 6.6 Types of contours are illustrated and their symbology is described in the Publications Symbols document for 1:24,000-scale USGS topographic maps.

are often laddered—that is, they are placed inline with the label on the adjacent labeled contours so that you can easily read the elevation values from one index contour to the next. On U.S. maps, the labels are also drawn relative to the bottom of the page rather than the terrain. On maps used in Great Britain and other countries, you will find the contour labels laddered as well, but they will always be oriented upslope, so in some places they will appear upside down on the map. On Swiss topographic maps, the contour labels will be randomly placed rather than laddered so as not to give the impression that the terrain has been altered by humans.

Several modifications to standard index and intermediate contours are used to portray special aspects of the land surface and to simplify map reading. **Supplementary contours** are placed in areas where elevation change is minimal. And if there is a lot of space between index and intermediate contours, these lines are added to indicate that elevation measurements have been made, even if they are few and far between. These lines often are used in floodplain areas where a slight change in relief might have a major impact on the stream channel and flooding pattern. Sometimes you will also see them in areas of rugged terrain to emphasize important or unique features. On USGS topographic maps, they are shown in a lighter tint than standard index and intermediate contours (figure 6.4).

Sometimes you will come across dashed segments of standard contours or will find dashed contours between the standard contours (figure 6.4). These lines are called **approximate** or **indefinite contours.** They are used to indicate the approximate location of contours where the information isn't reliable, usually in areas where the vegetative surface cover precludes economically contouring the ground so that the contours will meet National Map Accuracy Standards (see chapter 10 for more on map accuracy and these standards). Although these lines contribute valuable terrain information, they should be viewed with caution since they are intended to give a general impression of the terrain, not precise elevations.

To aid in identifying closed depressions or basin-like features, small right-angle ticks may be added to the down slope (or inside) of contours (figure 6.4). These **depression contours** help to focus the map user's attention when the depression is small relative to the size of the map sheet. Depression contours are especially useful in distinguishing small depressions from small hills because there is seldom enough space for the mapmaker to label the contours of these features with their elevation values. You will sometimes see a special type of depression contour called a **cut contour** where a roadway has been blasted through landscape, drastically lowering (cutting) the terrain (figure 6.6). Usually these are accompanied by **fill contours**, which show where the terrain was raised to support a road or railway grade (figure 6.6).

When depression contours are used, they are sometimes labeled, sometimes not. In either case, note that the first (outside) depression contour is always of the same elevation as the adjacent standard contour. Moreover, depression contours merely represent special cases of the standard contours on the map and thus share the same interval and elevation values.

In areas of high relief where there is insufficient space to show all the contours, carrying or feathering contours may be used. Sometimes you will see two or more contours merge into a single contour; these **carrying contours** are used to represent vertical or near-vertical topographic features such as cliffs and escarpments (figure 6.6). A feathering-out treatment, (**feathering**) is used to drop contours in areas of high relief where coalescence would cause the lines to bleed together (figure 6.6). The highest and lowest elevation contours are extended the farthest and dropped, then the next highest and lowest contours are extended and dropped, and so on.

If a contour is on an ice mass, like a glacier, the positional accuracy of the line will also be depicted as "indefinite" and the contour will be shown in blue. **Glacial contours** are those that represent the surface of an ice mass or permanent snow field at the date of the photography used to compile the feature (figure 6.7).

An uncommon type of contour is a **submerged** or **underwater contour.** It depicts the former river channel in an area that was inundated because of damming. These contours are printed in dark blue, black, or brown.

Isobaths

Isobaths, also called **depth contours** or **depth curves,** are lines of equal water depth below the MSL or MLLW datum. They are found on nautical charts and other **bathymetric maps** (maps that show water depths). The first isobaths appeared on European charts of river estuaries in the late 1500s. By the end of the eighteenth century, enough soundings were taken that isobaths could be drawn on a large number of nautical charts. In this century, electronic depth sounders have provided detailed information for drawing isobaths on virtually all nautical charts (figure 6.3). In addition, information on deep ocean depths collected by sonar methods and detailed analysis of artificial satellite data have allowed isobaths to be accurately drawn on bathymetric maps of the oceans and other deeper water bodies.

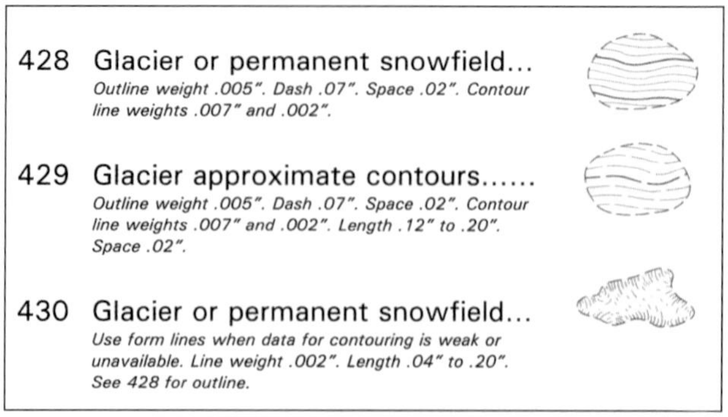

428	Glacier or permanent snowfield...
	Outline weight .005". Dash .07". Space .02". Contour line weights .007" and .002".
429	Glacier approximate contours......
	Outline weight .005". Dash .07". Space .02". Contour line weights .007" and .002". Length .12" to .20". Space .02".
430	Glacier or permanent snowfield...
	Use form lines when data for contouring is weak or unavailable. Line weight .002". Length .04" to .20". See 428 for outline.

Courtesy of the U.S. Geological Survey.

Figure 6.7   Blue is used instead of brown for contours that relate to glaciers and hydrographic features.

Isobaths on large-scale nautical charts differ from contours—isobaths are based on the MLW or MLLW datum and use uneven depth intervals. Charts produced by the U.S. National Ocean Service, for instance, show isobaths for 1, 2, 3, 5, and 10 fathoms, with greater depths shown in multiples of 10 fathoms, typically 20, 50, and 100. Shallow water isobaths appear on charts for the same reason that shallow water depths are shown in fathoms and feet—mariners need this information for safe coastal and harbor navigation.

Hypsometric tinting

Hypsometric tinting (also called **layer tinting** or hypsometric coloring) is a method of "coloring between contour lines" that enhances the relative relief cues for contours while maintaining the absolute portrayal of relief. On some maps made with this technique, contours or isobaths have a stepped appearance much like a layer cake. This is because the space between contours is given a distinct gray tone or color, called a **discrete hypsometric** or **layer tint.** For terrain surface maps, the colors often relate to the type of land cover found at different elevation zones, such as green for low-lying verdant valleys, brown for high-elevation treeless areas, and white for snowcapped peaks.

Visually enhancing the zones between contours using discrete hypsometric tinting helps you to see differences in relief. It also allows you to more easily determine the elevation or depth range for any location. The simplest discrete hypsometric tinting is on U.S. nautical charts, where a light blue tint is usually added to all water areas within the 5-fathom isobath (figure 6.3). When mapmakers use a series of blue tones to tint between isobaths, the usual rule is, "The darker the tone, the deeper the water." If the map is well designed, hypsometric tinting should give a visual impression that appropriately portrays the underwater surface being mapped.

With newer mapping capabilities, the abrupt change between hypsometric tints can be minimized by gradually merging one tint into the next, giving a smooth appearance to tonal gradation. This **continuous hypsometric** **tinting** (figure 6.8) is readily achieved using computer mapping software and **digital elevation model (DEM)** data, which we'll explore later in this chapter. Continuous hypsometric tinting is being used more and more frequently; you will see it on poster-sized wall maps that are more like art and in recreational atlases or gazetteers.

This technique uses a **color ramp,** which is a continuous succession of colors between a specified beginning and ending color. These can be strung together to make a progression across many colors. The key advantage of this method is that each elevation in the DEM grid is assigned a color corresponding to the exact elevation rather than a range of elevations as in discrete hypsometric tinting.

As with discrete hypsometric tinting, the colors selected are often assumed to relate to the ground cover typically found at various elevations (figure 6.9). The difference is that the colors vary so gradually that you cannot see the individual "steps" of the elevation ranges. For that reason, it is harder to determine an exact elevation or elevation range at any location on these maps. To find this information you will have to rely on contours, if they are also shown.

Figure 6.8 Continuous layer tinting creates a smooth progression of hypsometric tints for water depths and the land surface form on this bathymetric map of the Caribbean Sea.

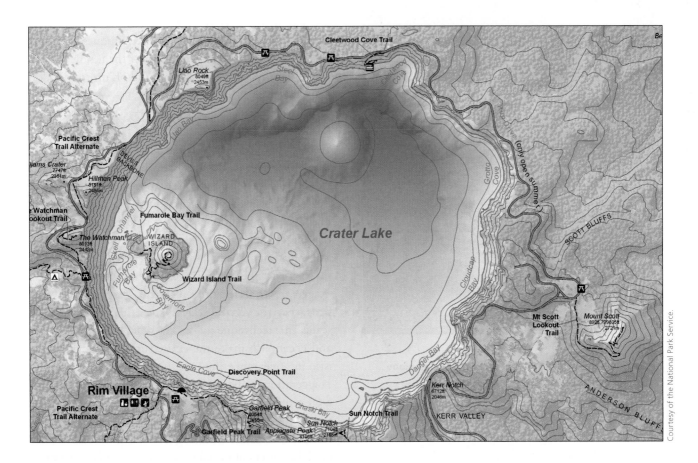

Figure 6.9 In addition to the contours, the elevation is shown with hypsometric tinting. The higher elevations are shown in white as snowcapped peaks. The continuous hypsometric tint progresses to brown for the rocky outcrop areas and then to green for the lower elevation valleys.

You will also find some maps, such as older wall maps, that show the highest elevation zone in dark brown or red, the intermediate zones in buff and light brown, and the lowest zone in dark green. The water in the valley bottoms is blue. These color progressions are based on the perceptual phenomenon that some colors such as blue appear to visually fall away from you, while others such as red visually rise toward you. Although this coloring scheme has been widely used, its visual effectiveness hasn't been satisfactorily proven.

Colors for hypsometric layer tinted maps aren't always effectively chosen by mapmakers, nor do darker tones always signify lower water depths, especially when coupled with relief shading (described later in this chapter), as they often are. Thus, it's especially important to check the legend before using maps that have hypsometric tinting.

Courtesy of Nutri-Systems Corp., bathymetric 20" globe.

Figure 6.10 Elevation variations on a raised relief globe have to be greatly exaggerated to give a reasonable impression of the landform.

RELATIVE RELIEF MAPPING METHODS

In our day-to-day life, we're usually concerned with the local range between high and low heights, or the **relative relief,** rather than the absolute elevation values. We often think of relief in terms of terrain features like hills and plains, mountains and valleys. Relative relief methods are used on many maps to give a visual three-dimensional effect that makes terrain features easy to see. One way to achieve the perception of realism is by depicting the landform surface in raised relief on a globe.

Planimetric perspective maps

Not to be confused with planimetric maps, which represent only the horizontal positions of features and not the vertical positions that topographic or bathymetric maps show, **planimetric perspective maps** give an overhead view of the mapped area. Showing elevation variation on maps that have a viewpoint as if you were looking straight down from an airplane is challenging—you would be able to see variations in the surface from the side in perspective view. Nonetheless, mapmakers have developed a number of clever techniques to do just this. We will take a look at a few of them now so that you will know how to read this kind of information on maps you use.

Raised relief globes

Since globes present the truest picture of the earth as a whole, we might conclude that **raised relief globes** provide the most realistic and useful portrayal of the vertical dimension. But there is a flaw in this reasoning. We saw in chapter 1 that if the earth were reduced to the size of a bowling ball (a common globe size), the earth would be "smoother" than the ball.

To create a realistic impression of relief, mapmakers exaggerate the elevation. **Vertical exaggeration** happens when the vertical scale is larger than the horizontal scale. This technique makes low relief landforms visible and it can also make the entire surface somewhat realistic in appearance (figure 6.10). In practice, a vertical exaggeration of about 20 to 1 is typical on relief globes of a 60 centimeter (24 inch) diameter. Therefore, in addition to the highly generalized landforms because of their small scale and the handling and storage inconvenience of globes, users of raised relief globes also face the problem of large vertical distortion. Makers of these globes also face the problem that they are costly to produce.

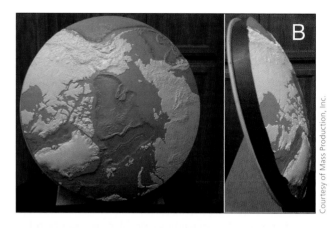

Figure 6.11 The layer-by-layer construction of relief models often leaves them with a stepped surface (A), although smoothing out the steps between contours creates a more realistic portrayal, as does showing the earth's curvature (B).

Relief models

We can minimize the problem of high vertical distortion by using physical **relief models** rather than globes. Relief models are in effect chunks of a giant globe, often constructed to show the curvature of the earth. They let us focus on one small portion of the earth at a larger scale than a raised relief globe, allowing relief to be shown with less vertical exaggeration than on globes. Some relief models are constructed with discrete elevation layers, much like discrete hypsometric tinting, but the areas between contour lines are flat planes at the same elevation (figure 6.11A). The stepped effect is almost always a result of the mapping technique, not the nature of the landform surface. Although layered landforms do occur in regions of terraced agriculture, open pit mining, and horizontal sedimentary beds of varying resistance to erosion, these landscapes are relatively rare.

More realistic models are constructed with a smooth, continuous landform surface (figure 6.11B). Of all maps, these relief models probably provide the clearest picture of the terrain, although the terrain is usually exaggerated vertically to create the clear picture. Unfortunately, like all physical models, they suffer the disadvantages of bulk, weight, and high cost of production. For these reasons, relief models are usually used only in permanent or semipermanent displays. They are frequently found in the lobbies of government and private agencies that deal with environmental problems; in city-regional planning exhibits; and in university geology, geography, and landscape architecture departments. People working on promotional schemes or research projects are especially fond of using relief models. Parks, urban renewal projects,

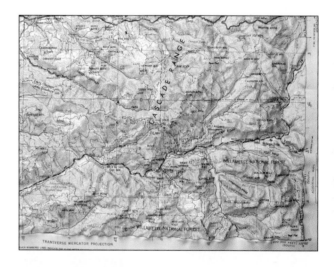

Figure 6.12 A raised relief topographic map is printed on a sheet of plastic molded to form a three-dimensional surface.

malls, and dam sites are favorite subjects. For these purposes, the inconvenience of models is offset by the true-to-life impression of the landscape they provide.

Raised relief topographic maps

There is one type of relief model that partially sidesteps the problems of weight and high cost. This is the **raised relief topographic map.** It is made by taking a flat topographic map, printing it on a sheet of plastic, and using heat to vacuum-form it into a three-dimensional model (figure 6.12). The resulting map isn't a curved piece of the globe, as are many relief models. It is really just a flat map sheet with an undulating landform surface.

Raised relief topographic maps are available from private mapping firms. They cost about 10 times more than a conventional map of the same size and area, and are relatively fragile and difficult to store. Since the flat plastic sheet must be stretched, some displacement of features on the raised surface is also common. But the realism of relief often compensates for these drawbacks. A more serious concern is that the horizontal position of geographical features may not be portrayed accurately. This is inevitable because of the displacement caused by stretching the flat map into a three-dimensional plastic model. The inaccuracy problem is compounded by the fact that this type of relief map is usually produced at medium and small scales so as to have wider sales appeal.

Hachures

On large-scale flat maps, the terrain is sometimes rendered with tiny, short lines called **hachures,** arranged so that they face downhill. Each hachure line lies in the direction of the steepest slope, showing the amount of slope with some accuracy. The best known hachuring method is called the **Lehmann system,** after its founder **Johann G. Lehmann,** a Saxon military officer who introduced the method to Europe in 1799.

In the Lehmann system, the thickness of hachures is varied—the steeper the slope, the wider the hachures (figure 6.13A). At times, the hachure technique has been so rigorously applied that you can derive numerical slope values by comparing a zone of uniform slope with the map legend. More commonly, only a general impression of steepness was intended.

Figure 6.13 With the Lehmann hachure method (A), relief is portrayed by increasing the width of hachures with increasing slope. The Dufour method (B) eliminates hachures on the northwest sides of hills to give a three-dimensional appearance to the terrain.

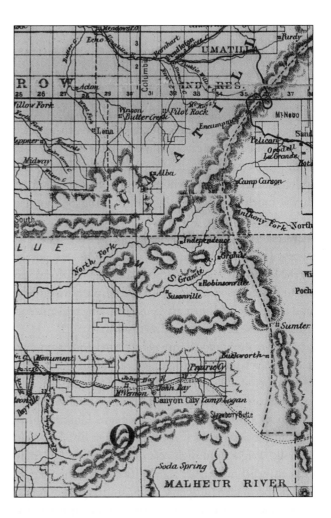

Figure 6.14 Hachures on nineteenth century small-scale maps like this piece of an 1887 county and township map of Oregon and Washington look like caterpillars crawling across the map.

Hachuring is a poor method when applied to small-scale maps, as it was in the nineteenth century (figure 6.14). The hachure lines for mountain ranges had to be so simplified and stylized that they look like hairy caterpillars crawling across the map. In fact, cartographers will refer to these as "hairy caterpillars." If you see these creatures on modern maps, realize that they're intended to show only the general location of hills and mountain ranges.

In the mid-nineteenth century the Lehmann system was largely replaced by the **partial hachuring system** of the Swiss cartographer **Guillaume H. Dufour.** In Dufour's method, hachures are eliminated on the northwest sides of hills on north-oriented maps (figure 6.13B). Eliminating hachures in this manner greatly improves the three-dimensional impression of relief and makes terrain features far easier to identify. The problem with Lehmann and Dufour hachures is that they are tedious to execute and obscure other map features. Hence, hachures are rarely used in the United States today, but the Dufour system is the predecessor of the relief shading method widely used throughout the world and can still be found on topographic maps made by the Swiss and other national mapping agencies.

Relief shading

Relief shading, also called **hill shading** or **shaded relief,** has been used on maps since the late nineteenth century to enhance the three-dimensional appearance of terrain features. We know from looking down at the earth from an airplane that patterns of light and shadow on hills give us the strongest impression of relief. Relief shading attempts to re-create these tonal variations on maps to give the same three-dimensional effect.

The principle underlying relief shading is that drawings of three-dimensional objects appear correct to us when we use an imaginary light source from the upper-left corner of the illustration. This is equally true of maps (figure 6.15, left). If the opposite lower-right light source position is used, objects will appear inverted. **Relief reversal** is exactly the opposite effect from what was intended—hills look like valleys and valley bottoms look like ridge tops (figure 6.15, right). The Dufour hachuring method was based on this principle and hence is a crude form of relief shading. Place names, of course, will usually provide an obvious clue to proper map orientation. Still, you should develop the habit of checking the map orientation before interpreting any relief shaded map.

On north-oriented maps, relief shading is based on an imaginary light source at a fixed position off the

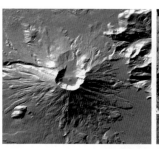

Courtesy of the U.S. Geological Survey.

Figure 6.15 Relief shading (left) of Mount St. Helens, Washington, using oblique illumination with a fixed light source from the northwest. A vertical aerial photograph (right) of the same area illustrates the differences between relief shading and shadowed terrain. Relief reversal is evident in the air photo because the light source (the sun) illuminates the terrain from the southeast.

northwest corner of the map (figure 6.15, left). The lightest shading occurs on northwest facing slopes, which are at right angles to the imaginary light source rays, and the lightest tones are used for those areas that have the highest slopes. The darkest shadows are cast over southeast facing slopes, with darker tones assigned to steeper slopes. Lightness and darkness, then, is determined not only by slope steepness but also by terrain orientation with respect to the light source.

Relief shading is not the same as the shadowing that appears on real earth features. For instance, compare a mid-morning vertical photograph of the terrain in figure 6.15 to its relief shaded representation. On the air photo, shadows cast by hills are equal in darkness throughout. These shadows also cross valleys to darken adjacent hillsides. In contrast, the relief shading varies in darkness within shaded areas and stops at the base of valleys.

Another problem with relief shaded maps is more serious than relief reversal. Relief features oriented roughly parallel to the imaginary light rays (northwest to southeast) aren't shaded enough to give proper relief cues. To circumvent this problem, mapmakers often let the imaginary light source "float" around to the map's west and north sides so that terrain with north–south and east–west orientation also receives sufficient shading. This use of a variable light source is probably the best way of all to show relative relief on a flat map. This can be achieved analytically with computer software or it can be generated through an artistic rendering (figure 6.16). The map may look as real as a picture of a physical relief model of the landform.

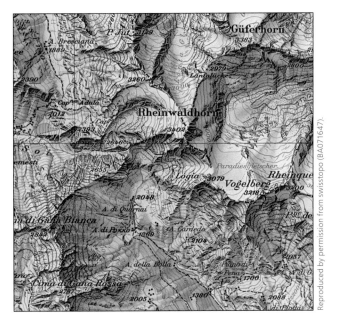

Figure 6.16 The use of artistic shading from the north to the west depending on the orientation of ridges gives an excellent impression of a continuous landform surface on this Swiss 1:100,000-scale topographic map. Rock outcrop drawings also enhance visual realism. If you turn this illustration upside down, relief reversal will cause the hills to look like valleys and the valleys to look like ridge tops.

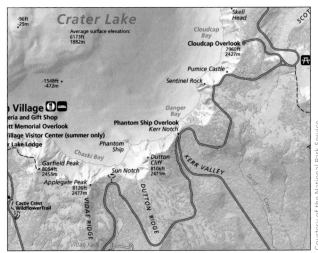

Figure 6.17 A forest texture has been added through bump mapping to this relief shaded tourist map of Crater Lake, Oregon.

Relief shading used to be a laborious manual chore that required a great deal of artistic skill and a thorough understanding of **geomorphology** (the study of landforms and terrain forming processes). The expense involved in producing shaded relief maps by hand could rarely be justified. Thus, despite the dramatic visualization relief shading affords, for practical reasons not many such maps were made. This situation has changed dramatically, however, since relief shading by computer is now easy and the data required are plentiful. Relief shading can be produced in seconds using computer mapping and GIS software and digital elevation model data.

Bump mapping

Recently, maps have been produced with surface textures added to relief shading, such as the textured relief shading for Crater Lake National Park seen in figure 6.17. The idea is to give a general indication of the surface texture or land cover, adding to the visual realism of the relief shading. In this case, a method called **bump mapping** is used that modifies the original DEM by adding elevation values around randomly scattered points that represent vegetation in the landscape. The relief shaded DEM then casts shadows for the vegetation as well as the terrain.

Oblique perspective maps

There are a number of methods that can be used to create maps that give us a bird's-eye view of the landscape. These **oblique perspective maps** have a more three-dimensional look to them than planimetric maps that portray the landscape from a vantage point directly above the area mapped. Oblique perspective views have strong intuitive appeal and a high level of readability. This accounts for their popularity for everyday maps. But there is a problem. On these maps, you can't place features at their true location. Positional displacement of a terrain feature occurs in direct proportion to the height of its symbol—the shift in horizontal position between the bottom and top of a tall mountain is much greater than for a small hill. The top, bottom, or middle of a hill can be placed in its correct horizontal position on the map, but not all three at once. Let's look at a number of ways that mapmakers provide oblique perspective views starting with terrain profiles and fishnet maps.

Terrain profiles and fishnet maps

Oblique perspective surfaces can be produced with sets of parallel lines called **terrain profiles** that follow the surface of the terrain. In figure 6.18, for example, parallel terrain profiles are drawn horizontally or vertically across the terrain surface. More typically, these are combined into profiles drawn across the landform at right angles to each other. The result is called a **fishnet map** or **wireframe map** because it resembles a net or wire mesh draped over the terrain.

More closely spaced profiles produce a smoother fishnet surface. When computer software was first able to create fishnet maps, they increased in popularity. However, more realistic representations of the terrain have since supplanted their use. You will find that many of the concepts relating to fishnet maps will also be applicable to other oblique perspective maps. We will continue our discussion of oblique perspective maps using fishnet diagrams as an illustration because they depict the terrain so clearly.

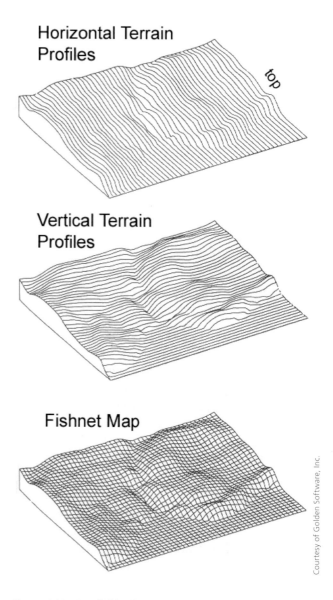

Horizontal Terrain Profiles

top

Vertical Terrain Profiles

Fishnet Map

Courtesy of Golden Software, Inc.

Figure 6.18 Parallel horizontal and vertical terrain profiles can be combined to make a fishnet map.

5 degree   15 degree

45 degree   75 degree

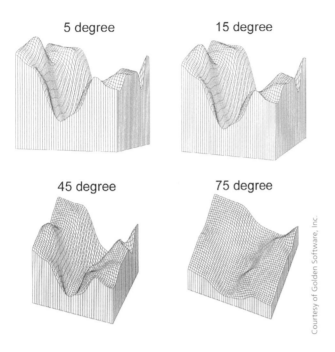

Courtesy of Golden Software, Inc.

Figure 6.19 Changing the vantage point viewing angle may drastically alter the terrain portrayal, as this sequence of 5-, 15-, 45-, and 75-degree viewing angles demonstrates.

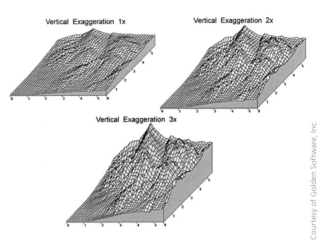

Vertical Exaggeration 1x   Vertical Exaggeration 2x

Vertical Exaggeration 3x

Courtesy of Golden Software, Inc.

Figure 6.20 Cartographers normally use vertical exaggeration to increase the three-dimensional appearance of oblique perspective maps and other three-dimensional perspective views of the terrain.

Mt. Olympus, Washington, USA, block terrain example from Map Illustrations Web site, www.mapillustrations.com.au.

Figure 6.21 The fishnet map that underlies this perspective view map of Mt. Olympus, Washington, is draped with contours and other topographic map information.

With oblique perspective maps, the quality of relief depiction depends on the vantage point set by the mapmaker. An oblique viewpoint around 45 degrees above horizontal is generally the most useful. Notice in figure 6.19 that as the vantage point viewing angle decreases, the problem of terrain blocking increases. To circumvent the blocking problem, you can sometimes "look behind the hill" if you use two maps, each with a different viewing direction. Computers now allow the generation of dynamic displays that you can tilt and turn as desired to see the landscape from multiple vantage points.

To create a realistic impression of relief, mapmakers exaggerate the vertical scale of most oblique perspective maps. A vertical exaggeration between 2 and 3 is common with large-scale oblique perspective maps (figure 6.20), while a 50-to-1 exaggeration is often used for small-scale maps of states or countries. Regions of little relief are vertically exaggerated more than regions of substantial relief. Since different degrees of vertical exaggeration can produce quite different relief impressions, you should make it a practice to check the legend to see if the vertical exaggeration factor has been indicated.

Perspective view regional maps
You will note that the oblique perspective maps above show bald terrain, denuded of land-cover features. With the aid of computer mapping software, mapmakers have

been able to overcome this problem. It is now common to drape relief shading over the underlying fishnet surface to create a more realistic **perspective view regional map** that approximates a block diagram, but leaves out the subsurface terrain characteristics (figure 6.21). Contours and other map symbols can be draped over the map as well for viewing on a computer screen or printed copy.

Another approach is to drape aerial photography or satellite imagery over the terrain surface (figure 6.22). Mapmakers also drape other information—everything from land-use categories to surface geology data—over surfaces. These combinations will only expand in the future as sophisticated surface rendering software allows three-dimensional digital trees and other realistic map

Figure 6.22 Landsat imagery has been draped over a DEM of Los Angeles and the San Gabriel Mountains.

symbols to be draped on the surface. And since all data are in digital form, you can view the terrain from any number of different vantage points (see "Dynamic Relief Portrayal" later in this chapter).

These types of displays can now be generated easily with computer mapping software, like ESRI's ArcScene and ArcGlobe (applications that are part of ArcGIS 3D Analyst). These maps are usually constructed on an oblique orthographic projection. More often than not, they include the horizon in the display.

Block diagrams

A **block diagram** portrays a piece of terrain as if it were cut out of the surface of the earth. The vertical sides of the block allow the underlying rock formations or other subsurface geologic information to be shown (figure 6.23). Block diagrams reach their highest degree of sophistication in their smooth, continuous surface form, where an attempt is made to give an artistic picture of the actual land surface. The natural appearance of the landform can be achieved in two ways—by using sketch lines to accentuate those terrain features that give distinct form to the landscape or by relief shading the surface. Either technique makes the terrain easily understandable.

Popular magazines and advertisements take advantage of the realistic terrain picture provided by block diagrams. The illustrations in geology, physical geography, and other environmental textbooks are also commonly of this type.

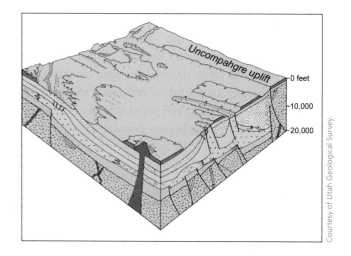

Figure 6.23 A line drawing block diagram showing terrain features and subsurface geology in the vicinity of the Uncompahgre uplift, Utah.

Figure 6.24 Landform drawings like this view of Mt. Bachelor, Oregon, represent one of the highest achievements in artistic rendering of the terrain surface.

Courtesy of Peter Powers.

Landscape drawings

Artistically rendered, oblique perspective maps, called **landscape drawings,** can provide nearly photorealistic terrain portrayals. Because of the immense amount of hand labor involved in their construction, the best examples are found for regions of special interest, such as national parks and popular mountain resorts. Landscape drawings have become popular as ski area posters like the Mt. Bachelor, Oregon, poster in figure 6.24.

Hillsigns

The stylized drawing of hills from an oblique perspective is one of the oldest ways to show relief. From the medieval period until the mid-1800s, hills and mountain ranges on both large- and small-scale maps were represented by crude line drawings of highly stylized hills called **hillsigns** (figure 6.25). These drawings look like conical mole hills or haystacks that may bear little resemblance to the actual features symbolized. Yet, even today, these simple drawings are an effective way to show the general nature of the terrain, particularly on medium-scale to small-scale maps. They are commonly found on recreational, advertising, and similar maps designed for the public, where only a general impression of landform character is needed (figure 6.26).

Figure 6.25 Hillsigns on old maps like this rendition of the California coastline resemble mole hills or haystacks.

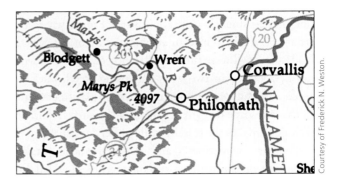

Courtesy of Frederick N. Weston.

Figure 6.26 A modern version of hillsigns on a section of a landform map of the Willamette Valley, Oregon.

COMBINING METHODS ON MAPS

Cartographers often combine different relief portrayal methods to enhance the three-dimensional appearance of the terrain while also making it possible for map users to determine elevations or depths. This often involves a combination of absolute and relative relief portrayal methods. There are lots of possible combinations; we'll take a look at two here—relief shading and contours and relief shading and hypsometric tinting.

Relief shading and contours

A skillful combination of contours and relief shading is one of the most effective relief portrayal methods. The USGS uses this technique to produce topographic quadrangles like the one in figure 6.27 that contain classic examples of different terrain features. Producers of commercial travel and recreation atlases have also created relief shaded topographic maps for entire states, many available in digital form.

Further enhancements are possible for maps combining relief shading and contours to show the terrain. Large-scale Swiss topographic maps as shown in figure 6.16 are noted for combining contours and relief shading with line drawings of rock outcrops and ridges. These manually produced maps are truly works of art, yet elevations can be determined easily from the contours.

Relief shading and hypsometric tinting

It is also common to see maps that have both relief shading and hypsometric tinting; indeed, it is unusual to find a map today that has hypsometric tinting without relief shading, primarily because both techniques can be achieved so easily using computers and digital elevation models (figure 6.28). If the mapmaker can create one effect, it is easy to add the other. The challenge is to correctly assign the colors in the hypsometric tint to the elevation values. Unknowledgeable mapmakers will sometimes apply a color ramp or progression that does not mimic the typical land cover, which as we stated earlier, is the intent of this method. This is especially problematic if the area being mapped is extremely large, such as a continent or the world. In these cases, the land cover varies greatly across elevation over the area. Be careful to refer to the legends for maps such as these.

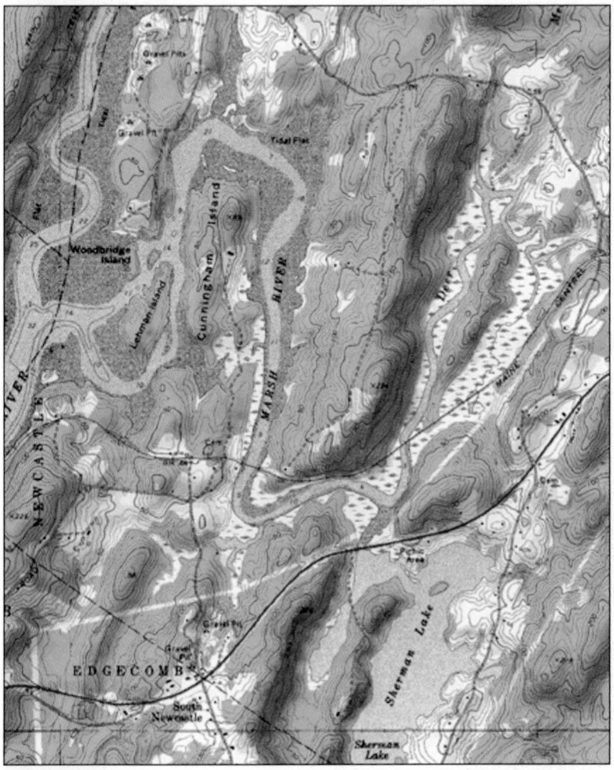

Courtesy of the U.S. Geological Survey.

Figure 6.27 Relief shaded contour maps are produced by the USGS for certain topographic map sheets such as this 1:24,000-scale map segment covering the Sherman Lake, Maine, area.

Courtesy of Tom Patterson, National Park Service.

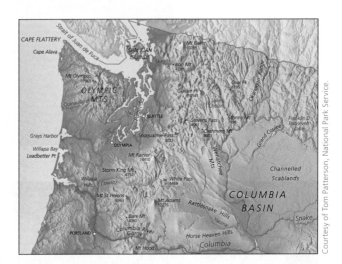

Figure 6.28 Both relief shading and hypsometric tinting have added to this section of a landform map of the United States that shows mountains in 3D perspective.

STEREOSCOPIC VIEWS

Stereoscopy is any technique capable of recording three-dimensional visual information or creating the illusion of depth in a photograph, map, or other two-dimensional image. The illusion of depth is created by presenting a slightly different image to each eye. Many three-dimensional displays use this method to convey depth to the reader, and maps are no exception.

Stereopairs

One way you can visualize the terrain in three-dimensions is by viewing a stereopair of relief shading or oblique perspective views (figure 6.29). A **stereopair** is a set of two maps from slightly different vantage points that, when viewed together, give the impression of a three-dimensional surface. This mimics the same thing our eyes do all the time—you can test this if you close first one eye then the other. The image you see shifts just slightly—together the two images allow us to see the world in 3D.

The stereovision mechanism, called a **stereoscope** (or **pocket stereoscope** for the portable version), is the same as that used to view photographic stereopairs, which are often used for air photo interpretation. In this case, you view relief shaded maps of the terrain surface instead. When you view these two maps with a stereoscope, your mind merges them into a single three-dimensional mental image of the terrain.

Stereoscopic polarization

A more technically complex way to display the terrain stereoscopically is to view a special computer monitor that alternates the two maps at least 30 times a second. With this method called **stereoscopic polarization** (or **stereoscopic projection**), the first map is displayed with horizontally polarized light, the second with vertically polarized light. Map viewers wear special goggles with polarizing filters that allow the right eye to see only the horizontally polarized map and the left eye the vertically polarized map, so that the terrain is seen stereoscopically. An added advantage

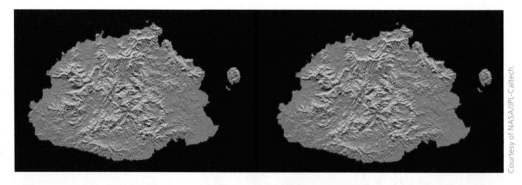

Figure 6.29 When you view a relief shaded map stereopair stereoscopically, you see a truly three-dimensional image of the terrain. Try viewing this stereopair of Fiji with a pocket stereoscope.

Courtesy of NASA/JPL-Caltech.

Courtesy of NASA/JPL-Caltech.

Figure 6.30 Anaglyph of a relief shaded map of Los Angeles and the San Gabriel mountains created from Shuttle Radar Topography Mission data. Try viewing this map with blue and red anaglyph glasses used to view 3D movies.

Figure 6.31 **This layer tinted map of the Hawaiian Islands will stand out vividly in 3D when viewed through Chromadepth glasses.**

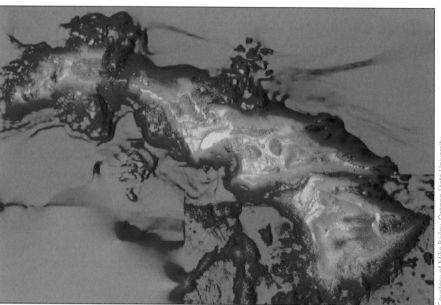

is that the viewer is able to pan, zoom, and otherwise manipulate the map.

Anaglyphs

A special form of stereopair is produced by printing the maps constructed from two vantage points in red and blue and then superimposing one upon the other. The resulting stereopair is called an **anaglyph.** When you view the anaglyph through special glasses equipped with red and blue lenses, you see the red map with one eye and the blue map with the other (figure 6.30). This allows you to see the map stereoscopically.

Chromadepth maps

Another method uses the colors in the rainbow (red, orange, yellow, green, blue, indigo, and violet) to create the impression of different heights within a display. You may have used inexpensive Chromadepth glasses to look in three dimensions at illustrations in comic books, childrens' textbooks, and perhaps printed or displayed maps such as figure 6.31. Chromadepth glasses, patented by Chromatek Inc., create a stereoscopic effect using holographic film lenses that combine refraction and diffraction of light to make each lens act like a thick glass prism. These glasses create the visual impression of colors in the visible spectrum being closer or farther from your eyes, with red in the foreground, blue in the background, and orange, yellow, green in intermediate positions. Terrain maps designed with a blue-to-red layer tint color progression will have the strongest three-dimensional effect.

Although stereopairs, stereoscopic polarization, anaglyphs, and Chromadepth maps are effective visually

and fun to look at, they are also somewhat impractical. They cost more to produce and require more viewing effort than standard relief shaded or oblique perspective maps. And because the relief impression is created in the brain, not on a sheet of paper, the image is ephemeral and not subject to analytical map use procedures. Despite these drawbacks, however, anaglyphs and other stereoscopic viewing methods are used extensively in certain earth sciences, such as geology. These stereoscopic views may also become more common in the future, since they are easily produced by computer.

DYNAMIC RELIEF PORTRAYAL

Computer technology has made it possible to put relief portrayal into motion. If relief portrayals from a sequence of vantage points are animated, you get the impression that you're flying and that the terrain is passing under and around you. The visualization of space and movement can be so realistic that you may actually feel pangs of airsickness! This **fly-through** effect occurs because your mind finds it easier to accept your body moving than the terrain moving.

Animated methods

Technological advancements have introduced the potential of creating **animated maps** that give you a sense that things are changing on them either over time, through space, or both. By viewing a motion picture that gives the impression of a camera flying over and around a region, you can gain a dramatic, dynamic impression of the landform. The effect is similar to that of viewing a physical model, such as the relief modes we saw earlier, from a number of positions. In both cases, we have the advantage of being able to change vantage points and, therefore, perspective (although in the cinema version the sequence of movement is preprogrammed). In fact, animated terrain maps are commonly made by essentially rotating a movie camera over and around a physical model of the terrain using computers. Simulators to train pilots and astronauts, for example, have used this dynamic mapping method.

Inexpensive video (television) cameras have brought new life to animated mapping in recent years. But the most important advance came with the advent of high-speed, digital computers and high-resolution color monitors. With the aid of computers and sufficient numerical data representing the terrain, there is no longer a need to photograph or videotape a physical model or the terrain surface itself. Instead, the images can be created from the digital database through a series of calculations that take into consideration the vantage point of the observer, the orientation of the surface, and the position of the illumination source. The images are then displayed on the monitor screen and may be recorded electronically in video mode or digitally for subsequent playback.

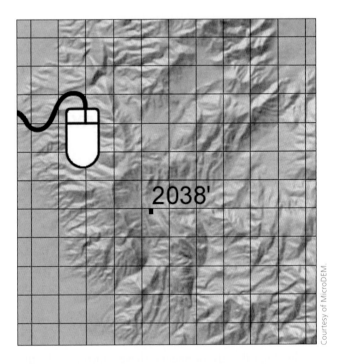

Courtesy of MicroDEM

Figure 6.32 **When a map displayed on an electronic screen is queried about the elevation at a point, the computer can search the database used to make the map to get the answer.**

Interactive methods

Interactive maps take animated methods one step further. While animation takes you over the terrain on the path chosen by the animator, interactive relief maps put you in control. You can call up any image onto the screen, in any sequence. If you operate the controls, you can simulate movement realistically from one vantage point to any other. You control the fly-through path using either a mouse or, better yet, a joystick with three axes of movement that are transmitted to a computer. You might start up high to get the overall view, then fly in closer to get a better look at features of special interest. You can view the terrain from as many heights and directions as needed to get a feel for the nature of the surface. You get the sensation of flying over the terrain with your continuously changing view in the fly-through effect.

Not only can you interactively control your movement over and about the terrain, you can also interact with data about the area mapped (figure 6.32). This type of interaction takes place in a series of steps. First, terrain data are displayed on an electronic screen using a relative relief technique, such as relief shading. Second, you "point" at the location whose elevation you are seeking. The pointing may be done by touching a touch-sensitive screen or by moving a mouse or similar electronic device.

Next, you "ask the map" what the elevation is at the indicated point or what a terrain profile is like along a given path. You may be able to do so through a keyboard, by verbal query, or by pointing to an electronic menu of questions displayed on the side of the screen. Finally, GIS software such as ArcGIS will note the location or path you indicated, search through the elevation data records used to make the map, perform the necessary computations, and provide you with the requested elevation data or profile.

Other datasets draped over the terrain can also be displayed and queried. With these "point and ask" portrayals, the map merely provides a graphic version of the landform, and you direct analytical questions to the underlying digital terrain data or overlying draped data. In other words, the map serves as a window on the data. By not burdening the map with the need to portray absolute relief information, the quality of visualization can be improved. At the same time, elevation values can be determined with greater precision than would be possible with a graphic portrayal alone.

DIGITAL ELEVATION MODEL DATA

Most of the relief portrayal methods discussed in this chapter are now carried out on computer mapping software that relies on **digital elevation model (DEM),** also sometimes called **digital terrain model (DTM)** data. As we saw earlier, relief shading, hypsometric tinting, fishnet mapping, and perspective view maps can now be made easily with DEM data. To understand these maps better so you know how to use them appropriately, it will be useful to take a closer look at the nature and various sources of some of the DEM data that are used to make these maps.

A DEM is a sample of elevations or depths taken on a regular grid. A DEM isn't a map in the sense we are using the term in this book, since a DEM is a digital database of elevation values. Thus, it's better to think of a DEM as a data source for mapping the terrain, rather than as a map. From this perspective, DEMs serve the same role as other forms of digital cartographic data, such as those generated by electronic image scanners and digitizing. Digital elevation models are tremendously important in modern mapping and map use because computers can easily carry out computations on a matrix of DEM data to create maps of the terrain in seconds.

ETOPO5, ETOPO2, and GLOBE

A number of government and private organizations are involved in creating regular grids of elevation values. You can download world maps of land and ocean topography made from the ETOPO5 or ETOPO2 global DEM distributed by the U.S. National Geophysical Data Center. **ETOPO5** is an equal-angle grid at a spatial resolution of 5 minutes (called arc-minutes) in latitude and longitude, meaning that each cell is given the average elevation of its 5-minute-by-5-minute quadrilateral on the ground (average elevation within four-ninths the area of a USGS 1:24,000-scale topographic map sheet). The **ETOPO2** DEM is a global equal-angle grid with a higher spatial resolution of 2 minutes, and hence is the dataset most used currently for global terrain mapping. The National Geophysical Data Center also distributes finer-resolution **GLOBE** 30-arc-second DEM data for the world's land areas. The U.S. Geological Survey distributes a similar **GTOPO30** grid. The GLOBE grid is global in extent, but grid cells in oceans are given a constant depth of 500 meters, which means "no available data."

Because of the convergence of meridians (explained in chapter 1), these equal-angle grid cells vary in shape and area from large, essentially square cells at the equator to small, tall, thin cells as the pole is approached. Consequently, equal-angle grids normally are not displayed in their raw form, but are transformed using a map projection such as the equirectangular, Mollweide elliptical, or Lambert azimuthal equal area (see chapter 3 for descriptions of these projections).

Near the equator, however, grid cells are close enough to square in shape that they can be faithfully displayed as maps composed of square pixels. For example, in figure 6.33 grid cells from ETOPO5, ETOPO2, and GLOBE DEMs covering a 5-degree-by-5-degree quadrilateral in west-central Ecuador are mapped at a display resolution of 100 pixels per inch. Notice the progressive

ETOPO5

ETOPO2

GLOBE
(30 sec.)

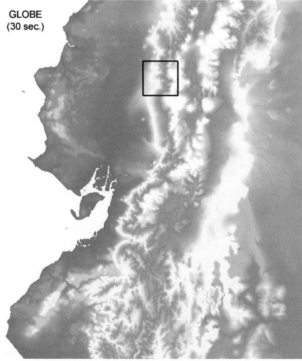

Figure 6.33 These continuous hypsometric tinted maps of a 5-degree-by-5-degree quadrilateral covering west-central Ecuador were created at a display resolution of 100 pixels per inch from the ETOPO5, ETOPO2, and GLOBE 30-second resolution DEMs. The area in the box is shown in figure 6.34.

increase in terrain detail as the spatial resolution of the grid increases. You would be correct in assuming that accuracy of elevations at particular points increases with finer spatial resolution, since elevations within smaller quadrilaterals on the ground are being averaged.

Shuttle radar topography mission

The highest resolution worldwide DEM is the 3-second dataset created from **Shuttle Radar Topography Mission (SRTM) data.** The SRTM space shuttle mission was an international project led by the U.S. National Geospatial-Intelligence Agency (NGA; formerly the National Imagery and Mapping Agency, or NIMA) and U.S. National Aeronautics and Space Administration (NASA). The SRTM radar device was carried on the space shuttle *Endeavour* from February 11 to 22, 2000, and acquired enough data during its 10 days of operation to create a near-global high-resolution database of the earth's topography.

The device had two radar antennae, one located in the shuttle's payload bay, and the other on the end of a 60-meter (200-foot) mast that extended from the payload bay once the shuttle was in space. These antennae collected radar return data used to determine land elevations. The shuttle's orbit allowed data to be collected between 60°N and 56°S latitudes. Elevations for land areas within these latitudes were determined within a spatial resolution of around 50 meters by 50 meters (150 feet by 150 feet). The horizontal accuracy of grid cells is within 20 meters (60 feet) of their true position, and the vertical accuracy of the elevations is +/–10 meters (32 feet) of their true average elevation.

You can download SRTM 3-arc-second resolution data (90-meter-by-90-meter cells at the equator) in 5-degree-by-5-degree tiles for any land area within the latitude range of the shuttle's orbit. The continuous layer tinted map in figure 6.34, for example, was created at a display resolution of 100 pixels per inch from a 0.5-degree-by-0.5-degree segment, 1 percent the area of an SRTM tile. Notice the tremendous increase in spatial detail compared to the GLOBE map segment of the same area. Elevation data at a 1 arc-second (approximately 30-meter-by-30-meter) resolution can be downloaded for the continental United States. Maps of interesting terrain features have been made from these higher resolution data. Most maps have been oblique perspective views, often draped with satellite images of similar spatial resolution such as the Landsat Thematic Mapper scene in figure 6.30.

Figure 6.34 **This continuous layer tinted map of the 0.5-degree-by-0.5-degree quadrilateral in west-central Ecuador outlined in figure 6.33 was created at a display resolution of 100 pixels per inch from the SRTM global 3-second resolution DEM.**

National Elevation Dataset

In addition to high-resolution SRTM data, you can obtain DEMs for the United States from the USGS **National Elevation Dataset (NED)** Web site. The NED was created by merging the highest resolution, best quality elevation data available across the United States into equal-angle grids covering the continental United States, Hawaii, Alaska, and the U.S. island territories. The grid resolution is 1 arc-second except for Alaska's 2-arc-second dataset. The shaded relief maps in figure 6.35 illustrate the terrain detail visible at these two spatial resolutions. The NED is not a static product, but one that is updated bimonthly to incorporate the best available DEM data. Currently, a ⅓-arc-second (approximately 10-meter-by-10-meter) resolution grid is being assembled that eventually will cover the continental United States seamlessly.

NED 1-arc-second DEM

NED 2-arc-second DEM

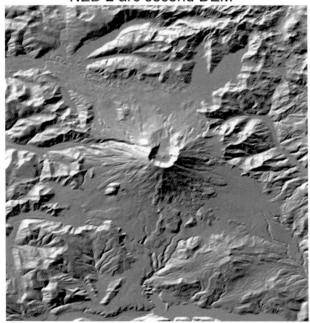

Figure 6.35 These two relief shaded maps of Mount St. Helens, Washington, illustrate the terrain detail inherent in the NED 1- and 2-arc-second DEMs.

Coastal Relief Model

The **Coastal Relief Model** distributed by the U.S. National Geophysical Data Center (NGDC) on CDs or over the Internet is the first DEM of the U.S. Coastal Zone, extending from the coastal state boundaries to as far offshore as NOS hydrographic data will support a 3-arc-second grid of the seafloor. This seaward limit in many areas reaches beyond the continental slope. The Coastal Relief Model contains data for the entire coastal zone of the conterminous United States, including Hawaii and Puerto Rico. Alaska and the Aleutian Islands will soon be added to the model, and eventually the Great Lakes and surrounding state coastal areas will be included as well. The bathymetric map shown in figure 6.36 was made from Coastal Relief Model data for the central California coast, including the seafloor just west of Monterey Bay. You can easily see submarine features such as Monterey Canyon and marine avalanches.

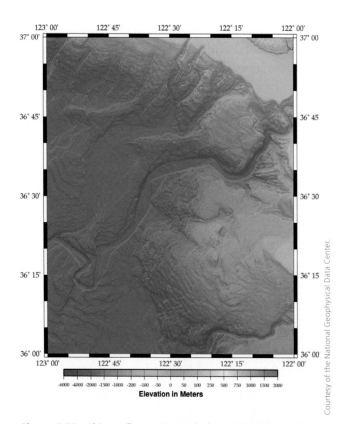

Courtesy of the National Geophysical Data Center.

Figure 6.36 This seafloor map made from NGDC Coastal Relief Model data shows submarine features in a 1-degree quadrilateral just to the west of Monterey Bay, California.

Lidar

Light detection and ranging (lidar) is a remote sensing system that uses rapid pulses of laser light striking the surface of the earth to determine elevations. The lidar laser pulse emitter and detector is mounted in the bottom of a remote sensing device, usually an airplane, along with a GPS receiver used to determine the aircraft's horizontal position and altitude. Part of each laser light pulse is reflected or scattered back to the detector. The distance from the detector to the spot where the pulse was reflected is determined by measuring the time it takes for each light pulse to travel to the spot and be reflected back, since the pulse return time is proportional to distance from the detector.

Current lidar systems are able to time and record the distances for more than one return per pulse, *if* the pulse can continue through a terrestrial feature, like vegetation. First, second, and third pulse returns may be from the forest canopy, understory, and bare ground in a wooded area. In urban areas, the first pulse return is used to measure the elevations of the tree canopy and building roofs, while the last return is assumed to be from the ground surface. Lidar datasets can be very large, often hundreds of thousands or even millions of points per square mile. The distance values for these points are then processed mathematically to create very

high-resolution DEMs of small areas, such as the 1-meter resolution DEM for the portion of West Linn, Oregon, displayed as a continuous layer tinted, relief shaded terrain map in figure 6.37. Notice the subtle differences in elevation that allow you to see streets, buildings, and excavations on the map.

As you can see, there are many ways we can learn about relief in the landscape from maps. These types of portrayals are important to map users. We not only find them intriguing, but they also help us move beyond the fundamental two-dimensional nature of a map. Because we experience relief in our environment every day, relief portrayal on maps is something that we are eager to pay attention to and try to learn from both explicitly or implicitly. We may view the portrayal of relief differently from the other symbols and elements on the map since it is something we can directly relate to our own experience. For that reason, we may also be more apt to find fault with or shortcomings in the way relief is portrayed on maps.

Through airline travel and images we see from space, we understand better than previous generations the three-dimensional nature of our world. Cartographers have tried innovative ways to capture this, through such clever approaches as bump mapping, anaglyphs, and Chromadepth maps. With the use of computers, fly-throughs allow an animated view of the landscape and, through interaction, users can now decide where and how high to fly over a digital terrain surface. More data at higher resolutions are helping cartographers in the portrayal of relief. And the methods are also being applied to places that have not been fully mapped before, such as ocean bottoms and other planets. Our understanding of the world around us, and even other worlds, is expanding through the techniques that mapmakers have used to help us understand the third dimension on maps. With a greater understanding of the techniques they use, you have a better ability to read the relief portrayed on a map.

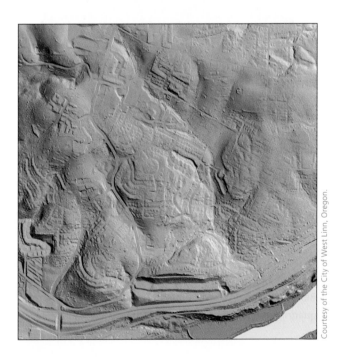

Courtesy of the City of West Linn, Oregon.

Figure 6.37 Streets, excavations, and building tops are visible in this continuous layer tinted and relief shaded map made from 1-meter resolution lidar data covering part of West Linn, Oregon.

SELECTED READINGS

Baldock, E. D. 1971. Cartographic relief portrayal. *International Yearbook of Cartography* 11: 75–8.

Barnes, D. 2002. Using ArcMap to enhance topographic presentation. *Cartographic Perspectives* 42 (Spring).

Castner, H. W., and R. Wheate. 1979. Reassessing the role played by shaded relief methods in Topographic Scale Maps. *The Cartographic Journal* 16: 77–85.

Curran, J. P. 1967. Cartographic relief portrayal. *The Cartographer* 4 (1) (June): 28–38.

Eyton, R. J. 2005. Unusual display of DEMs. *Cartographic Perspectives* 50: 7–23.

Grotch, S. L. 1983. Three-dimensional and stereoscopic graphics for scientific data display and analysis. *IEEE Computer Graphics and Applications* 3 (11): 31–43.

Hurni, L., B. Jenny, T. Dahinden, and E. Hutzler. 2001. Interactive analytical shading and cliff drawing: Advances in digital relief presentation for topographic mountain maps. In *Proceedings of the 20th International Cartographic Conference,* ICC 2001 Beijing.

Imhof, E. 1982. *Cartographic relief representation.* Translated and edited by H. J. Steward. New York: Walter de Gruyter.

Irwin, D. 1976. The historical development of terrain representation in American cartography. *International Yearbook of Cartography* 16: 70–83.

Jenny, B. 2001. An interactive approach to analytical relief shading. *Cartographica* 38 (1and 2): 67–75.

Kennelly, P. 2002. Hillshading with oriented halftones. *Cartographic Perspectives* 43: 25–42.

Kennelly, P., and A. J. Stewart. 2006. A uniform sky illumination model to enhance shading of terrain and urban areas. *Cartography and Geographic Information Science* 33 (1): 21–36.

Kraak, M. J. 1993. Cartographic terrain modeling in a three-dimensional GIS environment. *Cartography and Geographic Information Systems* 20: 13–8.

Kumler, M. P. 1995. *An intensive comparison of triangulated irregular networks (TINs) and digital elevation models (DEMs).* Toronto: University of Toronto Press, Inc.

Lobeck, A. K. 1958. *Block diagrams and other graphic methods used in geology and geography.* 2d ed. Amherst, Mass.: Emerson-Trussel Book Co.

Patterson, T. 1997. A desktop approach to shaded relief production. *Cartographic Perspectives,* NACIS 28: 38–40.

Petrie, G., and Kennie, T. J. M., eds. 1991. *Terrain modelling in surveying and civil engineering.* New York: McGraw-Hill, Inc.

Price, W. 2001. Relief presentation: Manual airbrushing combined with computer technology. *The Cartographic Journal* 38(1).

Robinson, A. H., et al. 1995. Portraying the land-surface form. In *Elements of cartography*, 6th ed., 527–48. New York: John Wiley & Sons, Inc.

Ryerson, C. C. 1984. Relief model symbolization. *The American Cartographer* 11(2): 160–64.

Schou, A. 1962. *The construction and drawing of block diagrams.* London: Thomas Nelson & Sons, Ltd.

Tait, A. 2002. Photoshop 6 tutorial: How to create basic colored shaded relief. *Cartographic Perspectives* 42 (Spring).

Watson, D. F. 1992. *Contouring.* New York: Pergamon Press.

Yoeli, P. 1983. Digital terrain models and their cartographic and cartometric utilization. *The Cartographic Journal* 20 (1): 17–23.

chapter
seven **QUALITATIVE THEMATIC MAPS**

7

Qualitative thematic maps

In the introductory chapter, we saw that maps can be divided into broad categories based on their function. For example, topographic quadrangles and world atlas sheets are used to locate features and learn the basic geography of a region. Nautical and aeronautical charts and road maps are used to plan travel routes. Some maps used in advertisements were made to help persuade you to accept some idea or take some action. In this chapter we will focus on a fourth category of maps—thematic maps.

Thematic maps emphasize a single theme or a few related themes. A map of different climate zones is a good example, as are maps of land cover, vegetation zones, species ranges, and population density. Basic geographical reference information will appear on a thematic map in order to provide the locational base for the reader, but the theme will stand out visually as the most important message of the map.

Thematic maps can show both qualitative and quantitative information. In chapter 8, we'll see the methods mapmakers use to portray **quantitative information,** or information that portrays a magnitude message, such as population density, annual rainfall, and stream flow. In this chapter, we'll focus on the ways mapmakers show **qualitative information,** or data that have classes varying in type but not quantity. Examples include land cover, zoning, and soils. First, we'll look at the basic types of qualitative information for point, line, and area features. Next, we'll examine the ways that mapmakers show a single point, line, or area feature. We'll also see examples of how mapmakers create **multivariate maps** that show the geographical relationships between two or more themes. Finally, we'll explore **dynamic qualitative thematic maps** that focus on changes in feature locations and attributes over time.

QUALITATIVE INFORMATION

Qualitative information tells you only what different things exist—lakes, rivers, roads, cities, farms, and so on. By contrast, **quantitative information** consists of data giving you the magnitudes of things, such as how large, wide, fast, or high they are. This simple dichotomy between qualitative and quantitative information pervades our descriptions of the environment. But to categorize all information as either qualitative or quantitative is needlessly restrictive, since scientists think of data as being at one of four basic measurement levels. A **measurement level** is a classification used to describe the nature of numerical information about features. The **nominal level** is associated with qualitative information, whereas quantitative data can be at the **ordinal, interval,** or **ratio** measurement levels (see chapter 8 for details about the three quantitative measurement levels).

Nominal-level information tells you simply which category (class) a feature belongs to. Features within a category are assumed to be relatively similar, and categories should be quite distinct. This is sometimes described as

"minimum within class variation and maximum between class variation." The different types of nominal-level data collected for point, line, and area features is the basis for qualitative thematic maps. In a geographic information system (GIS), *attributes* carry the descriptive information for the geographic *features*. For example, zoning and ownership are common nominal-level attributes used to describe parcel features in a dataset. Sometimes the classes are defined by numbers, such as the land capability classes in figure 7.1. These numbers do not represent a quantitative measure—rather they are merely labels attached to nominal-level information about the features. Can you see how it would be inappropriate to work with these numbers if you assumed they were quantities?

Point feature information

What is a point feature? This question is harder to answer than you might think because two types of point features are mapped using similar symbols. The first type of point feature is a *zero-dimensional* entity without width or area defined solely by its geographic location. The horizontal survey control points discussed in chapter 1 are an

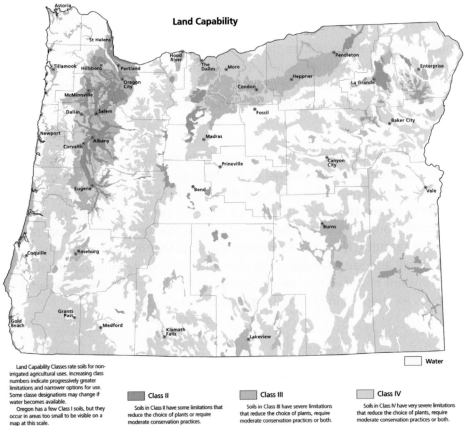

Figure 7.1   Land capability classes are designated by numbers that represent qualitative rather than quantitative differences between the categories.

From W. G. Loy and S. Allan. 2001. *Atlas of Oregon.* 2nd ed. Eugene, Ore.: University of Oregon.

example of zero-dimensional point features. Although the metal control point marker is actually a feature that takes up some area (albeit very little) on the ground, truly zero-dimensional geographic coordinates define the control point.

The second and far more common type of point feature is a two-dimensional feature that is *conceived of as a point for mapping purposes.* A large-scale thematic map for the parks department might show trees as areas by mapping the extent of the tree crown. The tree data might also include the species type—a nominal-level attribute. So a map could be made that showed trees as an area feature that is symbolized using colors to represent different species types. A smaller-scale map might show trees as points with different colors indicating the different tree species. In this case, trees that cover an area are represented as points for mapping purposes. This example shows you that features with areal extent may be mapped either as points or areas depending on the map's scale and purpose.

Data collection for point feature information

Data describing the locations and attributes of features conceived of as points for mapping purposes are collected in a variety of ways. In our tree mapping example, a Global Positioning System (GPS) receiver may have been used to **survey** (see chapter 5 for more on surveying) the location of each tree, and the species may have been determined by **field observation** (observing features *in situ*). It is also possible that the location of each tree was mapped on a field sketch, and the species type was determined through an interview with a park employee. Point data are also collected through **image interpretation** using photographic and digital images, such as air photos or satellite images, to identify objects and their attributes (see chapter 9 for more on image maps). Any of these methods could be used create a point feature in a GIS database with an x,y coordinate to identify its location and an attribute to carry the species information.

The accuracy of the information about the location and attributes of point features is an important consideration for you as the map user. In our tree mapping example, if a professional botanist trained in using GPS receivers collected the data, great faith can be placed in the accuracy of the point data in terms of both the location and the species attribute of the trees. If the same botanist used field observation to record the location and species, you can still be confident of the species attribute, but you should assume that the mapped position of each tree is only approximately correct. A **photogrammetrist** (a person trained to compile reliable data from aerial photographs) may have determined the location and species of each tree from large-scale air photographs. In this case, different tree species are usually determined with the aid of tree **identification keys** that show the typical appearance of each species on an aerial photo. Although photogrammetrically compiled locations can be very precise, identifying different types of trees from the typical appearance on aerial photos usually is less accurate than direct field observation. In this case, you can place more faith in the mapped position of the trees than in the species attributes.

Line feature information

Like point feature information, line feature information on maps can either relate to truly *one-dimensional* features with length and direction but no width, or features that are *conceived of as lines for mapping purposes.* Surveyed lines such as property, political, and administrative boundaries are true one-dimensional features. Lot boundaries, for example, are composed of straight or curved line segments whose endpoints are surveyed locations defined by coordinates. Most lines we see on maps, however, represent features that have width but are portrayed graphically as one-dimensional lines. We know that roads, for example, have standard design widths and that these features vary in width on the ground. However, roads are often shown with one-dimensional lines on maps. Surface composition, such as asphalt, concrete, gravel, and dirt, is an example of a nominal-level attribute for these features. On large-scale engineering plans, the road width is shown as the edge of pavement and on some maps it is shown as the **right-of-way** (real property designated for a specific purpose). On smaller-scale maps, mapmakers use line symbols that vary in width, color, pattern, or combinations of these to represent the location and attributes (such as width or surface composition) of roads.

Data collection for line feature information

Like point feature data, line feature information is commonly collected through a variety of methods, including surveys, field observation, and image interpretation. Let's consider a map of roads that shows differences in surface composition. Road locations may have been **digitized** (converted from paper to digital format) from the original engineering plans. The road surface type may have been entered as an attribute of the features. Alternatively, a person could drive each road with a GPS receiver on the car to digitize road segments and use field observation to note if the road surface composition is asphalt, concrete, gravel, or dirt. A photogrammetrist could digitize the roads as centerlines from air photos and then use a road cover interpretation key to record the surface composition.

As with our tree mapping example, the data collection method has an impact on both the locational accuracy of the features and their attributes. City and county public works or engineering departments often store road data collected using one or more of these methods in their GIS database. Other road attributes such as type (interstate, U.S. or state highway, ramp, and primary or secondary road) and direction (two-way or one-way) also are recorded. Mapmakers regularly use these digital data to create a variety of maps, including those designed specifically to show road surface composition.

Area feature information

Mapmakers commonly use two-dimensional area features to divide a region into areas that have some common qualitative attribute. The idea is that qualitative features within a certain category share some common trait, while differing significantly from the features in another category. The category is then mapped as if it were homogeneous within the data collection area, with no internal variation.

Data collection for area feature information

Mapmakers use qualitative area feature data collected by ground survey, image interpretation, or other methods such as census taking to determine the category for each data collection area. A **census** is a survey that collects data from all the members of a population, whether it's people, animals, businesses or other entities, within a defined space at a specified time. Often various statistics are noted for each member of the population, such as demographic, economic, and social attributes.

HOMOGENEITY

The qualitative feature data may give the mapmaker information about the actual degree of homogeneity within each feature. The features might be truly **homogeneous**—uniform in structure or composition throughout. For example, a stand of trees might prove to be entirely of the same species. The stand is a completely homogeneous qualitative area feature since all of its defining objects (trees) are of the same category (species) within the homogenous area (stand).

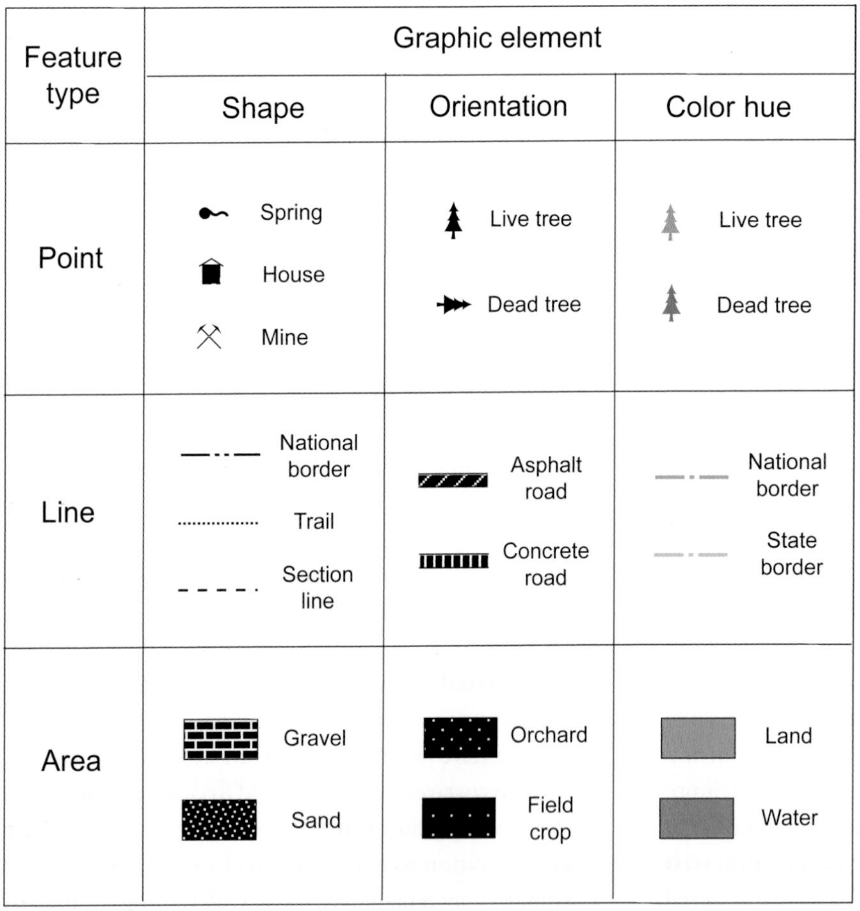

Feature type	Graphic element		
	Shape	Orientation	Color hue
Point	●⚊ Spring ⬛ House ⚒ Mine	🌲 Live tree ➤➤ Dead tree	🌲 Live tree 🌲 Dead tree
Line	—·—·— National border ·············· Trail — — — · Section line	▨▨▨ Asphalt road ▥▥▥ Concrete road	—·—· National border —·—· State border
Area	Gravel Sand	Orchard Field crop	Land Water

Figure 7.2 **The visual variables that naturally evoke qualitative differences among features are shape, orientation, and color hue. A shape repeated across an area or along a line creates the element of pattern. Orientation can also be used to create patterns within graphic marks.**

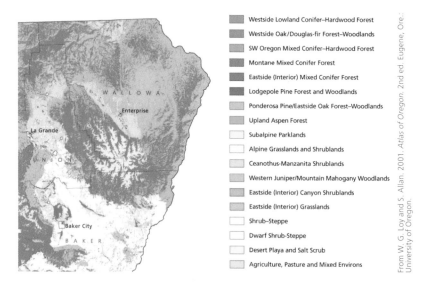

Figure 7.3 Color hue is used to differentiate most habitat classes on this map of the northwest Oregon, but color lightness is also used (note the symbols for the first few classes in the legend). It is useful to refer to the legend to understand that all the classes are at the nominal level, so the map shows only qualitative differences.

Westside Lowland Conifer–Hardwood Forest
Westside Oak/Douglas-fir Forest–Woodlands
SW Oregon Mixed Conifer–Hardwood Forest
Montane Mixed Conifer Forest
Eastside (Interior) Mixed Conifer Forest
Lodgepole Pine Forest and Woodlands
Ponderosa Pine/Eastside Oak Forest–Woodlands
Upland Aspen Forest
Subalpine Parklands
Alpine Grasslands and Shrublands
Ceanothus-Manzanita Shrublands
Western Juniper/Mountain Mahogany Woodlands
Eastside (Interior) Canyon Shrublands
Eastside (Interior) Grasslands
Shrub–Steppe
Dwarf Shrub-Steppe
Desert Playa and Salt Scrub
Agriculture, Pasture and Mixed Environs

From W. G. Loy and S. Allan. 2001. *Atlas of Oregon*. 2nd ed. Eugene, Ore.: University of Oregon.

The interesting thing about features mapped as though they were homogeneous is that most are actually not. In our tree example, although the trees within the stand may be the same species, there may be an understory of shrub, brush, or grass. At the edges of the stand, the trees may mix with other species creating a mixed species transition zone between two single species stands.

Mapmakers deal with this problem in several ways. They could decide to define the boundary between the two single species stands as the middle of the transition zone. Now neither of the stands are homogeneous area features, although they are mapped as being so. They may add a transition zone to the classification so that the map now includes a mixed class as well as the two single species classes. This solution increases the complexity of the map, but allows the single species stands and the transition zone to be mapped as homogeneous areas that more accurately reflect what's on the ground. They may then add a note to the map explaining that they've included the transition zone as part of one of the stands.

Let's take a look at an example of homogeneity within line features. In our road example, many roads are probably completely asphalt or totally concrete. When these homogeneous features are shown on the map as asphalt or concrete roads, the maps agree with what's on the ground. However, other features may have more than one surface type, such as a forest road that's mostly gravel with short sections of asphalt and other sections that are dirt. In this case, the composition of the majority of the road could be used as the category for the entire road segment. An alternative is to **segment** the features by breaking them at locations where the road surface type changes. The feasibility of this approach relates to the map scale (see chapter 2 for more on map scale). For example, you

would not analyze, navigate, or manage road segments based on what would be minute variations at a small map scale. However, roads that have many surface types but were classified into one category on a larger-scale map may cause travel and other problems for the unwary map reader. Without the appropriate surface composition information, map users may find themselves unable to plan for associated changes in access, traffic, or speed.

SINGLE THEME MAPS

The simplest and most common qualitative thematic maps are those that emphasize a single theme. To read these **single theme maps,** you must first determine what features have been shown with point, line, or area symbols. Cartographers use point, line, and area **graphic marks** on the map to represent geographic features. The graphic marks are symbolized using what mapmakers call **visual variables.** Certain visual variables work well for qualitative data because they do not impart a magnitude message; others work well for quantitative data because they do. Symbols on maps are easiest to read if the mapmaker has assigned the correct visual variables to the points, lines, or areas for the features shown.

The visual variables that evoke in our mind qualitative differences among features are **shape, orientation,** and **color hue,** or what we think of as the "color" of a symbol, such as red, green, or blue (figure 7.2). A shape can also be repeated along a line or across an area to create a pattern—we sometimes call this **line** or **area pattern.** Likewise, orientation can be used to create patterns within the graphic marks—this is called **pattern orientation.**

For example, a set of point symbols might be designed as circles with lines in them that are oriented in different directions. Each of these visual variables gives the impression that features are different in type or kind but not more or less in magnitude. Understanding the visual variables and how they are used in creating the graphic marks on maps helps you to correctly interpret the symbols on a map.

Keep in mind, however, that maps are not always designed as well as they might be. Thus, some other visual variable (size, texture, color value, or color intensity) may be used that is really most appropriate for showing quantitative differences (see chapter 8 for a discussion of quantitative visual variables). When maps are incorrectly made using the wrong visual variables, your first reaction will be that quantitative rather than qualitative information is being shown. The only way to keep from being misled is to read the legend to see what the symbols actually represent.

For example, on the map in figure 7.3, the mapmaker quite properly used color hue to differentiate most of the habitat classes. Nevertheless, for some categories color value was varied, which could easily give you

the impression that the categories differed in some quantitative way.

For this map, the legend clarifies any ambiguities in the symbols.

Point feature maps

Point feature maps have symbols that show the existence of something at a specific location. The word "point" has a rather loose interpretation here. It isn't being used in the strict mathematical sense of a one-dimensional figure but, rather, as the center of a circle, square, or some other map symbol representing a point feature found at this location, or as we saw earlier, representing a feature conceived of as a point for mapping purposes.

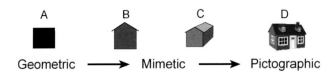

Figure 7.4   Qualitative point symbols range from geometric (A) to pictographic (D).

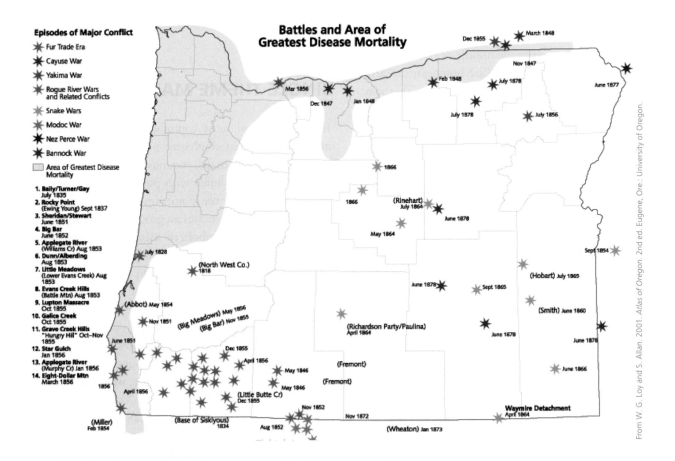

Figure 7.5   Simple geometric shapes are used to show different episodes of major conflict on this map. Without the legend, it would be impossible to tell what the symbols represent.

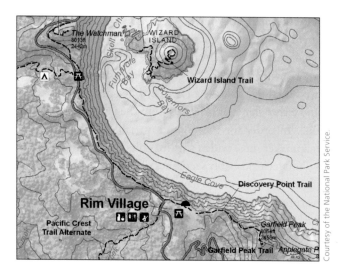

Figure 7.6 Mimetic symbols are used on this map to show the various attractions at different locations within Crater Lake National Park.

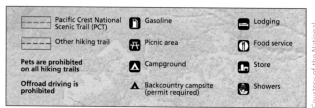

Figure 7.7 The legend for the map of Crater Lake.

Qualitative point features are usually represented with point symbols somewhere on a continuum between geometric and pictographic. **Geometric symbols** use simple shapes, such as squares, circles, and triangles to represent features (figure 7.4A). Because they are so simple, they usually require a legend to be interpreted correctly (figure 7.5). At the other end of the continuum are **pictographic symbols**, which are designed to look like miniature versions of the features they represent. Pictographic symbols are often used on landmark maps that require a reader to identify specific buildings or other landmarks at first sight. On these maps, small sketches of the buildings can sometimes be used to allow the reader to distinguish individual buildings immediately. These are more like example D in figure 7.4.

Somewhere in the middle of the continuum is a type of symbol called a **mimetic symbol**—one that "mimes" the thing it represents. Mimetic symbols are often created as a combination of geometric shapes such as a square with a triangle on top to represent a house, or they can be more complex, like a small cartoon of a particular type of building like a ranger station or museum. Because these symbols are intuitive, they are popular for mapping point features or labeling area or line features. You will find mimetic symbols on tourist maps, recreation maps, children's maps and maps on the Web. A good example is the push pin you see on Google maps.

A special type of mimetic symbol is the **standard symbol**, which is a symbol that is used as a standard in some mapping practice or for some mapping product. Sets of standard symbols have been created to show different

categories of transportation, recreation, and other activities. The U.S. National Park System uses these to depict visitor facilities in the national parks (figure 7.6).

Standard symbols have a more professional look but they still suffer problems associated with more pictographic symbols in general. They have to be relatively large for details to be apparent in crowded spaces. In addition, only a limited number of environmental features can be successfully symbolized since readers will become confused if there are too many mimetic symbols, and they will have to refer constantly to the legend to decipher the map. What, for example, is an obvious icon for a vista or an overlook? And what if there are 10 such cryptic symbols on your map? Although these symbols are intended to be intuitive at a glance, you'll often have to check the map legend to determine what is being symbolized (figure 7.7).

Mapmakers sometimes cleverly change the orientation or hue of a mimetic symbol to show two or more *attributes* of the point feature. Notice in figure 7.2 that the orientation or hue of the mimetic symbol for a coniferous tree can be changed to show the tree as alive (vertical or green) or dead (horizontal or brown). Mapmakers sometimes use both orientation and hue differences to make sure that you see the attributes of each feature.

The drawbacks of mimetic symbols are overcome by using geometric shapes such as circles, squares, triangles, and so forth (figure 7.4A). Although these symbols may look very abstract, they can be read correctly even when very small. They can also be used for larger categories having hundreds of individual features. This lets mapmakers pack more information into the map than they can with larger mimetic or pictographic symbols. Furthermore, since the correspondence between real world features and their geometric symbols is strictly arbitrary, any point feature can be represented this way. The greater level of abstraction embodied in geometric symbols increases their flexibility as feature display tools. It also requires close study of the legend to determine what is being symbolized.

Line feature maps

Mapmakers make **line feature maps** with qualitative line symbols showing different categories of linear features, such as roads, streams, or boundaries. As with point symbols, the word "line" isn't used here in the strict mathematical sense of a one-dimensional figure. Although line features may be truly one-dimensional (such as boundary lines), line feature maps can also be used to show features that are conceived of as lines for mapping purposes (such as a river, which has width in reality but is shown with a line on a map).

The lines used to symbolize linear features on maps have width as well as length, and the symbol width rarely corresponds directly with the feature width on the ground. For example, if you measured the width of a thin blue line that represents a stream on the map in figure 7.8,

it will usually be much wider than the width of the corresponding feature on the earth when its map width is converted to ground distance. Lines on the map are drawn wider to show the category using the appropriate visual variables—color hue, line shape (or pattern), and either shape or orientation of the pattern within the line. In order for these visual variables to be seen, the mapmaker must widen the lines on the map.

You have probably seen lines of different hues on maps (figure 7.9). Some hues have been standardized for certain features. For example, water features are usually shown with blue lines, boundaries are depicted with red lines, roads are drawn in black, and different types of contour lines are shown in brown. Other times hue carries a categorical message, as on the map of railroad ownership in figure 7.9. To read these kinds of maps correctly, you must refer to the legend.

Shape repetition (line pattern) is also commonly used to distinguish different categories of line features (figure 7.2). Again, there are some standards. For example, administrative boundaries are often shown using a variety of dashed line symbols, and railroads are commonly shown using a solid line with cross hatches that mimic the railroad ties. The individual marks repeated along the line usually are geometric, but they also can be mimetic. Mimetic shapes may be easier to discern than abstract geometric shapes, but they're more difficult to miniaturize and repeat along a line. Therefore, most patterned line symbols use repetitions of geometric shapes.

On many maps, color hue and line pattern are used together to allow the map to show a greater variety of lines. When you think about the number of line symbols on a map, this can be very helpful. For example, the lines on a map can be used to represent features that are linear (fences or walls), as well as networks (roads and railroads), the edges of area features (administrative boundaries), and the land surface form (contours). In addition, there may be lines to show the graticule, the edges of the mapped area, and other bounded areas on the page such as the legend or title. Having a variety of line symbols to choose from helps the cartographer create a map that you can more easily decipher.

Area feature maps

Qualitative **area feature maps** use area symbols to portray area features that are homogeneous within regions. On such thematic maps, area features are best symbolized using the visual variables that give the impression of differences in type or kind. These include color hue, pattern shape, and pattern orientation. For example, a mapmaker might use two hues to show the

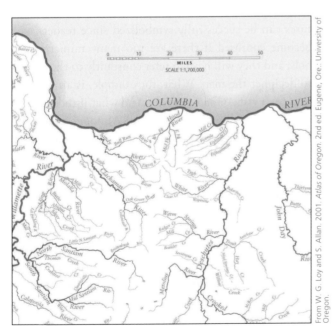

From W. G. Loy and S. Allan. 2001. *Atlas of Oregon.* 2nd ed. Eugene, Ore.: University of Oregon.

Figure 7.8 The lines on the portion of this map shown do not accurately reflect their width on the ground. The width of the line is used to show variations in the stream flow, not the actual width of the stream on the earth.

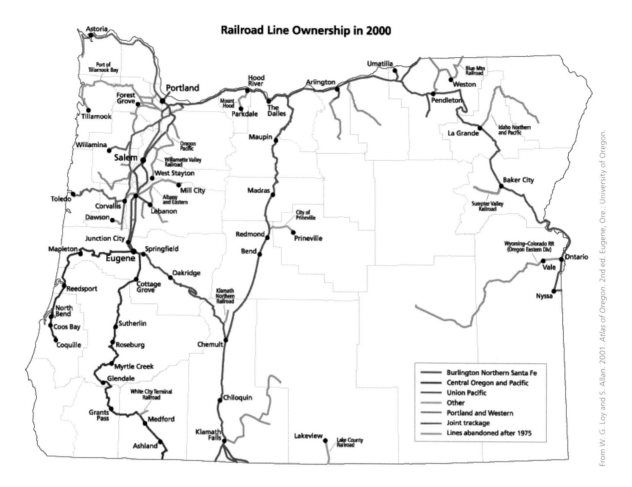

Figure 7.9   Different hues are used to distinguish ownership of railroad lines on this map.

states carried by the Democratic (blue) and Republican (red) party presidential candidate (figure 7.10). Alternatively, she could use a pattern of donkeys as mimetic symbols to fill the Democratic states and a pattern of elephants for the Republican ones. Or she could use a combination of pattern shape and hue (red elephants and blue donkeys) to make sure that you correctly see the category for each region.

If mapmakers instead use visual variables that give a magnitude impression, such as color lightness, color intensity, pattern texture, or size, you're likely to see quantitative rather than qualitative differences in the data. You may think that one symbol depicts more of something than another symbol, when this wasn't the mapmaker's intent. Look, for instance, at figure 7.11. This map shows different Level III ecoregions. Level IV ecoregions are shown using the same hue but varied in lightness and intensity. Without reading the legend, your first impression might be that the map is showing quantitative information as well as qualitative.

Figures 7.10 and 7.11 exemplify two basic types of qualitative thematic maps that differ in the kind of data

Figure 7.10   Area symbols can be used to distinguish regions from each other on the basis of attributes such as states won by the Democratic and Republican party presidential candidates.

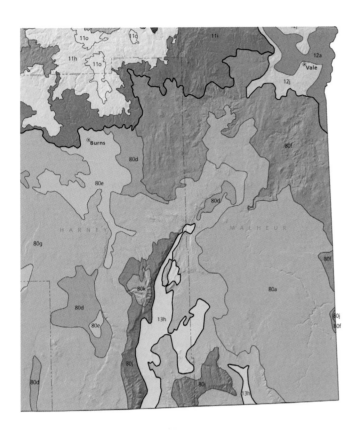

Deep Loess Foothills — 10i
Deschutes/John Day Canyon — 10k

11. Blue Mountains
John Day/Clarno Uplands — 11a
John Day/Clarno Highlands — 11b
Maritime-Influenced Zone — 11c
Melange — 11d
Wallowas/Seven Devils Mountains — 11e
Canyons and Dissected Highlands — 11f
Canyons and Dissected Uplands — 11g
Continental Zone Highlands — 11h
Continental Zone Foothills — 11i
Blue Mountains Basins — 11k
Mesic Forest Zone — 11l
Subalpine Zone — 11m
Deschutes River Valley — 11n
Cold Basins — 11o

12. Snake River Plain
Treasure Valley — 12a
Unwooded Alkaline Foothills — 12j

13. Central Basin and Range
Salt Desert Shrub Valleys — 13h

78. Klamath Mountains
Rogue/Illinois Valleys — 78a
Siskiyou Foothills — 78b
Umpqua Interior Foothills — 78c
Serpentine Siskiyous — 78d
Inland Siskiyous — 78e
Coastal Siskiyous — 78f
Klamath River Ridges — 78g

80. Northern Basin and Range
Dissected High Lava Plateau — 80a
Pluvial Lake Basins — 80d
High Desert Wetlands — 80e
Owyhee Uplands and Canyons — 80f
High Lava Plains — 80g
Semiarid Uplands — 80j
Partly Forested Mountains — 80k

From W. G. Loy and S. Allan. 2001. *Atlas of Oregon.* 2nd ed. Eugene, Ore.: University of Oregon.

Figure 7.11   Color hue is used to distinguish Level III ecoregions, and color lightness and color intensity are used to show the minor classes within the ecoregions.

collection areas being mapped, though both use color hue as the primary visual variable. The presidential election map in figure 7.10 is based on legislatively defined data collection areas (states). On the map, states are given one of two hues, depending on which candidate received the most votes. The areas are rightfully portrayed as homogeneous, since the candidate with the most votes receives all the state's electoral college votes. This type of map is often called a **categorical map**—a map that has polygons enclosing areas assumed to be uniform or areas to which a single description can apply.

The ecoregion map in figure 7.11 is an example of mapping area features that are inherently homogeneous in some way, such as having the same bedrock geology, soils, or vegetation. Though this map is compiled by letting the environmental data determine the boundary between classes (which differs markedly from one compiled using already defined administrative or other boundaries bearing no natural relation to the data) many mapmakers still call it a categorical map.

Notice that the different categories may be purely of one feature ("High Desert Wetlands") or of two or more intermixed features found in certain regions ("Owyhee Uplands and Canyons"). You can see that the mapmaker can define inherently homogeneous areas in many ways, so you should carefully read the map legend to understand what each category means.

MULTIVARIATE MAPS

Most qualitative thematic maps use a separate symbol to represent each category. But sometimes mapmakers show several attributes of the feature within the same symbol on a **multivariate map.** They accomplish this feat in two ways.

One method is to use a different visual variable to show each attribute. In theory, mapmakers could show four different qualitative attributes at once by varying the symbol's shape, hue, pattern shape, and pattern orientation. In practice, symbols showing even two or three feature attributes can be very difficult to read if they are not created carefully. You'll usually have little trouble telling when multivariate information has been mapped, because the symbols will appear more complex than those showing a single attribute.

With the second method of symbolizing multivariate information, mapmakers show a concept defined by a composite of attributes rather than raw data for a single theme. At first glance, many qualitative thematic maps seem to show a single theme when they may actually show multivariate information. This is the case with the ecoregion map in figure 7.11. The concept of an ecoregion is defined by temperature and precipitation ranges as well as typical vegetation, soils, geology, human influences on the landscape, and more. You may think

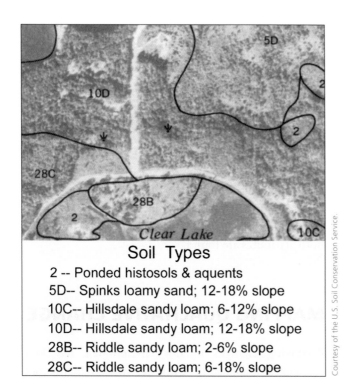

Figure 7.12 Symbols representing soil classes on a map signify areas on the ground where certain combinations of soil attributes exist.

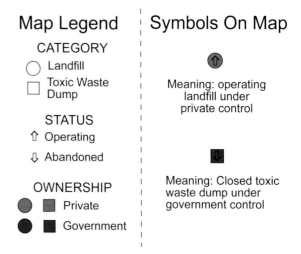

Figure 7.13 A great deal of attribute information can be shown on a map through the use of multivariate point symbols.

of ecoregions as a single theme, but each component of ecoregions plays a crucial part. Similarly, the concept of a soil class is defined by a set of soil attributes, including slope, depth, drainage, color, and texture (figure 7.12). You can probably think of a number of other examples of environmental phenomena that are determined by a composite of attributes.

Reading these kinds of maps can be tricky because the symbols on these multivariate thematic maps look exactly like those on single theme maps. Only the nature of the information symbolized, not the form of the symbols, has been changed. Therefore, it is essential to check the legend as your initial step in map reading.

Multivariate point symbols

Mapmakers like to use point symbols to show multivariate qualitative information because they can pack information into each symbol by combining several visual variables. The resulting symbol is sometimes referred to as a **glyph.** Some multivariate point symbols are pictographic, but most are geometric. For example, the symbols for landfills and dumps shown in figure 7.13 are circles and squares. Arrows in two opposite directions indicate the status of operation, and different hues represent ownership.

Mapmakers sometimes use the same visual variable more than once to create multivariate symbols. For instance, they may use two shapes, such as a star within a circle, which is a common way to show capitol cities. More often they will show two different attributes of a feature with two different qualitative visual variables as in figure 7.13. Although they should use visual variables with qualitative connotations to construct these symbols, they don't always do so. So a good rule of thumb is to check the legend to make sure you understand what the symbols mean.

Multivariate line symbols

You probably won't find many examples of maps showing multivariate qualitative linear data. One reason is that their options for visual variables are limited to using primarily color hue and pattern shape (a shape repeated along a line). But occasionally you'll find such maps.

For example, road maps may be enhanced with multivariate data. It's common to see highway maps with red lines for freeways, black lines for state highways, and gray lines for U.S. or state highways. Sometimes these single variable line symbols are augmented to show a second variable. For instance, dot patterns may be added alongside each line to show the scenic routes that can occur along any of the various road types.

You may also see multivariate line feature maps with line symbols wide enough to have different hues and line patterns within the two bounding lines. These are called **cased line symbols** as the interior line is bounded by a casing that is shown in a different color. You will find these symbols on many kinds of maps. For example, road categories might be shown by different hues, and the casing lines could be patterned as solid, dashed, or dotted lines to show where the roads are in tunnels or on bridges. These cased line symbols tend to cover a larger amount of map space, as we saw earlier with the stream lines on maps, so they are often used only for major roads.

Multivariate area symbols

Mapmakers can make qualitative multivariate maps for area features in a number of ways. One common method is to overlap two types of area symbols that are appropriate for nominal data. The most common visual variables they use are color hue for one attribute and pattern shape (repetition of shape within the polygon that represents the area) for the other attribute. For example, the map in figure 7.14 shows the type of ocean bottom off the Oregon coast by using light gray for sand, dark brown for bedrock, and light brown hue for silt and mud. Vertical and horizontal dashed line patterns are overlaid to show where crab or shrimp are harvested. The area patterns for shellfish type overlap to form "+" signs where both crab and shrimp are harvested. This map lets you see if there is any geographical relationship between the type of ocean bottom and the kind of shellfish harvested there.

On other maps, two different hues such as yellow and blue are used for two different categories so that their area of overlap is seen as the hue combination—in this

case, green. However, the overlap area may go unrecognized, because green is normally seen as a separate hue, not a mixture of yellow and blue. Another problem with this approach occurs if two hues whose combination is not easily recognized are used. Could you predict what hue should be used to show the overlap between an area symbolized with purple and another shown in green? In such cases, it's necessary to refer to the map legend to see what each hue represents. An alternative is to use alternating bands of the two hues within the overlap area—in this example, stripes of purple and green. These symbols are more easily seen as areas in which both attributes are found.

MAPPING QUALITATIVE CHANGE

Portraying changes in our environment has long challenged mapmakers. Until recently, changing phenomena weren't attractive candidates for cartographic representation because the data were difficult to find, the maps were more time-consuming to create, and the displays were more challenging for the map reader to understand. These limitations will become more evident as we discuss the approaches used on maps.

There are two types of qualitative change maps: those that show *change in the attribute* of features at a location over time, and those that show the *change in the location* of a feature over time. For both of these, there are a number of mapping methods that can be used, including small multiples, change maps, and superimposed maps.

Small multiples

To show qualitative change over time, mapmakers often use a method called **small multiples** in which the same basemap is used in a series but the data shown on them change (figure 7.15). This method can be used to map both change in type or attribute over time and change in location over time. Small-multiple maps allow you to easily become familiar with the geographic region shown on the basemap, so you can focus on the changes in the data between time periods.

Small-multiple maps require you to compare the distributions among the maps in order to see the change. To use these maps appropriately, you must first be able to decipher the type of map used in the series, and then you must be able to accurately interpret the changes among the maps. The opportunity for misinterpretation increases with every map added to the series.

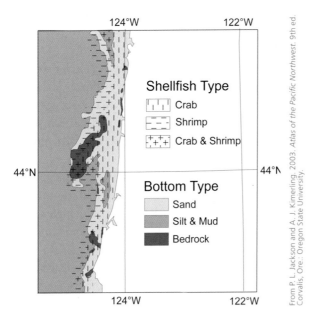

From P. L. Jackson and A. J. Kimerling. 2003. *Atlas of the Pacific Northwest.* 9th ed. Corvalis, Ore.: Oregon State University.

Figure 7.14 The overlap of different area symbols can be used to create a multivariate map showing places where different categories overlap.

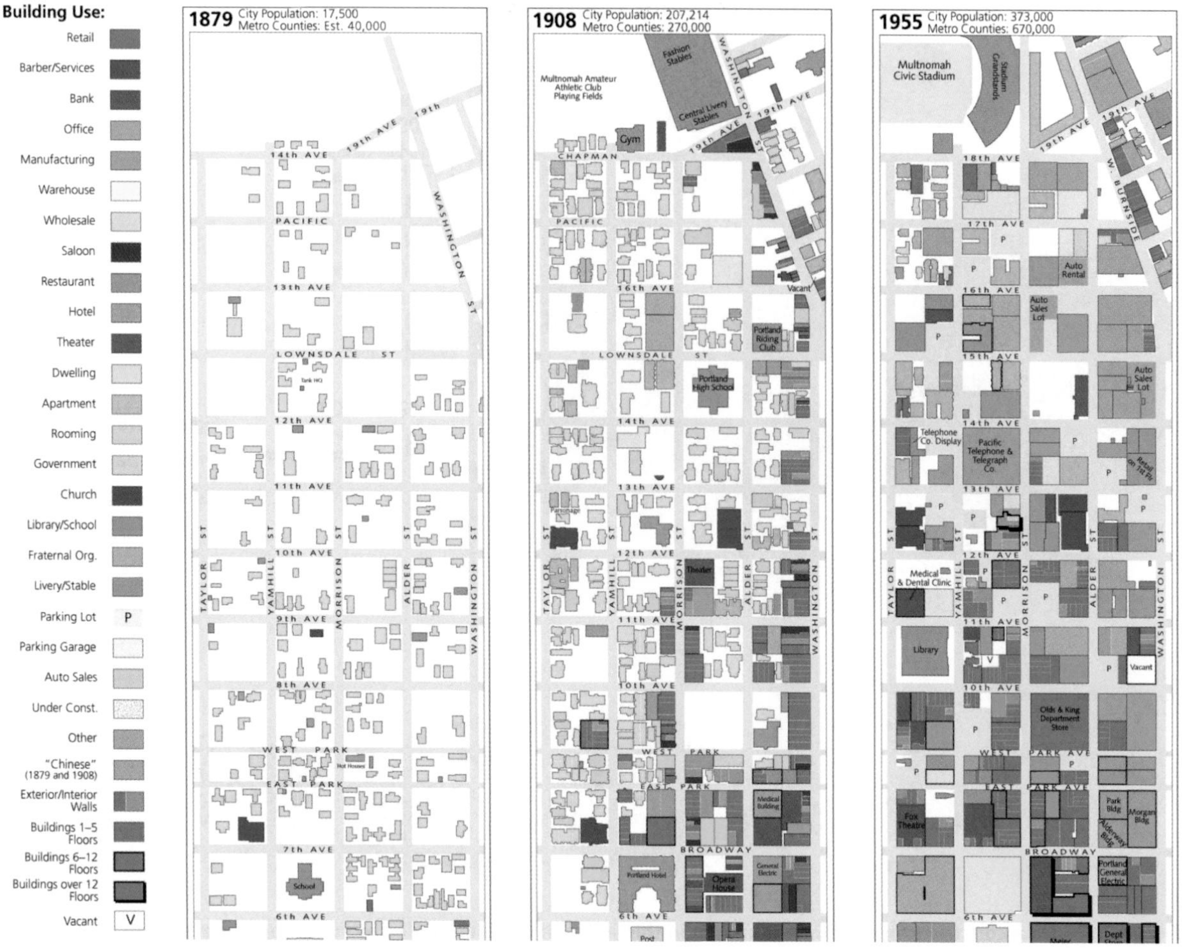

Figure 7.15 These small multiples are used to show qualitative attribute changes in the same location over time. Historical building use in Portland, Oregon, is shown on the same geographic base, allowing you to focus on changes in the number of buildings and their use.

Change maps

An alternative to small multiples is for the mapmaker to show only where change occurs. **Change maps** are used to show locations in which attributes have changed over a certain time period or where features have changed location over time. For instance, the red areas in figure 7.16 represent land-cover change from nonurban to urban from 1986 to 2002 in the Minneapolis–St. Paul, Minnesota, metropolitan area. From this figure, it's obvious that most of the nonurban-to-urban change occurred on the fringe of the core urban area.

Qualitative change maps can show changes for many categories simultaneously, in which case each category is given a separate map symbol. In the example above, only one change was uniquely symbolized—the change from urban to nonurban. The categories on the map could be expanded to include changes between other categories as well, such as forest to agriculture or forest

to urban. Each of these change categories would have a unique symbol (color, in this case.) To help readers, these maps are accompanied by before and after maps, or maps of single categories. As with small multiples, appropriate use of these maps requires that multiple maps are interpreted correctly.

Superimposed maps

Mapmakers can also show qualitative change by **superimposing** a map of one date on that of another. You can then see what changes have taken place since the earlier map was made. The maps in figure 7.17 show dam construction over time. Although each time step could be shown using a different symbol, in this series, only the most recent time step is symbolized uniquely and all other changes are aggregated over time. Notice that this series also makes use of small multiples.

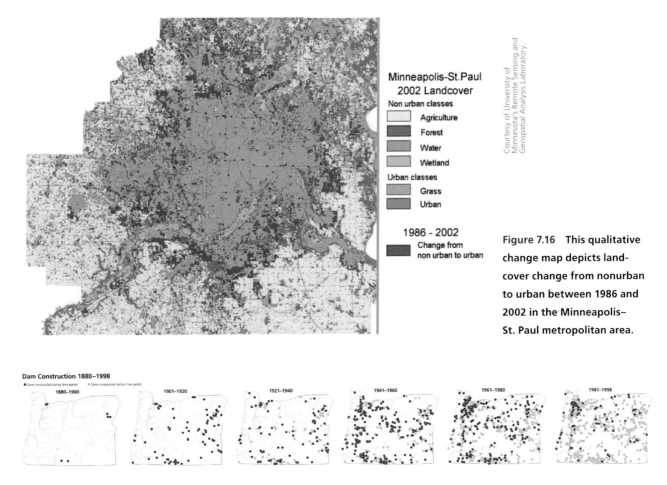

Courtesy of University of Minnesota's Remote Sensing and Geospatial Analysis Laboratory.

Figure 7.16   This qualitative change map depicts landcover change from nonurban to urban between 1986 and 2002 in the Minneapolis–St. Paul metropolitan area.

Figure 7.17   Dam construction over time is shown in this series of small multiples by superimposing the one time period over all previous time periods.   From W. G. Loy and S. Allan. 2001. *Atlas of Oregon*. 2nd ed. Eugene, Ore.: University of Oregon.

Superimposed maps require that the mapmaker use very clear symbology so that the reader can understand where the symbols are superimposed. For a succession of time intervals, you need to be able to be able to see each discrete time period as well as all of them combined in order to accurately study the long-term historical pattern of change in the region.

Change in attribute over time

The maps in figures 7.15, 7.16, and 7.17 are all examples of how *changes in attributes* over time can be displayed. Each method has advantages and disadvantages. Let's now see how the same methods can be used to show change in location over time.

Change in location over time

The second type of qualitative change map shows *change in the location of features* over time. Since point, linear, area, and three-dimensional features vary in the way they change location through time, we'll consider them separately.

Point symbols

Movements of point features are often mapped as **annotated route lines** tracing the paths along which the features move with explanatory labels, point symbols, and other indicators of the changes in location. For instance, figure 7.18 shows the westward route taken in 1805 by Lewis and Clark through present day Oregon and Washington on their historic journey to the Pacific Coast. Their westward movement is shown by a series of point symbols indicating their campsites beside the Snake and Columbia rivers from October 10 to November 10, 1805. The map appears to give you a full picture of their rate of travel, but notice that only 24 campsite symbols are on the map. Since no information is given about sites where they camped more than one night, your understanding of their actual westward rate of movement is incomplete.

Mapmakers can also use many individual point features to show routes or tracts, such as those followed by migrating birds or fish or the predicted or observed tracts of tornadoes and hurricanes. GPS receivers are

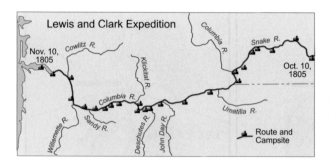

Figure 7.18 This map shows the westward route taken by the Lewis and Clark expedition through Oregon and Washington, indicating their campsites from October 10 to November 10, 1805.

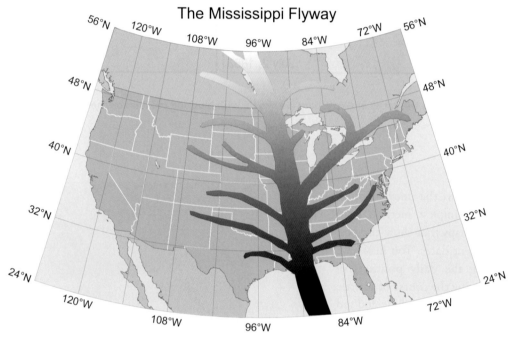

Figure 7.19 Route data are often generalized to broad paths or corridors, such as the Mississippi Flyway for migratory birds.

now being used to gather highly accurate information about the paths that individual animals, such as birds, whales, wolves, and even humans, take for migration, predation, and other activities. Satellite imagery can be used to more accurately identify the center of eyes of tornadoes and hurricanes. Since the exact route followed by each individual differs slightly, the route data are usually generalized to broad paths or **corridors of movement** for mapping purposes. Thus you might find a map of "Tornado Alley" or the "Mississippi Flyway" (figure 7.19) on which a composite of individual paths is depicted. Be sure you recognize that this type of movement map is intended to show only the general geographic pattern of movement over a certain time period.

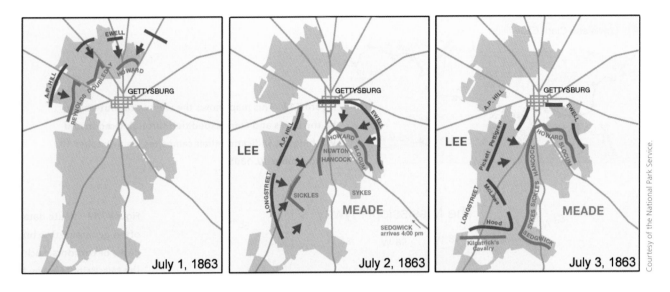

Courtesy of the National Park Service.

Figure 7.20 **The positions of troops during the Battle of Gettysburg are shown with blue and red lines on these three maps.**

Line symbols

There are several ways to show the movement of qualitative line features. The simplest method is to create a set of small-multiple maps that show the positions of the features on different dates. You may have seen military maps showing the daily positions of battle fronts or troop lines during the course of a battle. One example of these military maps uses blue and red lines to show the combined daily position of Union and Confederate troops, equipment, and control during the Civil War Battle of Gettysburg from July 1st to 3rd, 1863 (figure 7.20). Notice that this map series shows more than the advance or retreat of the same battalions over the three days—new troops that arrived and reserves that were called into battle are also mapped.

Another way to show change in the position of linear features over time is to superimpose the line features from different time periods on a single map. For instance, your TV weather channel may use a map showing the current and forecast position of the jet stream (figure 7.21) to help explain the changing weather conditions.

Area symbols

The movement of qualitative area features generally occurs at their edges along a linear front. In figure 7.22, the ash plume from the 1980 eruption of Mt. St. Helens in Washington state is shown as a series of **isochrones** (lines of equal time difference). The area of ash in the upper atmosphere at a given time is the same as the area within the single isochrone for that time period. Note

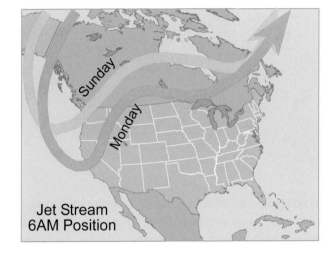

Figure 7.21 **This map shows superimposed qualitative linear features (the jet stream on different days) to give you a better understanding of a weather forecast.**

that most of the advance is at the front edge of the isochrones.

Changes over time in the location of qualitative area features can also be mapped. **Spatial diffusion** is the term we use to refer to changes over time *and* changing places—it is the transfer or movement of things, ideas, people from place to place. Figure 7.23, for example, shows the systematic areal expansion of gypsy moths in the northeastern United States from 1890 to 1971.

It is also possible to map qualitative area change on a change map. Consider a map of the tidal zone—this

Mount Mazama and Mount St. Helens Ash Fall

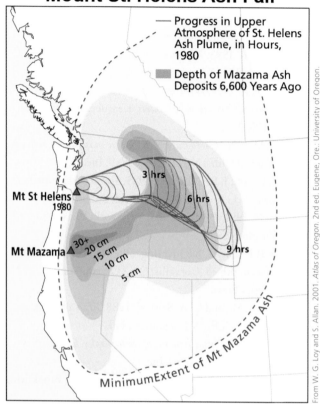

— Progress in Upper Atmosphere of St. Helens Ash Plume, in Hours, 1980

▨ Depth of Mazama Ash Deposits 6,600 Years Ago

From W. G. Loy and S. Allan. 2001. *Atlas of Oregon*. 2nd ed. Eugene, Ore.: University of Oregon.

Figure 7.22 This map shows the change in location over time of the ash plume from the 1980 eruption of Mt. St. Helens.

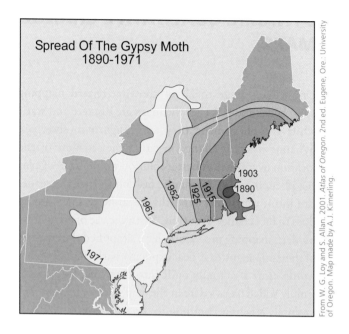

From W. G. Loy and S. Allan. 2001. *Atlas of Oregon*. 2nd ed. Eugene, Ore.: University of Oregon. Map made by A. J. Kimerling.

Figure 7.23 The change over time of an areal feature can be shown as an expanding or contracting front. This map depicts the spread of gypsy moths in the northeastern United States at various points in time from 1890 to 1971.

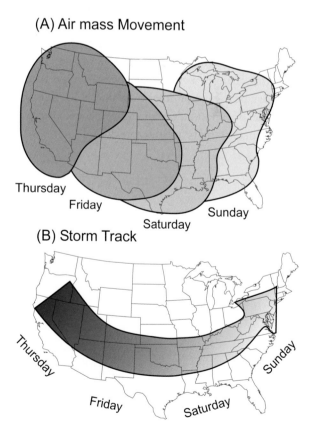

Figure 7.24 Movement of three-dimensional features such as air masses and storms are commonly depicted on maps by a series of area symbols (A) or a linear flow symbol (B).

kind of map would reflect the daily fluctuation in the area covered by seawater due to tidal changes. The map in figure 7.16 is a good example showing the diffusion of urban land cover over a 16-year period.

Three-dimensional features

It is also possible to see maps of the change in location of three-dimensional features, such as air masses. True volumetric change can be mapped with 3D models or animated 3D perspective views such as those discussed in chapter 6. On planimetric maps, the three-dimensional feature is commonly collapsed into a two-dimensional outline by projecting the two-dimensional feature boundary down or up onto the ground surface. The feature outlines for different dates are then superimposed on the basemap to show changes over time. This is a standard method of mapping air mass movement (figure 7.24A). Sometimes even greater cartographic abstraction is used, and the movement of the feature is shown only by a linear flow symbol (figure 7.24B) (see chapter 10 for more on cartographic abstraction).

DYNAMIC QUALITATIVE CHANGE MAPS

One disadvantage to all of the time-related mapping approaches above is that the maps are discrete and static while the phenomena they are designed to show are dynamic (there is change in either the attribute or the location over time). A better solution would be to display the phenomena in a dynamic fashion so that the reader can intuitively understand that what he or she is seeing does in fact change over time. With the use of computers, modern mapping procedures have simplified the creation of dynamic maps that deal explicitly with time. In other words, time is actually built into these maps—it is not simply added on as a date in the map's title or margin. For example, it is now common to show the continual movement of the jet stream and other weather elements in an animated time sequence.

The map reader may find these challenging to read if the maps are not displayed slowly enough or if they are not repeated often enough. The optimal situation is for the map reader to be able to have interactive control over the display of the map animation, which is not always the case.

These maps require a very prominent legend to indicate the changing time period. This can be shown as a label that changes, a clock that advances, pages torn off a calendar, or the use of other clever methods. But if the map reader does not understand the associated passage of time, the map could be misinterpreted.

SELECTED READINGS

Bertin, J. 1981. *Graphics and graphic information-processing.* New York: Walter de Gruyter.

Carnachan, R. 1993. *Wisconsin soil mapping, Guide 4.* Madison, Wis.: Wisconsin State Cartographer's Office.

Chaston, P. R. 1995. *Weather maps: How to read and interpret all basic weather charts.* Kearney, Mo.: Chaston Scientific, Inc.

Dent, B. D. 1996. *Cartography: Thematic map design.* 4th ed. Englewood Cliffs, N.J.: Prentice-Hall, Inc.

Hole, F. D., and J. B. Campbell. 1985. *Soil landscape analysis.* Totowa, N.J.: Rowman & Allanheld, Publishers.

Holmes, N. 1991. *Pictorial maps.* New York: Watson-Guptill Publications.

Jackson, P. L., and A. J. Kimerling. 2003. *Atlas of the Pacific Northwest.* 9th ed. Corvallis, Ore.: Oregon State University Press.

Monmonier, M., and G. A. Schnell. 1988. *Map appreciation.* Englewood Cliffs, N.J.: Prentice Hall.

Robinson, et al. 1995. *Elements of cartography.* 6th ed. New York: John Wiley & Sons, Inc.

Robinson, V., ed. 1996. *Geography and migration.* Brookfield, Vt.: Edward Elgar Publishing Co.

Saint-Martin, F. 1990. *Semiotics of visual language.* Bloomington, Ind.: University Press.

Tufte, E. R. 1997. *Visual explanations: Images and quantities, evidence and narrative.* Cheshire, Conn.: Graphics Press.

Wrigley, N. 1985. *Categorical data analysis for geographers and environmental scientists.* New York: Longman, Inc.

chapter
eight **QUANTITATIVE THEMATIC MAPS**

8

Quantitative thematic maps

In the previous chapter, we discussed qualitative thematic maps, which emphasize the location of different kinds of environmental features. Sometimes we want to know not only what and where but also how much of something exists at some location. In such cases, we turn to **quantitative thematic maps**—maps that show a single theme or a few related themes of **quantitative information** (information that portrays a magnitude message).

On quantitative thematic maps, cartographers use a variety of symbols to depict magnitude information telling you how many, large, wide, fast, high, or deep things are. The visual variables that inherently connote differences in magnitude are size, pattern texture, gray tone or color lightness, and color saturation (figure 8.1). A well-designed series of symbols using variations in one or more of these visual variables will appear to you as a progression of magnitudes from small to large or low to high.

Many types of quantitative thematic information can be portrayed using these visual variables. Data for a **single theme map,** such as population density for different cities or the precipitation recorded at various weather stations (figure 8.2), relate to one variable for the features mapped.

You can also find **multivariate maps** that are a composite of two or more related themes. Figure 8.3 shows two themes relating to economic losses due to earthquakes for Oregon counties. Notice that the two themes are displayed using different visual variables. The total estimated earthquake loss for each county is shown by the height of a narrow bar, and the lightness of a purple hue shows each county's earthquake loss ratio.

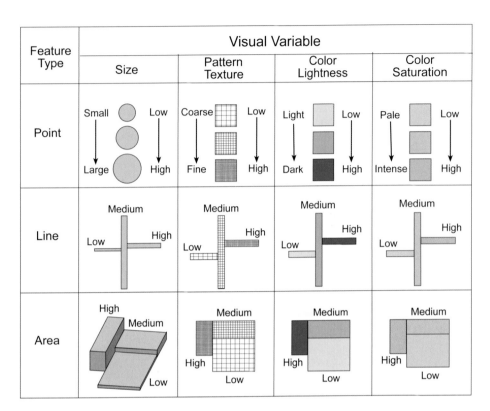

Figure 8.1 The visual variables that inherently impart a magnitude message are size, pattern texture, gray tone or color lightness, and color saturation.

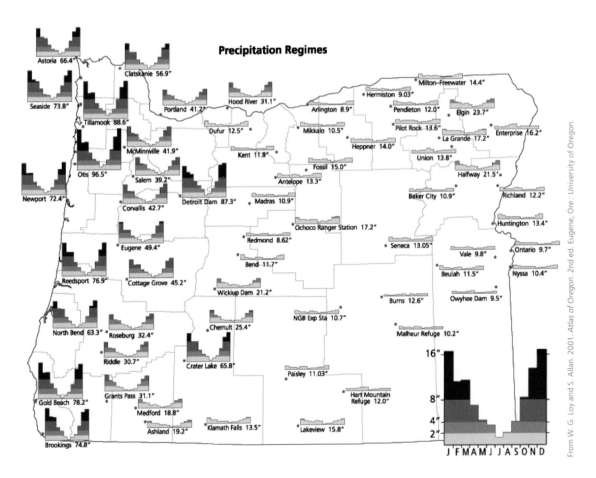

Figure 8.2 This quantitative thematic map shows precipitation regimes for a set of weather stations in the state of Oregon. Note that the size of the bar and the color lightness are used together to show how much rain is recorded for each month at each location.

QUANTITATIVE DATA

Data collectors have several choices. They can take physical measurements describing a theme at different locations within a region. They can collect data for every feature in a theme to obtain what is called a **population count.** The population count can then be mapped, such as in figure 8.4 where point symbols are used to represent each of the ski areas in Oregon.

The ski areas were symbolized to show the total number of day visits from 1999–2000.

In this case, the data were collected for every feature in the theme in order to produce this population count map. It is possible for cartographers to collect the data for each point feature of interest when the number of features is fairly small or the data can be easily collected for the entire population. The map of ski areas in figure 8.4 is a good example of data collection for a manageable number of features.

Another alternative is to collect data for only a portion of the region or population; this is called taking a **sample.** A map of household income in a set of cities could be made from surveys collected from each individual household. The entire population of households may have been surveyed, although more likely the map was made from a sample of households in each city. The key difference between a population count and a sample is that your goal with a population is identification of the characteristics, while the goal with a sample is making inferences about the characteristics of the population from which the sample was drawn.

Quantitative thematic maps can be created from all or part of the measurements taken for population counts or sample data. Figure 8.5 is a map of the types of high technology companies in the Portland, Oregon, metropolitan area—it only shows those companies with at least 300 employees, so some of the smaller high-tech companies are not included on the map.

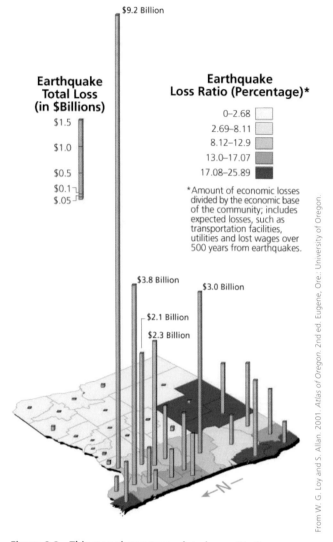

From W. G. Loy and S. Allan. 2001. *Atlas of Oregon.* 2nd ed. Eugene, Ore.: University of Oregon.

Figure 8.3   This map shows two related quantitative themes—color lightness is used to show the earthquake loss ratio, and the size (height) of the bars is used to show the earthquake loss estimates in dollars.

Figure 8.4   This map shows the number of day visits for every ski area in Oregon.

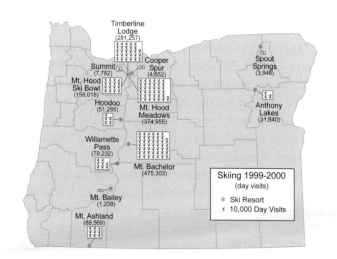

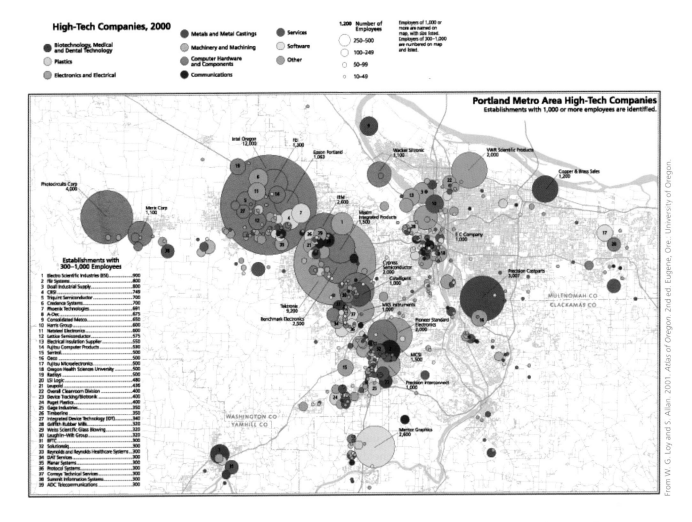

Figure 8.5   Only high-tech companies with at least 300 employees are shown on this map of the Portland metropolitan area.

Counts and measurements

Maps are made from a wide variety of counts and measurements. **Counts** are the total number of features, and **measurements** or **amounts** are the quantities associated with features. Counts and measurements can be collected for point, line, or area features. Let's look first at data that describe individual *point features*.

Figure 8.5 provides a good example of count data for point features. The features are the high-tech companies and the counts are the number of employees. For the largest companies, the number of employees is shown right on the map. For smaller companies, the number in the point symbol refers to the table at the left where the number of employees is listed.

As with point features, maps can also be made for counts or measurements that relate to line features. Counts within sections of a line feature are common. An example is the highway traffic map in figure 8.6 in which the traffic volume is recorded for each section

of the highway. Abrupt breaks where sections meet are your clue that each line segment is symbolized to show the number of vehicles that pass per section of road. An example for a river is the number of fish caught per river mile. Catch data from your state fish and game department may have been used to make maps showing fishing success in the river. The total number of fish for each river mile could also be estimated from counts taken at sample points using nets or other methods to trap or stun all the fish at each point.

Measurements of *line features* are recorded in several ways. Say you are looking at a map of water temperature along a river. It is possible that temperatures were measured continuously along the river from thermal infrared remote sensors that collected measurements every few minutes over the entire portion of the river that was mapped. Although the sensing device was operating continuously, the data from the sensor are usually recorded as values for grid cells of a certain spatial resolution that make up the thermal image. The temperature mapped at any point along the river may be an average of the grid cells perpendicular to the direction of flow, or the value for the grid cells at its center.

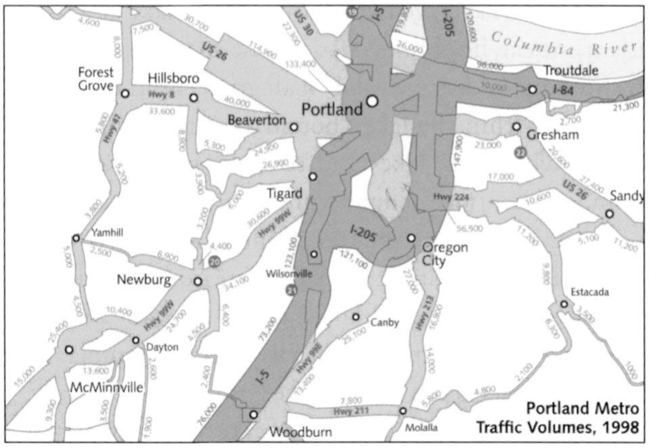

It is more likely that the river temperatures were obtained from thermometer measurements taken at a small number of *sampling sites* (locations where a measurement or measurements are taken) along the river's course. The map may only show the temperatures at these points, or the cartographer may have interpolated between the measured values to create a continuous dataset for the river. Only by looking at explanatory notes for the map can you tell how the data for the map were obtained. The map in figure 8.7 uses the same approach to display the mean streamflow for the rivers.

Area feature maps can also show counts or measurements. When you are more interested in understanding the spatial nature of a distribution of features than the location of individual features, a map showing counts of features within data collection units will best suit your needs. There are many different kinds of **data collection units**—natural or human-defined units that are used to divide the entire population or extent for the

Figure 8.6 This traffic volume map shows the number of vehicles that were recorded for sections of roads in the Portland, Oregon, area.

From W. G. Loy and S. Allan. 2001. Atlas of Oregon. 2nd ed. Eugene, Ore.: University of Oregon.

Figure 8.7 Average annual streamflow is measured at selected sampling stations along the rivers from which the flow at intermediate locations was interpolated.

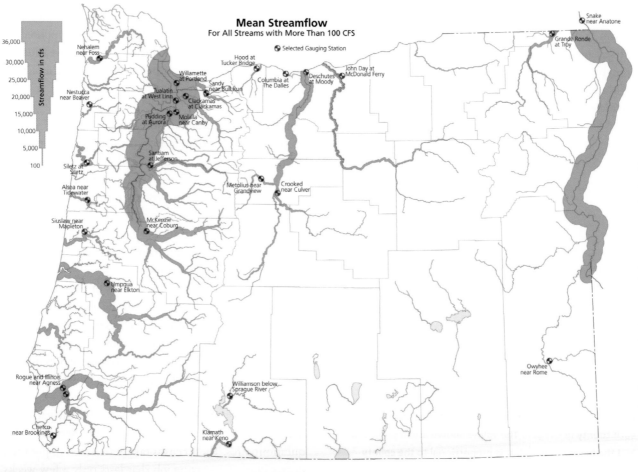

From W. G. Loy and S. Allan. 2001. Atlas of Oregon. 2nd ed. Eugene, Ore.: University of Oregon.

purposes of data collection. For example, enumeration districts are geographic areas assigned to a census taker, usually representing a specific portion of a city or county. Watersheds, ecoregions, climate zones, and stream reaches are but a few examples of others. Figure 8.8 is an example of population counts within the counties of Oregon. The count of individual features is called a **census.** Federal, state, and local government agencies conduct censuses to learn basic population characteristics. Private firms focus their census taking on product advertising and marketing. Census maps are produced to make it easier for you to visualize the spatial pattern of high and low counts. Such census maps will show you how many features were counted in each data collection area, but not where each feature was located at the time of the census.

Maps are made from a variety of census data. U.S. Census Bureau data collectors determine the number of people within households, from which the total population count within different data collection areas (blocks, block groups, census tracts, cities, counties, and states) can be determined by simple summation. The areas usually are irregular in shape and size, and maps made from population count data inherently show the irregularities among areas. Population counts can also be based on a grid of identical square or hexagonal cells covering a region. A variety of biological phenomena are counted within grid cells. Maps of biomass, animal density, insect infestation intensity, and many other themes are made from biological census data.

Maps can also be made for measurements of the characteristics of area features. Let's take a lake as an example. The average temperature of the lake at a particular time can be obtained from a thermal infrared image by finding the average value of all grid cells in this image that fall within the lake's boundary. **Limnologists** (scientists who study lakes and streams) may take a large number of surface temperature measurements at different sample points on the lake, from which they compute an average temperature. They then create a map showing average temperatures of lakes within a region from the average values for image grid cells or actual temperature measurements taken in each lake.

Another approach would be to take measurements at a small number of sampling sites within the lake. The map may only show the values at these points, or the cartographer may have averaged these values to create a representative value for each lake. The explanatory notes for the map can tell you how the mapmaker derived the values.

Map designers also can obtain data they think of as a **continuous surface** (a surface representing a geographic phenomenon that has continuously changing values) from remote sensing or direct measurement at a number of sample points. You can think of average annual precipitation across your state as a continuous surface (figure 8.9). The average value at each point in the state could be obtained from meteorological satellite imagery taken daily. But more often an average precipitation map is made by interpolating between averages computed for the weather stations within the state. Each weather station, of course, is a sample point from which the surface is inferred.

Count and measurement accuracy

It seems logical that a map showing the counts or measured values for a theme would be accurate. This is true in most cases, but not always. A number of errors may reduce the quality of counts or measurements. The errors are due to *instrumental, methodological,* and *human deficiencies* during data gathering. Such errors are usually well disguised on maps, sometimes on purpose and sometimes inadvertently.

Take the case of the U.S. Census of Population and Housing conducted each decade by the U.S. Census Bureau. Since every household in the country is supposed to be surveyed, maps produced from the counts will be faultless . . . or will they? A full head count of 300 million people is an immense job. Let's look at how the data are gathered.

Once every 10 years, the Census Bureau gathers a nationwide team to track down nonresponders to mailed questionnaires. These census takers are asked to put their hearts into a low-paying job that lasts only a few weeks.

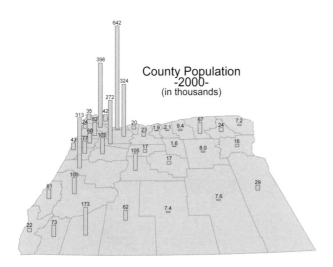

Figure 8.8 Population counts on this map are shown for each county in Oregon. The values shown are in thousands, and the height of each bar is proportional to the county population in the year 2000.

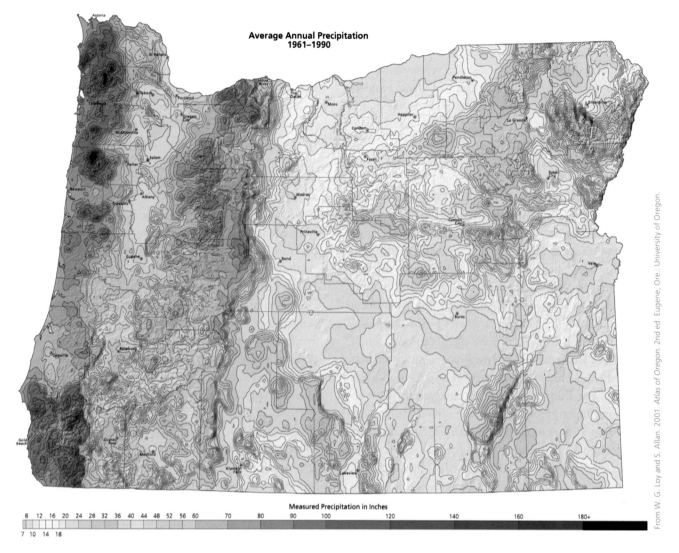

**Average Annual Precipitation
1961–1990**

Measured Precipitation in Inches

8 12 16 20 24 28 32 36 40 44 48 52 56 60   70   80   90   100   120   140   160   180+

7 10 14 18

Figure 8.9 This map of annual precipitation was created using point measurements of precipitation to produce a continuous surface from digital grid estimates of yearly averages.

From W. G. Loy and S. Allan. 2001. *Atlas of Oregon*. 2nd ed. Eugene, Ore.: University of Oregon.

They must brave strange neighborhoods, sometimes repeatedly. The people to be interviewed often aren't home, and when they are home they may be hostile or their dog might be. The people interviewed are supposed to respond truthfully even though this means telling secrets that could get them into trouble with their landlords, the welfare office, or local authorities. Finally, after all the information has been gathered, it still has to be processed, analyzed, and mapped.

You can see the potential for all three types of errors: instrumental (questionnaire), methodological (interview procedure), and human (between the interviewer and the respondent). Yet the Census Bureau has an electronic display and Web site that at any time will give you the "actual" number of people in the United States. Be your own judge of how accurate and current this population count is.

Physical measurements are subject to these same sources of error. Imagine determining the average weight of fish in a stream by netting at several locations and weighing each fish. The net may not be fine enough to capture all species (instrumental), the sample locations may not be representative of the entire stream (methodological), and the person doing the netting may be too distracted to perform the job thoroughly (human deficiency).

Spatial samples

As mentioned earlier, quantitative thematic maps are often based on a *spatial sample* of features rather than a full population count. The idea behind sampling is to use a small part of the population to find out what you want to know about the entire population. There are several reasons that samples are used to estimate what the entire population is like. For one thing, the time and cost involved in obtaining a full population count may be prohibitive, so only a sample can be taken. There are also continuous physical phenomena like temperature that exist everywhere and change in magnitude from moment to moment. It is impossible to measure their values constantly at every point on the earth, but for practical purposes a spatial sample of temperatures obtained from weather stations can be used to estimate the temperature at any location over a selected period of time.

Maps are made from several types of spatial samples (figure 8.10). Data for a map showing average temperature of this lake and others in the region may have been collected using a basic **sampling method** (the means of obtaining the sample from the total population). The simplest approach would be to travel by boat to a number of geographic positions and measure the temperature at each position—this is called a spatial **point sample.** A more difficult method is to have the boat slowly follow a **transect line** and obtain temperature measurements along the line either continuously or at a constant time interval. The third method is to navigate the boat to **quadrat** sampling areas (rectangular sampling units) predefined on a chart of the lake. With careful navigation, temperature readings can be taken within each quadrat.

The arrangement of sample points, transect lines, or quadrats can be random or systematic, as seen in figure 8.10. With **random spatial samples,** all locations are given an equal probability of being selected for data sampling. In contrast, **systematic spatial samples** are

arranged so as to collect data at regularly spaced distance intervals. With this rule, sampling is often done at a rectangular or triangular grid of points or quadrats, or along equally spaced parallel transect lines. With a **stratified spatial sample,** known characteristics of a feature distribution are used to guide data collection. This type of sample is obtained by first subdividing the geographic feature or region into zones (strata) possessing distinctive traits. Random or systematic samples are then taken so as to reflect the importance of each stratum. In our lake example, the sample points were stratified by water depth, since the shallow and deep portions of the lake were equally important to estimating the average temperature. Notice the same number of sample points is randomly placed in the deep and shallow sections of the lake, reflecting their equal importance.

Statisticians may argue about which form of sampling gives the best data for estimating a true average temperature, but one thing is certain—a higher density of randomly or systematically determined sample points, transect lines, or quadrats should give a better estimate, particularly if the sample is stratified in an appropriate manner.

Measurement levels

The qualitative data for the maps described in the previous chapter were at the nominal measurement level. Nominal-level data (also called categorical data) consist of categories used to distinguish different types of features within a map theme. Familiar nominal categories include different types of trees or roads, different nationalities and religions, or different soil-types. At the nominal level, there is no information about the relative size or importance of each category. Nominal-level data do not give the map reader a "magnitude message."

In contrast to qualitative data, the data for quantitative thematic maps are at the ordinal, interval, or ratio measurement levels. From an analytical perspective, these measurement distinctions are important. **Ordinal-level data** are ranked according to a "less than" to "greater than" system. How much more or less one class is than another isn't specified, since there are no numerical values. Examples of ordinal data include small, medium, and tall trees; minor and major highways; and low-to-high-density housing. Figure 8.11 indicates how each of the 40 cities in the Portland, Oregon, metro area ranks in terms of household income. Be careful here. Do not confuse the ranks, which appear as visually obscure numerical labels, with the population in 2000, which is shown by the eye-catching sizes of the symbols.

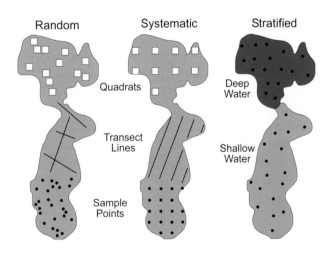

Figure 8.10 **Different spatial sampling methods have been used to find the average temperature of this lake. Sampling points, transect lines, and quadrats have been randomly and systematically located across the lake, and a point sample stratified by water depth was also taken.**

Household Income Ranked within Portland Metro Area
(40 incorporated cities and unincorporated areas)

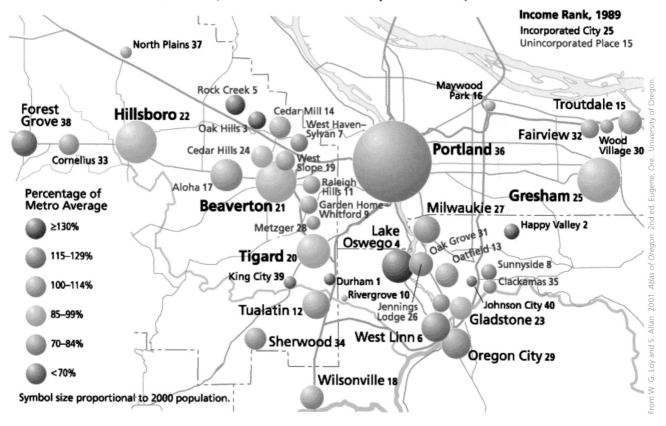

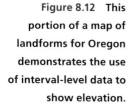

Figure 8.11 Labels indicate the ranks for Portland Metro Area 1989 household income. Point symbol sizes and colors relate to other population characteristics.

Figure 8.12 This portion of a map of landforms for Oregon demonstrates the use of interval-level data to show elevation.

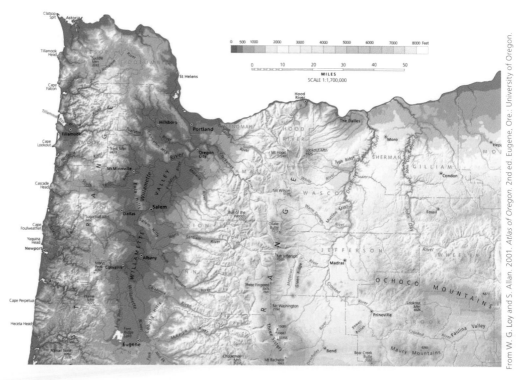

Interval-level data consist of numerical values on a magnitude scale that has an arbitrary zero point. Land elevations (figure 8.12) are an excellent example of interval-level data, since the zero datum is arbitrarily defined as mean sea level (see chapter 1 for more on the definition of mean sea level). To see how arbitrary mean sea level is, you only have to think about how sea level rises and falls in the past and present—the zero point shifts over time and doesn't denote the absence of elevation.

When looking at interval-level data on a map, you should realize that only the numerical intervals between classes are valid mathematically. Notice that the elevation classes are in 500-foot intervals. You can correctly conclude that the difference in elevation for each class is the same, since the elevation interval is the same. However, it is incorrect to say that an elevation of 2,000 feet is twice as high as an elevation of 1,000 feet. To further understand this concept, imagine sea level were to rise 999 feet so that the second elevation is now 1 foot and the first is 1,000 feet. The ratio of the elevation difference between the same two points is now 1,000 to 1, but this ratio does not mean the first elevation is 1,000 times as high as the second elevation.

Ratio-level data also consist of numerical values on a magnitude scale, but, in contrast to interval-level data, the zero point isn't arbitrary. Instead, the zero point denotes absence of the phenomenon. You will find many quantitative thematic maps showing ratio-level data. Themes such as population density (or any other density), annual precipitation, crime rate, tree heights, or temperature in degrees Kelvin have zero points that denote total absence.

In figure 8.13, there are a number of Oregon counties in which none of the population works in foreign-owned companies.

With ratio-level data, both the numerical intervals and the ratios between values are mathematically correct. When you look at a map like in figure 8.13 with classes such as 0.25–0.49, 0.50–0.75 and 0.75–0.99 for the percent of total employment, you can conclude that the density range is the same for all classes. You can also assume that 0.50 is two times more than 0.25, and 0.75 is three times more.

SINGLE THEME MAPS

The simplest quantitative thematic maps are those that show a single theme at the ordinal, interval, or ratio measurement level. It is convenient to subdivide map themes into point, line, and area features in order to explore how cartographers have devised special ways to create sets of quantitative point feature, line feature, and area feature symbols.

Point feature maps

Proportional or graduated symbols

To show quantitative information at specific points, mapmakers vary one or more visual variable to portray variations in the magnitudes of the attributes for the features. If they vary the symbol size in proportion to the magnitude for each feature on the map, the symbols are called either **graduated** or **proportional symbols**.

Proportional symbols are used to represent the exact data values because the visual variable (size in this case) of each symbol is scaled proportional to its data value (figure 8.14, center). A difficulty with proportional symbols arises when there are too many similar values, as the differences between symbols may become indistinguishable. In addition, the symbols for high values can become so large that they obscure the other symbols.

Another alternative is **graduated symbols**, in which the quantitative values are grouped into classes and all the features within a class are shown with the same symbol (figure 8.14, right). Although you can't tell the value of an individual feature, you can tell that its value is in a certain range. The category ranges can be either numerical (1–10, 10–20 , 20–30, and so on)—or they can be at the ordinal level (low, medium, high); the latter instance is called **range grading**.

Jobs in Foreign-Owned Companies

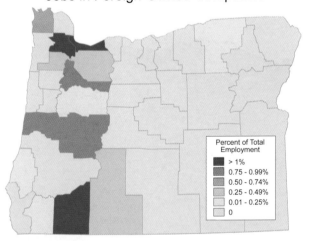

Percent of Total Employment

- > 1%
- 0.75 - 0.99%
- 0.50 - 0.74%
- 0.25 - 0.49%
- 0.01 - 0.25%
- 0

Figure 8.13 This map uses ratio-level data—the zero point is meaningful, and operations such as addition and subtraction as well as multiplication and division are meaningful.

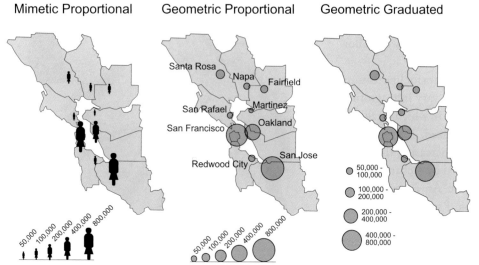

San Francisco Bay Area County Seat Population
-2000-

Mimetic Proportional   Geometric Proportional   Geometric Graduated

Figure 8.14 Proportional and graduated point symbols are commonly used to show quantitative information. Mimetic symbols (left) may look interesting, but simple geometric symbols (center and right) are usually easier to read. Graduated symbols (right) aggregate the data into a small number of classes to make the symbols easier to read.

Quantitative point symbols are usually **geometric,** such as circles, squares, or triangles, or **mimetic symbols**—miniature caricatures of the features they represent (see chapter 7 for more on these types of point symbols). If the point symbols are mimetic, they will usually be varied only in terms of size (figure 8.14, left). For example, if a human figure of one size indicates a city population of 100,000 people, the same figure eight times as large shows a city with 800,000 residents. If the point symbols are geometric forms like circles, the most common way of showing changes in magnitude is to vary their size. Pattern texture, color lightness, and color saturation may be varied as well to show different magnitudes. Darker shades of red, for instance, could indicate larger populations, so that the larger circles are also darker.

Reading geometric symbols isn't as straightforward as it might seem. The difficulty arises because of the way the human eye and brain work. The brain doesn't perceive signals from our eyes in a linear fashion; instead, the size of geometric symbols is progressively underestimated as the area or volume of the symbols increases. This discrepancy between the apparent size and absolute size of map symbols is minimal with respect to symbol height, is worse with respect to area, and becomes a major problem with three-dimensional symbols. We judge the magnitude of three-dimensional symbols by their area rather than their volume. Thus, the area that a sphere covers on the map, not its volume, is what you're likely to see (figure 8.11).

It's most difficult to read proportional point symbols when a continuous sequence of symbols has been used. The human eye simply doesn't function precisely enough

to differentiate between such slight variations in symbol size. The inevitable result is that the map user doesn't appreciate much of the effort that went into making the map. In fact, continuous gradation of magnitude symbols may actually contribute to map reading error because of the increased confusion it may cause.

Proportional point symbol reading difficulties are largely avoided when symbols are labeled, as in figures 8.4 and 8.8. Limiting the sizes to a small number of classes also helps, so graduated symbols are sometimes the better choice for cartographers. With this approach, the symbols are usually different enough in size that the eye can easily tell them apart (figure 8.14, right). Although information has been lost by reducing the magnitude data to a few classes, graduated symbol maps are generally the easiest to read of the single-variable quantitative point symbol maps.

Line feature maps

For line features, mapmakers show quantitative information associated with the features or how some values change along a line or between lines. The visual variables that can be used to show quantitative information for lines include size, color lightness, color saturation, and pattern texture. The best visual variable to use is size that would vary relative to the value being shown. Figures 8.7 and 8.15 provide excellent examples. On these maps, the more water that flows down a segment of a river, the wider that part of the line. The magnitude at any point along a line is difficult to read because the human eye isn't sharp enough to discriminate between slight changes in width. One solution is to divide the legend guide into a small number of classes. In the figure 8.15 legend, the thinnest line indicates 0–1 million gallons of water per minute, the next thinnest line 1–2 million gallons per minute, and so on. The map reader then must guess which legend class best fits the mapped stream flow at a particular location.

Another way to show quantitative data for line features is to represent different features that vary in magnitude so that they form a **visual hierarchy** in which categories of features are organized and prioritized. The visual hierarchy is created by systematically increasing the width (size) of the line, often accompanied by a change in the lightness or saturation of its color. An example is the road symbols found on the typical state highway map (figure 8.16). You will often find an ordinal-level hierarchy of expressways, U.S. highways, state highways, and county roads shown by progressively narrower lines.

Another technique that mapmakers use is to vary the textures, lightnesses, color saturations, or color hues as "fills" within cased line symbols. For instance, the degree of traffic congestion on central Seattle freeways and major highways is illustrated in figure 8.17 with the green, yellow, and red hues we associate with stop lights, plus black for stop-and-go traffic. As with other proportional line symbols, data are usually grouped into a small number of classes to simplify map reading. Note that although the primary visual variable here is color hue (which is more appropriate for qualitative data), you automatically associate red, yellow, and green colors with traffic (red = no speed, yellow = low speed, and green = higher speed).

A problem with using visual variables in the fill of a line symbol from your perspective as a map user is that the symbols are disproportionately wide, covering large areas on the map. This is necessary so that the variations in the fills can be seen. Hence, only a limited number of symbols can be used, and lines that are close together may have been displaced. For the map to make sense, the geographical base features may have been displaced as well to accommodate the line symbols.

Flow maps

Flow maps show ratio-level changes in magnitude along a line feature using a proportional line symbol called a **flow line.** On most flow maps, the line's width is made proportional to some magnitude, such as where the jobs provided to Oregon by foreign-owned companies come from in figure 8.18. The *direction of flow* is also important, and arrows are added to one end of the flow line to show flow direction. On many flow maps the mapmaker

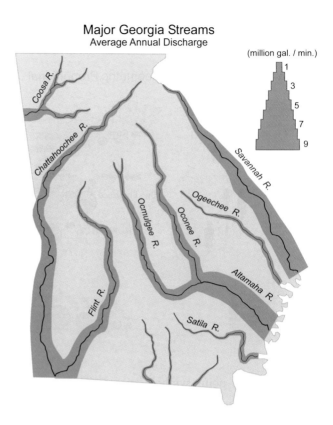

Major Georgia Streams
Average Annual Discharge

(million gal. / min.)

Figure 8.15   Average annual discharge for the major streams in Georgia is shown by varying the width of the lines.

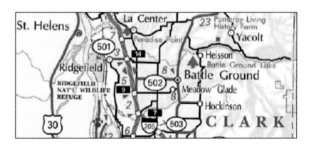

Figure 8.16   This section of the Washington State Department of Transportation highway map shows interstates, U.S. highways, state highways, and county roads by progressively narrower lines.

has to decide where to place the lines. Since they usually connect only a few places and are not related to the actual routes, the biggest challenge for mapmakers is to place them so that they do not overlap other flow lines.

Mapmakers may decide to add a different visual variable to flow lines that are made constant in width. Changes in amount of flow can be shown by varying the texture, color saturation, or color lightness of the lines. Say that a mapmaker wants to show how many artichokes are being freighted along a railway line from

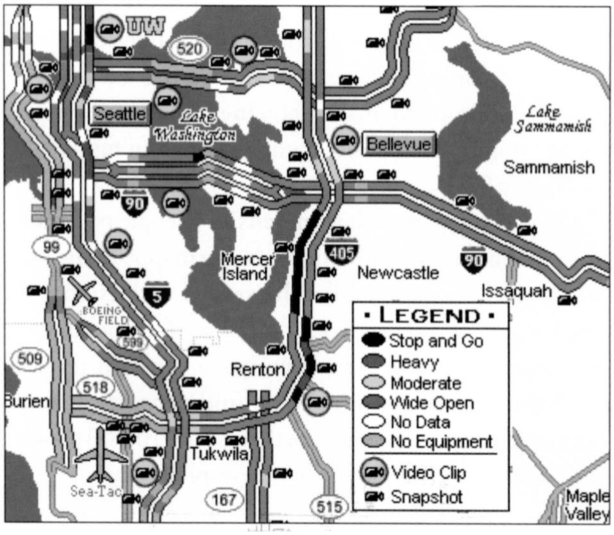

Courtesy of Washington State Department of Transportation.

Figure 8.17 This section of the Seattle traffic flow map from the Internet shows four ordinal levels of congestion from stop-and-go in black to wide open using three hues that connote different speeds.

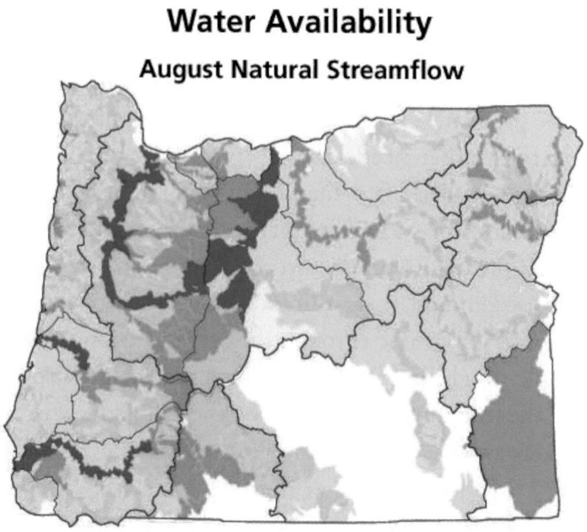

Water Availability
August Natural Streamflow

From W. G. Loy and S. Allan. 2001. *Atlas of Oregon.* 2nd ed. Eugene, Ore.: University of Oregon.

Figure 8.19 Watersheds are symbolized to show the water availability in August. Natural streamflow within each data collection unit is assumed to be homogeneous, with abrupt changes occurring at watershed boundaries. Although natural streamflow is not like this in reality, for mapping purposes these assumptions are made.

Figure 8.18 This flow map required the cartographer to decide where to draw the flow lines and how to symbolize them. On this map, the size of the line is proportional to the number of jobs provided to Oregon by foreign-owned companies.

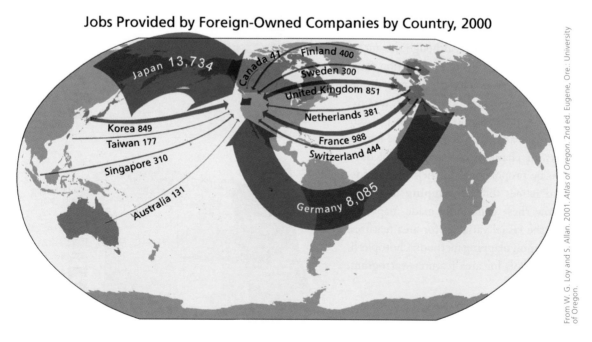

Jobs Provided by Foreign-Owned Companies by Country, 2000

From W. G. Loy and S. Allan. 2001. *Atlas of Oregon.* 2nd ed. Eugene, Ore.: University of Oregon.

Southern California to Chicago. One portion of the line could be shown three times darker than another to show that three times more artichokes are being shipped along it. The problem is that the lines have to be quite wide to make these differences clear.

When you read a flow map, focus your attention on the magnitude information, not on the flow line's precise location. Mapmakers often distort the basic reference geography to accommodate flow lines. A flow line showing the volume of ship traffic through the Strait of Gibraltar, for example, might be too wide to fit into the small space. Thus, the strait must be widened on the map; otherwise, it will look as if the ships are traveling over land.

Area feature maps

Quantitative area features vary in magnitude from place to place in patchwork-like fashion. For these themes, data collection areas have a constant magnitude within and abrupt changes at the boundaries. Some map themes related to human activity, such as tax rates by state, have a pronounced "stepped" character. More commonly, the quantity within each data collection area is conceived of as being distributed homogeneously so that it can be mapped, when in reality the quantity varies within each data collection area. The map in figure 8.19 is an example—the phenomenon being mapped (water availability) is more variable than the data collection units (watersheds) suggest. Population distribution is another example, since population patterns are a reflection of environmental and social factors, not of data collection areas. Population is not uniform within units and doesn't change abruptly at their boundaries, although the uncritical map user might get this idea from a population map.

We have seen that to show changes in magnitude on area feature maps, mapmakers vary the size, texture, lightness, or saturation of the symbol for the areas. Even more eye-catching maps are made by varying the height of individual areas, or changing the map area of data collection units to be proportional to their magnitude. There are standard names for the **mapping methods** associated with how these commonly made maps are created by varying the visual variables for area features. Let's look at five common mapping methods: choropleth, dasymetric, point symbols for area features, cartogram, and prism maps.

Choropleth maps

Many quantitative thematic maps of ordinal-, interval-, or ratio-level area data are made using the **choropleth mapping** method. The term comes from the Greek terms "choro" meaning place, space, or land, and "pleth"

meaning full, as in plethora. With this method, each data collection area is given a particular color lightness, color saturation, or pattern texture depending on its magnitude. Choropleth maps showing ratio-level Oregon county population density data are shown in figure 8.20 and figure 8.21.

On each map, **population density** (people per square mile) and not total population is being mapped. The mapmaker has **normalized** for the area of each county. In other words, the map is made to look as if the population is uniform throughout the county—a constant number of people per square mile. You'll find densities, percentages, rates (such as incidence of disease per 10,000 people), and other quantities on similar choropleth maps.

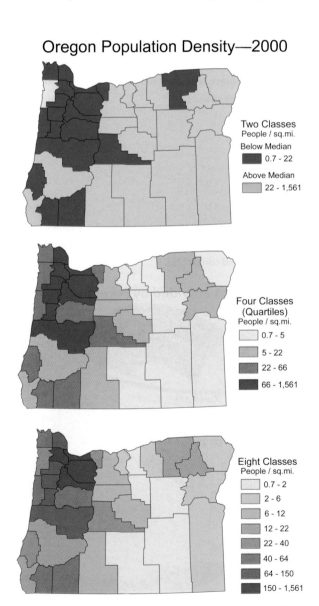

Figure 8.20 Choropleth maps of Oregon population density with two, four, and eight data classes show how the number of classes can change the appearance of the map.

Number of classes

Notice that the choropleth maps in figure 8.20 have two, four, and eight population density classes. You will likely see choropleth and other quantitative thematic maps having from two to eight classes for a theme. Progressive subdivision of the data range into classes simply involves reducing the numerical intervals between class limits. The simplest map has only two classes—above and below the median in our first example (figure 8.20, top). More information is shown if four classes (**quartiles**) or five classes (**quintiles**) are used. The map would show the

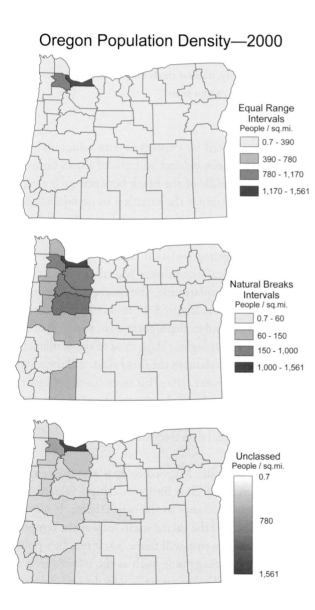

Oregon Population Density—2000

Equal Range
Intervals
People / sq.mi.

☐ 0.7 - 390
☐ 390 - 780
☐ 780 - 1,170
■ 1,170 - 1,561

Natural Breaks
Intervals
People / sq.mi.

☐ 0.7 - 60
☐ 60 - 150
☐ 150 - 1,000
■ 1,000 - 1,561

Unclassed
People / sq.mi.
0.7

780

1,561

Figure 8.21 Choropleth maps of Oregon population density that use equal range intervals (top), natural break intervals (middle), and the unclassed method (bottom) with brown tones proportional to county population densities show how the class interval selection method used can change the appearance of the map.

most information if the population density for each county were in its own class.

You can see that quantitative thematic maps made with different numbers of classes vary greatly in appearance. The number of classes has little to do with the nature of the mapped information; rather, you are looking at the mapmaker's arbitrary design decision. Depending on the number of classes used, the maps can look radically different, as illustrated in figure 8.20 and 8.21.

You may wonder how meaningful the map classes you see really are, particularly when the range of data values is divided into a small number of classes. Look again at the two-class population density map in figure 8.20. The "above median" and "below median" classes may tell you all you need to know, but the information content of the map is minimal. When only a few classes are used on the map, there's likely to be significant within-class variation that you can neither see nor assess.

Mapmakers increase the information content by using more classes, but this solution creates its own problems. Each additional class makes the graphic portrayal of the data more complex, as the eight-class map in figure 8.20 illustrates. When many classes are used, there is less within-class variability, but the visual complexity of the map makes it more difficult to read.

Class interval selection

The impression of magnitude variation you get from the map depends not only on the number of classes, but also on the method used to define the **class intervals** (the ranges for each numerical class). This is called the **classing method.** Let's look at the types of class intervals you are likely to see on choropleth and the other types of quantitative thematic maps discussed in this chapter.

Equal frequency intervals The three maps in figure 8.20 have two, four, and eight **equal frequency intervals,** in which the number of counties in each class are as close as possible to equal. The three maps give the impression of high population density throughout northwest Oregon and low population density in the southeast quarter of the state. Are these maps a realistic portrayal of the distribution of people in Oregon? To answer this question, you need to look carefully at the range for each class. Look again at the eight-class map. Values ranging from 150 to 1,561 people per square mile may not be particularly high population densities, since the average county density for the nation is a little under 100 people per square mile. The maps make much of Oregon appear too high in population density.

Equal frequency interval choropleth maps give the most faithful portrayal of the data when the range of values for each class is approximately the same. To have similar class ranges, the number of low, medium, and high values in the data must be about the same.

Equal range intervals A second way for the mapmaker to group quantitative data is by **equal range intervals** (figure 8.21, top). The range of data values is merely divided by the desired number of classes to obtain equal intervals. For example, dividing the 1,560 (1,561-0.7) Oregon population density range by 4 gives a constant interval of 390, or upper class limits of 390, 780, 1,170 and 1,561.

Equal range intervals are intuitively meaningful and easy to understand. Numerically constant intervals appeal to the same basic human data handling mechanism that makes percentage figures so attractive to us. Our minds are comfortable with the idea of segmenting numbers into equal fractional parts.

Equal range intervals produce the most meaningful map if an approximately equal number of data collection areas are in each class. For this to occur, data values must be equally distributed throughout their range. When data values are equally distributed across their range, choropleth maps made with equal frequency and equal range intervals should look identical.

The problem is that for many themes the data values are unevenly distributed across their range. Using Oregon population density as a typical example, there are many counties with a low population density and few with a high population density. In this situation, equal range intervals produce a strange map indeed. Most counties fall into one class, while some classes are empty, with no counties at all. Oregon population density appears to be uniformly low throughout the state, except for the two counties that contain Portland and much of its large suburban area. Although this map may be useful in showing the vast differences in county populations, it poorly communicates the actual variation in population density.

Natural break intervals Instead of defining classes according to equal data frequencies or ranges, the mapmaker may have established class limits at **natural breaks** in the distribution of data values. One way that mapmakers find natural breaks in a set of data is by creating a **frequency diagram (histogram)** for the data—a graphical display showing what proportion of features fall into each category.

An example of a frequency diagram is shown for Oregon population density in figure 8.22. The

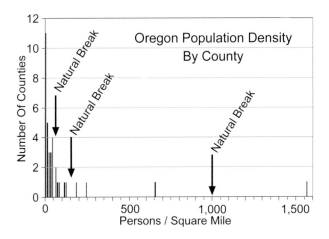

Figure 8.22 **On this frequency diagram of Oregon population density, natural clustering in the data and breaks between clusters can be seen.**

cartographer looked for clusters of data values with natural breaks between clusters and placed limits between classes in the middle of the break between each cluster. The idea is to minimize the variation in population density within each class while maximizing the variation between classes.

For the Oregon population density data shown in figure 8.22, natural breaks were placed at 60, 150, and 1,000 persons per square mile. The choropleth map made with these class limits (figure 8.21, middle) has the county containing Portland in the highest class, three suburban counties in the next highest class, most of the Willamette Valley south of Portland in the third class, and the rest of the state in the lowest class. This map, based on natural breaks in the data, shows the nature of population variation in Oregon better than the previous maps, which were based on equal data frequencies or ranges.

Critical value intervals Mapmakers may also use **critical values** to determine class limits. A critical value is one that has special relevance to the map's theme. It may be a physical aspect of the theme itself, such as the temperature below which a crop will freeze. Or it may be a politically defined dividing point, such as the income level at which counties fall below an artificially devised "poverty" line and are thus eligible for government assistance. If we defined median population density for Oregon as a critical value, the map at the top of figure 8.20 would be based on this critical value.

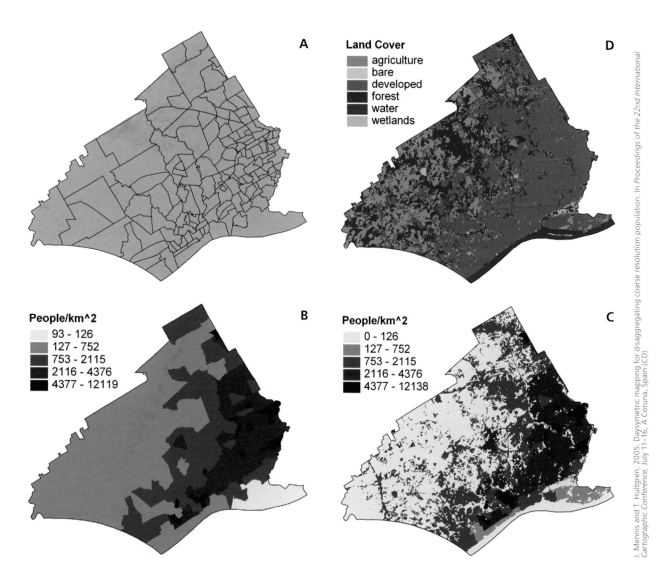

J. Mennis and T. Hultgren. 2005. Dasymetric mapping for disaggregating coarse resolution population. In *Proceedings of the 22nd International Cartographic Conference*, July 11–16, A Coruna, Spain (CD).

Figure 8.23 Dasymetric mapping is used to determine a more realistic representation of the population density in Delaware County, Pennsylvania (A). Remotely sensed data (D) were used to redistribute the population (B) to areas where the people were more likely to be living, then the densities were recomputed and the new areas were symbolized on the dasymetric map at the lower right (C).

Unclassed choropleth maps

On **unclassed choropleth maps,** each data collection area is given a lightness, saturation, or texture according to its magnitude. The unclassed choropleth map of Oregon population density (figure 8.21, bottom) has a continuum of brown tones ranging from white for the lowest density county to dark brown for the highest density. The intermediate population densities are linearly scaled between these two extremes. This map gives you an unbiased picture of Oregon population, since the mapmaker has not generalized the data by categorizing the values into a small number of classes.

The difficulty with reading unclassed choropleth maps is that unless the values are more or less evenly distributed

through the range of all values, the map will show most areas to be either high or low in value with only a few areas at the opposite extreme. The Oregon map is a classic example, with one county solid brown, one county medium brown, and the others a very light brown. The map looks similar to the equal range interval map, except that the discerning eye can see small variations among the light tones.

Dasymetric maps

On **dasymetric maps,** the mapped areas aren't political data collection units such as counties but rather areas of inherent homogeneity in the data. The idea is that each mapped area will have small internal magnitude variations, while there will be larger magnitude variations between mapped areas. In order for this to happen, the boundaries of the original data collection units are modified using related ancillary data and the attribute values are redistributed within the newly drawn units. As with choropleth maps, each homogeneous area can have a single value or an average value with little variation about the average for the area. The map of population density in Delaware County, Pennsylvania, provides a good example (figure 8.23).

In dasymetric mapping, the mapmaker starts with a map of the collection units. Ancillary data are used to help the cartographer adjust the boundaries of the units to more realistically represent homogeneous areas. Then the values are recomputed to reflect the addition or subtraction of values within the units.

As you might imagine, having access to different layers of information in a GIS greatly aids dasymetric mapping.

In the Delaware County example, the rural part of the county population is reassigned to the developed areas and the densities are recomputed. As a result, the population density in the western region decreases in the agricultural and forest areas, and in the eastern region the density decreases in the water and wetland areas. You can see that a dasymetric map with the homogeneous areas modified to reflect the data should give a more faithful representation of population density than a choropleth map showing the density class for each county.

Point symbols for area features

One of the more confusing quantitative thematic maps are those that use proportional point symbols to represent quantities within data collection areas. The problem is that you are led to think you're looking at data for a point feature when this isn't the case. The cartographer has placed the point symbol in the center of each data collection area and symbolized it using the same cartographic principles that apply to symbolizing point features (see *single theme point features* in this chapter).

The point symbols will be easiest to read when data collection areas are approximately equal in size on the map and when the range of data values is small. A map of California county population (figure 8.24) illustrates map reading difficulties when both the data range and the range of county areas are large. The map shows total county population with proportional squares placed at the center of each county. The very large population data range (from 1,500 to 9,000,000) results in a set of squares that range from barely visible to those covering several counties. Many of the squares extend past their county's

boundaries and overlap with the squares for neighboring counties. Reading the map is particularly difficult in the San Francisco Bay and Los Angeles regions, where small

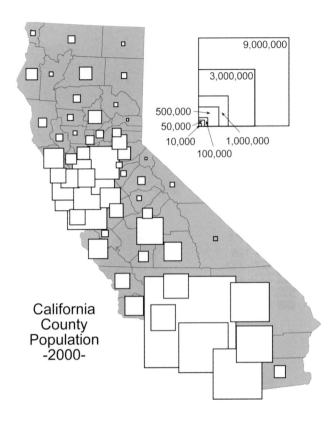

Figure 8.24   California population data from the 2000 census have been aggregated by county, with proportional squares placed at the center of each county to show the total county population.

counties and large populations have resulted in numerous symbol overlaps and county boundaries are hidden from view.

Cartograms

There may be times when you'd like to view an eye-catching portrayal of the relative magnitudes of area features rather than their exact spatial locations. At such times, you might turn to an odd-looking map called a **cartogram** or **area cartogram**. Mapmakers create cartograms by distorting the geographical size of data collection areas in proportion to their magnitudes. The size of each state or county, for example, might be made proportional to its population rather than its geographical area (figure 8.25). Although these maps are a little strange looking, they help overcome some of the problems that map readers have with point symbol and choropleth maps when there is a large range in the size of the areal mapping units.

To use a cartogram effectively, you must be able to compare sizes of data collection areas on the cartogram with the same areas shown on an equal area map projection (figure 8.25, top left). If the positions, shapes, and relative sizes of data collection areas in a region aren't familiar to you, a cartogram can be difficult to use because you'll need to compare two unfamiliar looking maps. For this reason, cartograms are most successful when they are used to map areal units that are familiar to the map reader, such as countries, states, and counties. This explains why "world by country," "country by state," or "state by county" cartograms are so common.

Ideally, when the size of the unit areas are made proportional to their magnitude rather than their geographical area, mapmakers retain as many spatial characteristics of conventional maps as possible. Preserving the shape of data collection areas will make it easier to compare a cartogram with a standard map. Preserving the **proximity** (nearness) and **contiguity** (boundary connectedness) to neighboring areas would also make the maps easier to compare. But shapes cannot be preserved without altering the proximity and contiguity of neighboring areas, and vice versa, so one or the other of these will be compromised in a cartogram. Let's look at these differences in cartograms.

Noncontiguous cartograms
Noncontiguous cartograms are the most commonly produced because they are the easiest to make. On these cartograms, the shape of areas is maintained at the expense of proximity and contiguity of neighboring areas. To create a noncontiguous cartogram for the population of California counties (figure 8.25, top right), the mapmaker enlarged or reduced each county in proportion to its 2000 population. The transformed data collection areas were then replaced as closely as possible in their relative geographical positions on the map. Preserving the shapes of data collection areas makes it easy to recognize and compare them with their counterparts on a conventional map. But as figure 8.25 also shows, proximity relations are only roughly maintained, and contiguity is sacrificed completely.

Pseudocontiguous cartograms
Pseudocontiguous cartograms appear contiguous at first glance but upon closer inspection turn out only to give the illusion of contiguity (figure 8.25, lower left). The transformed data collection areas share common boundaries, but the boundaries aren't the same as on a conventional map of the region.

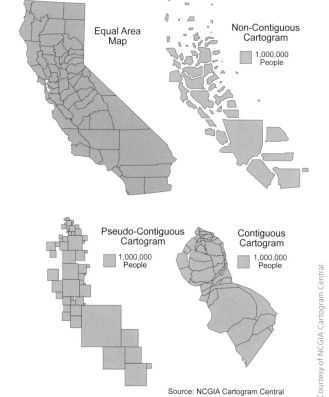

California County Population Cartograms
From 2000 Census Data

Equal Area Map

Non-Contiguous Cartogram

1,000,000 People

Pseudo-Contiguous Cartogram

1,000,000 People

Contiguous Cartogram

1,000,000 People

Source: NCGIA Cartogram Central

Courtesy of NCGIA Cartogram Central.

Figure 8.25 Noncontiguous, pseudocontiguous, and contiguous cartograms for California county population (based on 2000 census data).

A pseudocontiguous cartogram is made by transforming each data collection area into a simple geometrical shape proportional in size to the magnitude being shown. Rectangles are the shapes most commonly used. The new shapes are then arranged in what resembles their relative geographical position. When rectangles are used, the result is called a **rectangular cartogram.** By sacrificing shape and proximity, these cartograms can maintain a considerable degree of contiguity.

Pseudocontiguous cartograms may actually be no better than noncontiguous cartograms, and may even be worse because some map readers may believe them to be truly contiguous when in fact they aren't. The popularity of pseudocontiguous cartograms is probably better explained by the fact that they are easy to construct (only graph paper is needed) than by any map reading advantage they might possess.

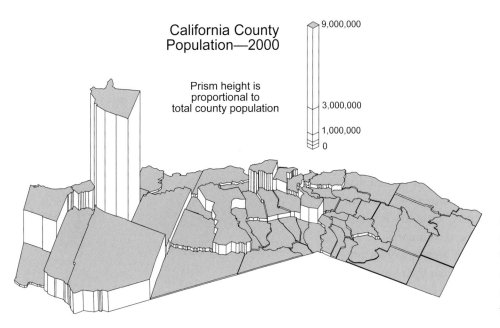

California County
Population—2000

Prism height is
proportional to
total county population

9,000,000

3,000,000

1,000,000

0

Figure 8.26 Three-dimensional prism map of California population by county. The height of each county is proportional to its total 2000 population.

Contiguous cartograms

Contiguous cartograms are by far the most interesting to look at. On these cartograms, the proximity and contiguity of neighboring areas are maintained, although this is accomplished at the expense of shape distortion. The shapes of regions are usually distorted in a subjective, apparently uncontrolled manner. Therefore, cartograms of the same theme created by different mapmakers would likely look dissimilar.

This lack of rigid geometrical mapping control is evident in the contiguous cartogram of California county population in figure 8.25, bottom right. Notice that while the shapes of some counties are fairly well preserved, other counties don't look like themselves at all. Yet, despite this shape distortion, the cartogram is effective; at only a glance, you can see precisely which counties have the greatest population. This would seem to indicate that the variable quality of shape preservation from one region to the next isn't always a major distraction. We appear able to accept a fair degree of shape distortion before areas become unrecognizable.

The biggest problem with contiguous cartograms is that they have in the past been difficult and time-consuming to make. To draw even a simple cartogram well requires a vast amount of labor, not to mention artistic talent. As a result, cartograms have been regarded more as a novelty than as valuable tools for environmental understanding.

But both problems—uncontrolled geometric distortion and difficulty of construction—have been solved by computer mapping software. It now takes little effort or artistic talent to create contiguous cartograms like

that for California population in figure 8.25 (and the other cartograms are even more straightforward to create). Although county shapes have been distorted, the distortion has been done in a recognizable way.

With increasing use of computer mapping software, cartograms of all types will become more available over time. In deciding whether to look at a cartogram or a conventional choropleth or point symbol map of a theme, you must decide if you prefer emphasis on the variation in magnitude among areas or on the region's geography. With conventional quantitative maps, you face the problem of decoding magnitude information from map symbols. Cartograms make these magnitudes obvious, but in distorting physical space, they force you to rely on your own familiarity with the geography of the region. The optimal solution, of course, is to use an equal area basemap, such as in figure 8.25, top left, as a reference for understanding the distorted geography seen on the cartogram.

Prism maps

A **prism map,** also called an **oblique stepped surface map,** is made by dividing the mapped region into data collection areas and showing the magnitude by varying the heights of the areas. Since the information to be mapped is usually collected in areas such as states, counties, or census tracts, it is easiest for mapmakers to divide the map into these same areas. Such a map for California population by county is shown in figure 8.26. Each county boundary has been raised vertically above the base level to a height proportionate to its total population. The resulting map looks like a three-dimensional, stepped surface.

Prism maps are visually impressive, but there are several problems with them. One drawback is that, although highs and lows are apparent, the exact height of the surface at a given location is difficult to determine. In addition, if the information that is mapped ranges greatly in magnitude, mapmakers may have transformed the data values to a more convenient mathematical form. For example, if the population of the United States by state were mapped as raw data, the few very populous states would be so much greater in magnitude than the majority of the low-population states that the map would show little magnitude variation between most states. But by mapping the square root of the state population, differences among the low-population states are exaggerated. Considering the impact of such data transformations on the appearance of a map, you should check the map legend and explanatory notes to see if the data values have been transformed by the mapmaker and, if so, in what manner.

Another potential drawback of prism maps is that they are drawn in oblique perspective. This means that the vantage point taken when the map was made is crucial to their appearance. A poor choice of viewpoint may cause important data collection areas to be obscured from view. For instance, the relatively low angle, east viewpoint used for the map in figure 8.26 caused two important counties, San Francisco and Santa Cruz (at the center of the top edge of the map), to be obscured.

Continuous surface maps

A fourth type of quantitative map shows a theme as a continuous surface. A **continuous surface map** is like a choropleth map in that it portrays the changing magnitude of some phenomenon from one place to another. It differs from a choropleth map in that the changes are gradual rather than abrupt. Temperature and landform elevation (figure 8.12) are examples of continuous surfaces that change gradually from place to place.

Continuous surface maps can be created from a sparse set of data values by interpolating between values in such a way that a smooth surface results. The average annual precipitation map in figure 8.9 was created by interpolating weather station precipitation data. Another example is a population surface map for a state created by interpolating between total population values placed at the centerpoint of each county.

Density distributions are also continuous surfaces. Rather than think of features as discrete entities, mapmakers can count how many are found in a small data collection area and assign that quantity to the center of each area. By performing this operation for all data col-

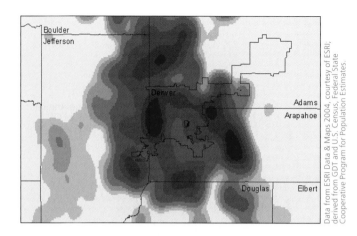

Figure 8.27 **The density of school children in the Denver area are mapped as a continuous density surface.**

lection areas, they create a continuous density surface (figure 8.27).

There are several ways to map a continuous surface. Many of the mapping methods were originally devised to portray the elevation surface—these are described in chapter 6. In this chapter we are focusing on the mapping of other types of continuous surfaces defined by other environmental or statistical quantitative data. The most important methods for mapping these continuous surfaces produce isoline (including isopleths and layer tinted isoline maps), dot density maps, and three-dimensional perspective maps. Let's look at each in turn.

Isoline maps

The most common method used to map continuous data surfaces is to connect points of a selected value with **isolines** (the prefix "iso" means equal). Isolines (also known as **isarithms**) have been given different names according to the type of information they show—**isotherms** are lines of equal temperature, **isobars** are lines of equal atmospheric pressure, **isohyets** are lines of equal precipitation, **isochrones** are lines of equal time difference (as shown in figure 8.30), and so on.

To create an isoline map, mapmakers must decide how many isolines to draw and how close together to place them. In other words, they must decide what the *interval* between successive isolines will be. Mapmakers can choose a **constant** or **variable interval** (see chapter 6 for more on intervals) for isolines. This choice will have a major effect on the map's appearance. It will also affect your map reading, so be sure to note whether the interval is the same across the map (constant) or whether it is larger or smaller in some places (variable).

In figure 8.28, constant interval isoline mapping has been used to show average annual hours of sunshine in the Pacific Northwest. Each isoline is labeled with its value so that regional variations in sunshine hours can be studied. A strong regional pattern is evident. Hours of sunshine at first decrease as you move from west to east across the region, with a rapid increase to the east of the Cascade Range.

In creating this map, the cartographers had to make a series of choices. They decided that 1,800 average annual hours of sunshine would be the lowest value shown, and they chose a constant isoline interval of 200 hours. Changing either of these factors will alter the appearance of the map. If more isolines were shown, for instance, you would see more detail than the seven isolines on the map, although the regional pattern of sunshine would remain the same.

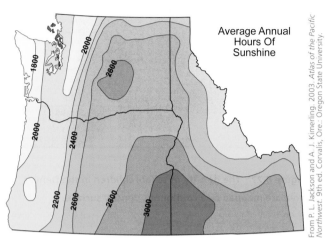

From P. L. Jackson and A. J. Kimerling. 2003. *Atlas of the Pacific Northwest.* 9th ed. Corvalis, Ore.: Oregon State University.

Figure 8.28 This isoline map shows average hours of sunshine received in the Pacific Northwest.

Isopleth maps

An isoline drawn across a density surface created from statistical data is called an **isopleth.** Isopleths look identical

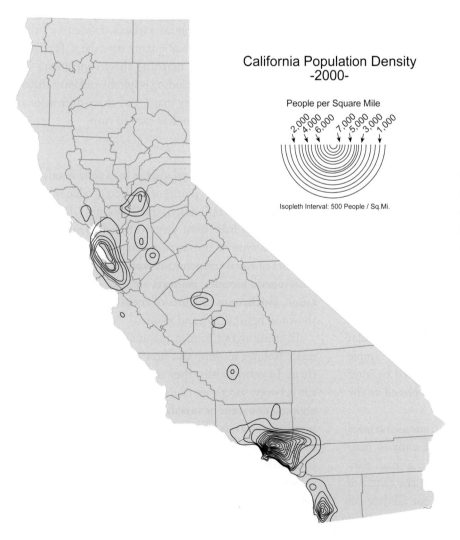

Figure 8.29 This isopleth map of California population density was created by drawing isopleth lines from a set of density values placed at the center of each county.

to standard isolines, but they differ in that they show a density or rate surface where the values can't physically exist at points. A common way to create the continuous surface for an isopleth map is to collect data by census units and then assign data value to the centroid (the center of each unit). Each value is divided by the census unit area, creating a density. The density values are then mapped using isopleths. The map in figure 8.29 is an example of an isopleth map—the population density of California was mapped by creating a continuous surface using data values at the centroids of the counties. You can tell this is an isopleth map because it's impossible, for instance, that 500 people per square mile can exist only along one line!

An isopleth map is at best an abstract, generalized representation of the data. Suppose, for example, that an isopleth shows the population density as 1,000 people per square mile. At a given position along the 1,000 isopleth, you may find no people, or you may find 5,000. The map is only intended to give a general impression of varying density over space. It is not to be analyzed location by location.

Layer tinted isoline maps

The way to read an isoline map is to ignore individual isolines and focus on the overall isoline pattern so that you can visualize the surface. Some people find this hard to do, however. To assist map users who have trouble focusing on the pattern rather than on separate isolines, some maps are made with a progression of gray tone or color lightnesses, color intensities, or textures added between isolines (see figure 8.28). This is called **layer tinting** (see chapter 6 for more on this technique). If the symbols are selected properly, you will see a magnitude progression from low to high, with the isolines seen as outlining different magnitude zones. The isolines often aren't labeled, which further encourages you to see the general pattern of highs and lows on the surface rather than to concentrate on individual isolines. If the lines aren't labeled, numerical range information for each, tint, intensity, or texture is found solely in the map legend.

One drawback of layer tinted isoline maps is that when only a small number of isolines are drawn, only a few tints will be used on the map and it will look highly generalized. Furthermore, even though a continuous surface is being mapped with isolines, the progression of tints may leave you with the false impression of a stepped surface or distribution. Fortunately, modern computer mapping software makes it possible for cartographers to create a larger number of isolines and a continuous lightness or saturation progression (see figure 8.9 as

an example). These modern layer tinted isoline maps enhance the impression of a continuous surface.

Dot density maps

Just as a series of isolines can be used to show a continuous surface, so can a set of point symbols. In fact, point symbols are the basis for one of the most effective ways of showing variations in density across a surface. The procedure is called **dot density mapping** (sometimes called *dot mapping,* but be careful not to confuse this with the point symbol mapping method in which each dot either represents the location of one feature or the magnitude of some phenomenon at a given position). What distinguishes dot mapping is that each dot represents greater than one feature, and the dots are then replicated to produce a varying dot density distribution.

When producing a dot density map, the map-maker repeats the same point symbol as many times as

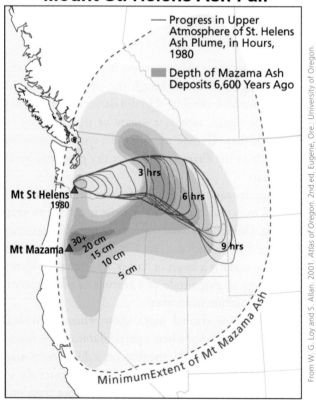

Figure 8.30 **Isochrones are used to map the progress of the Mt. St. Helen's ash plume over time. The depth of Mount Mazama ash is shown using isopachs, or lines of equal depth. Isopachs are enhanced by brown tints used to shade the areas between the lines. The layer tints help map readers to more easily see the continuous surface of ash depth.**

California Population
-2000-

Each dot represents
10,000 people

Figure 8.31 Dot density map of California population. Each dot represents 10,000 people in the surrounding area.

necessary to show the variations in density across the surface (figure 8.31). Although the circular dot is the most common symbol used in dot density mapping, any geometric figure, such as a square or triangle, may be used. The shape is irrelevant, since the symbol's meaning lies solely in the changes in dot density produced by its repetition.

An understanding of the difference between a point symbol map and a dot density map is crucial to map reading. On point symbol maps, each symbol represents only one feature, such as a city. Changing the form of the symbol—making it larger or darker, for instance—shows the change in magnitude of a feature or phenomenon from one location to another.

While point symbol maps show where individual features are located or where a particular feature is found with a given magnitude, the dots on a dot density map can't exist at the same locations as the features do in reality, because each symbol represents more than one feature. The aim of dot density maps isn't to give precise locational information but to present an image of changing density across the region. Thus, if the dot density map is well made, your eye won't be attracted to individual dots but to a general impression of changing spatial density. Not all dot density maps meet this criterion, however. The dots may be too large or too small in size, or too

few or too many in number. When this happens, you may receive a mistaken impression of the changes in density.

To create a dot density map, mapmakers first choose a **dot unit value** (the amount that each dot represents) and a **dot size** (the size of the symbols they will put on the map) for each dot. For example, they may decide, as they did for the California population map in figure 8.31, that each dot represents 10,000 people. Then they divide the number of features in the region by this unit value. The resulting number tells them how many dots to put on the map. Dot density maps usually include a legend, shown as a word statement, defining the unit value of a dot.

Dot density maps look simple, but they can be difficult maps to read if you want to determine exact values for particular locations. For one thing, as we've already seen, the human eye tends to under-perceive magnitudes. Psychological experiments have shown that as the density of dots increases, our estimates tend to fall below the actual density at an increasing rate. The result is that people viewing dot density maps typically get the impression that the range in dot density—and therefore the contrast in density from one region to another—is less than it really is. For this reason, you may find it necessary to compensate mentally for under-perception in your own density judgments to gain a true picture of the mapped distribution.

A well made dot density map will have variations in dot density that correspond to variations in magnitude within the data collection areas. Notice the within-county variations in dot density on the map in figure 8.31. To make this map, city population statistics were used along with maps showing the terrain, land use, and transportation features within each county to determine where people most likely lived. Dots placed according to this information are seen as smooth gradations from low to high population density. You should recognize this refinement of dot density mapping to be an interesting variation on the logic of the dasymetric technique discussed previously.

Incorrectly made dot density maps can be confusing, if not downright misleading. The more clustered the distribution, the more pronounced the potential discrepancy between reality and the dot density map. For example, a poorly created dot density map of California population might give the impression that population is scattered evenly within each county, with sharp breaks in density between counties. In reality, most of California's population is clustered within parts of the San Francisco Bay Area and Southern California counties. The classic mapping problem is San Bernardino County, the biggest county in California, which stretches from the eastern edge of the Los Angeles metropolitan area to the Arizona

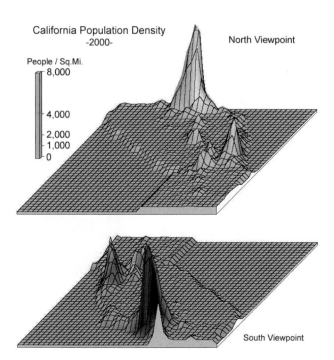

Figure 8.32 **3D perspective views of California population density from north (top) and south (bottom) viewpoints.**

and Nevada borders. If dots were placed evenly within the county, you would conclude that most of the Mojave Desert has a medium population density. As you see on the correctly made dot density map in figure 8.31, most of the Mojave Desert is unpopulated, and the vast majority of people live in the far southwestern corner of the county that's part of the Los Angeles metropolitan area.

3D perspective maps
A continuous surface can also be shown as a three-dimensional perspective map. If the mapmaker constructs closely spaced line profiles in two directions and in perspective view, you gain an impression not of individual lines but of a continuously varying 3D surface called a **fishnet** (figure 8.32). (See chapter 6 for more on fishnet maps.)

The fishnet map uses lines to depict the continuous surface. Notice how your attention is focused not on any one line or any one quadrilateral that is formed by the lines but on vertical undulations in the surface. It is actually the angle and length of sides of the quadrilaterals that depict the surface. Note, too, how differently this map shows the variability in California population density than does the prism map in figure 8.26.

Your ability to see all locations on the map is determined by the viewpoint and viewing angle selected by the mapmaker. In figure 8.32, the California population density surface is shown at a 30-degree angle above

the horizon from both a north (figure 8.32, top) and south (figure 8.32, bottom) viewpoint. Notice that different peaks and valleys in the surface are hidden from view on each map. Two or more maps are often required to see all parts of the surface depending on the data distribution. Animated 3D perspective maps are ideal for viewing the details of the continuous surface (see chapter 6 for more on these kinds of maps).

MULTIVARIATE MAPS

Sometimes mapmakers show more than one theme on a single **multivariate map** like figure 8.11. Mapmakers use four primary methods to show multivariate data on

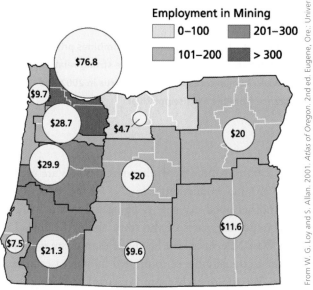

Figure 8.33 **By combining different mapping methods, two related variables—mining employment and production—can be shown on this one map.**

maps: combined mapping methods, multivariate point symbols, composite indices, and multiple displays.

Combined mapping methods

One way to show multiple quantitative themes on maps is to **combine mapping methods.** In figure 8.33, for example, two different mapping methods are combined to show related mining industry information. A choropleth map is used to show the number of people employed

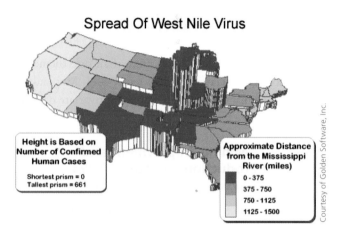

Figure 8.34 This multivariate map combines prism and choropleth mapping to show you the strong relationship between human cases of West Nile virus in 2002 and distance from the Mississippi River.

in the mining industry while point symbols are used to show mining production.

Another good example is figure 8.34, which shows two important parts of the story about the spread of the West Nile virus. The height of each prism is proportional to the number of confirmed human cases in each state. The top of each prism is colored according to its distance from the Mississippi River using a combination of two mapping methods. The map shows you the strong relationship between the incidence of the virus and proximity to the Mississippi River.

The advantages for map users of combining mapping methods are that the map is conceptually simple, useful for displaying a few variables (two or three), and useful for inspecting individual distributions. Limitations are that readability decreases as the number of variables increases and that it is difficult to convey the relative importance of the various types of information on the map.

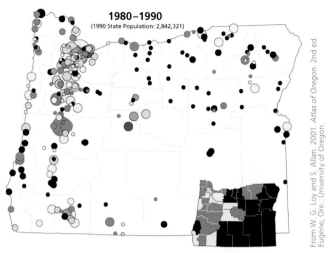

Figure 8.35 The growth rates of Oregon counties and cities are shown using color lightness and intensity variations within the circular point symbols. This relates to the population distributions, which are shown using symbol size.

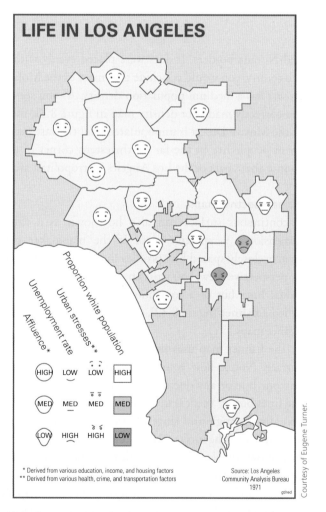

Figure 8.36 This map of life in Los Angeles shows four social status factors using glyphs called Chernoff faces.

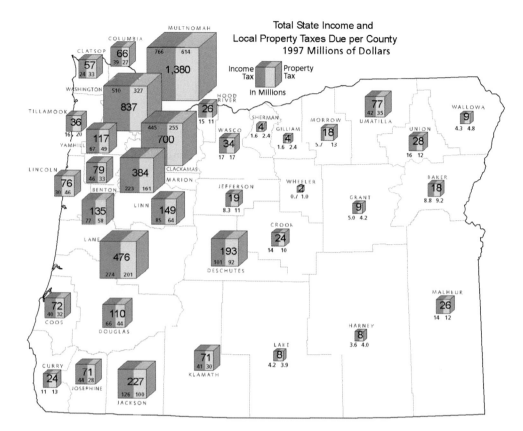

Total State Income and
Local Property Taxes Due per County
1997 Millions of Dollars

Figure 8.37 **Raw data values are mapped with segmented point symbols to show the taxes due by county for both income and property tax.**

Multivariate point symbols

Another method to show multivariate quantitative data on a single map is through the use of multivariate point symbols. Mapmakers can display two or more variables in a single point symbol, or they can segment a point symbol in order to show the relative magnitudes of subcategories of attributes for the features. Figure 8.35 illustrates the first approach of using a single point symbol to show two different types of quantitative information. Circle size represents population, while changes in color lightness and intensity within the symbol represent growth rate.

Maps can show several related categories of quantitative data with a single pictographic point symbol, sometimes referred to as a **glyph.** In the example in figure 8.36, the quality of life in Los Angeles is shown by different facial glyphs depicting four social status factors using a symbol (called a **Chernoff face**). The shape of the face shows levels of affluence; the mouth indicates unemployment rate; the eyes represent urban stress; and the tint of the symbol represents the proportion of the population that is white. The combined effect is so striking that you can almost sense how people in different sections of the city might feel. The reason you are so easily able to find similarities and differences is because it has been proven that humans can quickly, easily, and accurately read facial expressions.

The obvious advantage of looking at a single point symbol is that the one symbol simultaneously displays several variables, but there are visual limitations. Because the number of classes the human eye can distinguish is limited, the symbols are generally restricted to combinations of either two or three variables, with figure 8.36 as a rare exception. As noted above, humans can easily "disassemble" a human face into component parts (eyes, nose, mouth, etc.). This indicates that the best glyphs to use for multivariate point symbols should be easily recognizable shapes and icons.

Multivariate point symbols can also show magnitude information for categories within the attribute for a map theme. The total magnitude for each attribute is shown using point symbols that vary in size. Each symbol is then **segmented** (the parts show the relative magnitudes of subcategories of attributes) to show the different categories, often with different hues used to differentiate the categories. The different hues give you a clue that one of the variables mapped is qualitative and the other is quantitative.

You may see categories of a theme shown by bar graphs, pie graphs, or subdivisions of other regular geometrical shapes. One way this is done is to show the raw data for each subcategory. An example is the map in figure 8.37 in which the raw values for income tax and property tax

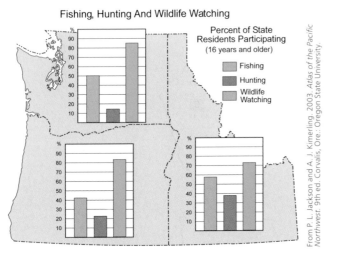

Figure 8.38 Bar graphs can be used to show categories of a theme such as major outdoor recreation activities. The graphs often extend outside the data collection areas because only the height of bars can be varied.

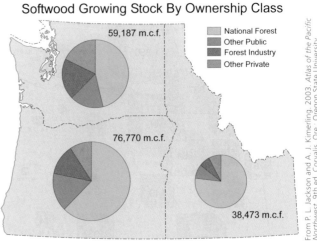

Figure 8.39 Pie graphs segment circles so that you can estimate the proportion of the total magnitude—softwood growing stock in this example—in categories such as landownership types.

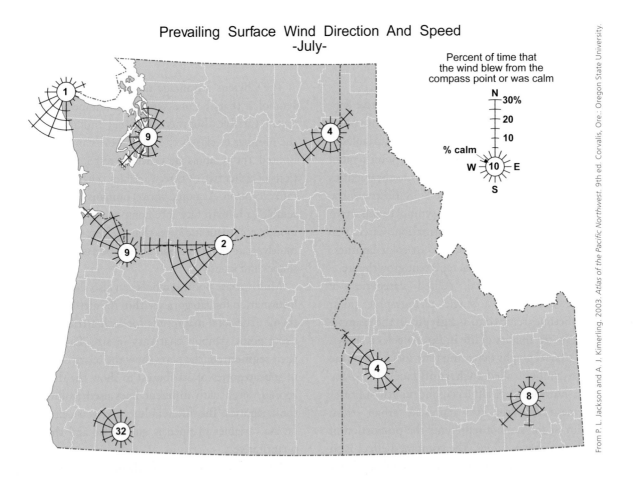

Figure 8.40 Multivariate maps can show categories of directional information, such as prevailing surface wind direction during July in the Pacific Northwest.

are shown in a single segmented cube symbol that reflects the total taxes due per county. Another way this is done is to subdivide each symbol into pieces or segments that show the proportions or percentages for subcategories (figure 8.38 and figure 8.39).

Another type of segmented multivariate point symbol shows you the percentage of time that a physical phenomenon such as wind comes from different directions. In figure 8.40 the wind roses show the July pattern of prevailing surface wind directions in the Pacific Northwest. You can see from the few weather stations shown on the map that certain areas (the Columbia River Gorge along the border between Oregon and Washington, for example) have a strong dominant wind direction relative to other parts of the region.

The use of segmented point symbols requires the ability to judge the relative length or area within each segment. Adjacent vertical bars are somewhat easier than segmented circles for most people to read and analyze with precision, because bars require only the estimation of relative height. The problem with bar graphs is that they often extend outside their data collection areas because only their height can be varied.

One advantage of multivariate point symbols is that they are good for the inspection of individual variables. It is best if the map reader is familiar with the form of point symbol that is used, such as pie charts, bars, wind roses, or human faces. A disadvantage is that it may be difficult to estimate and compare proportions, especially if multiple visual variables are used. Furthermore, the **visual field effect**—the modification of a symbol's appearance by nearby symbols—can alter the perception of a symbol. And it may be difficult to compare parts of symbols that are widely separated on the map, especially if there are

Figure 8.41 **This map shows categories of California city safety as defined by a safety index based on weighting the incidence of six basic crime categories.**

Data courtesy of Morgan Quitno Corporation.

many intervening symbols. Nonetheless, you may find these maps compelling, so you may be more inclined to spend time reading and analyzing them.

Composite indices

Composite data rather than raw data for each theme also can be shown. On **composite index maps,** several data variables are combined into a single numerical index. These maps are sometimes called *cartographic modeling* or *composite variable maps.*

More complex composite indexes involve the weighting of several factors. The categories of California city safety shown on the map in figure 8.41, for instance, involved creating a safety index by weighting the incidence of six basic crime categories. The data for the map were taken from the 2000 Morgan Quitno national awards for 322 cities. The six basic crime categories—murder, rape, robbery, aggravated assault, burglary, and motor vehicle theft—were used in a formula that measured how a particular city compared to the national average for each crime category. The outcomes for the six categories were weighted equally to obtain the safety index. The index values for the 80 cities on the map were sorted from low to high so that quartiles could be determined. The quartiles were given the names safest, safe, less safe, and least safe, and the city circles in each class were mapped by differences in gray-tone lightness.

At first glance, these circles look no different from circles showing a simple ratio—or even from circles showing only the size of cities. To completely understand what the circles show, you need to look carefully at the

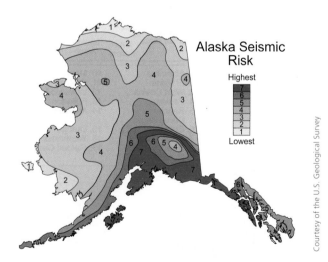

Figure 8.42   A seismic risk map for Alaska shows areas of lowest to highest risk. The risk index used takes into consideration a number of physical factors which, when combined, cause earthquake damage.

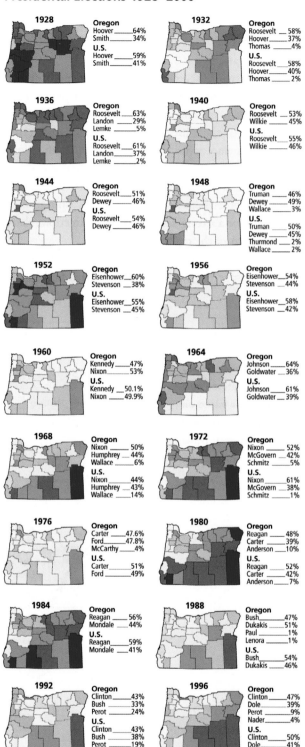

Presidental Elections 1928–2000

Figure 8.43   Small multiples are used to show this series of maps of the presidential election results in Oregon from 1928–1996.

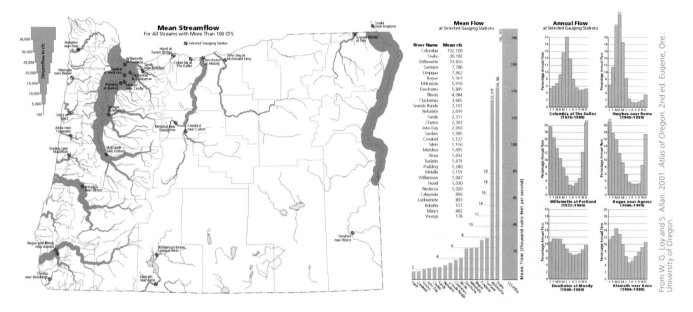

Figure 8.44 Complementary format displays are used to show the spatial (map) and temporal (graph) characteristics of streamflow in Oregon.

legend and explanatory notes on the map. For example, only California cities greater than 75,000 in population that reported the crime rate in all six categories appear on the map. In addition, the factors used to define the index and the weighting of factors may not be what you expected. You may have wanted to see a map on which murder and rape were weighted much higher than automobile theft, for example. You also may have been looking for a map that included additional city safety factors, such as earthquake or flooding frequency.

Multivariate maps of continuous areal phenomena area features commonly show ratios or indices rather than raw data. For instance, the seismic risk map for Alaska shown in figure 8.42 takes into consideration a number of physical factors that cause earthquake damage when combined. These factors are weighted and summed to give a seismic risk index that is used to define the seven risk categories from low to high, as shown on the map.

One advantage of composite index maps is that they are good for distinguishing patterns among variables. A limitation for the mapmaker, which may affect the quality of the map, is that it requires spatial data for each variable for the full extent of the study area. Often mapmakers are reduced to using data that are available rather than data that are appropriate, so it behooves you to study any notes that describe how the map was made.

Multiple displays

Another method mapmakers use to display multivariate data is **multiple displays.** Multiple displays can be generated in either constant or complementary formats. **Constant format displays,** as in figure 8.43, show a series of displays with the same graphic design structure that are used to depict changes in magnitude or type from multiple to multiple (map to map). This method is sometimes called **small multiples.** With this method, the consistency of design assures that your attention is directed toward changes in the data.

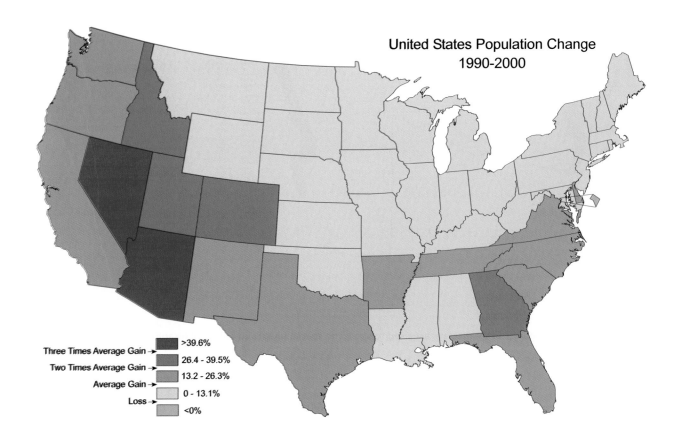

United States Population Change
1990-2000

Three Times Average Gain → >39.6%
Two Times Average Gain → 26.4 - 39.5%
Average Gain → 13.2 - 26.3%
0 - 13.1%
Loss → <0%

Figure 8.45 **This temporal change map shows the percent change in the conterminous United States by state between 1990 and 2000.**

A **complementary formats display,** as in figure 8.44, is the other multiple displays mapping method. With this method, the mapmaker combines maps with graphs, plots, tables, text, images, photographs, and other formats for the display of data.

One advantage of multiple displays is that they are better for comparing datasets than distinguishing among datasets, especially if there is a lot of complexity in the displays. If small multiples are used, they should be comparable, multivariate, small, high-density graphics that are based on a large data matrix and used to show shifts in relationships between variables. A limitation, especially for complementary format displays, is that the effectiveness is dependent to some degree on your ability and aptitude to understand each format.

QUANTITATIVE CHANGE MAPS

Quantitative change maps show the increase or decrease in an attribute value over a specified time period. This population change map in figure 8.45 was made by finding the percent change in state populations between 1990 and 2000 as reported by the Census Bureau. If a state had 10 million people in 1990 and 11 million 10 years later, the ratio would be $^{11}/_{10}$ million, a 10 percent increase.

You have to be careful when reading this type of map, because an equivalent percentage of change doesn't mean that the same number of features, people in this case, have been added or lost. If a populous state such as California increases its population by 15 percent, it is gaining far more people than if a sparsely populated state such as Nevada adds 50 percent to its population.

Quantitative change maps also can show **cyclical phenomena,** or phenomena that recur regularly. The climate of a region, for instance, can be described by a map composed of annual graphs showing the yearly cycle of temperature and precipitation at different locations (figure 8.46). The six annual graphs for Washington allow you to compare seasonal variations at the

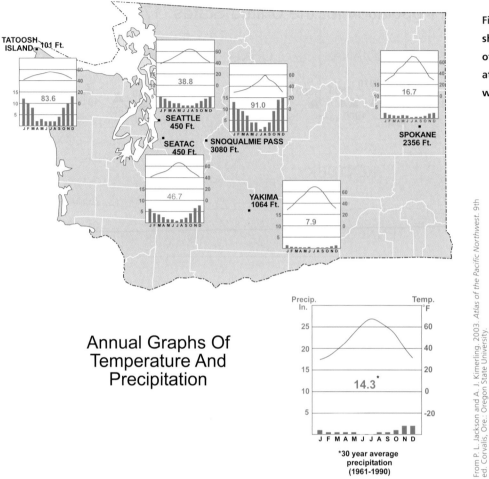

Annual Graphs Of Temperature And Precipitation

From P. L. Jackson and A. J. Kimerling. 2003. *Atlas of the Pacific Northwest*. 9th ed. Corvalis, Ore.: Oregon State University.

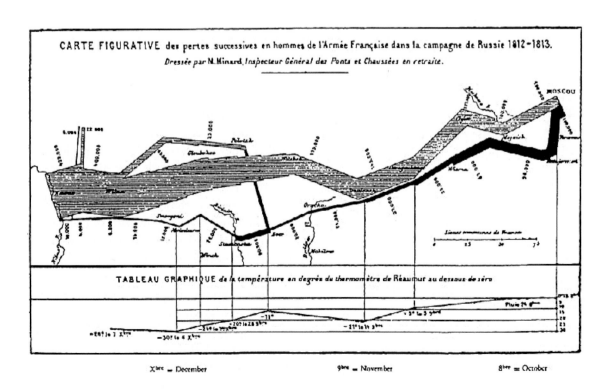

Figure 8.47 **One of the most famous quantitative time change maps was made in 1861 by Charles Minard. A sequence of flow lines is used to portray the disastrous losses suffered by Napoleon's army in the Russian campaign of 1812.**

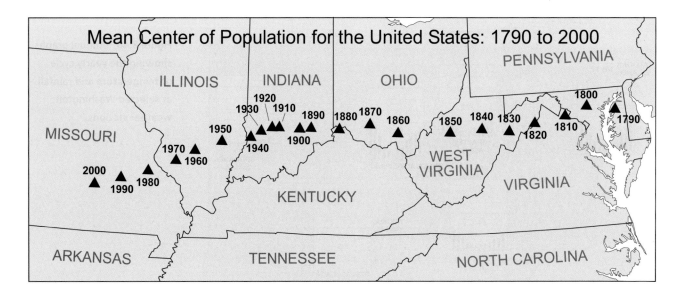

Figure 8.48 This map shows the continual westward shift in the mean center of population for the United States from 1790 to the present.

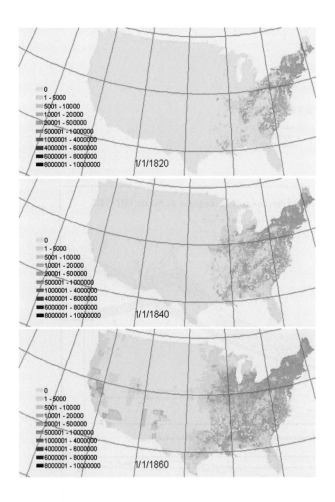

Figure 8.49 Time series maps of population density change in the United States 1820–1860.

different weather stations, or to study seasonal variations at a single station. Notice, for instance, the low values of monthly precipitation and the greater range of monthly temperature in the eastern half of Washington.

Time composite maps

Another way to show change over time is by superimposing data for several dates on a **time composite map.** You can then see exactly what changes have taken place over the time span covered by the map. Time composite maps are often used to trace the path along which some quantitative feature moved. One of the most famous quantitative time change maps was made in 1861 by **Charles Minard,** a French pioneer of the use of graphics in statistics and engineering. The map uses a sequence of flow lines to portray the disastrous losses suffered by Napoleon's army in the Russian campaign of 1812 (figure 8.47). Beginning at the Polish–Russian border, the gray flow line shows the size of the army at different positions and dates. The path of Napoleon's retreat from Moscow in the bitterly cold winter is depicted by the black flow line, below which is a graph of minimum temperature.

Time composite maps can also show the change in position of a geographical measure over time. A classic example is a map showing the position of the **mean center of population** as determined by the U.S. Census Bureau after tabulating the results of each census (figure 8.48). The Census Bureau describes this as "the point at which an imaginary, flat, weightless, and rigid map of the United States would balance perfectly if weights of identical value were placed on it so that each weight represented the location of one person on the date of the census."

Each triangle in figure 8.48 represents the population center of the United States calculated from each decade's census. The map shows the location of the population center calculated for each decade from 1790 to 2000. You can see that during the twentieth century, the mean center of population shifted southwest, from southcentral Indiana to its position in 2000 in southcentral Missouri.

Time series maps

A series of choropleth, prism, or other maps can used to show changes over time for the data collection areas on the map. In the three maps showing western U.S. population density at 20-year increments from 1820 to 1860 (figure 8.49), the choropleth method was used to create these visually striking **time series maps.** The three maps are actually snapshots taken from an animated map showing the continuous change in U.S. population density in 10-year increments from 1800 to 2000. You are likely to see an increasing number of animated quantitative maps as movie clips that can be downloaded from the Internet.

Another way to map a time series is by giving the impression that maps are being overlaid vertically while still allowing more than one map to be seen at the same time. This is an effective way to show a set of georeferenced surfaces. The 3D perspective views of the two precipitation surfaces for average December and June precipitation surfaces for western Oregon in figure 8.50 are shown from the same viewpoint and on the same vertical scale. Drawing a line vertically downward from the December surface shows you the corresponding point on the June surface. You can easily see that the amount of precipitation in western Oregon is much greater in December, and that the overall geographical pattern of precipitation is similar in December and June.

Attribute change maps

It is also possible to show a change in magnitude over time. Say you wanted to see the location of earthquakes of different magnitude. You could review a set of maps on which earthquakes in different magnitude classes are shown sequentially. This would allow you to see the location of the earthquakes relative to magnitude rather than time. Although you don't often see these types of maps they can be useful and may prove to be interesting and appropriate for some types of data.

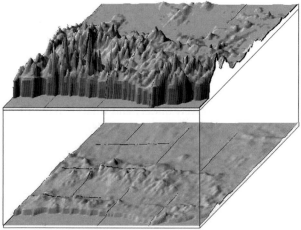

Western Oregon Seasonal Precipitation

Average December Precipitation

Average June Precipitation

Figure 8.50 December and June precipitation surfaces for western Oregon have been overlaid to show the large seasonal differences in rainfall.

DYNAMIC QUANTITATIVE CHANGE MAPS

As with qualitative change maps, a disadvantage of quantitative change maps is that the maps are discrete and static even though what is being mapped is dynamic. Displaying the phenomena in a dynamic fashion instead would allow you to intuitively understand what is changing on the maps. The same challenges apply as for dynamic maps showing qualitative data. You may find these hard to read if you do not have display or replay control, for example, allowing you enough time to see the maps and to review them so you can more fully understand them. Prominent legends and clear time passage indicators are a must. In most cases, these displays are pleasing to the eye and, if presented correctly, can help you better understand the dynamic nature of what is being mapped.

SELECTED READINGS

Bertin, J. 1983. *Semiology of graphics.* Madison, Wisc.: University of Wisconsin Press.

Dent, B. D. 1996. *Cartography: Thematic map design.* 4th ed. Dubuque, Iowa: Wm. C. Brown Publishers.

Jackson, P. L., and A. J. Kimerling. 2003. *Atlas of the Pacific Northwest.* 9th ed. Corvallis, Ore.: Oregon State University Press.

Mitchell, Andy. 1999. *The ESRI guide to GIS analysis –volume 1: Geographic patterns and relationships.* Redlands, Calif.: ESRI Press.

Robinson, A. H. 1967. The thematic maps of Charles Joseph Minard. *Imago Mundi* 21: 95–108.

Robinson, A. H., et al. 1995. *Elements of cartography.* 6th ed. New York: John Wiley & Sons, Inc.

Slocum, T. A., and S. L. Egbert. 1991. Cartographic data display. In *Geographical information systems: The microcomputer and modern cartography.* ed. Taylor, D.R.F., ed. New York: Pergamon Press.

Tyner, J. 1992. *Introduction to thematic cartography.* Englewood Cliffs, N.J.: Prentice-Hall.

Tufte, E. R. 1983. *The visual display of quantitative information.* Cheshire, Conn.: Graphics Press.

——. 1990. *Envisioning information.* Cheshire, Conn.: Graphics Press.

Wesson, R. L., et al. 1999. Probabilistic seismic hazard maps of Alaska. *U.S. Geological Survey Open-File Report 99–36.*

Wright, J. K. 1944. The terminology of certain map symbols. *The Geographical Review* 34 (4):654–55.

U.S. Census Bureau. http://ask.census.gov.

chapter
nine **IMAGE MAPS**

9

Image maps

Although people rely primarily on their eyes to learn about the environment, indirect experience is also important. Reports of distant places have traveled by word of mouth and other forms of communication since the beginning of humankind. But such methods of gathering data remotely provide information for only specific sites. In the last century, the significance of collecting images of the earth (and other planetary bodies) from a distance, called **remote sensing,** has grown enormously. We are now inundated with a vast array of remotely sensed images of our surroundings.

Remote sensing lets us observe features in the environment by using films or recording instruments **(sensors)** that are sensitive to the energy emitted or reflected from objects. Remote sensing opens mapping of both visible and invisible aspects of our environment. **Images** are produced from remote sensor data. In the past, the term "photo" was used instead of "image." But with today's technology, many images that look like photos really aren't. Thus, the more general terms "image" and "image map" are more appropriate.

Some images come directly from the sensor system. Others represent computer manipulations of the energy recorded by the sensor. These creations may contain artifacts characteristic of the photo-chemical or electronic processes involved. Many factors can influence the appearance of the resulting images, including the sensor vehicle's vantage point, the sensor's spectral sensitivity, the image's spatial resolution, the sensing instrument's technical quality, and atmospheric conditions.

Although remote sensor images are excellent at showing many aspects of the landscape, they may fail to depict others. Intangible features, such as political boundaries, aren't picked up on images unless physical features in the landscape happen to follow them or they happen to align with physical features. Such useful aids as geographical names, map scale, and reference grids are absent from raw images. Features on raw images aren't classified and identified in a key or legend. For these reasons, remote sensor images are often made more interpretable and useful by cartographically enhancing them with symbols, such as lines, words, numbers, and colors. These symbols are laid over the image base, produc-ing an image map. The **image map** may then be draped on a 3D terrain model. This chapter shows the ways in which images for these maps can be obtained. We'll look first at large-scale image maps created from **aerial photographs** (photographs of the ground taken from an aircraft, satellite, or other remote platform) that have been digitally processed into geometrically corrected **orthophotographs** (imagery in which distortion from the camera angle and topography has been removed).

AERIAL PHOTOGRAPHY

The development of cameras and photographic films began in the early 1800s. At first, the visible light portion of the spectrum was imaged as shades of gray. Later photographic films not only came close to duplicating the color-sensing capability of the human eye but also extended our imaging capability into the near-ultraviolet and near-infrared portions of the spectrum at both ends of the eye's sensitivity range (figure 9.1). In addition to broadening our image of the environment, these different films, when combined with filters for blocking out unwanted wavelengths, made it possible to image specific bands of visible light energy, like blue, green, or red, as well as the near-visible light energy, such as ultra-violet and infrared.

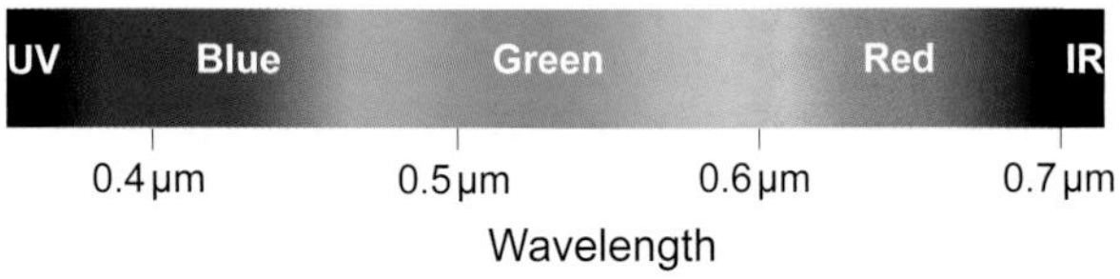

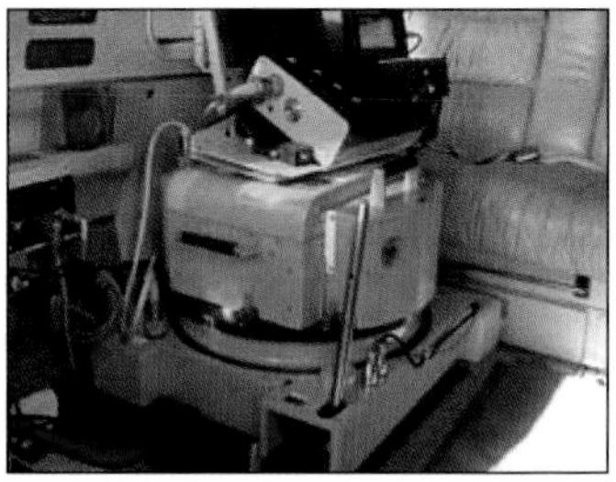

Figure 9.2 A typical aerial mapping camera placed vertically in an airplane.

Courtesy of the Washington State Department of Transportation.

Figure 9.1 Near-ultraviolet to near-infrared portion of the electromagnetic spectrum.

These advancements in imaging capability were coupled with the development of **platforms** (the vessels, crafts, or instruments from which the images are taken) that could be used to remotely image the environment. Airplanes, towers, balloons, and even birds have been used as remote sensing platforms. When the image is taken from above the ground with a camera, the result is an **aerial photograph,** or air photo for short.

What is imaged on an air photo depends on a number of factors. One is the type of camera used. In taking air photos for mapping purposes, photographers usually use the familiar kind of camera that produces individual pictures called **frames. Aerial mapping cameras** (figure 9.2) are specially designed so that they will expose large 9-inch-by-9-inch (23-centimeter-by-23-centimeter) frames on photographic film. Larger film makes it possible to image a larger area in detail on the photo. Complex, expensive camera lenses are used to minimize geometric distortion due to lens defects. The mapping camera is placed in a **gyrostabilized mount** (one that stabilizes side-to-side motion) that allows it to be pointed vertically downward at the ground.

Digital mapping cameras that record pictures electronically rather than on film are rapidly replacing traditional aerial mapping cameras. The digital mapping camera is large and uses the same high-quality lenses, but the imaging is done on a two-dimensional charge coupled device (CCD) array, not photographic film. The CCD device is similar to that used in your own digital camera, except that the array has thousands more rows and columns. Rolls of exposed film are replaced by a digital image memory hundreds of gigabytes in size that is capable of storing more than a thousand black-and-white or color **digital images** composed of pixels with numbers representing gray-scale or color shades equal in quality to air-photo frames on photographic film.

Large-scale air photos are obtained by flying the aircraft at around 10,000 feet (3,000 meters) above the ground along **flight lines** (paths that the aircraft follow), typically north–south or east–west (figure 9.3). Photos are taken along the flight line so that there is a 60-to-80-percent **overlap** (duplicated image of the ground in two successive air photos). Overlap between photos within a flight line is called **forwardlap.** Typically more than a single flight line is required to cover the area to be mapped, and adjacent flight lines are planned with a 20-to-30-percent **sidelap** to ensure that there are no gaps in the coverage.

Black-and-white photography

Conventional black-and-white aerial photography negatives are based on a film emulsion that records electromagnetic radiation in the 0.3 to 0.7 μm visible spectrum. Because of its sensitivity to visible light, black-and-white film is often called **panchromatic** (meaning "all-colors") film. Unfortunately, the shorter (0.3 to 0.4 μm) **near-ultraviolet (near-UV)** wavelengths are scattered by the atmosphere, requiring that a **UV haze filter** be placed over the camera lens to increase the clarity of the photo. These filters correct for UV effects (which cause images to look bluish and other colors to be modified) and eliminate haze (caused by dust particles in the air) from the photos. Positive photographic prints, as shown in figure 9.4, are made from the film negatives, or nowadays

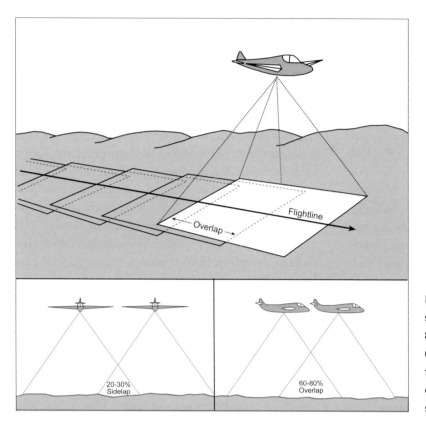

Figure 9.3 Adjacent air photos are taken so as to ensure duplicate imagery in 60 to 80 percent of the area along the flight line (forwardlap) and 20 to 30 percent between flight lines (sidelap) to ensure complete overlapping ground coverage (sometimes simply referred to as overlap).

from the digital aerial photography data for each frame. The greater the amount of visible light gathered by the camera lens onto the film emulsion, the lighter its tone on the final photographic print. Soil moisture content, surface roughness, and the natural color of objects all influence a feature's tone on a black-and-white photo. Moist soils, marshlands, and newly plowed fields tend to be darker than surrounding features, while human constructions such as roads and buildings tend to appear lighter.

Standard black-and-white photos have been put to many uses. The U.S. Forest Service uses them in making timber inventories and in mapping national forests. They are used by foreign mapping agencies and the U.S. Geological Survey (USGS) in the production of topographic maps in their quadrangle (1:24,000) series. The U.S. Natural Resources Conservation Service uses them as the base for mapping soils and agricultural activities. They are also used widely in transportation, recreation, and land-use planning purposes. Indeed, by far the largest amount of remote sensing has been done using standard black-and-white photography. This type of aerial photography is also used most often to create image maps.

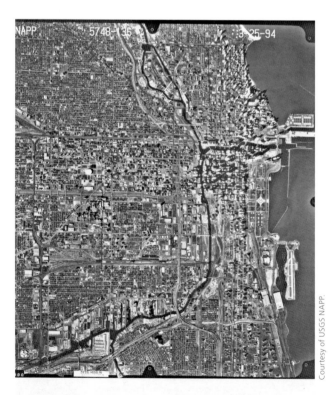

Figure 9.4 Photographic print (reduced from the original) made from a 9-inch-by-9-inch, black-and-white negative air photo of downtown Chicago, Illinois.

True-color photography

For a long time after its development in the 1930s, true-color film was little-used in mapping. For one thing, color film is more expensive than conventional black-and-white film. This is because it contains three separate emulsion layers sensitive to the blue, green, and red portions of the visible spectrum. Special film processing requirements add further to the cost. In addition, early color film had poor resolution capabilities.

Image clarity of color film has improved so dramatically in recent years, however, that it is comparable in quality to black-and-white film. In fact, color photos are usually easier to read than black-and-white photos because they capture the colors uniquely associated with special landscape features, such as the terra cotta color of the tile rooftops in figure 9.5. It is easier to distinguish subtle differences in colors than shades of gray.

Color is especially useful in revealing the condition of objects, such as the stage of a crop in its maturation cycle. For such applications as vegetation and soils classification, geologic mapping, and surface water studies, using color photos has proven simpler and more accurate than working from equivalent black-and-white images alone. Despite its extra cost, true-color film is rapidly growing in importance for understanding the environment. Of course, with digital cameras, the extra cost associated with color images is primarily related to printing.

Color infrared photography

During World War II, military researchers developed a special film that was sensitive to **near-infrared (near-IR)** wavelengths (0.7 to 0.9 μm) as well as to visible light (figure 9.6) called **color infrared (CIR) film.** One of the highest reflecting materials in the near-IR wavelength region is the cellular structure in the leaves of plants. Generally speaking, the healthier the vegetation, the higher the reflectance. This property turned out to be extremely useful for the military, since on a photograph produced with CIR film, artificial camouflage materials could be distinguished from live, healthy vegetation. Because CIR film is used for this purpose, it is sometimes called **camouflage detection film.** It is also known as CIR film because the spectral sensitivities of the dye layers in the film bear no relation to the natural colors of environmental features—the blue, green, and red emulsion layers record green, red, and near-IR radiation, respectively.

An environmental feature that absorbs a lot of near-IR energy, such as clear water, appears black on CIR photos. Features such as buildings and unhealthy vegetation absorb less near-IR energy and appear blue or blue gray. The most obvious feature on CIR photos is healthy vegetation, which appears bright magenta red rather than green due to its high near-IR reflectance. The cut vegetation, green paint, and rope netting used to conceal military installations are recorded in pinkish to bluish tones,

Courtesy of the North Carolina Department of Transportation.

Figure 9.5 **True-color air photo print, reduced in size from the 9-inch-by-9-inch original, of central Raleigh, North Carolina.**

Courtesy of USGS NAPP.

Figure 9.6 **CIR air photo print, reduced in size from the 9-inch-by-9-inch original, of central San Diego, California.**

in stark contrast to the background of bright reds produced by the surrounding healthy vegetation.

Although the first important applications of CIR film were in military reconnaissance, many other uses have been found. For vegetation studies, the film is now used for such mapping applications as crop inspection, tree growth inventories, and damage assessment of diseased flora. Plant diseases can often be detected on CIR photographs well before they would be visible to the unaided eye. Geologists have found CIR photos useful, too, in mapping near-surface structural features such as faults, fractures, and joints. These features can be detected because they often collect water, encouraging lusher vegetation growth than in the surrounding area. CIR film also enhances boundaries between soil and vegetation and between land and water, making it useful in mapping these features. In addition, the film is valuable for urban mapping because it shows a sharp contrast between vegetation and cultural features, and because the near-IR wavelengths penetrate smog easily.

Low- and high-altitude photography

The most detailed images are **low-altitude air photos** taken anywhere from just above the ground to around 1,500 feet above the surface. Towers and balloons provided convenient sensing platforms in the 1800s, but these devices were superseded in the twentieth century by light aircraft, particularly helicopters. These low-altitude photos (figure 9.7) usually cover a small ground area at a large map scale (see chapter 2 for more on map scale).

Acquiring low-altitude photographs at altitudes from 1,500 to 10,000 feet (500 to 3,000 meters) provides less environmental detail since the image scale is much smaller. But this type of low-altitude aerial photography has the

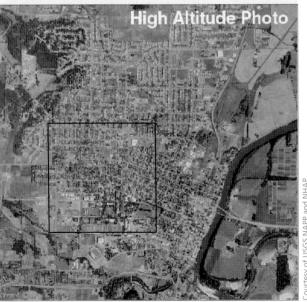

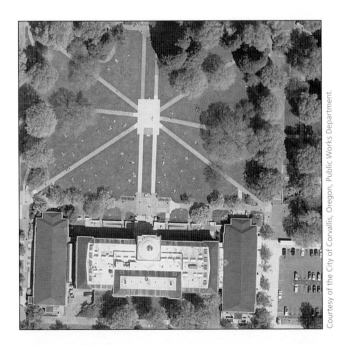

Figure 9.7 Low-altitude black-and-white air photo of the Memorial Union quad at Oregon State University, Corvallis, Oregon, taken from an altitude of around 1,000 feet (300 meters). Notice the students on the grass in the quad.

Courtesy of the City of Corvallis, Oregon, Public Works Department.

Courtesy of USGS NAPP and NHAP.

Figure 9.8 Portions of low-altitude (top) and high-altitude (bottom) black-and-white air photo of Corvallis, Oregon, (both reduced in scale from the original) taken at altitudes of around 10,000 and 40,000 feet (3,000 and 13,000 meters). The black box drawn on the high-altitude photo outlines the area covered by the low-altitude photo.

advantage that more ground area can be covered on a single photo. Black-and-white aerial photography of the type used to obtain basic elevation data for topographic mapping is usually taken from an altitude of 10,000 feet, almost 2 miles (3 kilometers) above the ground (figure 9.8, top). These photos are available in a standard 9-inch-by-9-inch frame format. Systematic coverage of the United States in this format began in the 1930s and has been repeated in many areas at intervals of 5 to 10 years. Other economically advanced countries have similar national aerial photography acquisition programs.

If there are advantages to taking pictures from two miles above the ground, then why not even higher? In 1987, the **National Aerial Photography Program (NAPP)** was established to develop a cloud-free aerial photography database of consistent scale and orientation and high image quality. The aim of this federal and state program was to provide complete coverage of the United States, updated every five to seven years. The photographs were taken from 20,000 feet (6,000 meters) along north–south flight lines through the west and east halves of USGS 7.5-minute topographic quadrangles. Ten 9-inch-by-9-inch photos at a 1:40,000 scale, each covering a 5-mile-by-5-mile (8-kilometer-by-8-kilometer) area on the ground, were needed for complete stereoscopic coverage of each quad (see chapter 6 for more on stereopairs).

It is now common to take high-altitude photographs from special aircraft flying at altitudes as high as 10 miles (16 kilometers). Examples are the U.S. National Aeronautics and Space Administration (NASA) ex-spy planes, such as the U-2 and SR-71 military reconnaissance aircraft. Black-and-white and CIR photography of the conterminous United States taken from these aircraft were acquired by the **National High Altitude Photography (NHAP) program,** which was operated by the USGS from 1980 to 1989. Each photograph was taken from an altitude of 40,000 feet (13,000 meters) and centered on a USGS 7.5-minute quadrangle. An illustration of the quality of this high-altitude aerial photography is provided by the photo centered on the Corvallis, Oregon, quadrangle, part of which is shown in figure 9.8, bottom. For most purposes, this air photo is equivalent in geometry to the standard topographic quadrangle of the area, although the photo shows far more ground detail.

The advantage of high-altitude photography is that a large ground area can be covered in a single photo. The high-quality photo coverage has many mapping applications. The damage caused by earthquakes, floods, and droughts, for instance, can be quickly monitored. Experiments involving forest resources, snow cover, crop yields, and many other environmental features are being

conducted to determine additional uses of high-altitude photography. The USGS used high-altitude photographs to make image maps and revise existing topographic quadrangles.

Geometric distortions on aerial photography

On vertical air photos, the scale of the photo will most likely be distorted radially away from its center. To provide a simple way to determine the photo center, precision aerial cameras contain fiducial marks placed at the midpoint of each edge (figure 9.9). **Fiducial marks** are small registration marks exposed on the film at the edges of a photograph. If you draw lines between opposite pairs of fiducial marks, these lines will intersect at the center of the photo, called the **principal point.** The point on the ground that was directly below the camera when the air photo was taken is called the **nadir.** The principal point on the photo is the nadir point on the ground only when the aircraft is flying parallel to the ground so that the camera is truly taking a **vertical photograph.**

The primary reason that radial scale distortion is characteristic of vertical photos is that objects of different heights are displaced radially away from the principal point of the photo because the photo is a central perspective view of the ground like what you would see if you looked straight down at the ground from the aircraft.

Courtesy of USGS NAPP.

Figure 9.9 **Format of a National Aerial Photography Program vertical air photo with the principal point and fiducial marks identified.**

Figure 9.10 Northeast quarter of a high-altitude vertical air photo of Chicago, Illinois, showing relief displacement of buildings away from the principal point (shown by a "+" mark and the letters PP at the lower left corner of the photo).

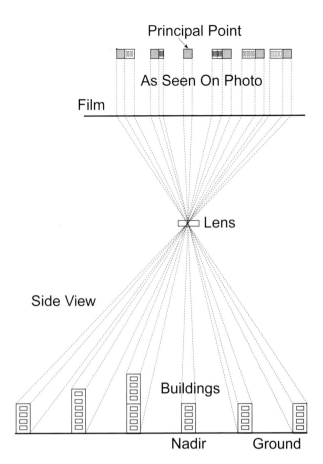

Figure 9.11 The relief displacement of features in central perspective air photos can be seen in this side view of buildings shown horizontally on the ground relative to the nadir and their appearance on the photo.

Relief displacement is the apparent "leaning out" of the top of a higher object on a vertical air photo. If the top of a feature is higher than the elevation of the nadir point, it will be displaced outward and imaged at a slightly larger scale. Inward displacement occurs with objects lying below average ground level. Notice how the sides and tops of tall buildings in downtown Chicago at the lower left corner and all along the lakeshore are displaced outward and radially away from the principal point of the photo (figure 9.10).

The geometry underlying this pattern of relief displacement is illustrated in figure 9.11. Notice the relation between the horizontal position of the six buildings on the ground at different distances from the nadir point and their associated appearance on the air photo. The greater the height of the buildings relative to the ground and the farther features are from the principal point of the photo, the greater the relief displacement. This pattern of relief displacement also holds true for hills, valleys, and any other topographic feature that varies in elevation.

The amount of scale variation and relief displacement on air photos is also influenced by the height of the camera above the ground and by the aircraft tilting the camera from vertical when the photo was taken, called **camera tilt.** The higher the flying height, the less the relief displacement and scale variation. Thus, vertical photographs taken by space shuttle astronauts exhibit so little distortion that they can be overlaid on topographic maps with only geometric discrepancies due to the visible curvature of the earth. In contrast, scale change and relief displacement is great on low-altitude vertical photos; features such as buildings and trees seem to be leaning out from the center of the photograph. If the region is hilly or if the camera is tilted when the photo is taken, scale distortion may be so great that the resulting image map will be difficult or impossible to use for mapping purposes. In such cases, these geometric distortions first have to be eliminated (if possible) from the air photos.

ORTHOPHOTOS AND ORTHOPHOTOMAPS

Orthophotos

Scale distortion in an air photo due to camera tilt can be removed by physically altering the geometry of the photo. The process is called **photo rectification** and is relatively simple to execute. **Rectified air photos** (those that have been georeferenced) still contain relief displacement due to difference in the heights of features (however, see chapter 5 for more on georeferencing). To remove relief displacements, it is necessary to turn the central perspective photo into an orthophoto. An **orthophoto** is a photograph that has been geometrically corrected through rectification so that the scale of the photograph is uniform and planimetrically corrected to remove distortion caused by camera optics, camera tilt, and differences in elevation. Rather than looking outward from the principal point to each feature, you will then be looking directly down on the landscape everywhere on the photo. Thus, all features on an orthophoto will appear in their true planimetric position.

Creating an orthophoto from aerial photography covering the area is a mathematically demanding process carried out by powerful digital computers. Prior to acquiring aerial photography of the area, ground control points determined by traditional surveying (see chapter 1 for more on horizontal control points) or collected by high accuracy Global Positioning System (GPS) receivers (see chapter 14 for more on GPS) must be identified and marked on the ground so that they will be visible on the photos. These control points are used to correct the photos for aircraft tilt, so that each photo will be oriented vertically to the ground and true or grid north (see chapter 12 for definitions of true and grid north). The mathematical corrections are made on digital scans of each photo, in which the photo is converted to a black-and-white or color image composed of several thousand rows and columns of picture elements **(pixels).**

Correction of scale differences and relief displacements in the center section of each photo is performed with the aid of digital elevation model (DEM) data covering the area (see chapter 6 for a description of DEM data). Digital computer software is used to geometrically match each scanned photo as best as possible with the DEM. The software then repositions each pixel on the scanned photo so as to remove both relief displacement and scale variation (figure 9.12). If in color, these planimetrically correct photo center sections are then color corrected to minimize the tonal and color differences between photos. Finally, the color-corrected photo center sections are pieced together into a planimetrically correct orthophoto that is constant in scale and placed on a map projection such as the transverse Mercator (see chapter 3) used for the UTM grid coordinate system (see chapter 4) covering the area. What cannot be corrected are the sides of the radially displaced buildings seen on the original photography; only the base of each building is in the correct geographic location.

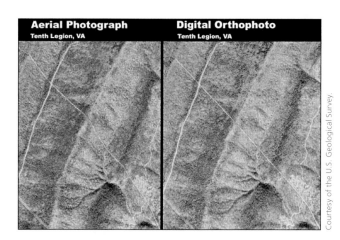

Figure 9.12   **Relief displacement and scale variation in the air photo of Tenth Legion, Virginia, (left) has been removed in the orthophoto of the same area (right). Notice the planimetrically correct straight powerline running over the hills, which appears curved on the air photo due to relief displacement.**

Courtesy of the U.S. Geological Survey.

Orthophotomaps

Orthophotos represent such an important breakthrough in image mapping technology that federal agencies in the United States cooperated with state and local governments, as well as the private sector. The orthophotos were used to create **orthophotomaps—** maps that use either an annotated or unannotated orthophoto as the base. Orthophotomap products developed by the USGS were derived from 1:40,000-scale black-and-white aerial photography that was scanned as images with pixels of approximately 1-meter spatial resolution (figure 9.13) to create orthophotos. These were used to produce 1:12,000-scale quarter quadrangles of 1:24,000-scale USGS topographic maps called **digital orthophotoquads (DOQs).** DOQs are available from a USGS online interactive map service called The National Map to display and manipulate on desktop computers. Also available are higher-resolution black-and-white, true-color, and CIR

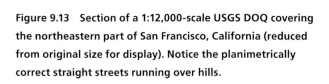

Figure 9.13   Section of a 1:12,000-scale USGS DOQ covering the northeastern part of San Francisco, California (reduced from original size for display). Notice the planimetrically correct straight streets running over hills.

Figure 9.14   Segment of the USGS high-resolution true-color orthophotograph of the Capitol Mall area in Washington, D.C.

Courtesy of the U.S. Geological Survey.

Courtesy of the U.S. Geological Survey.

orthophotos of major metropolitan areas, U.S. state capitals, and the national capitals as shown in the segment of the Washington, D.C., true-color DOQ in figure 9.14. These orthophotomaps serve as a uniform, high-quality base for a variety of mapping, geographical study, and planning activities.

Annotated orthophotomaps

Orthophotography provides you with environmental detail that cannot be portrayed with conventional map symbols for points, lines, and areas because of cartographic abstraction (see chapter 10). However, mapmakers can overlay conventional map symbols on an orthophoto to create an **annotated orthophotomap** that gives you more geographic information about the area. Annotated orthophotomaps are produced at different levels of graphic sophistication for a variety of map uses, as the four examples in figure 9.15 illustrate.

The map showing part of Okayama, Japan, that includes the Okayama Castle and Korakuen garden (figure 9.15, upper left) exemplifies how the mapmaker simply overlayed blue lines of different thicknesses and Kanji text on a color orthophoto to show you the extent of the original castle and garden area. On the Monroe, Oregon, street map (figure 9.15, upper right), the mapmaker overlaid white lines of constant thickness over all streets on the orthophoto, and then placed black street names over the white lines. School symbols and names have been added, as have railroads. Also notice that a gray line has been placed over rivers and streams,

and a transparent gray tone has been laid over the area covered by the city park.

The orthophotomap in figure 9.15, lower left comes from Lithuania's new national cadastre system (see chapter 5 for more on the cadastre) used for property ownership and taxation that includes cadastral maps downloadable from their Web site. Orthophotos serve as the basemap on which are overlaid point, line, and area symbols as well as labels to identify buildings, property boundaries, and corresponding tax lot numbers. Streets, street names, and water bodies are also shown with conventional map symbols to give map users additional locational information. The fourth example (figure 9.15, lower right) illustrates the use of an annotated orthophotomap in natural resource management by the state of Virginia Department of Forestry. Again, an orthophoto serves as the basemap, upon which different characteristics of forest land parcels are shown by overlaying transparent colors on the parcels and using white text to give the parcel identifier and acreage.

Historical feature boundary map from Ministry of Land, Infrastructure, and Transport of Japan; cadastral and forest management map courtesy of ESRI ArcNews.

Figure 9.15 Segments of annotated orthophotomaps used for delineating historical feature boundaries (A), street maps (B), cadastral mapping (C), and forest management (D).

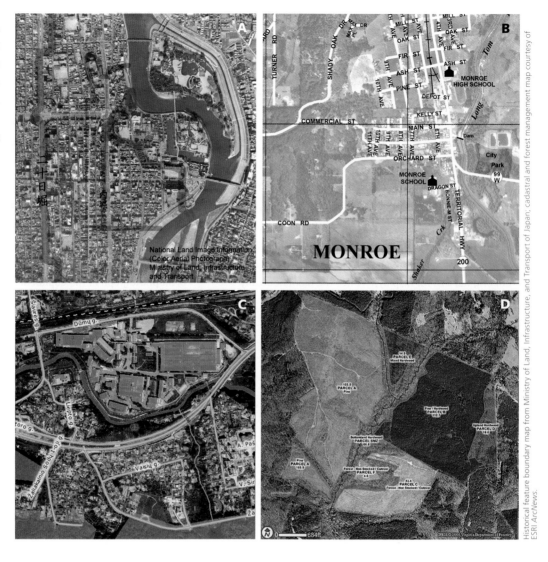

SATELLITE IMAGE MAPS

We mentioned previously that the higher the flying height, the less the relief displacement and scale variation on the images collected. Images obtained by orbiting satellites are at high enough altitudes that relief displacement of tall mountains is minimal and scale variation on the image is due to the earth's curvature. This scale variation can be corrected mathematically, and the resulting **orthoimage** (an image that has been processed to remove systematic geometric distortions) can be digitally processed so as to be on any map projection (see chapter 3 for more on map projection). We will use the term "orthoimage" to refer to a satellite image that has been corrected geometrically. Conventional map symbols can then be overlaid on the orthoimage to create a **satellite image map** of a particular environmental theme. Numerous satellites collect images of the earth and other planetary bodies, but we will focus on two major types—weather satellites and earth resources satellites.

Weather satellites

If you are in the United States, your local television or newspaper weather report probably uses one or more *weather satellite images* created from **Geostationary Operational Environmental Satellite (GOES)** imagery. Europeans have the **Meteosat system** with Meteosat-8 and Meteosat-9 over the Atlantic Ocean and Meteosat-6 and Meteosat-7 over the Indian Ocean. Russia operates the **Geostationary Operational Meteorological Satellite (GOMS)** (also known as Elektro) positioned over the equator south of Moscow, and China operates the **Feng-Yun** geostationary satellites. India also operates geostationary satellites carrying instruments for meteorological purposes.

Satellites can operate in several types of earth orbit. The most common orbits for environmental satellites are geostationary and polar. A satellite in **geostationary orbit** is always in the same position with respect to the rotating earth. This is also called the Clarke orbit because science fiction author Arthur C. Clarke first proposed in 1945 the idea of using a geostationary orbit for communications satellites. By orbiting at the same rate as the earth's rotation, in the same direction as the earth, the satellite appears stationary (synchronous) with respect to the rotation of the earth. Because a geostationary orbit must be in the same plane as the earth's rotation, it is always in the **equatorial plane.** Geostationary satellites provide a "big picture" view, enabling coverage of weather events. This is especially useful for monitoring severe local storms and tropical cyclones.

Satellites in polar orbit (sometimes called **near-polar orbit**) circle the earth at near-polar **inclination** (the angle between the equatorial plane and the satellite orbital plane—a true polar orbit has an inclination of 90 degrees). **Sun-synchronous orbit** is a special case of the polar orbit in which the satellite passes over the same part of the earth at roughly the same local time each day. This enables regular data collection at consistent times as well as long-term comparisons. The orbital plane of a sun-synchronous orbit must also rotate approximately one degree per day to keep pace with the earth's surface. Polar-orbiting satellites provide a more global view of the earth.

The broadest coverage imagery obtained from space is that from weather satellites such as GOES, positioned in a geostationary orbit above the equator at an altitude of 22,300 miles (35,680 kilometers). From fixed longitudes of 125°W and 75°W, respectively, the **GOES West** and **GOES East** satellites obtain the images of the western and eastern United States seen on television weather shows (figures 9.16 and 9.17). These two satellites record images of the western and eastern United States every 15 minutes and the entire hemisphere every three hours.

A single GOES image covers millions of square miles of ground surface, making it possible to see the cloud patterns over half of the United States instantaneously. By studying images taken every 15 minutes, we can easily see the detailed movement of clouds. This imagery has been used to produce very dramatic weather system videos that are commonly seen on TV weather broadcasts. The resolution of ground details on weather satellite imagery is particularly poor due to its extremely high altitude, however. Consequently, the use of the imagery is essentially restricted to broadscale atmospheric phenomena.

Weather satellite image maps for the conterminous United States are made by compositing the GOES West and East images collected in the visible and near-infrared regions of the electromagnetic spectrum, and in a third spectral region tuned to water vapor. **Compositing** is the process of combining multiple images to produce a single image of the area covered by the multiple images. Minimal geographic information (graticule lines, state boundaries, coastlines, and large lakes) is added to help you orient the map. The GOES *visible* satellite image map (figure 9.17, top) shows the amount of visible sunlight reflected back to the satellite sensor by clouds, the land surface, and water bodies. Thicker clouds reflect more light than thinner clouds, and hence appear lighter on the image. However, you cannot distinguish among low-, middle-, and high-level clouds, which is important for predicting the location and intensity of precipitation.

The *infrared* weather satellite image map (figure 9.17 middle) allows low-, middle-, and high-level clouds to be distinguished by their gray tone on the image. Warmer objects appear darker than colder objects on the image, so high-level cirrus or tall, highly reflective clouds such as thunderstorms with colder tops will appear white.

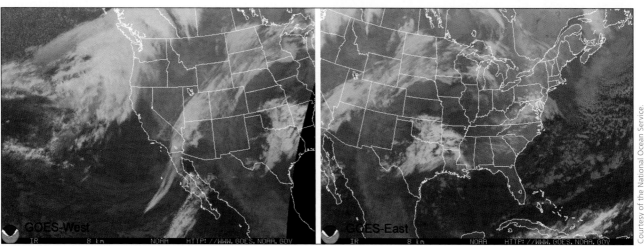

Courtesy of the National Ocean Service.

Figure 9.16   Examples of a NOAA GOES West and GOES East satellite image.

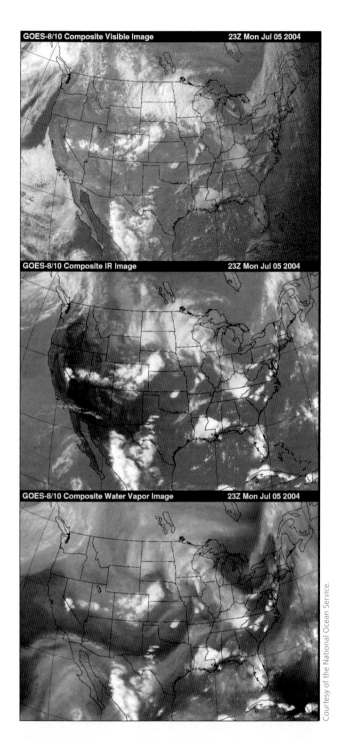

Courtesy of the National Ocean Service.

Figure 9.17 **Composite GOES weather satellite visible (top), infrared (middle), and water vapor (bottom) image maps of the conterminous United States allow you to see the cloud and moisture conditions at various times of day or night.**

On the GOES *water vapor* image map (figure 9.17 bottom), the lighter the cloud, the more moisture it contains. The white cloud plumes indicate moisture-laden thunderstorms, whereas the other clouds in the area are slightly darker, and hence contain less moisture. One of the most striking tonal differences between the visible and water vapor image is seen off the coast at the bottom left corner of the maps, where the dark gray, moisture-barren clouds are consistent with a thin layer of fog covering the ocean.

Weather satellite images are combined with conventional maps in several ways. Your evening weather show on television may include a map such as figure 9.18, top, in which clouds seen on the composite infrared image for the United States have been overlaid on a simple basemap with national and state boundaries. The two satellite images have been combined on a close cousin of the orthographic map projection (see chapter 3) that mimics how the earth would look to you from the position of the satellite in space. Coastlines and state boundaries are added to provide basic geographic location information.

The weather satellite image map is filled with complex point symbols that show values of several types of surface observations at hourly intervals for major meteorological stations across the country overlaid on a composite infrared satellite image. This particular map shows observed temperature, dew point, pressure, wind speed and direction, and cloud cover data for each meteorological station as recorded at 20Z on July 21, 2007. 20Z is 8PM "Zulu" time, the hour at the Greenwich prime meridian. Since the conterminous United States falls in time zones from six to nine hours west of the prime meridian, the map shows observations from 11AM on the west coast to 2PM on the east coast of the country. Locations of warm and cold fronts as determined from meteorological station observations are overlaid on the infrared image map, which is also color coded to show areas where precipitation is falling in different intensities based on information from weather radar across the country.

The weather map in figure 9.18, bottom, is similar to a weather map you have seen on television or in your newspaper. This map was made by overlaying the composite infrared satellite image on a relief-shaded basemap of the United States and vicinity. Color-coded precipitation intensities from ground weather radar are shown. The warm and cold fronts and **isobars** (lines of equal atmospheric pressure) shown on the map are the meteorologist's interpretation of surface air pressure variations, as well as frontal activity, across the country at a certain time. The meteorologist has also added yellow boxes where severe weather is predicted to occur.

Earth resources satellites

The electromagnetic spectrum is broad and rich in environmental information that can be sensed throughout its range. In the past several decades, we have learned to use images obtained simultaneously in different regions of the electromagnetic spectrum that we call **spectral bands,** each of which gives us valuable insights into the nature of our surroundings. **Multispectral** remote sensing devices are used to capture a ground scene in different spectral bands so that the resulting images are geometrically identical. Over the last three decades, artificial satellites have become the most important multispectral sensing platforms. If a coarse-spatial-resolution sensor is used, images are acquired several hundreds of miles above the earth's surface and cover a relatively large ground area at a small scale. If a high-resolution multispectral imaging device is used, it records

a small ground area at a very high spatial resolution. Most images we see come from what we call **earth resources satellites,** satellites launched with the primary mission of providing systematic, repetitive environmental system measurements, such as wind speed and direction, wave height, surface temperature, surface altitude, cloud cover, and atmospheric water vapor level. Let's examine imagery obtained from remote sensing devices on four of the earth resources satellites most commonly used for image maps—Landsat, SPOT, IKONOS, and QuickBird.

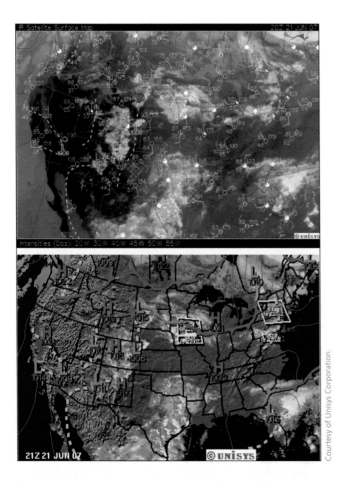

Courtesy of Unisys Corporation.

Figure 9.18 These two annotated weather satellite image maps of the conterminous United States show surface observation data placed over the composite infrared image (top) and a generalized current weather conditions map with clouds from the composite infrared image overlaid on a relief shaded basemap, along with generalized isobaric lines and locations of warm and cold fronts (bottom).

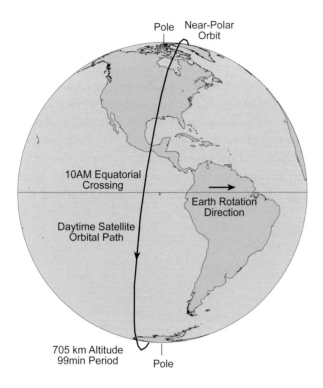

Figure 9.19 The sun-synchronous, near-polar orbit of Landsat 5 and 7 make it possible to provide nearly complete coverage of the earth's surface in 185-kilometer-wide swaths every 16 days.

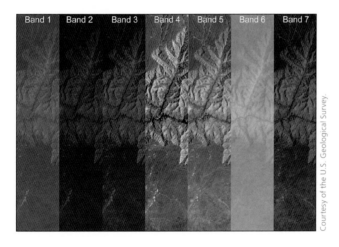

Figure 9.20 Images from bands 1 through 7 from the Landsat 7 ETM+ sensor provide different information about the environment, as seen in these images of the central portion of the Grand Canyon in Arizona.

Landsat

In 1965, NASA began the **Landsat** project to explore the potential for monitoring the earth from space. The first satellite, Landsat 1, was placed in orbit in 1972. Subsequent successful launches in the Landsat series took place in 1975 (Landsat 2), 1978 (Landsat 3), 1982 (Landsat 4), 1984 (Landsat 5), and 1999 (Landsat 7). Landsat 5 and 7 are currently in operation (Landsat 6, launched in 1993, failed to reach orbit), with remote onboard sensing instruments collecting hundreds of images of the earth every day. These images provide an excellent orthophoto base for satellite image maps made worldwide for applications in agriculture, geology, forestry, regional planning, global change research, and other fields.

Landsat satellites are in sun-synchronous near-polar circular orbits so that near-global coverage is possible due to the earth's rotation beneath the satellite (figure 9.19). Landsats 5 and 7 orbit the earth every 99 minutes from an altitude of 438 miles (705 kilometers), circling the earth roughly 14 times per day and passing over the same spot at the same time every 16 days. On each **descending orbit** (daylight pass from north to south), remote sensing instruments record electromagnetic radiation from the ground over a strip 185 kilometers wide. These data are used to create Landsat images covering a ground area 115 miles wide by 105 miles high (185 kilometers by 170 kilometers).

Most image maps are made today from images acquired by the **Enhanced Thematic Mapper Plus (ETM+)** sensor on Landsat 7. This sensor provides 30-meter-resolution images from the blue (band 1), green (band 2), red (band 3), near-IR (band 4), and mid-IR (bands 5 and 7) portions of the electromagnetic spectrum, as well as a 60-meter-resolution thermal infrared band (band 6) and a 15-meter-resolution panchromatic band (band 8). Digital pixel values for the thousands of rows and columns making up an image in each band are transmitted to ground receiving facilities where the data are processed into geometrically corrected images that can be ordered on one of several map projections. You can see in figure 9.20 how images of the same area appear differently in bands 1-7. Three of the eight Landsat ETM+ bands are often combined into composite images. The combination you use depends on what you want to see in the image. For example, the top composite image of Tampa Bay, Florida, at the top of figure 9.21 combines the blue, green, and red bands (1, 2, and 3) into a true-color appearing image, while the composite image at the bottom is a combination of the green, red, and near-IR bands (2, 3, and 4) that resembles CIR photography.

SPOT

Landsat isn't the only earth resources satellite program. The "commercialization of space" is well underway. The French space agency CNES, with the participation of Sweden and Belgium, has launched a series of land-resources satellites known as the **Systeme Probatoire** **d'Observation de la Terre (SPOT),** operated by the SPOT Image Corporation. SPOT 1 was launched in 1986, followed by SPOT 2 in 1990, SPOT 3 in 1993, SPOT 4 in 1998, and SPOT 5 in 2002. SPOT 3, 4, and 5 are still in operation, each placed in a sun-synchronous near-polar orbit at an altitude of 832 kilometers (517 miles) so that high-resolution 60-kilometer-by-60-kilometer vertical images can be obtained for the same area on the earth every 26 days.

The sensors aboard each SPOT satellite provide imagery in different bands and spatial resolutions, but used in combination, it is possible to obtain black-and-white, true-color, and CIR images at spatial resolutions of 2.5, 5, 10, and 20 meters. These images are created digitally from green, red, and near-IR data collected at a 10- or 20-meter spatial resolution and panchromatic data collected by SPOT 5 at 2.5- or 5-meter resolution. Color images at 2.5- and 5-meter resolution are created by using the panchromatic data to define the shape and texture of features in the images, and then coloring the image with the coarser-resolution green, red, and near-IR data. Sections of images from different parts of the earth are shown in figure 9.22 to illustrate the urban features distinguishable for Macau (20-meter resolution); Elche, Spain (10-meter resolution); Shaanxi, China (5-meter resolution); and Canberra, Australia (2.5-meter resolution). You can make satellite image maps at progressively larger map scales from 20-, 10-, 5-, and 2.5-meter resolution images like these, since all images are geometrically corrected to be constant in scale on a map projection surface.

Figure 9.21   Landsat ETM+ bands 1, 2, and 3 (top) and bands 2, 3, and 4 (bottom) are used in these composite images of Tampa Bay, Florida, to resemble true-color and CIR air photos, respectively.

Figure 9.22   True-color and CIR SPOT 3 and 5 image examples at the four available spatial resolutions: Macau at 20-meter resolution (upper left); Elche, Spain, at 10-meter resolution (upper right); Shaanxi, China, at 5-meter resolution (lower left); and Canberra, Australia, at 2.5-meter resolution (lower right).

IKONOS

New commercial satellite systems such as **IKONOS** (from the Greek word for "image") provide images from space at the same resolution as aerial photography. In 1999, the Satellite Imaging Corporation launched the IKONOS satellite into a sun-synchronous, near-polar orbit at an altitude of 423 miles (680 kilometers). This orbit and the ability to point the sensor away from vertical allows the same area to be imaged every three days in an 6.8-mile-wide (11-kilometer-wide) swath. The IKONOS sensor collects panchromatic images with a 2.6-foot (0.8-meter) spatial resolution (figure 9.23). In multispectral mode, blue, green, red, and near-IR images are collected at a 10.5-foot (3.2-meter) resolution to create both true-color and CIR images. An example is the true-color image of the Vatican in figure 9.24.

QuickBird

The DigitalGlobe corporation launched the **QuickBird** satellite in 2001 to obtain imagery at a slightly higher resolution than IKONOS imagery. QuickBird's lower 280-mile (450-kilometer) near-polar, sun-synchronous orbit allows vertical images of the same area to be acquired every 3.5 days in a swath 8 miles wide (13.5 kilometers wide). The sensor collects panchromatic data at a 2-foot (0.6-meter) spatial resolution when pointed vertically, and at a 2.5-meter resolution in multispectral mode. Panchromatic images such as figure 9.25 of the Washington Monument are virtually indistinguishable from large-scale aerial photography.

High-resolution imagery from IKONOS and QuickBird will open remote sensing to a wide range of new uses, including large-scale satellite image maps in color that are at the same spatial resolution as orthophotography.

Figure 9.24 IKONOS 3.2-meter resolution multispectral image of Vatican City, Rome, in Italy.

Figure 9.23 IKONOS satellite 0.8-meter resolution panchromatic image of the Great Pyramid at Giza, Egypt.

Figure 9.25 The QuickBird satellite's 0.6-meter resolution panchromatic image of the Washington Monument in Washington, D.C.

Examples

Let's now look at examples of image maps produced from images collected by these newer earth resources satellites as well as from the older Landsat and SPOT satellites.

Conventional map linework, area coloring, and lettering is commonly overlaid on geometrically corrected earth resources satellite images that serve as a locational reference base for the qualitative or quantitative information added to the image map. Although we can't possibly inventory all the possible examples of these satellite image maps, we can take a closer look at some representative samples. Let's focus on the three examples in figure 9.26, which will give you an idea of the variety of satellite image maps produced today.

The image map in figure 9.26, top, is a segment from a Benton County, Oregon, satellite image road map. The mapmaker made the county look green by printing the image from a single Landsat 5 Thematic Mapper band in green instead of in black-and-white ink. By doing this, roads, railroads, and lettering could be printed in black, rivers and the wildlife refuge could be shown with translucent gray tones, and U.S. Public Land Survey System section lines (see chapter 4) could be overlaid in white.

Very realistic satellite image maps, as in figure 9.26, middle, showing earthquakes and faults for the San Francisco Bay Area, result from combining relief shading created from DEM data with a mosaic of several Landsat 7 EMT+ color composite images. Notice how the relief shading helps you understand how the pattern of fault lines in black and red and the different size orange dots showing earthquake epicenters and magnitudes are related to the surface terrain.

The section of the flood extent map for the Kirulo, Bulgaria, area shown in figure 9.26, bottom is an example of an image map made by overlaying a transparent light blue color for the flooded area on a SPOT 5 CIR composite base image. Notice how you can see the flooded features on the ground, since they appear to be covered by clear blue water. Also note how the normal course of the river has been shown in dark blue, allowing you to see the degree to which the river overflowed its banks.

Figure 9.26 These three satellite image maps illustrate how mapmakers overlay conventional map symbols on satellite images to create satellite image maps. The top and middle maps are based on Landsat Thematic Mapper single-band and true-color images, respectively; the bottom map has a SPOT 5 CIR image as a base.

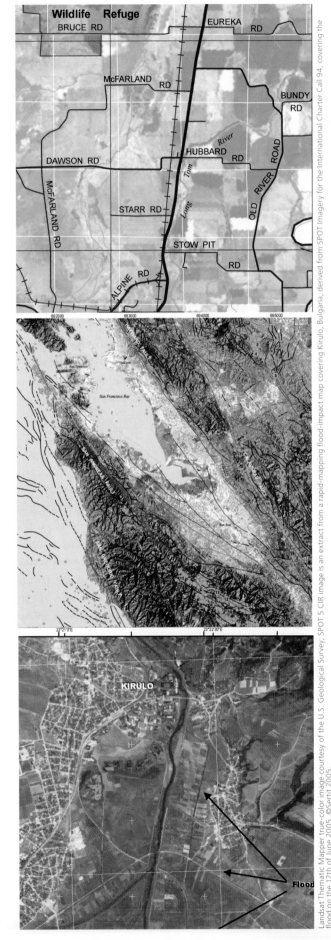

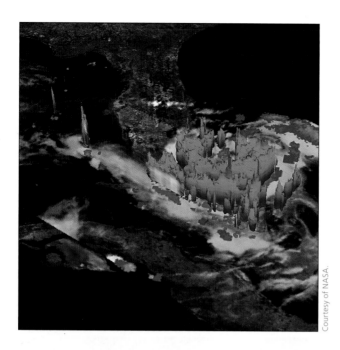

Courtesy of NASA.

Figure 9.27 **This oblique perspective view of the satellite image map of Cuba and Florida is overlaid with Hurricane Katrina data showing the horizontal distribution and vertical height of precipitation on August 25, 2005.**

A further enhancement to satellite image maps is achieved by draping the images over a DEM-created surface (see chapter 6 for more on relief portrayal) to create an oblique perspective view satellite image map (figure 9.27). Thematic data can also be overlaid on this base, as can conventional map symbols. This example shows a satellite image map of Cuba and Florida overlaid with precipitation extent and height data for Hurricane Katrina. By showing the precipitation data as a 3D surface over the satellite image, meteorologists were able to better visualize the predicted path and three-dimensional structure of the storm as it neared land.

DYNAMIC IMAGE MAPS

Dynamic image maps go beyond traditional static displays to include animated sequences of images shown in a predefined order, and they often allow the map reader to interact with map symbols, so that symbols are linked to text, pictures, or video clips. Animated image maps can be created to give the impression that you are flying over and around an area. You will increasingly see these fly-throughs (see chapter 6 for more on fly-throughs) in a variety of visual presentations, including news programs, training simulators, and video games. Let's look at a widely used program, ArcGlobe, that illustrates how interaction and animation can be applied to image maps created instantly on personal computers.

ArcGlobe

Many of you may have taken virtual flights to exotic locales by viewing ArcGlobe animated satellite and orthophoto image maps. **ArcGlobe** is an animated **virtual globe** program (a 3D software model or representation of the earth) developed by ESRI. Like the Microsoft Virtual Earth image in figure 9.28 and Google Earth, ArcGlobe works by continually superimposing digital images of the earth's surface on an orthographic-like map projection of a surface derived from DEM data. The DEM allows you to see the imagery of the earth in 3D perspective at different vertical exaggeration levels that you can select.

The 3D perspective image map in figure 9.28 is one frame from a fly-through that you could take over Manhattan from southwest to northeast. Flight controls superimposed on the image map allow you to continuously change the viewing distance, horizontal view direction, and vertical viewing angle. When virtually flying over large cities such as New York, you may also be able to turn on and off a 3D display of building outlines that give you a general idea of how the buildings would look from your ever-changing viewpoint.

In ArcGlobe you can also add your own data to the image map and make it available to others on the Internet. The resulting virtual fly-through is superimposed with map symbols showing your data. Many companies and organizations have taken advantage of this do-it-yourself capability, greatly expanding the types of information you can look at and query using these products. Undoubtedly others will mimic this capability in their own Internet applications as well.

Courtesy of Microsoft Virtual Earth.

Figure 9.28 This 3D perspective map of Manhattan, New York City, allows a virtual overflight of the skyline, even zooming down to the cityscape and soaring through the steel and concrete canyons between buildings.

The future of image maps looks bright and very interesting. We have only discussed the aerial photography and satellite imagery used most commonly today to create image maps, but many other meteorological, earth resources, and military reconnaissance satellites have been and will continue to be placed in orbit. Both static and dynamic image maps will continue to be made from imagery collected by the diverse and ever-expanding suite of remote sensing devices.

The possibilities for dynamic image maps seem limitless, as data about the earth are being collected at an accelerating rate using devices such as GPS to quickly and accurately determine the ground positions of features. These geographically referenced points, lines, and areas are increasingly stored and manipulated in geographic information systems (GIS) that, in turn, will be increasingly linked to dynamic image map systems like Microsoft Virtual Earth, Google Earth, or ArcGIS Explorer. For example, ArcGIS Explorer allows you to view imagery, topographic maps, physical features, shaded relief, historical maps, street maps and other maps and images for the entire world seamlessly at different levels of detail. You can combine these global maps and images with your own local cartographic data to create custom maps. You can also perform various map analysis tasks on the maps and images using ArcGIS functions such as determining what is visible from a particular point (see chapter 16 for more on visibility analysis). Incorporating these ArcGIS functions with a dynamic image map system allows you to answer geographic questions about the maps you generate and share the results with others. These dynamic image map systems may be married with new computer designs like Microsoft's Surface Computer and spherical display systems such as OmniGlobe, creating a revolution in how you will use and benefit from both traditional and image maps.

SELECTED READINGS

Barrett, E. C., and L. F. Curtis. 1992. *Introduction to environmental remote sensing.* 3d ed. London: Chapman & Hall.

Campanella, R. 1996. High-resolution satellite Imagery for business. *Business Geographics* (March): 36–9.

Campbell, J. B. 1996. *Introduction to remote sensing.* 2d ed. New York: The Guilford Press.

Carleton, A. M. 1991. *Satellite remote sensing in climatology.* Boca Raton, Fla.: CRC Press, Inc.

Ciciarelli, J. A. 1991. *Practical guide to aerial photography: With an introduction to surveying.* New York: Van Nostrand Reinhold.

Cook, W. J. 1996. Ahead of the weather: New technologies let forecasters make faster, more accurate predictions. *U.S. News & World Report* (April 29): 55–7.

Corbley, K. P. 1997. Applications of high-resolution imagery. *Geo Info Systems* (May): 36–40.

——. 1997. Multispectral imagery: Identifying more than meets the eye. *Geo Info Systems* (June): 38–43.

Dickinson, G. C. 1979. *Maps and air photographs.* 2nd ed. New York: John Wiley & Sons, Inc.

Evans, D. L., et al. 1994. Earth from the sky. *Scientific American* (December): 70–5.

Falkner, E. 1994. *Aerial mapping.* Boca Raton, Fla.: CRC Press, Inc.

Hamit, F. 1996. Where GOES has gone: NOAA's weather satellite imagery and GIS-marketed. *Advanced Imaging* (November): 60–4.

Lillesand, T. M., and R. W. Kiefer. 1994. *Remote sensing and image interpretation.* 3d ed. New York: John Wiley & Sons, Inc.

Office of Technology Assessment, U.S. Congress. 1993. *The future of remote sensing from space: Civilian satellite systems and applications.* Washington, D.C.: U.S. Gov. Printing Office.

chapter
ten **MAP ACCURACY AND UNCERTAINTY**

10

Map accuracy and uncertainty

Mapping, like architecture, is an example of functional design. Unlike an artist's representation of the environment in which geometric liberties are taken in order to convey an idea or emotion, a map is expected to be true to the location and nature of our surroundings. Indeed, our willingness to let maps "stand in" for the environment is because of this expected adherence to reality. Figure 10.1 illustrates the process that mapmakers use to transform data collected (through surveys, by GPS, from remote sensing, and more) about the geographical environment to a map. Maintaining the highest possible fidelity in the transformation process assures that the map is an accurate representation of reality. The map user also plays a part in the transformation process through reading, analysis, and interpretation, and high fidelity must also be maintained in these activities. For example, the user is responsible for checking the map against reality as features may have changed from the time the map was created.

We will use the term **map accuracy** to refer to the closeness of the map representation to the geographic phenomena. Map accuracy turns out to be a complex topic. There are a number of issues that relate to the map accuracy, including uncertainty, error, bias, precision, accuracy, scale, and quality. You will see some of these terms used interchangeably—understanding the subtle differences can help you use maps more appropriately, so we'll examine each in turn. But first we'll look at the basic characteristics of maps to see how maps differ from other representations of the environment, like architectural drawings, paintings, or photographs. At the end of this chapter we'll look at various aspects of map accuracy and how accuracy information is conveyed to the map user.

BASIC CHARACTERISTICS OF MAPS

Maps are a graphic representation of a geographical setting; stated another way, they are a collection of symbols used to represent a portion of the earth. In the introduction, we talked about the basic characteristics of maps that give them their "mapness." In this chapter, we can frame our discussion of map accuracy with the basic characteristics of maps because that will give an indication of how we can expect them to adhere to reality. All maps are concerned with two primary elements—*locations* and *attributes*. All maps are *reductions of reality* and therefore require *scale* conversions (see chapter 2 for more on scale). All maps are *transformations of space* involving applications of *map projections* and *coordinate systems* (see chapters 3 and 4 for more on these). All also are *abstractions of reality* and therefore require *generalization* of the geographic information. In addition, all maps use *signs* and *symbolism;* this is called **cartographic symbolization** (see chapters 7 and 8 for more on symbolism). In previous chapters we discussed accuracy issues relating to map projections and symbolism. In this chapter, we will look in more detail at scale and generalization.

Given the basic characteristics of maps, we can begin to understand how the faithfulness of their representation of reality can be compromised. Maps must maintain their connection to—and present a useful representation of—reality. This requirement has to be balanced against scale, projection, generalization, and symbolism requirements. As you can imagine, this is impossible without

sacrificing some degree of representational quality. It is in your interest as a map user to understand issues that relate to how accurately maps adhere to reality.

SCALE AND ACCURACY

As we learned earlier, all maps are reductions of reality. This means that it is impossible to show all the real-world detail on a map. If we did, this would essentially be a map with a scale of 1:1, which is the same as the world we live in! The ability to show detail on a map is determined by its scale because that dictates the amount of space available. Recall from chapter 2 that map scale is a ratio between distances on the map and their corresponding earth distances. When features are displayed in smaller spaces, distances between the symbols and lengths of features are reduced. Features may become too small to see. Patterns may disappear as features coalesce. There is less space to add additional helpful information such as labels. Visual chaos can result. Cartographers have to make careful decisions about the kind, number, and representation of features on the map. These decisions are governed in large part by the map scale (the map purpose, audience, or technical constraints also play roles).

Ideally the level of detail on a map is appropriate to the scale. A large-scale map of 1:1,000 will show less area with more detail than a smaller-scale map of 1:250,000. Because all maps are abstractions of reality, they require a reduction of information content in order for features to be **legible** (able to be seen and recognized). When the

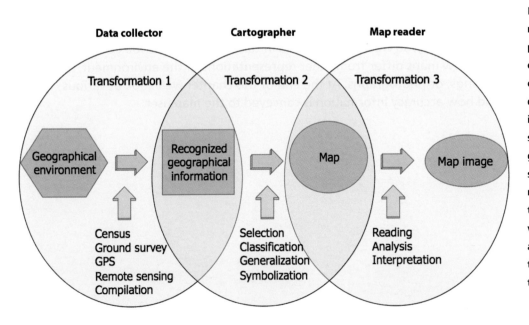

Figure 10.1 The map transformation process begins with the collection of data about our environment. These data are transformed into a map through selection, classification, generalization, and symbolization. The map user completes the transformation process when he or she reads, analyzes, or interprets the information on the map.

world around you is reduced to the scale of a map, the challenge is to remove unnecessary detail while preserving the basic geometrical form and spatial positioning of the simplified features. This is how the level of detail is made to fit the map scale.

CARTOGRAPHIC ABSTRACTION

Cartographic abstraction is the process of transforming data that have been collected about our environment into a graphical representation of features and attributes relevant to the purpose of the map (the second transformation in figure 10.1). Much of the power of maps lies in cartographic abstraction—salient aspects of the environment are preserved so that we can focus on patterns that may not be apparent when there is too much complexity or an overwhelming amount of information. To promote understanding about the information that is shown graphically, maps often portray the basic or universal character of the environment rather than individual features and unique attributes.

To transform geographic data to a map, cartographers rely on selection, generalization, classification, and symbolism. They decide what type of—and how much—information to portray (selection). In addition, they eliminate or deemphasize unwanted or unneeded detail (generalization and classification). Finally, they make appropriate choices about how the information will be shown graphically (symbolization). Let's look at each of these in more detail.

Cartographic selection
Cartographic selection refers to deciding which classes of features to show on the map. Based on the map's purpose and scale the cartographer will choose the relevant information to include on the map and determine what should be left off. Selection decisions relate to the themes of information on the map (hydrography, transportation, boundaries, physiography, cultural features, and so on), as well as the features within the themes. For example, on a large-scale topographic map, you will usually find **perennial streams** (those that flow all year long), **intermittent streams** (those that flow a large portion of the time but cease to flow occasionally or seasonally), and **ephemeral streams** (those that flow only during and immediately after periods of rainfall or snowmelt). On smaller-scale maps, the hydrographic theme will be included, but ephemeral and even intermittent steams may be eliminated.

There are also special-case features that will either be retained or eliminated on a case-by-case basis. For example, at a particular scale, all intermittent streams might be eliminated; however, one that has cultural significance might be retained. You could imagine that an intermittent stream passing through the center of a town and around which other features have been strategically located might be retained on a map on which all other intermittent streams have been omitted. Physiographic features like hills or mountains below a certain elevation might be eliminated, except ones that have some special cultural significance. Roads of a certain class might be eliminated except those that provide some important connection for motorists. These examples show you it is useful to be aware that there may be exceptions in the decisions that relate to the selection of whole classes of features.

Cartographic selection for general reference maps is much different than selection for thematic maps. For reference maps, the challenge the mapmaker faces is to decide which classes of features are of greatest interest and use to a wide variety of readers. For example, many people will use topographic maps for a variety of purposes, so these maps necessarily carry a lot of information, only some of which may be relevant for a particular map use task. With thematic maps, the challenge for the mapmaker is to decide which features to include as locational reference information. Only the base locational information relevant to the theme of the map will be included because superfluous information may distract you or cause you to misinterpret the map's message. While it is the responsibility of the mapmaker to choose the themes and features wisely, it is the map reader's responsibility to understand that only a limited selection of all possible features is shown on any map.

Cartographic generalization
Cartographic generalization refers to reducing the amount of information on a map through change to the geometric representation of features. The degree of generalization is proportional to a feature's spatial detail, and inversely proportional to map scale (figure 10.2). Even environmental features that are well defined, such as roads, rivers, or coastlines, don't get mapped in all possible detail because a map cannot carry all that information in the space allowed. Linear features and edges of areal features are smoothed on the map. We call this loss of detail **line generalization** for features like boundaries, rivers, and roads. There are also other types of generalization—we will look at these shortly.

1:24,000

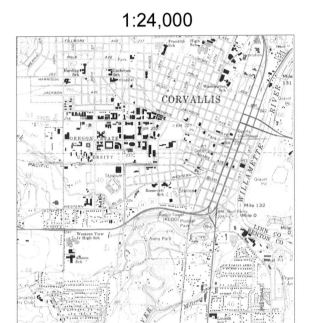

1:62,500

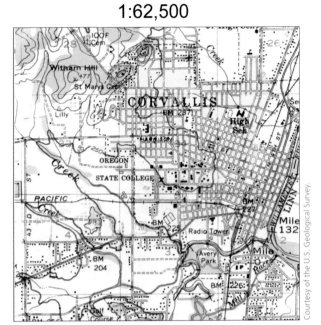

Courtesy of the U.S. Geological Survey.

Figure 10.2 The degree of line generalization is proportional to a feature's spatial detail and inversely proportional to map scale. Notice the smoothing of curvilinear features as the map scale is reduced from 1:24,000 to 1:62,500. Rectilinear features exhibit less sensitivity to generalization through scale change.

Generalization also can cause displacement of features as smoothing cuts off natural irregularities or causes features to overlap. But is this shift of a linear feature's position an example of positional error? For example, as a map is reduced in scale, two linear features next to each other, like a river and a road, will begin to coalesce. Since we know that a map should graphically communicate the geographical setting, the real message to the map reader should be that the features are next to each other in reality, not on top of each other. To preserve the integrity of the spatial relationships on the map, the cartographer has to displace one or both of the features. Is a map of the world on a postage stamp best critiqued in terms of its generalization errors? We think not. To say generalization is error misses the point. You might better think in terms of generalization effects, not generalization errors. By concentrating on generalization effects, attention is focused on the necessity of cartographic generalization rather than on the inevitable positional distortions.

Generalization in mapping is a complex topic and entire books have been written on the subject. Because there is so much literature available on generalization, and because there are many ways to approach the subject, we will severely limit our discussion here and instead provide a broad overview of generalization as it relates to map use. As the map reader, you want to be able to

assess some of the most common geometric changes that affect the accuracy of maps. Generalization can relate to either raster or vector representations—we'll look at both, starting with vector generalization.

Vector generalization operations

There are a variety of vector generalization methods, and no standard exists to supply you with a complete and agreed upon set. The set of vector operations shown in figure 10.3 is fairly comprehensive, and most mapmakers would agree it includes the most commonly used operations.

To see how these can be used, we will look at a study of a map that was reduced in scale from 1:5,000 to 1:50,000. The map at the top left of figure 10.4 shows the features with all their detail at the **source scale** (the scale at which the data were originally derived from their sources). The map at the larger 1:5,000 scale requires a number of generalization operations to still be readable at the smaller 1:50,000 scale, as shown in figure 10.4, top right. Streets originally represented with casings (lines of a different color than the fill) need to be collapsed to a single line. Complex building polygons need to be replaced with regular polygons. Some polygons need to be collapsed to point features. And groups of similarly shaped polygons need to be replaced with a simpler representation, such as a set of similar point symbols.

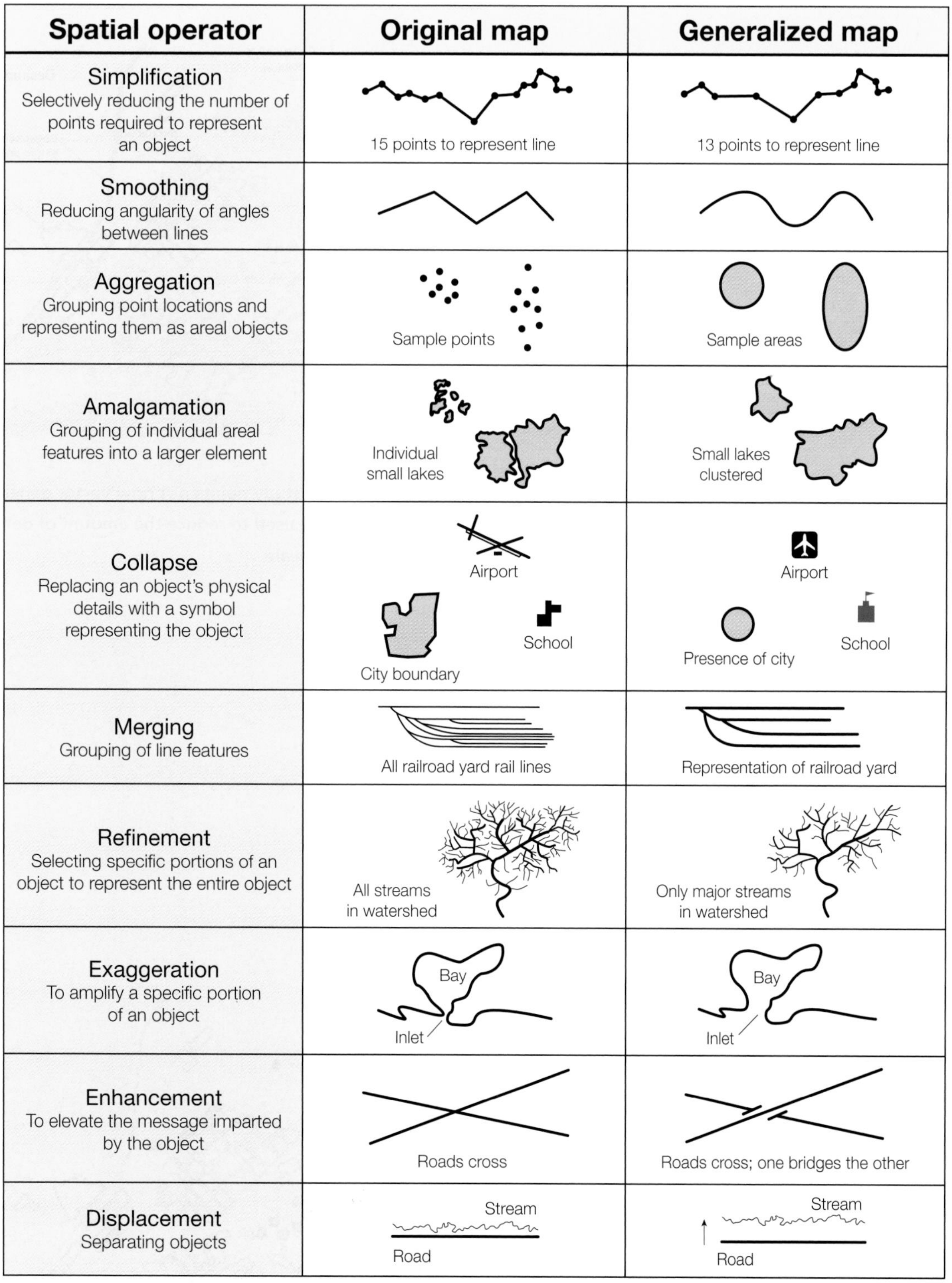

Spatial operator	Original map	Generalized map
Simplification Selectively reducing the number of points required to represent an object	15 points to represent line	13 points to represent line
Smoothing Reducing angularity of angles between lines		
Aggregation Grouping point locations and representing them as areal objects	Sample points	Sample areas
Amalgamation Grouping of individual areal features into a larger element	Individual small lakes	Small lakes clustered
Collapse Replacing an object's physical details with a symbol representing the object	Airport / School / City boundary	Airport / School / Presence of city
Merging Grouping of line features	All railroad yard rail lines	Representation of railroad yard
Refinement Selecting specific portions of an object to represent the entire object	All streams in watershed	Only major streams in watershed
Exaggeration To amplify a specific portion of an object	Bay / Inlet	Bay / Inlet
Enhancement To elevate the message imparted by the object	Roads cross	Roads cross; one bridges the other
Displacement Separating objects	Stream / Road	Stream / Road

Figure 10.3 **The most common vector generalization operations are illustrated in the examples above.**

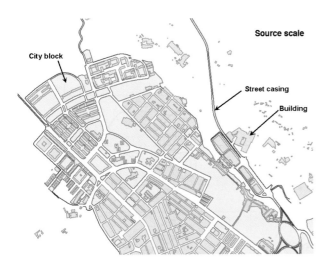

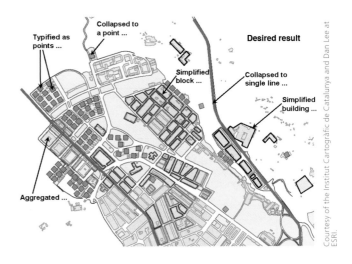

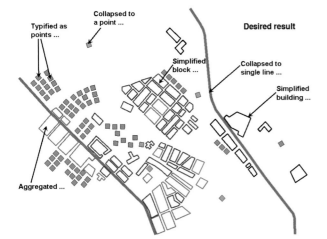

Figure 10.4 This study points out how vector generalization operations can be used to reduce the amount of detail for a map at a smaller scale.

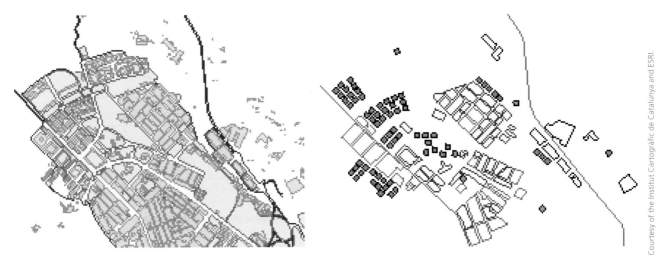

Figure 10.5 The final result of the vector generalization operations is a map that is more appropriate for use at a smaller scale.

Figure 10.7   After applying raster generalization operations, many of the smaller groups of cells have disappeared. Single cells were eliminated using a majority filter or by shrinking and expanding classes of cells.

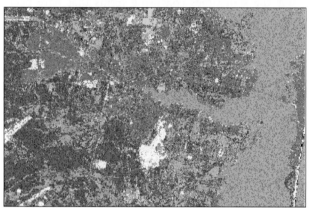

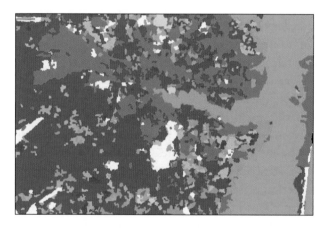

Figure 10.6   The raw satellite image at the top was first classified into an image with a select set of land-cover classes (shown at the bottom). Notice the many small, isolated single cells or groups of cells throughout the image.

Figure 10.8   Raster generalization operations were used to create this generalized land-cover map from the satellite image at the top of figure 10.6.

The final result is a map with an amount of detail that is appropriate to show at a reduced size relating to the smaller scale (figure 10.4, bottom). The maps in figure 10.5 are reduced to the final scale. In the map at the left, you can see how using the original source data produces a result that is very difficult to decipher. The map at the right is much easier to read.

Raster generalization operations

Generalization operations for raster data are used to either clean up small errors in the data or reduce unnecessary detail. The errors may be unclassified data originating from a satellite image, unnecessary lines or text originating from a scanned paper map, or artifacts from the conversion of the data from some other raster format. For example, a thinning function that thins linear features in a raster can be used to clean up scanned paper maps. Figures 10.6, 10.7, and 10.8 illustrate how raster generalization operations were used to eliminate errors and detail in a classified remotely sensed image.

Classification of an image into categories such as water, residential, hardwoods, conifers, and so on results in a jagged, unrealistic representation of the boundaries. This is often the result of various limitations of classification. In the classified image, a *single* cell may be misclassified as different from the cells surrounding it. In reality, the cell normally should belong in the category of the cells surrounding it. These types of errors can be smoothed using raster generalization operations, as shown in figure 10.7. Single, misclassified cells can be generalized by assigning them the value that appears most in its immediate neighborhood using a **majority filter function.** The boundaries between **zones** (cells with the same value) can be smoothed. Then, larger zones can be manipulated to invade smaller zones by expanding and shrinking their boundaries.

It may also be desirable to eliminate *groups* of cells smaller than a certain size. The smallest allowable size for a group of cells in a raster image is the **minimum mapping unit** (**MMU**). Groups below the specified size may be misclassified or too small to be represented as a readable group in the image. Eliminating these groups by combining them with the other classes results in a more generalized land-cover map (figure 10.8).

Classification

Classification can be defined as ordering, scaling, or grouping into classes that simplify the features and their attributes. The goal is to **typify** (provide a typical example of the essential characteristics of) the features and their attributes. In the process, the original features or attributes may be given a modified representation or replaced entirely with a different representation. Although we talked about classification in detail in chapter 8, we will point out a few important aspects of classification as they relate to cartographic generalization here.

Classification of qualitative attributes is achieved by grouping similar features. In figure 10.3, we can see that this could involve the aggregation of points into polygonal features or the amalgamation of areas into composite polygonal features. Classification of quantitative attribute values for map features can be done by numerical or statistical classification, which is what we talked about in chapter 8 relative to choropleth maps. For mapmakers, this requires a high level of expertise to classify features and attributes meaningfully. Mapmakers make better judgments about classification if they are trained to understand quantitative distributions, statistical measures, and classification methods.

Symbolization

Symbolization is the third part of the cartographic abstraction process. Once cartographers have selected, classified, and generalized the environmental data, they symbolize it so the map communicates some geographic "reality" to the map users. Of course, the operations of selection, classification, generalization, and symbolization are more interrelated than this sounds: mapmaking is more of an iterative process than a linear one. Cartographers continually refine the results they obtain through the application of the cartographic abstraction processes.

Symbolization is the use of visual variables to represent the data attributes (see chapters 7 and 8 for more on visual variables). Generalization can be applied to symbolism and will have implications for map accuracy. Cartographers generalize by symbolization in two ways.

First, they can reduce the level of measurement of the attribute value. For example, ratio-level rainfall data can be reduced from classes of 0–2 inches, 2–4 inches, and 4–8 inches to ordinal-level classes of light, medium, and heavy. Second, they can change the conception of the feature's dimensionality. For example, a city, whose boundary is a polygonal feature in reality might be reduced to a point for mapping at smaller scales (see chapter 7 for more on point features).

Because we explored symbolization in detail in chapters 7 and 8, we will move on at this point to talk about the sometimes confusing terms that have been used in relation to map accuracy and uncertainty.

UNCERTAINTY AND ERROR

Uncertainty relates to determination of what is missing in the map representation; it is a relative measure of the discrepancy between reality and its representation. **Error** is a value we use to measure the uncertainty. For features, the **positional error** is measured as the distance between the coordinates of a map feature and its location on earth. The greater the attribute and positional errors, the larger the set of potential values the true value lies within. So we are not really sure where the feature is or how it should be described; this is what we mean by uncertainty.

ERROR AND BIAS

Once we understand what error is, we can examine its relation to bias. **Bias,** whether intentional or not, is a *systematic distortion* of the representation as opposed to a *random error*. For example, consider lines on a map that represent streams digitized from an air photo. The streams that pass under dense tree canopies likely have some error in their representations because of the difficulty or inability to see them in the image when they were digitized. This is *random error*. If, however, all the streams are offset in a particular direction because of misregistration of the air photo to the control points (see chapter 9 for more on air photos and control points) or because a datum was used that is different from the rest of the information on the map (see chapter 1 for more on datums), then there is systematic distortion in the representation rather than random error. This is *bias*.

PRECISION AND ACCURACY

The words "accuracy" and "precision" have different definitions. We have said that map accuracy relates to how well the measured map coordinates conform to the true coordinates of a position. It sounds simple, but what criteria do we use to establish true coordinates? Obviously, "truth" in this case is defined relative to some agreed-upon standard. Examples of standards include the selection of an agreed-upon datum or an allowable horizontal distance from "true" coordinates.

The level of accuracy that you require on a map will vary for different purposes. For example, if you are trying to find the approximate extent of a national forest so that you can hike its trails, you could easily use a topographic map with confidence. However, if you wanted to know exactly where your property lines are so that you can put up a new fence, you would need a more accurate map, so a better choice would be a surveyor's plat map (see chapter 5 for more on these types of maps).

You will sometimes see the accuracy of a map described in terms of horizontal and vertical accuracy, but you may also see or need information about its attribute, conceptual, and logical accuracy. Although we will discuss each of these in more detail in the next part of this chapter, to illustrate positional accuracy (a measurement of the variation between mapped and true positions), table 10.1 shows the statistical standards set by the U.S. Geological Survey (USGS) for the maps it produces and uses.

    1:1,200 ± 3.33 feet
    1:2,400 ± 6.67 feet
    1:4,800 ± 13.33 feet
    1:10,000 ± 27.78 feet
    1:12,000 ± 33.33 feet
    1:24,000 ± 40.00 feet
    1:63,360 ± 105.60 feet
    1:100,000 ± 166.67 feet

Table 10.1 Accuracy standards for various scale USGS topographic maps.

The way to interpret these standards is that a point or a line on a map representing a clearly defined feature on the ground has a "probable" location somewhere within the stated standard. So on a 1:24,000-scale map, a feature on the map will be represented within 40 feet of its true location on the ground.

A related concept is precision. **Precision** is the amount of detail used to report a measurement. For example, for locational error a measurement of 10.625 is more precise than 10.6, but may or may not be a more accurate measure of the error. **Attribute error** is the misreporting of the characteristics of the feature. Precision in the description of feature attributes is distinct from its accuracy. A vegetation polygon with an attribute "mixed forest + some shrub + bare ground" is more precise than "mixed forest," but it may not be more accurate.

Precision has three meanings:

1. The number of significant digits reported for a measurement
2. The repeatability of measurements or the agreement among measurements
3. The rigor and sophistication of the measuring process

Achieving high precision for locational information means that features are positioned to within a fraction of a measurement unit of their true location. Highly precise attribute information means that the descriptive characteristics of features are captured in great detail.

As with accuracy, the level of precision required for particular map uses varies greatly. Surveying projects for road construction require very precise maps so that measurements can be made to fractions of an inch or centimeter. Demographic analyses based on census data do not require such exact measurements, so there can be less precision. For example, it would not make sense to talk about fractions of the number of people who live in a city block.

It is important to understand that precise data may be inaccurate and accurate data may be imprecise. Maps made with data collected using high-grade GPS units may still be inaccurate because of problems such as collecting data for the wrong area or about the wrong features (see chapter 14 for more on GPS and maps). In this case, the data used for mapping may be very precise but there are obvious errors that greatly diminish the accuracy of the information used to make the map. Conversely, highly accurate data may be imprecise. Recall our land-cover classification example. The data were generalized in order to increase the accuracy of the classification, but this resulted in a map with less precision.

False accuracy is reporting your map use results at a higher level of accuracy than it is possible to justify from the map. **False precision** follows along the same line— it is the reporting of results at a higher level of precision than it is possible to get from the map. If locations on a map are only asserted to be within 3 feet of their true position, it makes no sense to measure locations to a tenth of foot. You should only use a map to the level of accuracy and precision for which it was made.

There is a tendency for online maps to assume a greater level of accuracy or precision than the map supports. The ability to zoom to any scale gives users the impression that the underlying data increases or decreases in precision or accuracy. While this is sometimes the case (more highly precise and accurate datasets are used to make maps at larger scales), most uninformed map users will assume that the maps they see at all scales carry the same level of precision and accuracy. The real level of accuracy and precision are tied to the source map scale; they do not change if the user zooms in and out.

Highly accurate and highly precise data for mapping can be costly to compile, so these qualities are sometimes sacrificed in the interest of time or money when the map is made. You should always try to determine the accuracy of a map, especially if you are using it to make precise measurements or accurate decisions.

TYPES OF MAP ACCURACY

Rather than treat map accuracy as a single issue, we'll approach the subject from differing perspectives, including positional accuracy, attribute accuracy, conceptual accuracy, logical consistency, and temporal accuracy.

Positional accuracy

We depend on maps to show us a useful representation of where things are located. In many cases, location means the horizontal position of an object—its latitude and longitude, for example. But, since we live in a three-dimensional world, location also has a vertical component. Thus, we speak of the elevation of the land surface or the depth of a rock stratum. Any discussion of positional accuracy, therefore, is with reference to *horizontal position, vertical position,* or *both*. It is important to consider which are relevant to the map you are using. For example, a prism map of population, as in figure 8.26 (see chapter 8 for more on prism maps) requires only horizontal positional accuracy because the vertical aspect of the map is used for relative measurements. For a topographic map, on the other hand, both horizontal and vertical positional accuracy are important because absolute distance and height measurements may be made using these maps.

Topographic maps and many other maps produced in the United States include the marginal notation, "This map complies with National Map Accuracy Standards." To learn what this statement really means, read the explanation in box 10.1. Here you'll find that, as far as

U.S. National Map Accuracy Standards (and Canadian Provincial Government standards) are concerned, both horizontal and vertical map accuracy are measured statistically in terms of how far off a set of tested points are from their true position on the map.

Notice that horizontal and vertical accuracy pertains only to "well-defined points," such as surveyed triangulation points and bench marks. You will see that acceptable accuracy decreases progressively as map scale decreases — for maps of scales larger than 1:20,000, not more than 10 percent of the points tested shall be in error on the map by more than 1/30 inch (0.08 centimeters), whereas for maps at scales of 1:20,000 or smaller the error in position on the map can be no more than one-fiftieth of an inch (0.05 centimeters). Putting this in terms of error on the ground, one-fiftieth of an inch on a 1:24,000-scale topographic map represents 40 feet (12.2 meters) on the ground. This is also shown in table 10.1.

Vertical map accuracy is assessed similarly to horizontal accuracy — not more than 10 percent of the elevations tested shall be in error by more than plus or minus one-half the contour interval used on the map. For example, on a map with a contour interval of 20 feet (6 meters), the map must correctly show 90 percent of all points tested to within 10 feet (3 meters) of the actual elevation.

In 1987, the American Society of Photogrammetry and Remote Sensing (ASPRS) developed an alternative spatial accuracy standard for large-scale topographic maps, engineering plans, and other detailed maps of the ground produced at scales of 1:20,000 and larger. Two key differences from the National Map Accuracy Standards are (1) the allowable horizontal and vertical error is in feet on the ground, and (2) accuracy is defined in terms of an allowable root mean square error that varies with map scale. The Ordnance Survey topographic mapping agency in Great Britain and government topographic mapping organizations in other nations also use this type of map accuracy standard.

Root mean square error (RMSE) is defined as the square root of the average of the squared discrepancies in position *(d)* of well-defined points *(n)* determined from the map and compared with higher accuracy surveyed locations of each point:

$$RMSE = \sqrt{\sum_i^n d_i^{\,2} / n}$$

The maximum allowable horizontal RMSE for class 1 maps of the highest accuracy varies in a linear manner with map scale, set in English units as 10 feet for 1:12,000-scale maps, 5 feet for 1:6,000-scale maps, and

With a view to the utmost economy and expedition in producing maps that fulfill not only the broad needs for standard or principal maps, but also the reasonable particular needs of individual agencies, standards of accuracy for published maps are defined as follows:

1. Horizontal accuracy. For maps on publication scales larger than 1:20,000, not more than 10 percent of the points tested shall be in error by more than $^1/_{30}$ inch, measured on the publication scale; for maps on publication scales of 1:20,000 or smaller, $^1/_{50}$ inch. These limits of accuracy shall apply in all cases to positions of well-defined points only. Well-defined points are those that are easily visible or recoverable on the ground, such as the following: monuments or markers, such as bench marks, property boundary monuments; intersections of roads, railroads, etc.; corners of large buildings or structures (or center points of small buildings); etc. In general, what is well defined will also be determined by what is plottable on the scale of the map within $^1/_{100}$ inch. Thus, while the intersection of two roads or property lines meeting at right angles would come within a sensible interpretation, identification of the intersection of such lines meeting at an acute angle would obviously not be practicable within $^1/_{100}$ inch. Similarly, features not identifiable upon the ground within close limits are not to be considered as test points within the limits quoted, even though their positions may be scaled closely upon the map. In this class would come timber lines, soil boundaries, etc.

2. Vertical accuracy, as applied to contour maps on all publication scales, shall be such that not more than 10 percent of the elevations tested shall be in error by more than one-half the contour interval. In checking elevations taken from the map, the apparent vertical error may be decreased by assuming a horizontal displacement within the permissible horizontal error for a map of that scale.

3. The accuracy of any map may be tested by comparing the positions of points whose locations or elevations are shown upon it with corresponding positions as determined by surveys of a higher accuracy. Tests shall be made by the producing agency, which shall also determine which of its maps are to be tested, and the extent of such testing.

4. Published maps meeting these accuracy requirements shall note this fact on their legends, as follows: "This map complies with National Map Accuracy Standards."

5. Published maps whose errors exceed those aforestated shall omit from their legends all mention of standard accuracy.

6. When a published map is a considerable enlargement of a map drawing (manuscript) or of a published map, that fact shall be stated in the legend. For example, "This map is an enlargement of a 1:20,000 scale map drawing," or "This map is an enlargement of a 1:24,000 scale published map."

7. To facilitate ready interchange and use of basic information for map construction among all federal map making agencies, manuscript maps and published maps, wherever economically feasible and consistent with the uses to which the map is to be put, shall conform to latitude and longitude boundaries, being 15 minutes, 7.5 minutes, or 3.75 minutes of latitude and longitude in areal extent.

Issued June 10, 1941 U.S. BUREAU OF THE BUDGET
Revised April 26, 1943
Revised June 17, 1947
Source: Thompson, *Maps for America*, p. 104

Box 10.1   United States National Map Accuracy Standards.

so on. In metric units, the error tolerances are defined as 5 meters for 1:20,000-scale maps, 2.5 meters for 1:10,000-scale maps, and so on down to 0.0125 meters for 1:50-scale maps. Vertical accuracy for class 1 maps is similarly defined by an RMSE in the elevations of well-defined points of no more than one-third the contour interval used on the map. Maps of lower accuracy, called class 2 maps, have allowable RMSEs twice that of class 1, and still lower accuracy class 3 maps have allowable RMSEs three times as great as class 1.

Since only a few points on a map are actually tested to see if the map meets National or ASPRS Map Accuracy Standards, the accuracy of most well-defined points on the map likely is unknown. And, since map accuracy for these standards is defined as a maximum allowable percentage of points exceeding an error tolerance, the standards say nothing about the horizontal or vertical accuracy you can expect to find with respect to a specific location that is not one of the "well-defined points" used in the accuracy assessment.

Many of the intricate vegetation patterns existing in nature cannot be depicted exactly by line drawings. It is therefore necessary in some places to omit less-important scattered growth and to generalize complex outlines.

Types

The term "woodland" is generally used loosely to designate all vegetation represented on topographic maps. For mapping purposes, vegetation is divided into six types, symbolized as shown, and defined as follows:

- Woodland (woods-brushwood): An area of normally dry land containing tree cover or brush that is potential tree cover. The growth must be at least 6 feet (2 meters) tall and dense enough to afford cover for troops.

- Scrub: An area covered with low-growing or stunted perennial vegetation, such as cactus, mesquite, or sagebrush; common to arid regions and usually not mixed with trees.

- Orchard: A planting of evenly spaced trees or tall bushes that bear fruit or nuts. Plantings of citrus and nut trees, commonly called groves, are included in this type.

- Vineyard: A planting of grapevines, usually supported and arranged in evenly spaced rows. Other kinds of cultivated climbing plants, such as berry vines and hops, are called vineyards for mapping purposes.

- Mangrove: A dense, almost impenetrable growth of tropical maritime trees with aerial roots. Mangrove thrives where the movement of tidewater is minimal—in shallow bays and deltas, and along riverbanks.

- Wooded marsh: An area of normally wet land with tree cover or brush that is potential tree cover.

Density

Woods, brushwood, and scrub are mapped if the growth is thick enough to provide cover for troops or to impede foot travel. This condition is considered to exist if density of the vegetative cover is 20 percent or more. Growth that meets the minimum density requirement is estimated as follows: If the average open space distance between the crowns is equal to the average crown diameter, the density of the vegetative cover is 20 percent.

This criterion is not a hard-and-fast rule, however, because 20 percent crown density cannot be determined accurately if there are irregularly scattered trees and gradual transitions from the wooded to the cleared areas. Therefore, where such growth occurs, the minimum density requirement varies between 20 and 35 percent, and the woodland boundary is drawn where there is a noticeable change in density. A crown density of 35 percent exists if the average open space between the crowns is equal to one-half the average crown diameter.

Orchards and vineyards are shown regardless of crown density. Mangrove, by definition, is dense, almost impenetrable growth; crown density is not a factor in mapping mangrove boundaries.

Areas

On 7.5- and 15-minute maps, woodland areas covering 1 acre (0.4 hectares) or more are shown regardless of shape. This area requirement applies both to individual tracts of vegetation and to areas of one type within or adjoining another type. Narrow strips of vegetation and isolated tracts covering areas smaller than the specified minimum are shown only if they are considered to be landmarks. Accordingly, shelterbelts and small patches of trees in arid or semiarid regions are shown, whereas single rows of trees or bushes along fences, roads, or perennial streams are not mapped.

Clearings

The minimum area specified for woodland cover on 7.5- and 15-minute maps—1 acre (0.4 hectares)—also applies to clearings within woodland. Isolated clearings smaller than the specified minimum are shown if they are considered to be landmarks.

Clearings along mapped linear features, such as power transmission lines, telephone lines, pipelines, roads, and railroads, are shown if the break in woodland cover is 100 feet (30 meters) or more wide. The minimum symbol width for a clearing in which a linear feature is shown is 100 feet at map scale. Clearings wider than 100 feet are mapped to scale.

Landmark linear clearings 40 feet (12 meters) or more wide, in which no feature is mapped, are shown to scale. Firebreaks are shown and labeled if they are 20 feet (6 meters) or more wide and do not adjoin or coincide with other cultural features. The minimum symbol width for a firebreak clearing is 40 feet at map scale; firebreaks wider than 40 feet are shown to scale.

Source: Thompson, *Maps for America*, pp. 70–71

Box 10.2 Vegetation on standard topographic maps.

If you feel map accuracy standards are a bit vague about how close a map feature is to its ground location, you're on even shakier ground with maps that don't meet the accuracy standards. There is simply no way to ascertain the accuracy of data taken from these maps. We can expect, however, that the larger the map scale, the more reliable the plotted positions of features will be.

In contrast to tangible objects with clearly defined boundaries, some features, such as soils, wetlands, and forests, have boundaries that are transitional and difficult to find in the field. To appreciate what we mean by these types of features, read the material concerning vegetation mapping on standard topographic maps provided in boxes 10.2 and 10.3. Notice that the definitions of vegetation type, density, vegetated areas, and clearings are all human conceptualizations. The intricacy of the pattern to be mapped, the map scale, and the importance assigned to features all enter into the cartographer's judgment of how the mapping should be carried out.

The accuracy of conceptual features such as the boundaries of forested areas on topographic maps means something quite different from the accuracy of discrete features, such as roads, on the same map. As we have seen, positional accuracy is a measurement of the variation between the location of a mapped feature and its true position. It is dependent on the type of feature being mapped. Well-defined features, such as roads, buildings, and property lines, can be placed on a map with a high level of accuracy. Features with boundaries that are less distinct, such as vegetation, soils, and precipitation zones, lack sharp boundaries in nature and may reflect errors introduced by the cartographer's interpretation.

Attribute accuracy

As we learned earlier, all maps are concerned with two primary elements—locations and attributes. So far, we have primarily talked about accuracy and precision in relation to locations, but we can also talk about these in relation to attributes. **Attribute accuracy** is the fidelity in the description of characteristics of the geographic features. Attributes can vary greatly in accuracy and in precision. Accurate attributes describe geographic features with little variation. Precise attributes describe the geographic features in great detail.

To appreciate what we mean by attribute accuracy, read the material concerning vegetation mapping on standard topographic maps provided in boxes 10.2 and 10.3. To mitigate attribute accuracy problems, cartographers and data collectors usually try to define these characteristics explicitly through quantifiable measures. For example, an identification key (figure 10.9) used to determine the attributes for the vegetation polygon data compiled from air photos can be constructed to provide successively greater precision at each level of description

- Clearly defined woodland boundaries are plotted with standard accuracy, the same as any other well-defined planimetric feature. If there are gradual changes from wooded to cleared areas, the outlines are plotted to indicate the limits of growth meeting the minimum density requirement. If the growth occurs in intricate patterns, the outlines show the general shapes of the wooded areas. Outlines representing these ill-defined or irregular limits of vegetative cover are considered to be approximate because they do not necessarily represent lines that can be accurately identified on the ground. The outline of a tract of tall, dense timber represents the centerline of the bounding row of trees rather than the outside limits of the branches or the shadow line.

- In large tracts of dense evergreen timber, sharp dividing lines between different tree heights may be shown with the fence- and field-line symbol. Published maps containing fence-line symbols that represent fences and other landmark lines in wooded areas bear the following statement in the tailored legend: "Fine red dashed lines indicate selected fence, field, or landmark lines where generally visible on aerial photographs. This information is unchecked."

- Woodland is not shown in urban-tint areas, but it is shown where appropriate in areas surrounded by urban tint if such areas are equivalent to or larger than the average city block.

- Mangrove is shown on the published map with the standard mangrove pattern and the green woodland tint. Breaks in the mangrove cover usually indicate water channels that provide routes for penetrating the dense growth.

Source: Thompson, *Maps for America*, pp. 71–72

Box 10.3 Woodland boundary accuracy on standard topographic maps

(see chapter 7 for more on identification keys). If the features cannot be described with the greatest amount of detail, the interpreter falls back to a lower precision description. You can well imagine that the intricacy of the pattern to be mapped, the map scale, and the importance assigned to features all enter into the cartographer's judgment of how the assignment of vegetation attributes should be carried out.

As with positional accuracy, standards are applied to assure a minimum level of accuracy. For example, a standard might state that in 90 percent of cases tested, the correct category from among those defined must be chosen. In order to determine if a map meets these standards, ancillary source data with greater confidence is used. A statistically significant number of randomly chosen features is selected and checked for attribute (and positional) accuracy.

Conceptual accuracy

Conceptual accuracy means using the correct data to represent the real-world feature or phenomena on the map. It is the responsibility of the mapmaker to try always to present an accurate message. Sometimes error, lack of knowledge, or lack of attention will impede the

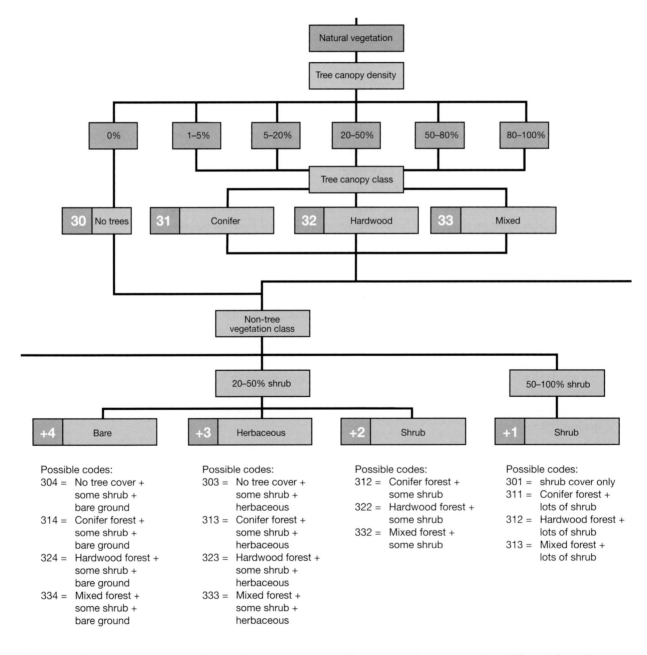

Figure 10.9 **This portion of a key used for air photo interpretation. The key provides more precise attribute information at successive levels of description.**

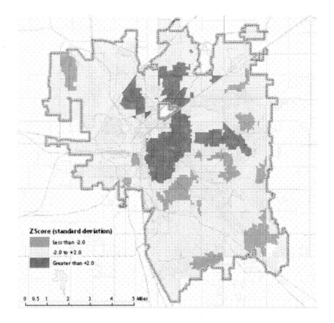

Figure 10.10 In the map at the left, the reader is led to assume that the cyan- and orange-colored areas are also important results of the hot-spot analysis. Since this analysis only identifies areas that have more clustering than expected (hot spots) or less (cold spots), the representation is not conceptually accurate. In the map at the right, this problem is corrected.

process. The map reader also shares in the responsibility of evaluating the map's message. This is illustrated in figure 10.10, which shows the results of a **hot spot analysis** used to identify clustering in sets of points. The analysis shows where there are clusters of more than the expected number of points and where there are less than the expected number. The results are based on the standard deviation of a metric that is statistically significant at a 95 percent confidence level (that is, values greater than 1.96 and less than –1.96). Greater than 1.96 means there is significantly more clustering than expected (hot spot); less than –1.96 means there is significantly less clustering than expected (cold spot). From this description, we can see that the analysis identifies only two significant classes (hot and cold spots); all other values are insignificant. Look at the map at the left in figure 10.10. Note that there are hot and cold spots, but there are also "warm" and "cool" spots as shown by the orange and cyan colors, respectively. This is inconsistent with the results of the analyses, so that map does not make conceptual sense. On the map at the right, the problem was corrected; only the hot and cold spots are indicated.

Logical consistency

Logical consistency is the internal consistency of the representation—it helps to ensure that the proper relationships are conveyed. The map at the top of figure 10.11 is difficult to read because the colors and widths of the lines and the way that they connect do not make sense (that is, a single line feature has internal inconsistencies). The map at the bottom has been corrected so that the features and their connections are easy to understand.

Temporal accuracy

Another important aspect of maps is their **temporal accuracy** or how well they reflect the temporal nature of the mapped features. An outdated map can lead to serious confusion, inaccurate measurements, and erroneous conclusions. For instance, if a new interchange was constructed after information was collected for your road map, you may miss the correct exit and become lost. Mapmakers sometimes try to extend a map's practical life by mapping features such as interchanges that have been planned but not yet built. The problem is that if the proposed roads are never constructed or aren't completed on schedule, then the map may be more confusing than it is helpful.

The question, then, is how dated can a map be and still be accurate enough to be useful? As usual, there is no one answer. You probably can tolerate different degrees of temporal inaccuracy with respect to different features and in different circumstances. The utility of a map depends on both the amount of change and your ability to imagine the change between the state of the environment when the map was made and the reality that now exists. Most of all, your tolerance for temporal inaccuracy depends on

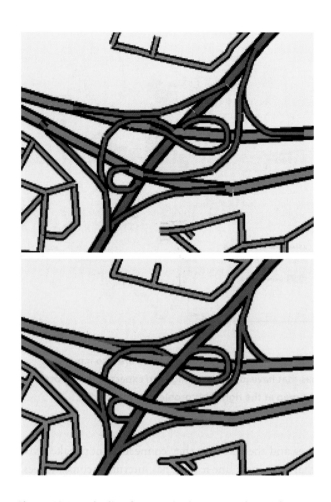

Figure 10.11 **The line features in the map at the top have internal inconsistencies; the map at the bottom makes more sense logically—the features and their connections are easy to understand.**

the consequences of arriving at a wrong conclusion about the environment through map use. Sometimes an outdated map may merely be inconvenient; at other times, the consequences of its use may be more serious. Can you imagine the consequences of using an outdated map in marine or air navigation? It's up to the user to determine that he or she has the latest information and is not fooled by a map that purports to be up to date but isn't. Experience will teach you that you always have to be ready to do reality checks with any map or chart.

Sometimes the temporal accuracy of a map is reduced because maps of a certain region haven't been produced or revised for a long time. Cartographers tend to give higher priority to mapping hitherto unmapped regions than to updating maps of areas that have already been mapped. Their logic is that it's easier for the user to cope with an outdated map than no map at all, assuming the reader will realize the representation is dated. At other times, the trouble is that, although updated maps have

been made, you don't happen to have one. Who hasn't known the frustration of trying to use a 20-year-old street or road map unearthed from the depths of the glove compartment? These outdated maps can cause no end of confusion. If you are such a skilled map user that you instantly spot changes in the environment and know what decisive action to take, you may be able to manage with an old map. But most people caught in this situation could better spend their effort locating a new map.

There are a number of ways to think about the temporal accuracy of maps and the features on them. We'll look at four: currency, mapping period, elapsed time, and the temporal stability of features.

Currency

One of the first things you should do when you pick up any map is to look for indicators of the map's datedness, or **currency.** The "up-to-dateness" of the map tells you if it really reflects the present time. Or, if you are interested in historical landscapes, the currency can tell you if the map accurately represents your time period of interest.

You may run into trouble finding currency information, for many maps aren't dated. At times the date is left off inadvertently. At other times, mapmakers deliberately avoid adding dates so that maps won't appear outdated through long revision cycles. Even if there is a date, it may be difficult to find or written in a cryptic code (table 10.2). If you must use an outdated map, look for changes between the map and the environment it depicts. You may need to extrapolate in areas where there are obvious indicators of change. Your need for action must be carefully weighed against the variations you find between the map and the environment.

From the examples above, you can see that date definitions are also subject to generalization, as they can range from less to more specific.

There are a couple of indicators that will give you clues to the currency of the map. The first, more specific indicator is the compilation date for the data used to make the map. The other, more often shown indicator is the date that the map was completed.

Date notation	Example
YYYY-MM-DD	1995-12-31
YYYYMMDD	19951231
YYYY-MM	1995-12
YYYY	1995
YYYY-Week#	1995-W52
YYYYWeek#	1995W52
YYYY-Week#-Day#	1995-W52-2
YYYYWeek#Day#	1995W522
YYYY-Day of the Year	1995-365
YYYYDay of the Year	1995365
Time notation	**Example**
hh:mm:ss	23:59:59
hhmmss	235959
hh:mm	23:59
hhmm	2359
hh	23
hh:mm:ss.fractions of a second	23:59:59.9942
hhmmss.fractions of a second	235959.9942
Date and time	**Example**
YYYYMMDDT(ime)hhmmss	19951231T235959

Table 10.2   **International standard date and time notation.**

Data compilation date

The growing use of data makes the issue of currency in mapping more complicated and potentially more specific. Digital data maps are usually compiled from information collected at different times and sometimes from maps of different dates. If the source date of all the digital data used to compile the map is recorded in its description or **metadata** (we will talk about this more later in the chapter), then it is possible to determine the currency of all aspects of the map.

Map completion date

In the absence of information about the data used to make the map, the date of completion at least tells you the latest possible compilation date for any of the data the mapmaker used. On many maps the completion date is clearly shown. The date on remote sensing images and air photos used for image maps (see chapter 9 for more on image maps) tends to be the most specific. Most dates on standard low-altitude photographs, for instance, are given to the nearest day. Dates on conventional maps are usually less specific. As you would suspect, the features shown on most maps don't represent data gathered at one instant in time, so the date is recorded less specifically to

reflect this temporal inexactness. The completion date may be only one of several dates shown for the map. Dates for revisions or additions to the map may also be indicated.

Notes on USGS topographic quadrangles, for example, tell when photographs were taken and when field checking of the map compilation was completed. Maps that use census information usually give the date of data collection as well as the date of completion.

Mapping period

Another way to think about temporal accuracy is the time it took to complete the map. **Mapping period** is the time between initial data collection and final map printing. If the mapping period is lengthy and if mapped phenomena change rapidly, the map can be well out of date before it is even completed.

The completion date isn't always given on a map—the length of the mapping period almost never is, so you usually have to infer it, if that is even possible. Look at the mapping notes on the topographic quadrangle in figure 10.12. The publication and map editing date is 1992, but the map was compiled from 1985 aerial photography. The mapping period for this map was seven years. This illustrates that a new map may, in a sense, be old if the mapping period was long and the environment mapped is dynamic.

Confusion over mapping period sometimes arises where least expected. Image maps are a good example. It is true that individual air photos give you an instantaneous

Courtesy of the U.S. Geological Survey.

Produced by the United States Geological Survey
Control by USGS and NOS/NOAA
Compiled from aerial photographs taken 1985
Field checked 1987. Map edited 1992
North American Datum of 1927 (NAD 27). Projection and
1000-meter grid, Universal Transverse Mercator, zone 6
10 000-foot grid ticks based on Alaska Coordinate System, zone 3
The difference between NAD 27 and North American Datum of
1983 (NAD 83) for 7.5 minute intersections is given in USGS
Bulletin 1875. The NAD 83 is shown by dashed corner ticks
There may be private inholdings within the boundaries of the
National or State reservations shown on this map
Swamps, as portrayed, indicate only the wetter areas, usually of low
relief, as interpreted from aerial photographs

Figure 10.12   **It is good practice to pay attention to the dates printed on maps. The publication date on this USGS topographic quadrangle is 1992, but the date of the aerial photography used to compile the map is 1985, so the mapping period was seven years.**

view of the environment. But image maps covering large ground areas at smaller scales are often created by taking a number of photos and then putting them together to make an orthophotomap (see chapter 9 for more on orthophotomaps). The photos used to make the orthophotomap may have been gathered over a period of hours or days, or—in the case of image maps made up of several satellite images—weeks or months. Usually, however, image maps have a shorter mapping period than conventional maps. This means that image maps are generally better than conventional maps at showing the current state of the environment at the date of publication.

The mapping period of some maps is now being reduced to almost the same temporal lag as image maps. Using computers and automated recording stations that continuously stream data to the mapping system, maps can be completed only minutes after information is collected. You may have seen examples of traffic maps that show current travel conditions. Computer maps of a short-lived or rapidly changing feature can be created while the feature still exists—hurricane tracking maps are an example.

Elapsed time

Another way to consider the temporal accuracy of maps is the **elapsed time,** or the length of time between its date of completion and its date of use. A traffic map that was current when you used it to gauge yesterday's rush hour conditions will be of little use to you today. The faster the environmental features on the map change, the shorter the acceptable elapsed time since the map was completed.

With the advent of ever more powerful computer mapping systems, mapping period and elapsed time are collapsing—map data are collected, maps are produced, and maps are used within decreasing amounts of time. It is not hard to imagine a time when the whole map transformation process will appear to be instantaneous.

Temporal stability of features

Another thing that contributes to errors on maps is the **temporal stability** of environmental features—that is, how long something stays the same over time. Different features on the same map change at different rates. As a result, some features will hardly have changed since the mapmaker began compiling it, while others will be vastly different or may have disappeared altogether. Thus, we can say that some features are more sensitive than others to temporal instability because they are short-term or intermittent. Roads are temporarily closed, for example, because of routine roadbed and bridge maintenance.

Rivers swollen by spring runoff or storm-induced flooding may be unnavigable for short periods. Certain soils may become impassable to farm equipment or recreational vehicles during spring break-up and periods of rainfall.

To read maps effectively, we must either grasp the effects of these—and similar—short-term conditions, or we must hope that the cartographer embedded information about the temporal stability of the phenomena on the map through symbology or annotation. For example, the USGS defines streams in part by their temporal nature as shown in this excerpt from the "Standards for USGS and USDA Forest Service Single Edition Quadrangle Maps":

STREAM/RIVER—A body of flowing water.

Characteristics
Show the following STREAM/RIVERS based on the portion of the year they contain water:

Intermittent	Contains water for only part of the year, but more than just after rain storms and at snowmelt.
Perennial	Contains water throughout the year, except for infrequent periods of severe drought.

In figure 10.13, you can see the symbology for these types of streams is different, so the way the features appear on the map gives you a clue to their temporal stability.

SOURCES OF ERROR

Now that we have a better understanding of accuracy issues that relate to mapped features and their attributes, it might help to know how errors come to be on maps. For this, we can turn to a discussion of sources of error.

There are many sources of error that may affect the quality of a map. Some are quite obvious, but others can be more difficult to discern. There is an inherent imprecision in cartography that begins with the projection process and its necessary distortion of the data, and imprecision may continue throughout the mapping process. Recognition of error and its sources will help map readers be more judicious when using maps.

Map error may be introduced deliberately. For example, the cartographer may purposely leave features off the map for design or copyright reasons. More often errors are introduced accidentally. For example, incomplete knowledge of the subject being mapped, lack of time, personal bias on the part of the cartographer, or technical clumsiness or ineptness can result in errors on maps.

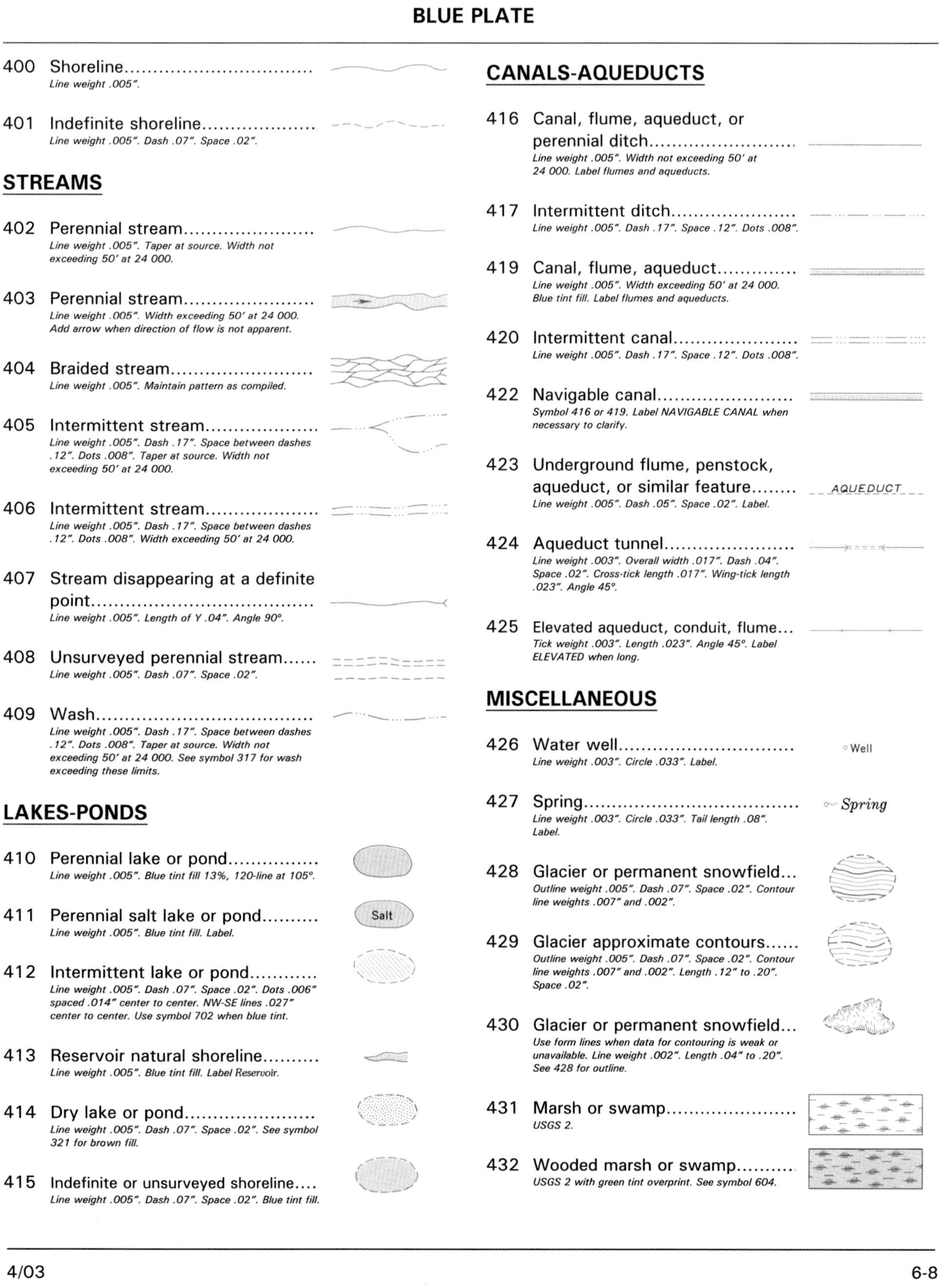

BLUE PLATE

400 Shoreline...............................
Line weight .005".

401 Indefinite shoreline.....................
Line weight .005". Dash .07". Space .02".

STREAMS

402 Perennial stream.......................
Line weight .005". Taper at source. Width not exceeding 50' at 24 000.

403 Perennial stream.......................
Line weight .005". Width exceeding 50' at 24 000. Add arrow when direction of flow is not apparent.

404 Braided stream.........................
Line weight .005". Maintain pattern as compiled.

405 Intermittent stream....................
Line weight .005". Dash .17". Space between dashes .12". Dots .008". Taper at source. Width not exceeding 50' at 24 000.

406 Intermittent stream....................
Line weight .005". Dash .17". Space between dashes .12". Dots .008". Width exceeding 50' at 24 000.

407 Stream disappearing at a definite point.................................
Line weight .005". Length of Y .04". Angle 90°.

408 Unsurveyed perennial stream......
Line weight .005". Dash .07". Space .02".

409 Wash...................................
Line weight .005". Dash .17". Space between dashes .12". Dots .008". Taper at source. Width not exceeding 50' at 24 000. See symbol 317 for wash exceeding these limits.

LAKES-PONDS

410 Perennial lake or pond...............
Line weight .005". Blue tint fill 13%, 120-line at 105°.

411 Perennial salt lake or pond..........
Line weight .005". Blue tint fill. Label.

412 Intermittent lake or pond............
Line weight .005". Dash .07". Space .02". Dots .006" spaced .014" center to center. NW-SE lines .027" center to center. Use symbol 702 when blue tint.

413 Reservoir natural shoreline..........
Line weight .005". Blue tint fill. Label Reservoir.

414 Dry lake or pond......................
Line weight .005". Dash .07". Space .02". See symbol 321 for brown fill.

415 Indefinite or unsurveyed shoreline....
Line weight .005". Dash .07". Space .02". Blue tint fill.

CANALS-AQUEDUCTS

416 Canal, flume, aqueduct, or perennial ditch..........................
Line weight .005". Width not exceeding 50' at 24 000. Label flumes and aqueducts.

417 Intermittent ditch.....................
Line weight .005". Dash .17". Space .12". Dots .008".

419 Canal, flume, aqueduct.............
Line weight .005". Width exceeding 50' at 24 000. Blue tint fill. Label flumes and aqueducts.

420 Intermittent canal.....................
Line weight .005". Dash .17". Space .12". Dots .008".

422 Navigable canal.......................
Symbol 416 or 419. Label NAVIGABLE CANAL when necessary to clarify.

423 Underground flume, penstock, aqueduct, or similar feature........
Line weight .005". Dash .05". Space .02". Label.

424 Aqueduct tunnel......................
Line weight .003". Overall width .017". Dash .04". Space .02". Cross-tick length .017". Wing-tick length .023". Angle 45°.

425 Elevated aqueduct, conduit, flume...
Tick weight .003". Length .023". Angle 45°. Label ELEVATED when long.

MISCELLANEOUS

426 Water well..............................
Line weight .003". Circle .033". Label.

427 Spring....................................
Line weight .003". Circle .033". Tail length .08". Label.

428 Glacier or permanent snowfield...
Outline weight .005". Dash .07". Space .02". Contour line weights .007" and .002".

429 Glacier approximate contours......
Outline weight .005". Dash .07". Space .02". Contour line weights .007" and .002". Length .12" to .20". Space .02".

430 Glacier or permanent snowfield...
Use form lines when data for contouring is weak or unavailable. Line weight .002". Length .04" to .20". See 428 for outline.

431 Marsh or swamp.......................
USGS 2.

432 Wooded marsh or swamp..........
USGS 2 with green tint overprint. See symbol 604.

Figure 10.13 This page out of the USGS document "Standards for 1:24,000 and 1:25,000 Scale Quadrangle Maps" shows the symbology for various hydrographic features. The symbology used helps map readers to understand the temporal stability of some of the features.

P. A. Burrough (1986) in chapter 6 of *Principles of Geographic Information Systems* talks about two types of sources of error in geographic information systems (GIS): obvious sources of error, errors arising from natural variation or from the original data, and errors arising through processing. Generally errors of the first two types are easier to detect than those of the third because errors arising through processing can be more subtle. We will use Burroughs's categories to talk about the source of errors on maps, but we will also add one other—factual errors.

Factual errors

Sometimes features are left off a map by mistake—a lake or town may have been overlooked by the cartographer. In other cases, features found on the map don't exist in the environment. Sometimes a name or symbol is misplaced on the map, or the wrong symbol or name is used. At other times a feature has disappeared from the environment, but its map symbol persists. Still other factual errors occur when a feature is misclassified by its map symbol, as when a railroad is shown as a road. These are all examples of **factual errors.** Thorough map editing can minimize errors of this type, but since mapmakers rely on data from diverse sources, mistakes can be made at many junctures in the map transformation process. If factual errors were known, they would be corrected as a matter of professional pride. In the event that errors do slip through, cartographers rely on feedback from map users to keep from repeating mistakes in subsequent map editions.

Obvious sources of error

Obvious sources of error relate in large part to the data used to compile the map—obviously if inappropriate data are used, the map will be less accurate. Obvious sources of error include the use of data that are out of date or incomplete (do not cover the entire area or do not include all the required features or attributes). Often the desired data may not exist and *surrogate* data have to be used instead. A valid relationship must exist between the surrogate and the phenomenon it is substituted for, but even then error can creep in because data about the phenomenon are not collected directly. Even if the right data do exist, lack of access or excessive cost can also result in the need to use surrogate data.

Errors resulting from natural variations or from the original data

Sometimes error is introduced because it was "hidden" from the cartographer—this includes error resulting from natural variation in the phenomena being mapped. Although these sources of error may not be as obvious, careful checking will reveal their influence on the map. To understand errors resulting from the original data we can think back to positional accuracy, attribute accuracy, conceptual accuracy, logical consistency, and temporal accuracy. Variations in data may be caused by measurement error introduced by faulty observation, biased observers, or miscalibrated or inappropriate equipment. There may also be a natural variation in the data being collected, a variation that may not be detected during collection. In any case, if the errors do not lead to unexpected results, their detection may be extremely difficult.

Errors arising through processing

Processing errors are the most difficult to detect by map users and must be specifically looked for. This requires knowledge of the data and mapping system used to create the map. These errors are more subtle, and they can potentially happen in multiple ways; therefore, they are usually more insidious, especially if they occur in multiple sets of data used to create the map.

One type of processing error is **numerical error.** Examples of numerical error include computer processing that rounds off operation results or faulty processors that compromise the data. These will affect other processing operations, such as conversion of scale, map projection, or conversion of data format (for example, raster to vector format or vice versa). **Topological errors** can occur when multiple layers of maps that are not uniform are overlayed, resulting in small lines or polygons that are unique features when they should be part of larger adjacent features. Even **physiological errors** of the mapmaker (tiredness, poor eyesight, shakiness, and so forth) can introduce error.

COMMUNICATING ACCURACY AND UNCERTAINTY

Maps should be documented as to accuracy and precision. A report of the map's pedigree recounts the steps in the map transformation process. Knowing the datasets and the processes used to compile them, you can tell if unreliable, unknown, or poor-quality data— or questionable or unsound mapping processes—have resulted in questionable or unknown map accuracy. Many agencies now provide data quality reports on how their maps were compiled and how the datasets used to make them were processed.

Metadata

In terms of mapping, **metadata** is a term for information describing the different sets of data used to create maps. Professional cartographers and map interpreters often look at **data quality reports** that are part of the metadata files for the digital datasets they are using. In the United States, a data quality report is a required part of the **Spatial Data Transfer Standard (SDTS)** developed under the leadership of the USGS. The idea behind the SDTS is that all datasets used in mapping and GIS will have a metadata file that contains a data quality report. Three elements of the data quality report are of particular interest to map users: lineage, positional accuracy, and attribute accuracy.

The SDTS defines **lineage** as the source material used to create the digital datasets for mapping, as well as any mathematical transformations of the source material. You will learn who was responsible for creating the source material, the date of original source material collection on the ground and of information used in later updates, and the map scale and projection used if the source material was in map form. The dates of source material collection allow you to assess the currency of the map you are using since you can easily determine the mapping period and the elapsed time from the date of source material collection.

The *positional accuracy* section of the data quality report gives the type of positional accuracy assessment carried out on the map, and the degree to which the map complies with a standard such as the U.S. National Map Accuracy or ASPRS standards based on root mean square error. You may also find that instead of these positional accuracy assessment standards based on an independent source of higher accuracy, a *deductive estimate* of accuracy based on the mapmaker's knowledge of errors occurring at each step in the map production process was used.

Attribute accuracy can be assessed in several ways. Deductive estimates of misclassification errors based on the mapmaker's professional experience are acceptable, although tests of accuracy based on comparison with ground sample sites of known categories are preferred. The idea is to determine the degree of correspondence between the categories of ground sample sites and the categories shown on the map at the same location. You may also see the accuracy of area feature categories on the map estimated by digitally overlaying a more detailed larger-scale map of higher accuracy. For example, the accuracy of land-cover categories on a 1:50,000-scale map could be assessed by overlaying a more detailed 1:24,000-scale land-cover map to determine the percent correspondence between the two maps for each land-cover category. To do this, classes like single-family and high-density residential shown on the 1:24,000-scale map may have been combined into the general residential category shown on the 1:50,000-scale map being tested for attribute accuracy.

Symbols and notations

The most direct way cartographers communicate varying degrees of map accuracy to users is with symbols that indicate differences in accuracy. For example, they may use dashed instead of solid line symbols for less accurately known boundaries (figure 10.14A). On USGS topographic maps, you'll find dotted lines used for less accurately positioned supplemental contours and form lines (see chapter 6 for more on different types of contours). And in figure 10.13, you can see that the topographic map symbol for indefinite shorelines is a dashed blue line rather than a solid blue line. Another map design strategy is to make the clarity or sharpness of symbols increase with higher-accuracy data. Less accurately located features are portrayed with blurred or fuzzy symbols. For example, a transitional boundary, like the extent of a wetland or climate zone, might be shown with a semitransparent line symbol that fades out to the background color along its edges.

A method almost as direct as using map symbols for inaccurate information is to add notes describing areas of lower accuracy on the map. A note such as "This region was not field checked" serves this purpose (figure 10.14B). On aeronautical charts, it's common to find notations of this sort, especially to warn that the general magnetic declination information given on the map may be subject to local disturbance.

Legend disclaimer

Sometimes a note in the map legend is the only indicator of map accuracy. The statement, "This map meets National Map Accuracy Standards," is an example of a **legend disclaimer.** You will now know what such a statement means. Remember, there are no rules saying that you must be warned of an inaccurate map by a specific statement in the legend, so this kind of statement will not always be there to help you.

A legend disclaimer is commonly found on commercial maps that use government maps as a base. For example, someone may enhance nautical charts with direction and distance information and then market this value-added product (value-added means that the map or chart has been enhanced with extra features). You're likely to find the statement, "This map is for reference only and should not be used for navigational purposes." The aim, of course, is to avoid lawsuits resulting from accidents ascribed to the map. It's a curious kind of message, however, since the map is clearly intended for navigational use.

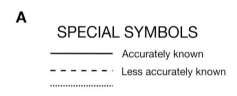

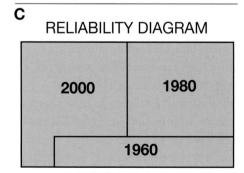

Figure 10.14 **The accuracy of mapped features is often indicated by dotted or dashed symbols (A), legend notes (B), or reliability diagrams (C).**

The most common types of legend notes tell you the date of map production and source of the data. The date is especially useful if the mapped region is undergoing rapid change. The source of data gives a hint of its reliability, since some data gathering organizations have better reputations than others for doing professional quality work. Data from promotional and private survey groups should be viewed with suspicion.

Reliability diagram

The third way to communicate map accuracy is through a **reliability diagram.** This is a simple outline map showing variation in source data used to produce the map (figure 10.14C). The diagram may appear in the legend or as an inset near the map margin. The date and source of the data used in mapping are usually provided.

LIABILITY ISSUES

Map accuracy is a concern for both the mapmaker and map user in a society quick to sue. You compound your problems if you are both mapmaker and user, as is often the case when working with digital maps. The question is to what degree do you place yourself at legal risk when making and using maps? Let's consider map liability issues from the perspective of both maker and user.

Mapmaker responsibility

The interpretation of liability law makes the responsibility of the mapmaker quite clear. In essence, "every reasonable effort" must be made to ensure map or cartographic database quality. What is reasonable is judged by contemporary professional standards.

Additionally, the mapmaker must inform or warn map users of potential problems and hazards in using the product. We discussed ways this can be done in the previous section on communicating map accuracy.

Map user responsibility

The legal responsibility of the map user is less well defined than that of the mapmaker. But the same advice holds true—"every reasonable effort" must be made to ensure high-quality map use. This is especially true if careless map use on your part has the potential to harm someone else. If your business is to market map products, then your responsibility to conduct yourself professionally is increased.

Speaking ethically rather than legally, you probably have no business using maps if low skill on your part could harm others or the environment. You especially have an obligation to understand that the map does not and cannot always show the whole truth. We discussed the many ways in which you can evaluate the truth of maps throughout this chapter.

Many subjects that seem simple turn out, upon closer inspection, to be surprisingly complex. This is the case with map accuracy. The more you think about the topic, the more facets you discover. Knowing, as you do now, the various issues that relate to map accuracy will lead you to use maps more judiciously in the future.

As we've noted, maps are graphic representations of the environment, and graphic representation by its very nature involves distortion. This is the paradox of cartography. We're using a distorted representation to make accurate decisions about spatial issues. Inaccuracies are the price we must pay to obtain the convenience of using maps. Since accuracy problems can't be eliminated—or we would have no maps—we must learn to live with them. Each time you use a map, you should evaluate its accuracy in light of the task at hand. You'll want to ask, "Can I live with the accuracy of this map in this situation?" Or, "What special care should I take if I use this map for this purpose?"

SELECTED READINGS

Bolstad, Paul. 2005. *GIS fundamentals: A first text on geographic information systems.* 2d ed. White Bear Lake, Minn.: Eider Press.

Branscomb, A. 2002. Uncertainty and error. In *Willamette River Basin Atlas.* 2d ed., 156–57. http://www.fsl.orst.edu/pnwerc/wrb/Atlas_web_compressed/PDFtoc.html.

Buckner, B. 1997. The nature of measurement, part II: Mistakes and errors. *Professional Surveyor* (April): 19–22.

——. 1997. The nature of measurement, part IV: Precision and accuracy. *Professional Surveyor* (July–August): 49–52.

Burrough, P. A. 1990. *Principles of geographical information systems for land resource assessment.* Oxford: Clarendon Press.

Chang, Kang-tsung. 2006. *Introduction to geographic information systems,* 3d. ed. Boston: McGraw-Hill.

Chrisman, N. R. 1991. A diagnostic test for error in categorical maps. In *Proceedings of AUTO-CARTO 10,* 330–48. Baltimore: American Congress on Surveying and Mapping.

——. 2002. *Exploring geographic information systems.* 2d ed. New York: John Wiley & Sons, Inc.

Congalton, R. G., and K. Green. 1997. *Assessing the accuracy of remotely sensed data: Principles and practices.* Boca Raton: Lewis Publishers.

Frans, J. M., et al. 1994. Visualization of data quality. In *Visualization in modern cartography,* ed. A. MacEachren and D. R. F. Taylor, 313–31. New York: Pergamon Press.

Guptill, S. C., and J. L. Morrison, eds. 1995. *Elements of spatial data quality.* London: Elsevier Science Ltd.

Hopkins, L. D. 1977. Methods of generating land suitability maps: A comparative evaluation. *Journal of American Institute of Planners* 43 (4): 386–98.

Lo, C. P., and Albert K. W. Yeung. 2002. *Concepts and techniques of geographic information systems.* Upper Saddle River, N.J.: Prentice Hall.

Lodwick, W. A., W. Monson, and L. Svoboda. 1990. Attribute error and sensitivity analysis of map operations in geographical information systems: Suitability analysis. *International Journal of Geographical Information Systems* 4 (4): 413–28.

Longley, Paul A., Michael F. Goodchild, David J. Maguire, and David W. Rhind. 2005. *Geographic information systems and science.* 2d ed. Hoboken, N.J.: John Wiley & Sons, Inc.

MacEachren, A. M. 1994. S*ome truth with maps: Primer on symbolism and design.* Washington, D.C.: Association of American Geographers

Merchant, D. C. 1987. Spatial accuracy specification for large scale topographic maps. *Photogrammetric Engineering and Remote Sensing* 53 (7): 958–61.

Monmonier, M. 1996. *How to lie with maps.* 2d ed. Chicago: University of Chicago Press.

Mowrer, H. T. 1999. Accuracy (re)assurance: Selling uncertainty assessment to the uncertain. In *Spatial accuracy assessment: land information uncertainty in natural resources,* ed. Kim Lowell and Annick Jaton, 3–10. Chelsea, Mich.: Ann Arbor Press.

Slocum, Terry A., Robert B. McMaster, Fritz C. Kessler, and Hugh H. Howard. 2005. *Thematic cartography and geographic visualization.* 2d ed. Upper Saddle River, N.J.: Reason-Prentice Hall.

Thompson, M. M. 1988. *Maps for America: Cartographic products of the United States Geological Survey and others.* 3d ed. Washington, D.C.: U.S. Government Printing Office.

Tobler, Waldo. 1979. A transformational view of cartography. *The American Cartographer* 1 (2): 101–6 .

USGS. 2003. National Mapping Program Standards. http://nationalmap.gov/gio/standards/.

In your study of map reading in part I, you've learned what you might expect to find on a map and gained an appreciation for the mapmaking process. This is necessary background for the second phase of map use: analysis. Here your goal is to analyze and describe the spatial structure of—and relationships among—features on maps.

Sometimes, you can carry out spatial analysis directly in the environment. Often, this is done more easily and inexpensively by analyzing features on maps. A map cuts through the confusion of environmental detail and complexity, making spatial relationships easier to see. Yet even a map doesn't make everything apparent at a glance. The purpose of map analysis, therefore, is to reduce what might appear to be a muddle of information on a map to some sort of order you can understand and describe to other people.

It is possible to do this visually—to view mapped information and describe it by saying, for example, "This area looks hilly" or "That pattern is complex" or "There seems to be a strong correlation between those variables." Traditionally, map analysis has been performed in just such a way. There are problems, however, with a visual approach to map analysis. First of all, such terms as "hilly," "complex," or "correlated" are subjective and vague. They are merely estimates based on personal experience. Two or more people looking at the same map would probably use different terms to describe it. You would also have a hard time conjuring up an image of the landscape on the basis of their nebulous descriptions. What picture comes to mind, for instance, if someone says that a hillside is "steep"? Your mental image of steepness may be quite different from that of the person sitting next to you. How do you know that the judgment of the person who described the hill as steep wasn't biased by his poor physical condition and the

hill isn't steep at all? Furthermore, all these problems with visual analysis are compounded as patterns grow more complex, as details on maps become more subtle, and as the number of mapped features increases.

Obviously, then, if you're going to extract information from a map so that someone else will understand what you have discovered, you need an objective way to describe mapped phenomena. By "objective," we mean repeatable: Two people looking at the same map pattern would describe it in the same way, so you could be sure that their descriptions were trustworthy. Such objectivity is provided by quantitative analysis, in which you use numbers rather than words and replace visual estimation with counting, measurement, and mathematical pattern comparison. These activities let you convert mapped information to numerical descriptions in a rigorous way. You may be satisfied simply with raw numbers—the area and depth of a lake, the length of its shoreline, or the number of houses along it. But it's often more interesting and useful to combine and manipulate raw numbers to obtain more sophisticated information. You might, for instance, want to add up the number of people in a city or state and divide it by the area to get an understanding of the population density (the number of people per square mile or kilometer). You could then compare this figure with the population density for other cities or states. There are hundreds of similar ways to combine, compare, and manipulate quantitative spatial information.

In theory, quantitative analysis is strictly repeatable. Using the same map and analytical procedures, each map user should arrive at the same conclusions. But there's no limit to the number of mathematical methods from which to choose. Moreover, the choice of best or most appropriate method is by no means obvious, even for someone

Part Two
Map Analysis

knowledgeable about quantitative analysis. This means that even with rigorous analysis, there's no guarantee that two or more people will come to identical conclusions when working from the same map.

In general, however, variations in conclusions are far less with quantitative than with visual analysis. You must decide whether the added objectivity and precision gained by using quantitative methods will be great enough to warrant the extra effort required. In many situations, verbal descriptions and estimates based on simple observation are all you need. You can capitalize on the advantages of quantitative procedures more easily, of course, by letting computers do the mathematical work for you.

In theory, too, quantitative methods are absolute and precise. In practice, however, there are several potential sources of error. As we saw in our discussion of map reading in part I, the map itself is not error-free. The tools, materials, and techniques of the data collector and the mapmaker lead to many distortions of mapped information. Even if it were possible to make a perfect map, map analysts would still introduce their own errors. Some of these errors are random and can be minimized only by exercising great care during analysis. Other potential errors are systematic. For example, they might be caused by inaccuracies in the data collection methods or our tools of analysis. Others are random and can be attributed to factors such as human bias and the fact that human vision has resolution limits. You can compensate for both types of systematic error once you know they exist.

Two other cautionary notes are warranted. First, map analysis is based on the assumption that we're working with Euclidian physical space, but the earth is basically spherical and thus non-Euclidian. This problem has plagued nearly everyone who has tried to conduct analytical studies based on map information. The second caution is that map analysis gives you descriptions, not explanations or interpretations. Analyzing a map's geometry is designed to facilitate map interpretation, not to substitute for it. Map analysis merely converts the complex pattern of symbols to a usable form.

A fascinating thing about map analysis is that you can, in a sense, get more out of a map than was put into it. When mapmakers show a few features in proper spatial relationship, they allow you to determine all sorts of things—directions, distances, area, densities, and so on—that they may not have had specifically in mind. This is one of the beauties of map analysis. It can make complex geographic relations more readily understandable.

The following discussion of map analysis is divided into eight chapters. The first two describe how to determine distances (chapter 11) and directions (chapter 12) from maps. In chapter 13, we discuss how to use distance and direction information in position finding and route planning. Chapter 14 deals with using the Global Positioning System in route planning and navigation. Chapter 15 explores various ways to measure the surface area, volume, and shape of features on maps. In chapter 16, we show how various aspects of the terrain and other continuous surfaces can be analyzed. In chapter 17, we look at spatial pattern analysis of features found on a map, and in chapter 18 we move our focus to ways of analyzing spatial associations among patterns.

As we move from map reading to analysis, we're shifting our attention from the theory behind maps to their practical use. It is here that the real fun of maps begins. However beautiful a map may be in theory and in design, it is at its most beautiful when it is being used.

chapter eleven DISTANCE FINDING

11

Distance finding

In our fast-moving society, we sometimes overlook the importance of distance. However, knowing the distance between places can be very important. Travel time and fuel consumption are closely tied to distance traveled, for instance, so distance measurement is crucial when planning hikes, wilderness road trips, and voyages by sea or air.

As we discuss this important aspect of map use, keep in mind that you can interpret "distance" in two ways. When working with maps, we usually think of the **physical distance**, distance measured in standard units such as feet or meters, between places. Since we assume that the map shows locations correctly, we expect a close relation between map and ground distance. We measure map distance with a ruler—usually along the shortest practical route for travel—and convert our measurement to ground distance in kilometers or miles.

In our everyday lives, however, it's also natural to think not of physical but of functional distance, distance measured as an expenditure of cost, time, or energy. Most of us have little notion of how long a kilometer or mile is. Instead, we view distance in terms of the time or energy we must expend to get from here to there. We ask not "How many miles?" but "How long does it take?" or "How hard is it to get there?" Although a few maps do provide information on functional distance, most are based on physical distance. Our first concern, therefore, is to learn to determine physical distance from maps. Then we'll see how we can use special maps to determine functional distance.

PHYSICAL DISTANCE

We measure physical distance using standard systems of measurement (a set of units that can be used to specify anything that can be measured). Some of these units are based on the size of the earth, as we will see below, while others are purely arbitrary. One widely used system based on arbitrary units is the system of **imperial units,** also known as **English units.** This system is sometimes referred to as "foot-pound-second" after the base units of length, mass, and time. This system is used in the United States, United Kingdom, Canada, Australia, India, Malaysia, New Zealand, Republic of Ireland, and other countries. To show how arbitrary this system is, the **yard** (three feet, or 36 inches) was decreed by King Henry I to be the distance from the tip of his nose to the end of his thumb with arm outstretched! When you use the English system to work with maps, you usually measure the ground distance in miles and map distance in inches.

When people talk about a mile, they usually mean a **statute mile** (1,760 yards or 5,280 feet), which is the commonly used distance of a mile on the ground, sometimes referred to as a **land mile.** And when we use the term "mile" in this book, unless otherwise noted, we are referring to this familiar statute mile. But as you use maps, you'll discover other types of miles. Unlike the arbitrary statute mile, these are based on the earth's circumference and represent a specific part of a degree. One such measure is the **geographical mile** (6,087.1 feet), which is one minute of longitude along the equator. The British **admiralty mile,** originally related to the length on the surface of the earth just south of Great Britain, was chosen to be 800 feet longer than a statute mile. The widely used **international nautical mile** (6,076.1 feet or 1,852 meters) is one minute of latitude on a perfect sphere whose surface area is equal to the surface area of the ellipsoidal earth (see chapter 1 for more information on the earth's shape and circumference).

Although not as familiar to most of us, the international nautical mile is important because it serves as the standard unit of distance in water and air navigation. It also provides the basis for the mariner's **knot,** which is the velocity of one nautical mile per hour. Knots are used in computing the speed of ships, planes, and wind.

In contrast to the English system, all distance units in the **metric system** were originally based on the earth's circumference. The **meter** (39.37 inches, which is slightly longer than the yard), represents a physically meaningful distance on the earth's surface. A meter was first defined as one ten-millionth of the distance from the equator to north pole along a meridian. Today the meter is defined relative to the speed of light and is $1/299,792,458$th of the distance light travels in a vacuum in one second.

When you use the metric system to work with maps, you'll usually measure ground distance in kilometers and map distance in centimeters. Conveniently, each kilometer equals 1,000 meters, and each centimeter is one-hundredth of a meter. Indeed, convenience is the metric system's great advantage. It's much easier to use than the English system because all units are powers of 10. The overwhelming convenience and worldwide use of the metric system is hard to ignore. The United States is one of the few nonmetric countries, and it is drifting toward the metric standard for its official government maps.

You should be familiar with both the metric and English systems, because there are situations in which one is more convenient than the other. This means you'll often have to convert from one system to the other. Table C.1 in appendix C will help you make these conversions. All these distance units have little real meaning in themselves, of course. They take on significance only when we use them to specify the distance between geographical locations.

Determining distance

When you need distance information, it's usually impractical to measure distance between places in the environment directly. The solution is to turn to maps. The relationship between maps and the environment is so close that measurements made on the right type of map can be nearly as accurate as ground measurements of distance.

Using maps to measure distance isn't always straightforward, however. A map is always smaller in scale than the environment it depicts. Map distances of centimeters or inches represent ground distances of kilometers or miles. To convert map to ground distance, you need to know the map's scale (see chapter 2 for more about map scale). An understanding of scale is vital in using maps to measure distances.

There are two ways to use a large-scale map to find distances between places. You can use the map's scale to measure the distance, or you can compute the distance using simple algebra applied to grid coordinates. Let's look at each of these methods.

Determining distance by physical measurement

Recall that there are three ways map scale is commonly expressed on maps—scale bars, representative fractions (RF), and verbal statements (see chapter 2 for more on map scale). The easiest way to figure distance is to use the *scale bar,* the ruler-like markings found on most maps. To

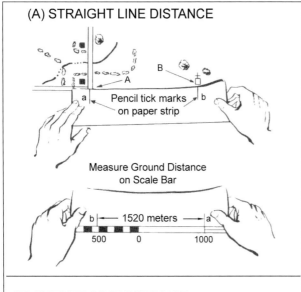

(A) STRAIGHT LINE DISTANCE

B

A

a   Pencil tick marks   b
on paper strip

Measure Ground Distance
on Scale Bar

b ◄— 1520 meters —► a

500   0   1000

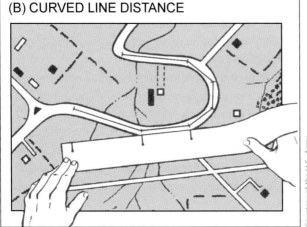

(B) CURVED LINE DISTANCE

Courtesy of the U.S. Army.

Figure 11.1   You can use the scale bar to compute the ground distance between two map features along a straight line (A) or a curved line (B).

do so, mark the distance between two map locations with ticks on a marker, like a strip of paper or a bit of string. Then place the marker with your distance on it along the scale bar (figure 11.1A). All you have to do is use the scale bar like a ruler to measure the distance. If the marker is longer than the scale bar, mark one full scale bar length on the marker and then reposition it against the scale bar and measure again. The distance value you obtain may not be very accurate, however, especially if you try to determine the length of a winding route (figure 11.1B).

You can also compute ground distance using the *representative fraction (RF)* printed on most large-scale maps. Simply measure the distance between two points with a ruler and then multiply the number of inches or centimeters by the denominator of the RF. For instance,

if the RF is 1:24,000 and the distance between points is 3.3 inches, then

$$24,000 \times 3.3 \ inches = 79,200 \ inches$$

You'll want to convert this distance to more familiar ground units such as miles or kilometers:

$$79,200 \ inches \div 63,360 \ inches/mile = 1.25 \ miles$$

or

$$1.25 \ miles \times 1.609 \ kilometers/mile = 2.01 \ kilometers$$

You can buy special map rulers that have scale bars for common map scales. MapTools rulers (figure 11.2, top), for instance, have rulers in meters, statute miles, minutes, and seconds for map scales ranging from 1:24,000 to 1:500,000. You simply place the ruler for your map scale on the map and directly read the ground distance between two points. Corner rulers like the one in figure 11.2, bottom are meant to be used with maps at common topographic map scales. Notice that all rulers are metric and graded in tenths of kilometers.

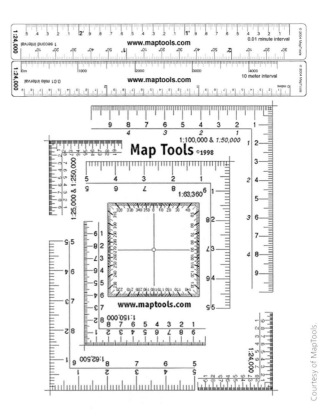

Courtesy of MapTools

Figure 11.2   Map rulers like the MapTools 1:24,000-scale ruler at the top allow you to measure distances in different units of measurement. The metric corner ruler at the bottom is designed for common topographic map scales.

You can speed your measurement of curved lines with a **map measurer.** This device consists of a wheel and one or more circular distance dials (figure 11.3A). Simply set the needle to zero, then roll the wheel along the desired path. The dials will give you the map distance in inches or centimeters. To convert this value to ground distance, multiply it by the denominator of the map's RF. Many map measurers have dials that directly give the ground distance at several common map scales. On the distance measurer in figure 11.3A, for example, there are dials for 1:25,000-, 1:50,000-, 1:75,000-, and 1:100,000-scale maps.

More sophisticated (and expensive) map measurers (figure 11.3B) have digital displays and built-in functions that let you enter the map scale before you make measurements. Ground distances are displayed, and you can connect the device to your personal computer so that you can download measurements.

Map measurers work best when the path between the two points is relatively smooth. Keeping the small wheel on a winding road or stream is a real exercise in finger control. If the wheel slips or binds on tight curves, errors result.

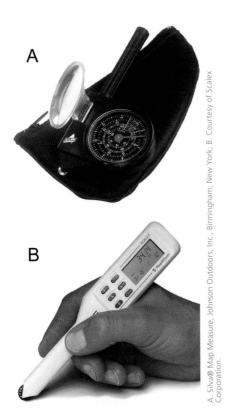

A. Silva® Map Measure, Johnson Outdoors, Inc., Birmingham, New York; B. Courtesy of Scalex Corporation.

Figure 11.3 Mechanical measuring devices, such as the Silva Mechanical Map Measure (A) and Scalex MapWheel (B), can make accurate distance measurements on maps in ground units, especially if the route isn't intricately detailed.

Determining distance by coordinates

Distance measurements become tedious if you must do many of them. If you wanted to know the distance from each of 10 houses to each of the other nine, you would have to make 45 calculations—quite a chore. But there is an alternative. If you use *grid coordinates* to compute the distances, you need only figure out the coordinates of the 10 houses and then use them over and over in your distance computations (see chapter 4 for more on grid coordinates). You can save time and effort if you have access to a computerized mapping system. In that case, you simply use the mapping software to process the coordinates of the locations and compute the distances. The computer will calculate these distances based on grid coordinates.

You can also use your GPS receiver as a convenient distance calculator. To do so, you merely enter the coordinates of the two points, press a few buttons, and the receiver will display the distance between the two points (see chapter 14 for more on GPS). Again, these distances are based on grid coordinates.

Such digital procedures eliminate the need to know a great deal about either mathematics or computers. You'll feel more comfortable with the techniques, however, if you understand the basic mathematical process behind them. We'll work through some coordinate distance computations so that you'll know how they are done, although you may never have to perform them yourself.

Grid coordinates

Assume that the distance you want to measure is the hypotenuse of a right triangle. The **Pythagorean theorem** tells us that if *a, b,* and *c* represent the sides of a right triangle, and *c* is the hypotenuse, then $c^2 = a^2 + b^2$ (figure 11.4A). Changing *c* to *d*, which will represent distance, we restate the Pythagorean theorem as the following:

$$d = \sqrt{a^2 + b^2}$$

or

$$d = \sqrt{(x_2 - x_1)^2 + (y_2 - y_1)^2}$$

where (x_1, y_1) and (x_2, y_2) are the easting and northing coordinates of the locations between which the distance is to be determined (see chapter 4 for more on eastings and northings.) This formula is called the *distance theorem.* It states that you can determine the distance between any two points on your map by taking the square root of the sum of the squares of the differences in the *x* and *y* grid coordinates of the two points (figure 11.4B).

Figure 11.4 You can compute the distance between two points from their coordinates using the distance theorem, the grid coordinate version of the Pythagorean theorem.

(A) PYTHAGOREAN THEOREM

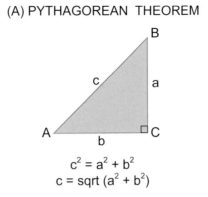

$$c^2 = a^2 + b^2$$
$$c = \text{sqrt}\,(a^2 + b^2)$$

(B) DISTANCE THEOREM

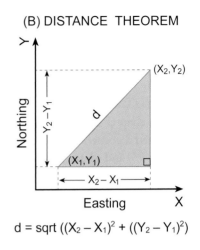

$$d = \text{sqrt}\,((X_2 - X_1)^2 + ((Y_2 - Y_1)^2)$$

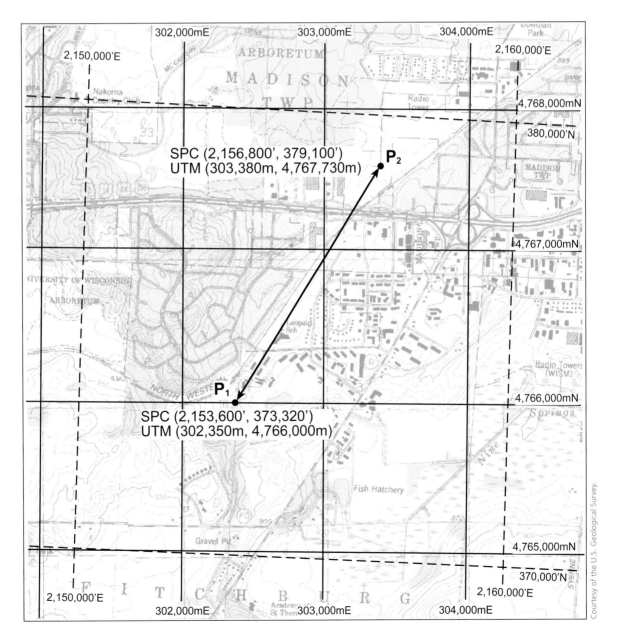

Figure 11.5 Grid coordinates are used to compute the distance between two points on the Madison West, Wisconsin, quadrangle (see text for explanation).

The usefulness of the distance theorem is demonstrated in figure 11.5, using data derived from universal transverse Mercator (UTM) and state plane coordinate (SPC) grid coordinate systems found on the Madison West, Wisconsin, 7.5-minute topographic quadrangle (UTM and SPC grid coordinate systems are discussed in chapter 4). Let's say we want to find the distance between a corner of the University of Wisconsin Arboretum (P_2) and a building to the southwest (P_1).

1. Extend the marginal ticks on the map to form boundaries around the UTM and SPC grid cells in which the two places are located. On the map in figure 11.5, we used solid black lines to extend the UTM tics and dashed black lines to extend the SPC ticks.

2. Determine the UTM and SPC easting (x) and northing (y) for each point (as explained in chapter 4):

$$P_1 \begin{cases} SPC \begin{cases} x_1 = 2,153,600' \\ y_1 = 373,320' \end{cases} \\ UTM \begin{cases} x_1 = 302,350\text{m} \\ y_1 = 4,766,000\text{m} \end{cases} \end{cases}$$

and

$$P_2 \begin{cases} SPC \begin{cases} x_2 = 2,156,800' \\ y_2 = 379,100' \end{cases} \\ UTM \begin{cases} x_2 = 303,380\text{m} \\ y_2 = 4,767,730\text{m} \end{cases} \end{cases}$$

3. Now transfer the numerical values determined in step 2 to the distance formula:

$$\begin{aligned} d_{spc} &= \sqrt{(2,156,800 - 2,153,600)^2 + (379,100 - 373,320)^2} \\ &= \sqrt{3,200^2 + 5,780^2} \\ &= \sqrt{10,240,000 + 33,408,400} \\ &= \sqrt{43,648,400} \\ &= 6,607 \text{ feet or } 2,014 \text{ meters} \end{aligned}$$

or

$$\begin{aligned} d_{UTM} &= \sqrt{(303,380 - 302,350)^2 + (4,767,730 - 4,766,000)^2} \\ &= \sqrt{1,030^2 + 1,730^2} \\ &= \sqrt{1,060,900 + 2,992,900} \\ &= \sqrt{4,053,800} \\ &= 6,604 \text{ feet or } 2,013 \text{ meters} \end{aligned}$$

The computed distance from the SPC grid coordinates is 6,607 feet or 2,014 meters. UTM grid coordinates yielded a distance of 6,604 feet or 2,013 meters—a discrepancy of only three feet between the two computations. These computed distances also compare favorably with the distance we can obtain by using the map scale. The measured map distance is 3.3 inches, which, at a map scale of 1:24,000, yields a ground distance of *3.3 inches × 24,000 = 79,200 inches ÷ 12 inches/feet = 6,600 feet.*

There's no doubt that using grid coordinates to find distances is a simple, accurate method. Its disadvantage is that you can use it only for relatively small regions because both locations must lie within a single grid zone. Your use of SPC grid coordinates is therefore limited to areas the size of a small state, while UTM coordinates are restricted to zones of 6 degrees longitude.

You might think you could solve this problem by extending the grid coordinates over larger regions, but grids of greater extent decrease the accuracy of distance computations. Larger grids incorporate greater scale distortion, since the earth curves progressively away from a plane surface. So this is not a feasible solution.

Figure 11.6 You can find the distance between widely separated places using spherical coordinates.

Spherical coordinates

Fortunately, there is a solution to computing longer distances. You can find the distance between any two places on earth by using spherical geographic coordinates. To find the distance between two locations, you need only know the latitude and longitude of each and use basic spherical trigonometry. GPS receivers perform this task automatically, but it's good to understand the procedure used.

Suppose you want to know how far it is from Seattle to Miami along the great circle route (the shortest path between two places on the globe) as shown in figure 11.6 (see chapter 1 for more on great circles). Follow these steps:

1. Look up the latitude and longitude of the two cities. You can use table C.4 in appendix C to help you find the distances between the world's major cities. This table shows you that Seattle is located at 47°36′N, 122°20′W, and Miami is located at 25°45′N, 80°11′W.

2. Form a triangle, the sides of which are arcs of great circles. Construct this spherical triangle by connecting Seattle and Miami with the arc of a great circle. Extend arcs of great circles along meridians from each city to the north pole. These are called **meridional arcs** (thicker lines in figure 11.6).

3. Consider what you know about this spherical triangle. You know that the distance from equator to pole is 90 degrees. Thus, you can calculate side *b* by subtracting the latitude of Seattle from 90 degrees, or

$$b = 90° – 47°36′ = 42°24′$$

Similarly, side *c* can be calculated like this:

$$c = 90° – 25°45′ = 64°15′$$

Angle *A* is the difference in longitude between the two cities:

$$122°20′ \ (Seattle) – 80°11′ \ (Miami) = 42°09′$$

4. Convert the angles determined in step 3 to decimal degrees:

$$A = 42°09′ = 42.15°$$
$$b = 42°24′ = 42.40°$$
$$c = 64°15′ = 65.25°$$

5. Then determine the sines and cosines for *A, b,* and *c:*

$$sin(b) = sin(42.40°) = 0.67430$$
$$sin(c) = sin(64.25°) = 0.90070$$
$$cos(A) = cos(42.15°) = 0.74139$$
$$cos(b) = cos(42.40°) = 0.73846$$
$$cos(c) = cos(64.25°) = 0.43445$$

6. Transfer the numerical values from step 5 to the Law of Cosines from spherical trigonometry, which can be used for calculating one side of a triangle when the angle opposite and the other two sides are known:

$$cos(a)= cos(b) × cos(c) + sin(b) × sin(c) × cos(A)$$
$$= (0.73846) × (0. 43445)+(0. 67430) × (0.90070)$$
$$× (0.74137)$$
$$= 0.32082 + 0.45028$$
$$= 0.77110$$
$$a = cos^{-1} (0.77110) = 39.547° = 39°32′50″$$

7. Convert this angular distance to ground distance by finding the proportion of a circle of the earth's circumference spanned by the angle, using the following equation:

$$Distance = a°/360° × circumference$$
$$= 39.547°/360° × 24,874 \ miles$$
$$= 2,732 \ miles$$

or

$$Distance = a°/360° × circumference$$
$$= 39.547°/360° × 40,030 \ kilometers$$
$$= 4,397 \ kilometers$$

The great circle route from Seattle to Miami is 2,732 miles or 4,397 kilometers.

You'll have to adjust the trigonometric equation if the two cities fall in different hemispheres. This situation occurs when the great circle route crosses the 180° meridian, the prime meridian (0°), or the equator. If one city is north and the other south of the equator you'll have to add, rather than subtract, the latitude of the southern hemisphere city to 90 degrees to obtain its meridional arc. This is illustrated for São Paulo and Rome in figure 11.6.

When both cities lie south of the equator, you can either add both latitudes to 90 degrees or use the south rather than the north pole as a meridional reference point. If one city is east and the other west of the prime meridian, like São Paulo and Rome, you add rather than subtract their respective longitudes to obtain angle A. Except for these slight changes, you compute spherical distance just as you did when both cities fell in the same hemispherical quadrant.

Finding complex line and perimeter distance

You've now seen how to measure or compute distances along straight lines and great circles. You can also find the length of a more complex line or the perimeter of the boundary line for an irregular area by breaking the line into short, straight segments. You then either physically measure or compute line segment lengths using the equations presented above for grid and spherical coordinates. Finally, you add the segment lengths to obtain the total distance (figure 11.7A).

You can use the same method to determine the length of a smooth curved line. All you do is approximate the curve with a series of straight lines (figure 11.7B). You must decide how long to make the line segments, of course, and this decision will affect the accuracy of your results. Although shorter line segments lead to less error, they also require more measurements or computations if the segments are of different lengths. You must reach a balance between accuracy and effort.

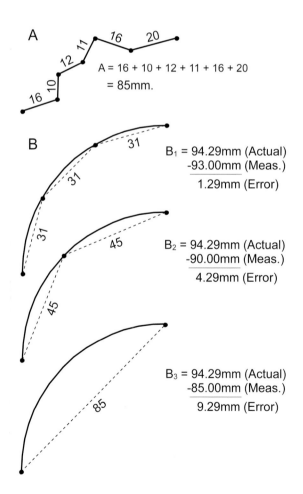

Error factors

No matter how you determine distance from maps, some error is bound to slip into your results. The methods you use, the judgments you make, or the calculations you do may be faulty. These are called *external errors,* since you impose them on your map use from the outside. To minimize external errors, use only proven techniques and instruments and take care to avoid mistakes in computation.

But even if you make no external errors, your final distance figure is still likely to be inaccurate to some degree. The reason is found in the nature of the map itself. A variety of distance distortions are built into the process of transforming reality to a map (see chapter 10 for more on map accuracy). Fortunately, you can compensate to some degree for these *internal errors* (errors introduced in the mapmaking process) if you're aware of their existence. Let's look at some of these internal errors and see how you can compensate for them.

Slope error

Your first problem is that you normally measure or calculate ground distance as if the surface were flat. This map distance is always shorter than true ground distance—except in the rare case of perfectly flat terrain. As figure 11.8A shows, the steeper the slope, the greater the **slope error,** which is the difference between the true ground and measured map distance caused by slope.

To compensate for this slope error, recall the Pythagorean theorem we discussed earlier. Consider figure 11.8B. To compute ground distance *c* between points *A* and *B,* use the map distance *(b)* and the elevation distance *(a),* both of which you can determine from

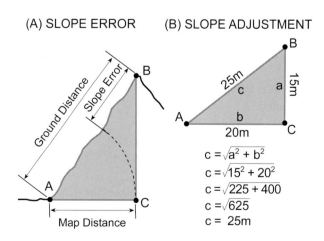

Figure 11.7 You can figure out the distance along a complex line if you divide it into short, straight segments **(A).** You can also approximate curved lines with straight line segments **(B).**

Figure 11.8 When measuring the distance between two points *(A and C)* on the map, it may be necessary to adjust for slope error.

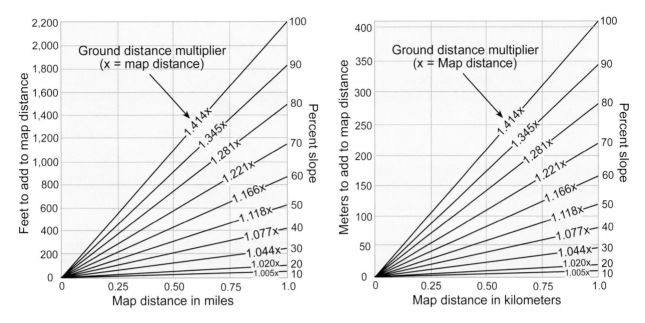

Figure 11.9 Use the feet or meter slope error values on the left vertical axis or the ground distance multiplier to convert the map distance to ground distance (horizontal axis).

a topographic map. For the example, in figure 11.8, the discrepancy between horizontal and ground distance is five meters.

Slope error increases with increasing map distance as well as steeper slopes. The two graphs in figure 11.9 give you the amount of slope error in feet and meters for different map distances as the slope percentage increases. From the right-hand graph, for instance, you can see that if the map distance you measure is one kilometer and the slope you find from the contours on the map is 40 percent, the actual ground distance you'll travel is 77 meters longer, or 1,077 meters. Notice that you can find the amount of slope error either from the values on the left vertical axis of each graph, or by multiplying the map distance by the **ground distance multiplier** coefficients for the percent slope lines on the graph drawn at 10-percent increments. For example, a map distance of 650 meters along a steep slope of 60 percent would be a ground distance of 650 × 1.166, or 758 meters.

Smoothing error
A second type of distance error can be traced to the smoothing of line symbols on maps. When showing linear features such as roads and rivers, mapmakers straighten curves and smooth irregularities (see chapter 10 for more on generalization). As a result, the measured route along a highway, railroad, or hiking trail may be shorter than the ground distance. Likewise, map distance along a shore or down a stream may mislead a map reader in a canoe, a kayak, or in a boat moving along a shoreline.

This **smoothing error** has something in common with slope error—both add to the measured map distance. A ground distance route may be the same length or longer than the map distance estimate, but it can never be shorter. Unless the ground is completely flat and the route perfectly straight, your computed distance will always fall short of true ground distance.

Although elevation information may be available to help you overcome slope error, you have no way of knowing how much mapmakers have smoothed linear features. There are, however, a few useful guidelines. One good rule to remember is that the smaller the map scale, the more smoothing mapmakers use. Thus, the discrepancy between map and ground distance increases as map scale decreases. This means you can minimize the problem if you use large-scale maps.

It's not quite that simple, however. Even on large-scale maps, some roads, rivers, and boundaries will be more smoothed than others. The more detailed and intricate the feature, the more it will be simplified. County roads, for instance, usually will be more smoothed than freeways, because they are more convoluted to start with. The map legend is rarely of much help in sorting out the effects of this **variable smoothing** of line symbols (that is, the varying levels of line generalization applied to different features on the map). Intuition is your best guide. You would naturally expect highways to curve more (and therefore be more smoothed by the mapmaker) than railroads, streams to be more sinuous than roads, and state highways to be less straight than freeways. Additionally, linear features in rugged, rocky regions will be more irregular (and thus more smoothed on the map) than the same features in flat, sandy regions. As you use your common sense and anticipate these differences within the same map, you'll be able to estimate ground distance better.

Scale variation

The larger the mapped region, the greater the **scale variation** across the map. Since no map projection maintains correct scale throughout, it is important to determine the extent to which it varies on a map. If you are measuring distances on small-scale maps of continents or hemispheres, it is good advice to trust the scale only near the centerpoint or standard parallels of the map or along meridian lines, since the ground distance between degrees of latitude along meridians is a constant 69.2 miles per degree on the sphere (see chapter 1 for more on the spacing of meridians and parallels).

Since distances are such an important aspect of our environment, aren't there any small-scale maps designed to keep distances correct? The answer is yes and no. Spherical globes faithfully represent distances, except for small errors introduced by the earth being ellipsoidal in shape. But for a small-scale map projection to preserve distance relations perfectly, it would have to show the shortest spherical distance between any two points as a straight line—and do so at true scale. That's a mathematical impossibility on a flat surface. We can, however, make measurements on map projections that have true distances along certain lines.

Recall from chapter 3 that the *azimuthal equidistant projection* transforms great circles passing through the projection center into straight lines on the map. Distances from all points to the projection's center are true to scale, but distances between all other points are incorrect. This creates a problem—to use the azimuthal equidistant projection to find distances from different starting points, you will need a new map centered on each starting point.

Disproportionate symbol error

A fourth source of error in figuring distance arises because most map features aren't symbolized in correct scale. If they were, the symbols wouldn't be large enough to see. Take, for instance, a road 30 feet (9 meters) wide. It might be shown by a thin line that is 0.01 inches wide (0.025 centimeters) on a map with a scale of 1:125,000. This would give it an effective ground width of 105 feet (32 meters)!

Disproportionate symbol error occurs when the size of the symbol used to map the feature is not proportional

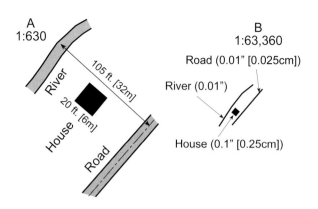

Figure 11.10 Disproportionate map symbols can have a major distorting effect on distance determinations between closely spaced features, as these 1:630- and 1:63,360-scale ways of depicting the same three features demonstrate.

to its ground size. This error becomes more extreme when a feature such as a political boundary, which has no width, is shown by a symbol, which has width on the map. And the trouble becomes worse as map scale decreases. There's simply not enough space to include everything. As a result, cartographers overemphasize features selected for the map. By the very nature of symbolism, depicting one feature can lead to excluding, distorting, or displacing its neighbor (see chapter 10 for more on generalization and symbolization).

Consider the case of a house on a 105-foot (32-meter) lot between a road and a river (figure 11.10A). If the river and road were both symbolized by lines 0.01 inches (0.025 centimeters) wide at a map scale of 1:63,360, they alone would use up 52.8 feet (16.1 meters), or half the space available. There wouldn't be enough room for the building symbol, which at 0.1 inch by 0.1 inch (0.025 centimeter by 0.025 centimeter) would itself take up 528 feet (161 meters)—five times the ground distance between the center of the river and road.

To make all three features visibly distinct, the mapmaker displaces them, or "pulls them apart." Such positional **displacement** is essential to retain the relative spatial relations between features—we can't have the road and river running through the house. But the result is that distance accuracy is compromised (see chapter 10 for more on accuracy).

Be wary, then, of any distance measurements involving closely placed map symbols. If, in figure 11.10B, you computed the distance between the centers of the river and road from the map, you would come up with a ground figure of 686 feet (209 meters), while the two are really 105 feet apart. If you didn't know about disproportionate map symbols and positional modification, you might even try to measure the distance across the river and find that the river is 52.8 feet (16.1 meters) wide when it is actually only 10 feet (3 meters) across.

Map users who recognize the problems of disproportionate symbols on conventional maps sometimes overlook the fact that the same thing occurs on air photos and remotely sensed images (see chapter 9 for more on images). Contrary to what you might think, a photograph doesn't show all environmental features in correct proportion. Since photography is based on reflected light, a shiny, highly reflective object is imaged disproportionately large on a photograph. Mapping teams have made use of this fact for years. Before taking photos, they place highly reflective targets on the ground. These targets stand out on air photos, helping the mapmakers locate known positions and correct any distance distortions when using the photos to compile topographic and other maps.

If these occurrences seem to defy logic, even more baffling is the fact that linear features on photos appear to be far larger than their natural size. Powerlines are often visible on a photo even though the large transmission towers holding them up are not. Railroad tracks can be seen when a car can't. On satellite images with a scanner cell size of 200 by 200 feet (61 by 61 meters), there is no logical way that a 60-foot-wide (18-meter-wide) road would show up—yet it does! These examples should remind us to keep disproportionate symbol error in mind when we make measurements on image maps as well as conventional maps.

Dimensional instability error

Unless maps are printed on a dimensionally stable material such as plastic, glass, or metal, they will stretch and shrink with changes in temperature and humidity. A map the size of a U.S. Geological Survey (USGS) 1:24,000-scale topographic quadrangle, if printed on paper (22 inches wide by 27 inches high), can change as much as 1 percent in length or width when moved from a cool and dry to a hot and wet environment. This **dimensional instability** caused by the stretching and shrinking of the medium on which the map is produced introduces errors in your distance measurements, especially if you're measuring long distances on the map.

Like slope, smoothing, scale, and symbol error, dimensional instability error can play havoc with distance calculations. No matter how precisely you compute distance from maps, no matter how carefully you wheel your mechanical distance finder along wiggles in the road, your final distance figure will be only approximate.

Premeasured ground distances

All the types of measurement errors discussed above occur because map distance isn't the same as ground distance. Such problems are eliminated if someone physically measures distances on the ground and puts the distance figures on the map. That's just what has been done on some maps—road maps in particular. *Premeasured ground distances* are an invaluable aid in determining route length and choosing between routes. Let's look at some of the ways that ground distances have been added to maps.

Distance segments

On some highway maps, ground distances are added between such places as towns or highway intersections using **distance segments** (figure 11.11A). Routes are divided into segments, with the distance along each segment labeled. To find the distance between two points, you merely add up all the distances for the segments along the route.

If your route is very long, of course, you'll soon tire of adding numbers. To help you out, mapmakers may provide a second level of distance numbers. Second-level distance segments are longer than the first and have as their end points major cities and intersections. These special points are shown by such symbols as stars or arrowheads (figure 11.11B). The symbols and segment distances are usually printed larger or in a different color to reduce confusion with the first-level figures.

Distance insets

To show the distance between widely separated points, mapmakers often include **distance insets.** Insets are usually double-ended arrows with names of features, such as cities, and distances between them (figure 11.11C). Insets

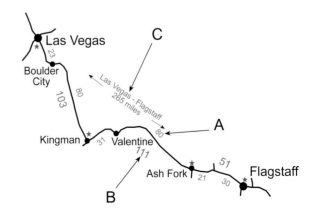

Figure 11.11 Distance segments (A), second-level distance segments, and distance insets (C) are often incorporated into the design of road maps to help in determining route length.

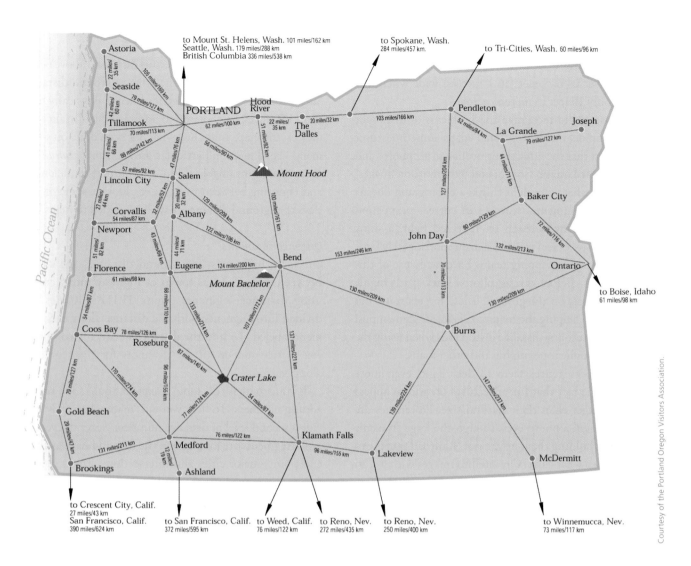

Figure 11.12 **A mileage map, such as this one for Oregon, shows ground distances between selected places in the state and to selected places in neighboring states.**

can be useful, though they may be hard to locate and are far from complete. All too often you'll find that the map-maker did not include the length of the route of interest to you.

Mileage maps

In addition to segment distances and distance insets, a **mileage map** can be used to indicate the distances between places (figure 11.12). On such a map, routes between major locations are shown by straight lines that are labeled with the distance along each route. These lines aren't meant to duplicate real roads but only to show that one place is connected to another by a route. Consequently, only the miles and kilometers printed for each line are correct, not the relative lengths or locations of the straight lines connecting the places. Notice that many of the cities and physical features on the map in figure 11.12 are not in their actual geographic locations,

but have been shifted to make the map more legible.

The main disadvantage of mileage maps is their size. They are usually relegated to a corner of a larger road map and are so small that most cities must be left off. Further-more, you have the same problem you did with segment numbers—you must add a lot of numbers together for longer routes and the more numbers you add, the more time consuming the task is and the more room there is for error. To find the distance from Portland to Ashland in figure 11.12, for instance, you must add six numbers.

Distance tables

The problem of having to add a great many numbers is avoided if distance figures are arranged in tables. There are two types of **distance tables,** or tables that identify the premeasured ground distance between a select set of locations. On a **rectangular distance table,** key loca-tions are listed alphabetically along the top and side to

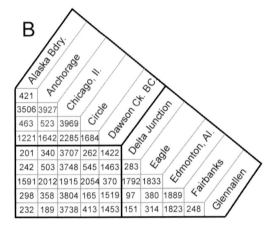

A	Alaska Bdry.	Anchorage	Chicago, Il.	Circle	Dawson Ck.	Delta Junct.	Eagle	Edmonton, Al.	Fairbanks	Glennallen
Alaska Bdry.		421	3506	463	1221	201	242	1591	298	232
Anchorage	421		3927	523	1642	340	503	2012	358	189
Chicago, Il.	3506	3927		3969	2285	3707	3748	1915	3804	3738
Circle	463	523	3969		1684	262	545	2054	165	413
Dawson Ck. B.C.	1221	1642	2285	1684		1422	1463	370	1519	1453
Delta Junction	201	340	3707	262	1422		283	1792	97	151
Eagle	242	503	3748	545	1463	283		1833	380	314
Edmonton, Alta.	1591	2012	1915	2054	370	1792	1833		1889	1823
Fairbanks	298	358	3804	165	1519	97	380	1889		248
Glennallen	232	189	3738	413	1453	151	314	1823	248	

Figure 11.13 Rectangular (A) and triangular (B) distance tables for Alaska.

create a row-column table of distances. To find the distance between two cities, look up one name along the side and the other along the top. Then find the cell where they intersect. To find the distance from Chicago, Illinois, to Fairbanks, Alaska, in figure 11.13A, locate Chicago in the left margin and Fairbanks on the top, and read the row-column intersection as 3,804 miles. Or look up Fairbanks on the left and Chicago on the top—the result will be the same.

Notice that the rectangular distance table is 50 percent redundant. To allow you to look up either city name in either margin, the table's upper-right half is the same as the lower-left half. To avoid this duplication and save space, mapmakers often use a **triangular distance table** instead. On this table, place names appear only once. In figure 11.13B, you can find the distance between Chicago and Fairbanks by locating Chicago along the edge and reading down the column until it intersects the row for Fairbanks. The distance again is 3,804 miles.

As with route distances and mileage maps, distance tables aren't all-inclusive. You may find one of the places you're looking for but not the other, or both may be missing from the table. On such occasions, you can sometimes arrive at an approximate figure by looking up the distance between nearby places.

Distance databases

The modern way to determine the distance between cities and other significant geographical places, such as state and national parks, recreation areas, and historical sites, is to access a **distance database** holding this information. Route-finding software for your personal computer may come bundled with these databases. Both sources often have the option of giving you either distances "as the crow flies" (straight-line distances) or actual road mileage distances.

Geographic information system (GIS) software such as ESRI's ArcGIS software uses the coordinate method described earlier in the chapter to compute the lengths of linear features from the grid coordinates that define each line. Point location databases, as described above, can also be entered into a GIS, and routes between locations can easily be computed along with their distances.

FUNCTIONAL DISTANCE

Physical distances tell only a part of the story. Consider the following example:

> In Los Angeles, with everybody traveling by car on freeways, nobody talks about "miles" anymore, they just say "that's four minutes from here," "that's twenty minutes from here," and so on. The actual straight-line distance doesn't matter. It may be faster to go by a curved route. All anybody cares about is the time (Wolfe and McLuhan 1967, 38).

There's more to distance than physical miles on the ground or measured miles on the map. Physical distance may not be as germane to our lives as functional distance, or distance measured as an expenditure of time, effort, or cost. As the above example illustrates, functional distance depends on many factors—mode of travel, driver circumstances, road conditions—the list is endless.

It's sad to say, but mapmakers have found it difficult to respond effectively to our concern with functional distance. In part, their shortsightedness is at fault. They have concentrated on making maps ever more geographically accurate, thereby satisfying the demands of engineers, land surveyors, military strategists, and others concerned only with physical distance. In the process, they often overlook the day-to-day needs of the rest of us.

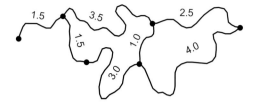

(A) Route segment labels (hours)

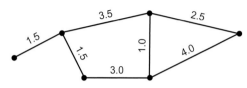

(B) Travel-time diagram (hours)

Figure 11.14 **Functional distances may appear as route segment labels (A) or on a travel-time diagram (B).**

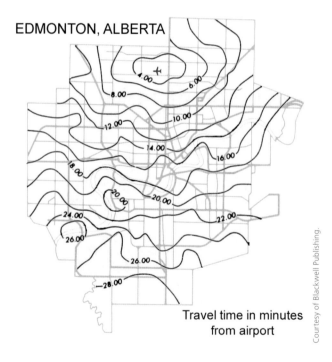

EDMONTON, ALBERTA

Travel time in minutes from airport

Courtesy of Blackwell Publishing.

Figure 11.15 **Isochrone map of Edmonton, Alberta, showing lines of equal travel time in minutes from the airport.**

But partly the fault rests with the nature of functional distance. While physical distance is always the same, distance measured in time or energy depends on many factors. Maps based on physical distance, therefore, aren't only simpler and less costly to make but are also more general in purpose. They can be used by anyone at any time. Maps showing functional distance may be more meaningful for specific purposes, but to be effective they must be tailor-made for selected situations.

The most common functional distance added to maps is **travel time,** or the time it takes to travel by standard means between locations, because it's the easiest to measure and show. Mapmakers can determine average travel times simply by studying speed limits along a route. They provide this information on maps in several ways. Like mileage maps, travel time maps are made by adding functional distance values as supplementary information for route segments on conventional maps. They can also make maps showing only functional distances. Another approach is to show travel time with isochrones.

Route segments

One method to add functional time to a map, as shown in figure 11.14A, is to give the travel time for each **route segment.** This is similar to the approach we saw earlier in which distances were added to route segments. You find your total travel time by summing the numbers of the segments, just as you do to determine physical distance.

Travel-time maps

The travel time between major locations can also be shown by straight lines labeled with the time it takes to travel along each route (figure 11.14B). These are called **travel-time maps.** As with mileage maps, the lines only to show that places are connected, so only the travel times shown for each line are correct, not the lengths or locations of the lines and sometimes not even the places. This procedure is used in road atlases, trip maps produced by automobile clubs, and maps in hiking guidebooks. Again, find the total functional time by adding the numbers for each segment.

Isochrones

A third mapping method is to give travel-time information continuously in all directions from a central point. Travel times are usually shown by lines of equal time distance, called **isochrones** as shown in figure 11.15 (see chapter 8 for more on isolines). Isochrones are arranged outward from the central point, with the spacing between them directly proportional to travel time. They'll be closer together in congested traffic areas, and farther apart in open areas.

This last way of showing travel times is more restrictive than the first two, since only time to and from the central point is meaningful. But all three methods are useful, especially if they are combined with physical distance measures on the same map. In these cases, you are given the best of both worlds—you can use physical distance units to compute gas mileage and travel time figures to estimate arrival times, travel fatigue, and other such factors.

All travel time figures are approximate, of course. They will vary with the driver, weather, time of day, route, vehicle, and more. Maps provide average figures, so you have to consider the circumstances of special cases.

If maps showing travel times are few, those giving other types of functional distance information are even harder to find. Such details as travel comfort, energy consumption, and route safety would all be of interest to map users. The core of the problem is that this kind of information is difficult to gather. And here again, average figures aren't always helpful in individual cases. GPS is starting to change this. You will often find that your GPS unit will tell you the functional distance in time travel as well as the physical distance on the ground. This allows you to consider your needs before you even start your journey—how much water and food to take hiking, how long between stopping for gas, and when you will want to stop for a rest area can all be estimated in advance of your departure. Once en route, GPS also allows you to modify decisions about your timing and the distances you travel.

As widespread automated mapping becomes more practical, however, we can envision maps being tailor-made for a particular person and trip. You'll give a computer all the information related to your journey and be presented with your own special and truly functional map. Or you can interact with a navigation unit while en route to get the functional and physical distances you need.

Functional distance by inference

In the meantime, as long as maps of functional distance remain somewhat scarce, the burden lies with you, the map user, to translate physical into functional distance. You could do this by searching out the information elsewhere and adding it to your maps, but such a process is usually too involved to be worth the effort. Your alternative is to determine functional distances through a process of **inference** in which you derive conclusions based on what you already know. You can conjure up a great deal of information from past experience stored in your mental maps.

You probably use the inference process frequently without realizing it. When figuring the length of a trip, for instance, you take traffic conditions into account. You know that to drive across town will take longer during rush hour than at other times. Holiday and weekend traffic is equally predictable, so you leave earlier or stay longer to shorten your travel time.

You can probably think of many other functional distance measures. Travel time is merely the starting point for a person with imagination. By drawing on your experience and intuition, in combination with careful map study, you can make many inferences that broaden your use of functional distance. Your imagination can add more to maps than cartographers can show.

SELECTED READINGS

Atwill, L. 1997. What's up (and down) at the USGS. *Field & Stream* (May): 54–5.

Bovy, P. H. L., and E. Stern. 1990. *Route choice: Wayfinding in transport networks.* Boston: Kluwer Academic Publishers.

Buttenfield, B. P., and R. B. McMaster. 1991. *Map generalization: Making rules for knowledge representation.* Essex: Longman Scientific & Technical.

FICCDC. 1988. The proposed standard for digital cartographic data. *The American Cartographer* 15 (1): 9–142.

Maling, D. H. 1989. The methods of measuring distance. In *Measurement from maps,* 30–52. New York: Pergamon Press.

McMaster, R. B., and K. S. Shea. 1992. *Generalization in digital cartography.* Washington, D.C.: Association of American Geographers.

Muehrcke, P. C. 1978. Functional map use. *Journal of Geography* 77 (7): 254–62.

Muller, J. C. 1978. The mapping of travel time in Edmonton, Alberta. *The Canadian Geographer* 22: 195–210.

Muller, J. C., J. P. Lagrange, and R. Weibel, eds. 1995. *GIS and generalization.* London: Taylor & Francis.

Monmonier, M. S. 1977. *Maps, distortion and meaning.* Washington, D.C.: Association of American Geographers.

Olsson, G. 1965. *Distance and human interaction.* Bibliography Series No. 2. Philadelphia: Regional Science Research Institute.

Peters, A. B. 1984. Distance-related maps. *The American Cartographer* 11 (2): 119–31.

Watson, J. W. 1955. Geography: A discipline in distance. *Scottish Geographical Magazine* 71: 1–13.

Witthuhn, B. O. 1979. Distance: An extraordinary spatial concept. *Journal of Geography* 78 (5): 177–81.

Wolfe, T., and M. McLuhan. 1967. *McLuhan hot & cool.* New York: Dial Press.

chapter twelve   DIRECTION FINDING AND COMPASSES

12

Direction finding and compasses

Some people like to think the true outdoors people among us have somehow managed to retain the inborn sense of direction that animals seem to have. If this is true, however, how do we account for the many animals as well as famed backwoods guides who've become lost? A better explanation for the direction-knowing powers of animals, "primitive" people, and guides is that they have learned to be more observant than the rest of us. This is good news, for it means that anyone can learn to tell direction and, with practice, share the distinction of having a "sense of direction."

If we say that certain people have a "sense of direction," what do we mean? That they always know where the North Star is? Or the north pole? Or that they can always find their way home? Direction, by definition, can only be determined with reference to something. The reference point may be near at hand or far away, concrete or abstract. This reference point, whether it is some object or some known position, establishes a **reference line,** sometimes called a **baseline,** between you and it. Direction is measured relative to this reference line.

In its simplest form, direction is determined egocentrically, that is, it is self-centered. Your reference line is established by the way you are facing. You go left or right, "this way" or "that way," straight ahead or back, up or down the road—all in relation to an imaginary line pointing out from the front of your body.

A common form of egocentric direction is to use a symbolic clock face. You are assumed to be located at its center, facing 12:00. Your reference line is the line projected from your position straight ahead to 12:00. Now say that you want to find the direction to a distant object. The line from you to the object is called the **direction line.** Direction is given in hourly angular units as the difference between the reference line and the direction line. Something ahead of you to the right might be at 2:00, directly to your right would be at 3:00, behind you at 6:00, and directly to your left at 9:00 (figure 12.1A).

GEOGRAPHICAL DIRECTION SYSTEMS

Your direction finding ability will improve if you learn to think **geocentrically** in terms of a *geographical direction system*. In a geographical system, as with the clock face, direction is measured in angular units of a circle (figure 12.1B) with north at the top. In other words, north is equivalent to 12:00 on the clock face. This is probably why we tend to think of north as "up" and why north is customarily placed at the top of maps. Oddly enough, the convention of orienting maps with north at the top is also used in the southern hemisphere— most likely because early European settlers carried the custom there.

The north reference point is useful because the reference line is no longer oriented to your own body, as it is with egocentric directional methods. No matter which way you turn, north remains the same. To make this system valuable in direction finding, all you need do is find some reference line on the earth's surface that ties north to the ground. The direction to a distant object can then be given as the angle between the north reference line and your direction line.

There is no single north reference line used on all maps, because there are actually three types of north. Each has its advantages and disadvantages as a reference point, and each is best suited for certain purposes. The reference point used on most maps is true north.

True north

True (or **geographical**) **north** is a fixed location on the earth—the north pole of the axis of earth rotation. A great circle line from any point on earth to the north pole—that is, a meridian—is known as a **true north**

reference line. Thus, any meridian can serve as your reference line in finding true north.

The advantage of using true north to reference direction is that you can find it in the field without using any special instruments. You can determine direction simply by making reference to natural features, much as animals do. People have used the sun and other stars as directional reference points since they first began observing nature. North was probably chosen as the reference point on maps because there are such good celestial ways to help us find it.

One of the oldest and most reliable ways to find true north is to locate the **North Star (Polaris)**. It is positioned in the northern hemisphere sky less than one degree away from the **north celestial pole** (the spot in the sky directly above the north pole). Finding Polaris is a simple matter because it is a relatively bright star conveniently located at the tip of the handle of the Little Dipper (figure 12.2). Because the earth is rotating on its axis, all the stars seem to move in concentric paths around Polaris (figure 12.3). We could hardly ask for a better true north reference point.

You can use the sun as well as the stars to tell direction. For thousands of years, people have found the direction of true north by noting the position of the sun at noon (or at 1:00 PM daylight saving time). At other times of day, you can use the shadow cast by an object, and a bit of patience, to find true north. Here's how.

The first step is to find an object that casts a well-defined shadow on level ground (figure 12.4A). Next, mark the spot at which the top of the shadow touches the ground. After waiting 30 minutes or so, again mark the tip of the shadow. A line drawn between your two marks is a true east–west line. Since the sun rises in the east and sets in the west, the shadow travels from west to east. Therefore, your first mark will be on the west end of

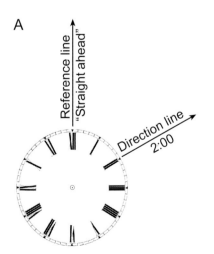

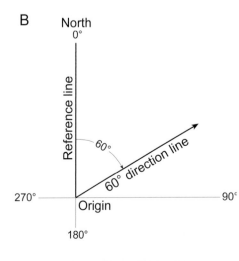

Figure 12.1 Direction is defined as the angular distance from a reference line. Sometimes a clock face is used as a directional reference system (A), but more commonly direction is measured in degrees from north (B).

the line and your second mark on the east end. With west to your left and east on your right, draw a perpendicular line. True north is at the top of the line.

The direction to true north can be determined in a few seconds if you are wearing a watch with hour and minute hands. Simply hold the watch horizontally at eye level, and then turn the watch until the hour hand points to the sun (figure 12.4B). Now picture a spot on the dial halfway between the hour hand and 12:00 (1:00 for

daylight saving time). In the northern hemisphere, a line from the center of the dial through that spot will point to true south, so the opposite direction is true north. In the southern hemisphere, the line points to true north. This method works best if you are located at the center of a time zone where the sun truly faces south at noon.

The problem with relying on the sun or Polaris as a reference point for true north is that the sun is visible only during the day and Polaris only at night. Also, they can be hidden by clouds. To overcome these difficulties, people have created their own "artificial stars" in the form of navigation satellites. The constellation of satellites developed and operated by the U.S. Department of Defense is called the **Global Positioning System (GPS)**, though this is now being augmented by regional systems developed by other countries (see chapter 14 for more on GPS).

Since GPS satellites transmit signals to the earth continuously on a 24-hour basis, we now have an all weather, day or night ability to determine the direction of the true north reference line. It couldn't be easier. You merely turn on your GPS receiver, enter the latitude and longitude of a distant feature, and in seconds the GPS receiver gives you the direction to the feature with respect to true north.

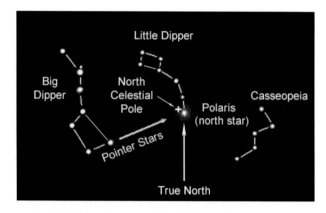

Figure 12.2 True north can easily be determined by observing the stars in the northern hemisphere sky at night to find the North Star (Polaris).

Courtesy of Jerry Pool.

Figure 12.3 A time-lapse photo of the night sky in the northern hemisphere captures the apparent circular movement of the stars around the north celestial pole and Polaris, the North Star. In fact, the earth, not the stars, is rotating.

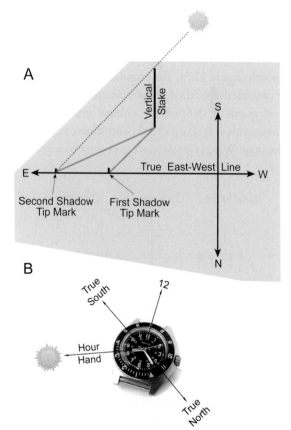

Figure 12.4 True north can be determined in the daytime by using shadows cast by the sun in conjunction with markers placed on the ground (A) or with a clock face (B).

Grid north

True north isn't always the most useful direction reference line. On maps that have a grid coordinate system overprinted (see chapter 4), you will find a second north—**grid north,** which is the northerly or zero direction indicated by the datum of the grid used on the map. Unlike true north, it doesn't refer to any geographical place. It is purely artificial, established for the convenience of those who use maps to measure or compute directions. Rectangular grids are often superimposed on maps to make it easier to use a protractor when measuring directions. The vertical lines on these grids point to grid north.

The universal transverse Mercator (UTM) grid system, state plane coordinate (SPC) grid, British National Grid, and Swiss Coordinate System (see chapter 4 for more on these grid coordinate systems) are standardized systems in which the lines of constant easting running from top to bottom are oriented to grid north. Grid north lines usually aren't in the same orientation as true north meridians on the map. Grid north lines are straight and parallel, whereas true north meridians converge toward the north and south poles. If a map is centered on a vertical meridian, the grid line in the center of the map will also be vertical. It will point to true north only if the meridian is the central meridian for the UTM or SPC zone. The farther east or west the mapped area is from the center of the grid zone, the greater the difference will be between grid and true north (figure 12.5). The angular difference between these two norths is called *grid declination.*

Grid north is defined by a single grid coordinate system because it is relative to the coordinate system used for the map. When two or more grid systems occur on a

single map, the system selected for showing grid north in the legend or margin may be difficult to determine. For example, on a standard topographic quadrangle for the United States published by the U.S. Geological Survey (USGS), the grid lines on the SPC system probably won't be at the same angle as the grid lines on the UTM system. The angular difference between these two grid norths on the Madison West, Wisconsin, quadrangle is illustrated in figure 12.6.

Clearly, using a map with several grid orientations can be confusing. Fortunately, this confusion is avoided for UGSG topographic maps and other large-scale U.S. map products because UTM grid lines are used as the standard grid north. Check the map notes in the margin to determine grid north for maps that use other grid coordinate systems.

Magnetic north

True north is valuable to mapmakers and GPS users, and grid north is a helpful aid to direction measurement and calculation. The third north—**magnetic north**—is convenient for the map user in the field who has a magnetic compass. Magnetic north is the direction that your compass needle points to at the earth's magnetic pole.

The reason for using magnetic north as the directional reference point is its global utility. In effect, the earth is a giant magnet, with its magnetic field running roughly north and south (figure 12.7) with a difference of approximately 11 degrees between the magnetic and geographic poles. A magnet also has a magnetic field running between its north and south poles. The magnetic property that makes a compass work is that when two magnets are put together, their like poles repel and unlike poles attract.

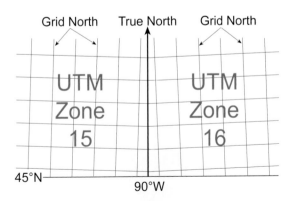

Figure 12.5 There can be a considerable angular difference between true north and grid north away from the center of a grid zone, such as the UTM grid coordinate system shown here. The maximum grid declination is at the edge of zones.

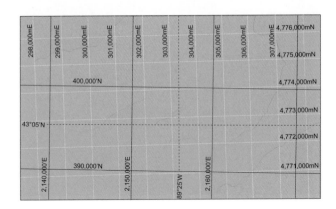

Figure 12.6 Grid north differs substantially from true north (indicated by the dashed line for the 89°25′ meridian) for the UTM (blue) and SPC (black) grid lines in the Madison, Wisconsin, area. The near-vertical UTM easting lines are the standard grid north on USGS topographic maps of the area.

Thus, a freely moving magnetized needle will align itself with the earth's magnetic field. The needle's north pole will point to the earth's north magnetic pole. If this end of the needle is marked distinctly, say with red paint or with the letter N, it will show us the magnetic reference line at our location.

Theoretically, the needle of a magnetic compass will point to the magnetic north pole from any place on the earth's surface. In practice, however, because of local magnetic disturbances, it doesn't always work out this way, as we'll see later in this chapter.

Magnetic declination

We've already seen that grid north usually differs slightly from true north. Greater angular differences occur with magnetic north. The north pole of the earth's magnetic field is currently located in the Nunavut Territory of Canada (west of Ellesmere Island), approximately 1,300 kilometers (800 miles) south of the true north pole (figure 12.8). This large distance between the positions of the two poles means that the true north and magnetic north reference lines rarely coincide, so the magnetic compass needle rarely points to true north. **Magnetic declination** (called **compass variation** on charts) is the angular difference between true and magnetic north.

The magnetic declination is easy enough to determine that maps are available that show **isogonic lines,** or lines of constant angular difference between the two norths (see chapter 8 for more on isolines). Such an **isogonic map** is shown in figure 12.9. Note that a line of "no declination" (the solid black line labeled 0°) runs through the east central part of the United States. At any position along this **agonic line,** the true and magnetic north poles are aligned and the compass needle points to true north. In contrast, magnetic declination exceeds 22° in the northwestern and northeastern corners of the country.

At the extreme northwest, the compass needle points to the east of true north and has **easterly declination,** while in the northeast the needle points to the west of true north and has **westerly declination.** If you live in the east central United States, you'll have little problem with magnetic declination for everyday purposes. But when you use a compass in the northwest or northeast parts of the country, you can't ignore the large difference between true and magnetic north. Compass users in the Midwest may be blithely unaware of magnetic declination, but if they try to use their compass in Washington or Maine they could be in for a shock.

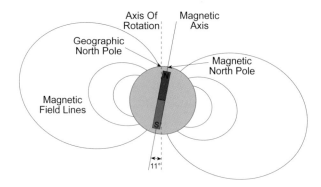

Figure 12.7 **The earth's magnetic field resembles that of a simple magnet. There is a difference of approximately 11 degrees between the magnetic and geographic poles.**

Annual change

To make matters worse, the magnetic declination for a given location changes slightly from year to year, as the dashed lines of equal annual change in minutes show in figure 12.9. The magnetic poles wander across the earth's surface, albeit in a somewhat predictable manner (figure 12.8). This is why the date shown on a topographic map or nautical chart is a critical piece of information for map users. If the map is used a decade or so after it was created, the declination may have changed significantly.

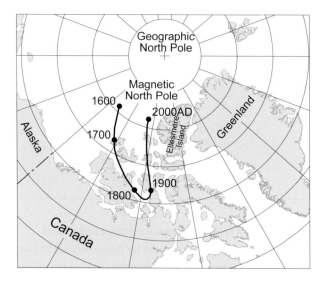

Figure 12.8 **This map shows that magnetic north and true north rarely coincide at a location and that the magnetic north pole shifts greatly over time. In 2000, when this map was made, the magnetic north pole was located about 1,300 kilometers (800 miles) south of the geographic north pole.**

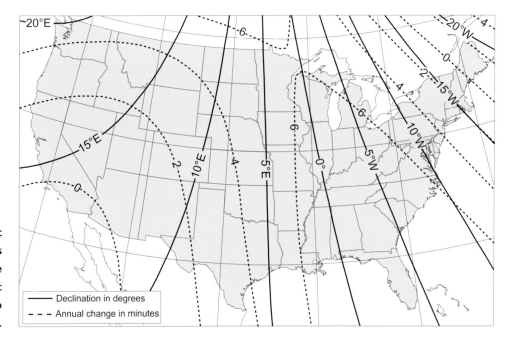

Figure 12.9 Isogonic map of the conterminous United States (1995). The zero-declination agonic line is slowly shifting to the west.

Nautical charts, which strive to give up-to-date direction information, tell exactly how much the variation is changing each year and in what direction. Charts for the United States give the **annual increase or decrease** in minutes of a degree per year. Canadian charts indicate the amount and direction of annual change (13'W, for example), then leave it to you to figure out if this is an increase or a decrease in declination.

A 1994 nautical chart of Vancouver, British Columbia, for instance, shows a compass variation of 20°30'E with an annual change of 7'W. The westerly annual change is in the opposite direction of the east compass variation, so the annual change is a decrease. By 2000, the 7'W annual change had accumulated over six years to 42'W, so that the variation was 20°30'– 0°42', or 19°48'.

Declination diagram

Declination is important to keep in mind, especially when you're in the field or navigating a boat or airplane. It will, of course, be more critical to take note of declination at some times than others, depending on how much accuracy is needed. Surveyors and other professionals who work with precise directional information must always be conscious of declination. You can think of other examples as well. For instance, someone building a solar house would also want a precise determination of true north so that the structure could be aligned as effectively as possible. The rest of us can sometimes disregard declination; other times we should not.

Large-scale nautical and aeronautical charts are designed to be used with a compass in a boat or aircraft.

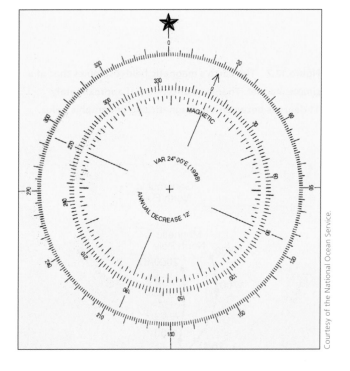

Courtesy of the National Ocean Service.

Figure 12.10 A typical compass variation (declination) diagram from a U.S. nautical chart for Prince William Sound, Alaska. The diagram is positioned on the chart so that 0° on the outer circle shows true north, while 0° on the inner circle shows magnetic north. Compass variation and its annual change are also indicated around the center of the diagram.

Compass variation, called **declination** for these uses, is shown on a circular **compass rose** printed in one or more places on these charts (figure 12.10). The outer circle of the compass rose is oriented to true north, while the inner ring is oriented to magnetic north. The angular difference between the zero-degree-north points on the two rings indicates the compass variation or declination at the time the chart was made. Notice that the annual change in compass variation is also shown near the center of the diagram along with the date the chart was made so that the navigator can determine the cumulative declination since the date indicated.

The standard topographic quadrangle series published by the USGS also provides us with declination information, since they are designed for use with a compass in the field. On these maps a **declination diagram**—a diagram that shows the angular relationship, represented by prongs, among grid, magnetic, and true norths—is printed at the lower left margin of the map. Notice that the true, magnetic, and grid north reference lines are marked at the top with a star, the letters "MN" and an arrow, and the letters "GN," respectively (figure 12.11). Always use the declination values printed on the diagram. Don't measure the angles on the diagram with a protractor because small angular differences are often exaggerated on the declination diagram to make them more discernible.

The angular differences among the three north reference lines, taken from the Madison West, Wisconsin, USGS 1:24,000-scale quad, are illustrated in figure 12.11. Notice that UTM grid declination and magnetic north declination are given for the center of the

sheet. Declination in areas toward the sides of the map could be slightly different. The larger the area covered by the map, the greater the change in declination from one side of the map to the other.

By comparing the three reference lines in the declination diagram, you can tell where the region covered by the map is located relative to the central meridian of the UTM grid and the isogonic chart of the country in figure 12.9. If the north–south UTM grid lines tilt to the east, you know that the mapped region lies to the east of the central meridian of the UTM zone. If the magnetic north lines slant to the east, you know that your mapped region is west of the agonic line. The reverse is true, of course, if the UTM or magnetic north lines slant to the west. In the declination diagram shown in figure 12.11, for example, UTM grid lines tilt west and magnetic north lines east. Consequently, you can tell that the Madison West quadrangle is located west of both the agonic line and the central meridian of the UTM zone.

Compass direction systems

Our systems for giving names or numerical values to directions come from the various ways we have for identifying directions on different kinds of compasses. Three compass direction systems—compass points, azimuths, and bearings—are of particular importance because of their widespread use in navigation, scientific work, and land surveying.

Compass points

The oldest compass direction system is the use of **compass points** (figure 12.12), which are directions

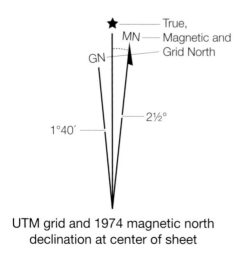

UTM grid and 1974 magnetic north
declination at center of sheet

Courtesy of the U.S. Geological Survey.

Figure 12.11   Declination diagram showing relations among true, grid, and magnetic north reference lines on the Madison West, Wisconsin, USGS 1:24,000-scale quadrangle.

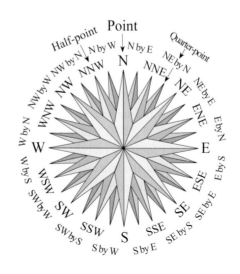

Figure 12.12   There are 32 points on the standard compass card: 8 points, 8 half-points, and 16 quarter-points.

indicated by arrows or prongs on the graphical face of the compass (called the **compass card**). Early mariners, who used the winds to find their way, devised compass points. The first mariner's compass card had eight points, representing the directions of the principal winds. But eight directions, sailors found, weren't exact enough. So they split their compass card further—first into eight additional **half-points** and later into sixteen more **quarter-points.**

The mariner's compass card thus came to have 32 points. The card, which looks something like a 32-petaled flower, is also called a **compass rose.** It is very similar to the compass rose we saw for navigational charts (figure 12.10) but directions are indicated by compass points rather than degrees. Unlike the circular compass on a navigational chart, the points on a mariner's compass card are named using a standard lettering system for the common directions. Figure 12.12 shows how it's done. The sizes of the compass rose points establish the priority system used to name direction. North, northeast, east, southeast, south, southwest, west, and northwest have first priority; the eight half-points have second priority; and the 16 quarter-points have third priority. A compass rose reading is therefore of the form NE (verbally stated as "northeast"), ENE (stated as "east-northeast"), and NE by E (stated as "northeast by east"). Each of these terms, of course, has a direct numerical counterpart in degrees, since half-points are 45 degrees apart and quarter-points are 22.5 degrees apart.

In modern times, the mariner's compass point system has been largely replaced by azimuth and bearing readings. Not only are azimuths and bearings less awkward and confusing to use, they are far more accurate.

Azimuths

The most common system of defining compass directions is the use of azimuths. An **azimuth** is the horizontal angle measured in degrees clockwise from a north reference line to a direction line (figure 12.13). Azimuths range from 0 degrees to 360 degrees and are written either in degrees, minutes, and seconds (45°22′30″) or decimal degrees (45.375°) (see chapter 1 for more on converting between degrees, minutes, seconds, and decimal degrees). A **back azimuth** is the opposite direction (180 degrees) from a given azimuth (in chapter 13, you will see how azimuths and back azimuths are used in position finding). The rule for determining back azimuths is to add 180 degrees if the azimuth to the feature is less than 180 degrees, and to subtract 180 degrees from the azimuth if it is greater than 180 degrees.

Azimuths are named according to which north reference line is used. Thus, there are **true, magnetic,** and **grid azimuths.** Compass roses for nautical and aeronautical charts (figure 12.10) have both true and magnetic north azimuth circles to simplify navigation by either true or magnetic north directions.

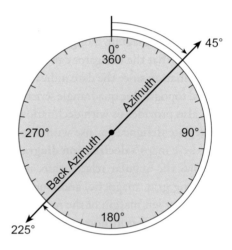

Figure 12.13 Azimuths and back azimuths are measured in degrees clockwise from a north reference line at the top of the azimuth circle.

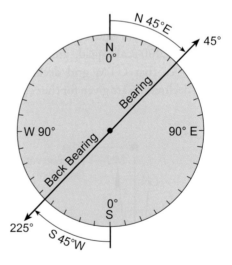

Figure 12.14 Bearings and back bearings range from 0 to 90 degrees and are measured from north or south toward east or west.

Bearings

The third compass direction system is the use of **bearings** (figure 12.14). Like azimuths, bearings are horizontal angles given in degrees. The difference is that azimuth readings go through the entire circle from 0 to 360 degrees, while bearing readings range only from 0 to 90 degrees. Bearings are measured clockwise (eastward) or counterclockwise (westward) from either a north or south reference line, whichever is closer to the direction line. To avoid ambiguity with this method, it's essential to give both the reference line (north or south) and an orientation (east or west) in addition to the angular measure in degrees. Therefore, bearings are written as N30°E (meaning 30 degrees east of north), S25°W (meaning 25 degrees west of south), and so on.

A compass with a **bearing card** (as shown in figure 12.14) is frequently referred to as a **surveyor's compass** because it has long been preferred as a surveying instrument. Surveyors like the bearing system because specifying the opposite direction (called a **back bearing**) is simply a matter of changing letters. For example, the back bearing of N45°E is S45°W. But since surveyors often must convert bearings to azimuths in their record keeping, it's common for bearings to be augmented with azimuths on the compass card (you will learn more about the use of bearings and back bearings in position finding in chapter 13).

Conversions

With three systems for specifying direction, sooner or later you'll want to convert one type of compass reading to another. When you do so, remember the declination diagram discussed earlier. Also consider the mathematical relationships among the three systems of specifying direction. For instance, a true azimuth of 85 degrees converts to a true bearing of N85°E, and both may be *roughly* described by the "east" compass point.

The declination information provided on large-scale topographic maps and nautical charts lets you convert true, magnetic, and grid azimuths from one form to another. To understand declination information more easily, it's a good idea to make an enlarged sketch of the declination diagram and an approximately correct direction line. For example, in figure 12.15A we sketched a direction line for a 65-degree true azimuth in a locale with a 17°E magnetic declination and 3°W grid declination. By marking the two declinations and the true azimuth with arcs, you can easily see that the magnetic azimuth for this direction must be 65 degrees minus the 17-degree magnetic declination, or 48 degrees. Similarly, you can see that the grid declination must be added to the

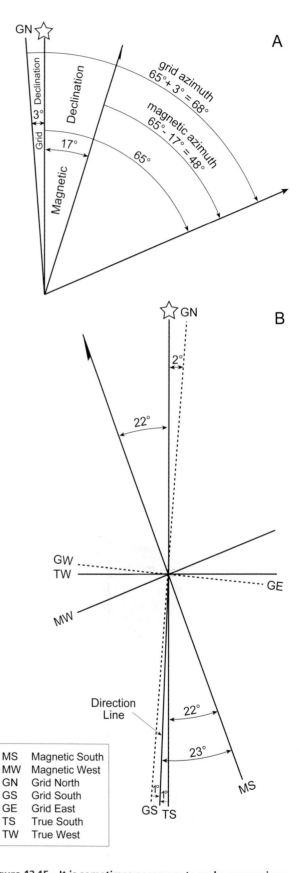

Figure 12.15 It is sometimes necessary to make conversions from one direction system to another. Conversions of azimuths (A) and bearings (B) are described in the text and illustrated in these figures.

true azimuth when converting to a grid azimuth. In this example, the grid azimuth is 65 degrees plus 3 degrees, or 68 degrees.

Converting bearings to and from true, magnetic, and grid north reference lines is more complex, but a sketch of the angles involved again clarifies what needs to be added and subtracted. In figure 12.15B, we have sketched a magnetic bearing of S23°W at a locale with a magnetic declination of 22°W and grid declination of 2°E. The diagram includes both magnetic and true south and the 22-degree angular difference between the two. Notice that the magnetic bearing is 1 degree west of true south, so the true bearing must be S1°W.

We should also consider **grid bearing notation,** or the system of expressing grid bearings. When expressing bearings, keep in mind that they can also be relative to a grid or magnetic south direction line. Figure 12.15B shows magnetic and grid south, and the 2-degree difference between them. The direction line for the magnetic bearing is 1 degree to the east of grid south, so the grid bearing must be S1°E.

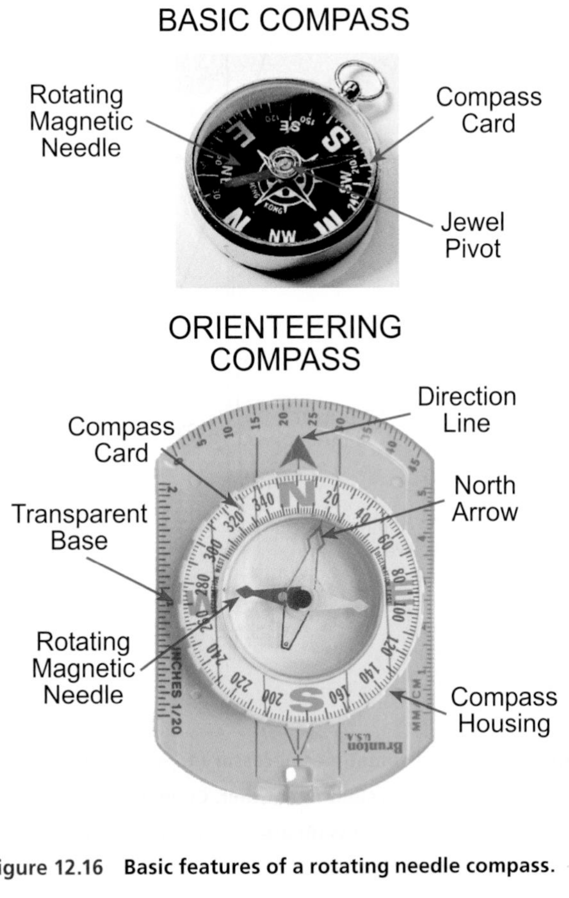

MAGNETIC COMPASSES

One of the most useful aids to navigation and map work is the **magnetic compass.** This clever device has three main parts—a needle or **floating disk,** which has been magnetized so that it will align itself with the earth's magnetic field; a **jewel pivot** or fluid bowl, which allows the needle or disk to float freely; and a dial, called a **compass card,** marked with directions (figure 12.16). The case that encloses all these parts is known as the **compass housing.** Some compasses also have long or round levels for aligning the compass with horizontal (figure 12.17). Sometimes **sights** and a **viewing mirror** are added to help you determine angles more accurately (figure 12.19).

Types of compasses

Compass quality varies greatly, usually in direct relation to the price. You may be confused by the wide range in prices and designs, but it helps if you understand that there are only four basic types of compasses. For convenience, we'll refer to these as rotating needle, rotating card, reversed card, and electronic compasses.

Rotating needle

Rotating needle compasses have the direction system inscribed clockwise on the compass card (figure 12.16, top). The compass needle is balanced on a center pin and rotates independently of the compass card. This design has the advantage that you can make either magnetic or

BASIC COMPASS

Rotating Magnetic Needle

Compass Card

Jewel Pivot

ORIENTEERING COMPASS

Compass Card

Direction Line

Transparent Base

North Arrow

Rotating Magnetic Needle

Compass Housing

Courtesy of Brunton, Inc.

Figure 12.16 Basic features of a rotating needle compass.

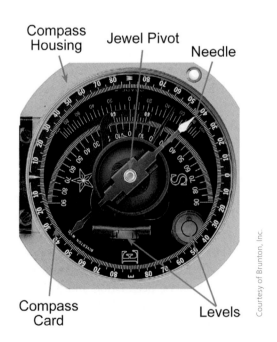

Compass Housing

Jewel Pivot

Needle

Compass Card

Levels

Courtesy of Brunton, Inc.

Figure 12.17 Basic components of a Brunton magnetic compass.

true readings. To find magnetic north, rotate the compass card slowly until north on the dial is lined up with the needle. With your compass set this way, you can take a sighting on any object and the reading on the compass will be relative to magnetic north.

Suppose, for instance, that you want to find the direction to a distant tree. With the needle stabilized on north, face the tree and project a visual direction line (**sight line**) from the center of the compass to the tree. You will now be able to read from the dial the number of degrees between magnetic north and the tree.

Your reading will, of course, be approximate. A more precise reading is possible if a sighting aid has been added to the compass. Sighting aids are of two forms. One type, resembling a standard raised gun sight, is mounted on opposite sides of the housing like the compass in figure 12.19. Instead of peering over your compass at the tree, you simply rotate the sighting aid until you can see the tree through it. You can then read the angle between the sight line and magnetic north. Readings are thus made easier and more accurate. Most rotating needle compasses don't include these raised sights, however, because they increase the cost and bulk of the compass, and because approximate readings are usually good enough.

A second form of sighting aid is an orienting arrow along the direction line as found on **orienteering compasses.** On these compasses, the housing (consisting of a compass card, a north arrow, and a rotating magnetic needle) is mounted on a transparent rectangular base (figure 12.16, bottom). These compasses are used along with maps in **orienteering** to find your way across unfamiliar terrain. To use an orienteering compass, you merely align the direction line, which is etched into the base, with your destination, and then rotate the compass housing until the orienting arrow is aligned with the magnetic needle. Now read the direction on the compass card beneath the sight line. This is the magnetic reading for the direction to your destination.

As we mentioned, an advantage of rotating needle compasses is that they allow you to make true as well as magnetic readings. To find true rather than magnetic north, first look at a declination diagram or compass rose on the map or chart to find the local magnetic declination. Then rotate your compass card east or west until the needle points to the amount of declination for your area. If the local declination is 5°E, you will line up the needle with 5°E on the dial. Now when you take a sighting to an object, you will obtain a true reading, because you have compensated for magnetic declination. Therefore, a rotating needle compass is easy to use with maps when a true reading is most useful, as well as in the field when a magnetic reading is of the most use.

Rotating card

On a **rotating card compass,** the magnetic needle and compass card are joined and work as a single unit floating in a bowl. Consequently, you don't have to bother with aligning the needle and compass card, as you did with the rotating needle compass. As soon as the disk stabilizes

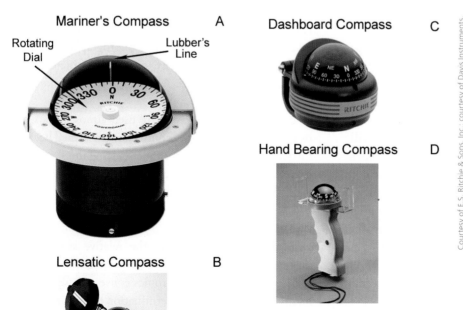

Courtesy of E.S. Ritchie & Sons, Inc.; courtesy of Davis Instruments.

Figure 12.18 Rotating card compasses.

and the needle points to north, you're set to make your sightings. Magnetic readings are thus simplified—an advantage when you are in the field without maps or charts. But the rotating card compass points only to magnetic north, not to true north. To obtain a true reading, you'll have to add (or subtract) local declination to (or from) your reading.

The **mariner's compass** is probably the most widely used rotating card compass. Since medieval times, mariners have placed a floating magnetized compass card in a bowl of clear liquid (figure 12.18A). A vertical pin in the middle of the bowl centers and stabilizes the compass card while allowing it to turn freely to magnetic north with changes in boat direction. The compass is suspended in such a way that it remains horizontal as the ship pitches and rolls in the water. A **lubber's line** parallel to the centerline of the ship is marked on the edge of the bowl to allow directions to be read relative to the direction of travel.

Lensatic compasses, available as military surplus items, also fall into the rotating card group (figure 12.18B), as do the common **dashboard compasses** with floating dials (figure 12.18C). Mariners usually have onboard a floating dial **hand bearing compass** (figure 12.18D). The compass may be mounted on a pistol grip and have a vertical lubber's line and gun sights, all to help you align the compass quickly and accurately.

Reversed card

With either a rotating needle or rotating card compass, you'll have to figure out the angle between true north and the object on which you are sighting. A **reversed card compass** like the Brunton compass in figure 12.17 directly displays this angle for you. You'll have to pay more for such convenience, of course. But for some people, like surveyors, hydrologists, and others who make much use of maps and compasses, the ease of making readings is worth the extra cost.

Reversed card compasses, like rotating needle compasses, have a freely rotating needle independent from the compass card. They are almost identical to rotating needle compasses, except that they include a number of refinements that make them more useful. The first thing you'll notice about a reversed card compass is that the direction system seems to be backward—west and east are reversed on the compass card (figure 12.19). The manufacturer hasn't made a mistake—there is a good reason for this design.

To visualize the logic behind the reversed direction system, imagine that you are facing north and that the needle on your compass is pointing north. Now turn

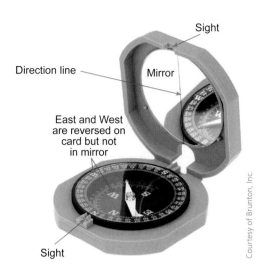

Figure 12.19 Reversed card compass.

slowly and align the sight on the compass with the object whose direction you are seeking. Although the needle actually remains stationary, it will look as though it is moving around the compass card in the opposite direction from which you move. But since west and east have been switched on the compass card, the direction will appear correct in the mirror that you look at to read the azimuth.

The reversed card compass is the most sophisticated, expensive, and versatile of the three types we've discussed so far. Like the rotating needle compass, it can be set for local magnetic declination, thereby permitting true readings. While rotating needle compasses sometimes include a sighting aid, reversed card compasses always do. More than one direction system is often incorporated onto the compass card to make several different types of readings possible. Many professionals use a tripod with reversed card compasses, just as photographers do with very expensive and sensitive cameras, to increase the accuracy of the readings. All these features make for a very useful compass.

Reversed card compasses are used almost exclusively by professionals. **Forester compasses** (so named because foresters use them to plan timber cutting lines) fall into this class. The best known example of a forester compass is the Brunton Pocket Transit (figures 12.17 and 12.19).

Digital

Digital compasses measure the relative strengths of magnetic fields passing through two magneto-inductive sensors. Sophisticated electronics convert these measurements into a continuous determination of the direction of the earth's magnetic field, which is magnetic north. The instrument then computes the azimuth between magnetic north and your straight-ahead direction of travel.

Figure 12.20 **Digital compasses with digital displays, including a watch-compass combination.**

Electronic compasses use digital displays to show headings (figure 12.20). Most allow you to enter the magnetic declination so that you can determine and display true azimuths as well.

Another form of electronic compass is the gyrocompass used on ships and aircraft. A gyrocompass finds true north by using a motor-operated gyroscope whose rotating axis is held parallel to the axis of the earth's rotation and thus points to true north, not the earth's magnetic pole. Since gyrocompasses are not based on magnetism, they are not affected by iron or other ferrous metals in the ship or aircraft.

Compass deviation

Notice on the isogonic map (figure 12.9) how in certain areas the isogonic lines appear to curve rather than form a straight line to the magnetic pole. This compass deviation occurs when the compass direction is strongly affected by **local anomalies.** These anomalies may be caused by nonmagnetic forces, such as regional variation in earth density, or by additional local magnetism, such as that produced by magnetic ore bodies. Localized magnetic anomalies that shift the compass needle 30 degrees in less than a mile exist on the earth but cannot be plotted on the small-scale isogonic map. The exact location of these anomalies may be described in notes printed on a navigational chart covering the area.

Compass deviation isn't always so predictable, though. It may also be caused by **local disturbances,** such as power lines, thunderstorms, and iron objects. Since this second source of deviation won't show up on isogonic maps, you must constantly watch for it. Try to keep away from known disturbances and be on the lookout for unknown ones, which may make your compass needle pull "off" north or act erratically.

DIRECTION FINDING ON LARGE-SCALE MAPS

Determining the azimuths or bearings of direction lines is the basis for several position-finding methods that we discuss in chapter 13. You can accurately measure the azimuth or bearing of a direction line if you plot its endpoints on a large-scale map designed for navigation purposes. Topographic quadrangles and nautical and aeronautical charts are good examples, since all are made on conformal map projections that preserve the geometry of the azimuth and bearing systems on the earth's surface.

A straightedge and protractor are all you need to measure these angles. A commonly used straightedge for map reading is a T square, which is a rule with a short perpendicular crosspiece at one end. This is used together with a **protractor**, which divides a circle into equal angular intervals, usually degrees. You are probably already familiar with semicircular and circular protractors used in basic geometry. You can use these for direction finding, along with semicircular, triangular, square, and course-plotting protractors specially made for measuring directions on maps (figure 12.21). All protractors have the **angle scale** along their edge and an **index mark** at the center of the protractor circle.

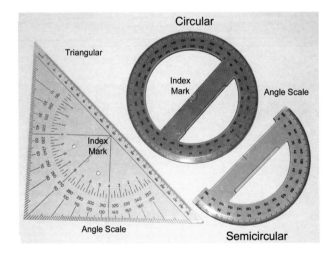

Figure 12.21 **Types of protractors used to measure azimuths and bearings from maps.**

Topographic quadrangles

To determine the true azimuth for the direction line between points *A* and *B* on a topographic quadrangle, use the following procedure with a semicircular or rectangular protractor (see figure 12.22):

1. Place the map on a rectangular table so that the bottom edge of the map is parallel with the bottom of the table. Align the bottom of the map with the top edge of a T square placed along the left edge of the table. The map is now oriented to true north.

2. Lightly draw the direction line on the map.

3. Place the protractor along the top edge of the T square, then move the T square vertically and the protractor horizontally so that the index mark is over the start of the direction line.

4. Read the true azimuth from the angle scale on the protractor.

You can use the same procedure to determine a grid azimuth, except that you must first orient the map to grid north. To do so, slightly rotate the map on the table until either a northing line or edge ticks for the same northing are aligned parallel to the bottom of the map. If UTM easting lines are printed on the map, you can align the index mark and the 90° mark on the protractor with one of these vertical lines. Now move the index mark to the start of the direction line, and you can read the grid azimuth from the protractor.

Magnetic azimuths are best measured by finding the true or grid azimuth for the direction line, then converting to a magnetic azimuth using the procedure described earlier in this chapter. You may be tempted to orient the map to the magnetic reference line on the declination diagram and then measure the magnetic azimuth directly with your protractor. Don't do this! Remember

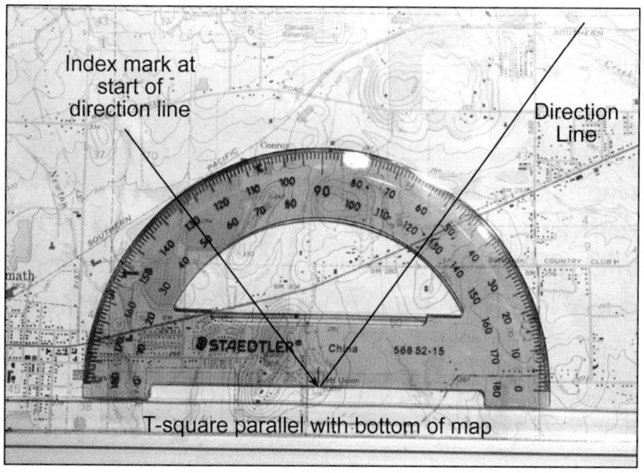

Figure 12.22 Finding the true azimuth of a direction line on a topographic map. See the text for the steps in the measurement procedure.

Courtesy of the U.S. Geological Survey.

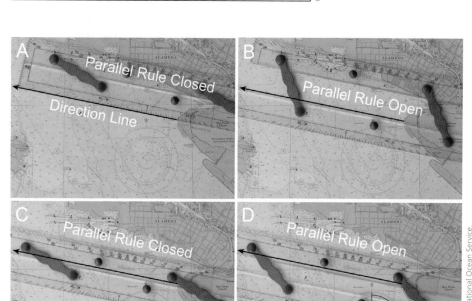

Figure 12.23 Finding the true or magnetic azimuth of a direction line by using a parallel rule and a compass rose on a nautical chart. See the text for the steps in the measurement procedure.

Courtesy of the National Ocean Service.

that the declination angles are usually so small that the angles between the reference lines are exaggerated to make them distinguishable. Only the declination values printed on the map are correct.

Nautical charts

Accurately determining directions on nautical charts is a critical task in marine navigation. Centuries ago chart-makers simplified the task by putting one or more compass roses on each chart. The compass rose on a nautical chart has an outer circular angle scale for true azimuths and an inner scale for magnetic azimuths (figure 12.10). Together with a compass rose, marine navigators use a special straightedge, called a **parallel rule,** to determine directions (figure 12.23). A parallel rule consists of two straightedges connected by metal bars that allow the straightedges to remain parallel as they are separated.

To operate the parallel rule, first push the two straightedges together, and align the top or bottom edge with the direction line (figure 12.23A). Holding the bottom straightedge firmly in place, separate the straightedges so that the top straightedge is closer to the compass rose (figure 12.23B). (You would reverse this walking process if the compass rose were above the parallel rule.) Next, firmly hold the top straightedge and move the bottom straightedge until it closes the gap between it and the top straightedge (figure 12.23C). Repeat this holding and moving procedure until you can place the top or bottom of the straightedge at the center of the compass rose (figure 12.23D). The parallel rule should be parallel to the direction line so that you can read the true or magnetic azimuth from the outer or inner scale on the compass rose.

Computing directions from grid coordinates

You can compute a grid azimuth if you know the grid coordinates (see chapter 4 for more on grid coordinates) for the endpoints of the direction line. The computation is done with simple trigonometry, as illustrated in figure 12.24. You first need to determine in which *quadrant* the direction line lies. If the beginning point has easting and northing grid coordinates (E_1, N_1) and the ending point is at (E_2, N_2), the rule for quadrants is the following:

> Quadrant I: E2 > E1 and N2 > N1
> Quadrant II: E2 < E1 and N2 > N1
> Quadrant III: E2 < E1 and N2 < N1
> Quadrant IV: E2 > E1 and N2 < N1

Once you know the quadrant, you can figure out the azimuth of the direction line. This is done by forming a right triangle. Angle α is calculated by finding its trigonometric tangent. Remember that the tangent of an angle in a right triangle is the ratio of the opposite and adjacent sides to the angle (rise over run). The lengths of the opposite and adjacent sides are the differences of the eastings and northings for the endpoints of the direction line. Slightly different equations are needed for each quadrant, since the difference in eastings defines the opposite side in quadrants I and III, and the adjacent side in quadrants II and IV. The beginning and ending northings and eastings for the direction line must also be reversed in some quadrants to always have positive distances for the opposite and adjacent sides.

After you find angle α, you need to add 90, 180, or 270 degrees for quadrants IV, III, and II, respectively, in order to convert the angle to a grid azimuth. An example of this calculation is to find the grid azimuth for a direction line from UTM (easting, northing) coordinate (345,630mE, 4,335,480mN) to coordinate (353,287mE, 4,308,592mN). Since the second easting is greater than the first, and the second northing is less than the first, the direction line lies in quadrant IV.

Knowing the quadrant, you can now solve the quadrant IV azimuth equation with your handheld calculator or a software program to do the following calculations:

$$\alpha = \tan^{-1}\left(\frac{(N_1 - N_2)}{(E_2 - E_1)} \right)$$

$$\alpha = \tan^{-1}\left(\frac{(4,335,480 - 4,308,592)}{(353,287 - 345,630)} \right)$$

$$\alpha = \tan^{-1}\left(\frac{(26,888)}{(7,657)} \right) = \tan^{-1}(3.511558)$$

$$\alpha = 74.1°$$
$$\text{Azimuth} = 74.1° + 90° = 164.1°$$

Geographic information system (GIS) software such as ArcGIS uses calculations from grid coordinates to compute the azimuths between pairs of points, line segments, and perimeters of area features. Grid, true, or magnetic azimuths are not typically stored in the GIS database, but it is possible to find Internet services that will compute these based on the date the map was published and the location of magnetic north at that time.

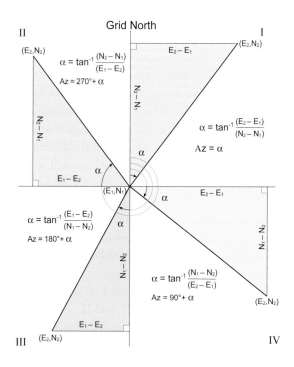

Figure 12.24 Trigonometric basis for computing grid azimuths from grid coordinates.

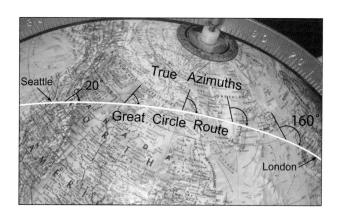

Figure 12.25 True azimuths along a great circle route are shown correctly on a globe.

DIRECTION FINDING ON SMALL-SCALE MAPS

"Fly north to get east" was Charles Lindbergh's philosophy when he reached Japan by flying over the north pole. Determining direction on a round earth, as every ship and plane pilot knows, is a far different matter from direction finding over short distances.

Until now, we've defined direction as the angle between a direction line and reference line. But when dealing with directions for points over hundreds of miles or kilometers apart, the flat earth methods we have discussed so far must be modified to account for earth's sphericity. On the spherical earth, there are two types of direction lines—**great circle routes** or the shortest path between two points on the globe, and **rhumb lines** of constant compass direction (see chapter 1 for more on great circles). Let's look at both these types of direction.

Great circle routes

Directions along the great circle route between two points on the earth's surface are true azimuths because they are measured relative to true north. What makes great circle routes most valuable to navigators is that they are the shortest possible routes between two locations on the spherical earth. To minimize travel time,

navigators traveling long distances would ideally follow a great circle route.

Measuring directions on a globe

One way to measure true azimuths along a great circle route is to use a globe (figure 12.25), since all true azimuths along the path are correct. You can hold a string tightly against a globe to find the great circle route between two locales. A friend can mark the angles from meridians to the great circle, and you can measure these on the globe with your protractor. You will notice that the great circle route between Seattle, Washington, and London, England, for example, has continuously increasing true azimuths from around 20° at Seattle to around 160° as London is approached. This procedure is time-consuming and difficult to do accurately, especially since the protractor and straightedge will not lie flat on the globe. Flat maps are used to make better measurements.

Measuring directions on the gnomonic map projection

The only map projection on which all great circles project as straight lines is the *gnomonic projection* (see chapter 3 for more on this projection). You can use the gnomonic projection to find the true azimuth at a point along the great circle path between two distant cities like Seattle and London (figure 12.26). Here's how:

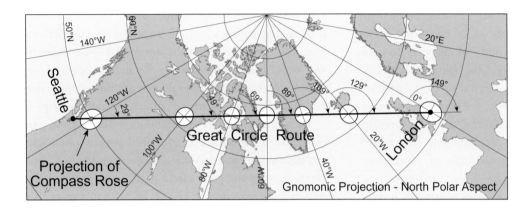

Figure 12.26 The great circle route from Seattle to London is a straight line on the gnomonic map projection. Circular "compass roses" at points where the route crosses major meridians are projected as ellipses, showing the distortion of directions that makes it possible to measure true azimuths only roughly along the route.

1. Draw the straight-line great circle route between Seattle and London on the gnomonic projection.

2. Find the westernmost meridian that the great circle route intersects (120°W in this example). Mark the point of intersection with a dot (a red dot on our illustration).

3. Place your protractor so that the index mark is on the dot and the zero-degree mark on the angle scale is aligned with the meridian.

4. The clockwise angle from the meridian to the great circle line is the true azimuth (29° at the 120°W meridian).

5. Now repeat these steps for each major meridian.

Notice the projections of simplified "compass roses" (ellipses on which the north–south and east–west directions are indicated) at the intersection of meridians and the great circle route—all are ellipses! Your measurement with the circular protractor is being done on elliptical direction dials that change in shape along the route. You could make accurate direction measurements if you had an elliptical protractor that would let you constantly change its elongation, but no such device exists. Hence, your protractor only allows rough measurements of true azimuths.

Rhumb line directions

When using a compass to navigate, navigators may soon grow weary of continually making directional measurements or computations and steering their craft to keep exactly on a great circle route. They can alleviate these problems by following a **rhumb line,** a direction line that is extended so that it crosses each meridian at the same angle. Rhumb lines are routes of constant compass readings and are therefore extremely useful to navigators. If

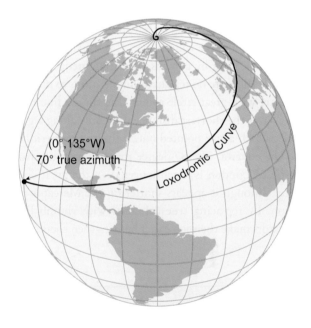

Figure 12.27 A rhumb line heading of 70 degrees starting at the equator will trace out a loxodromic curve as it converges on the north pole.

they follow one of these lines, navigators can maintain a course without constantly figuring out new headings and making turns. They merely check the compass to be sure that they are crossing each meridian at the angle of their rhumb line. Their path may curve, but their compass direction won't change.

Rhumb lines running east–west along the equator or north–south along a meridian return eventually to their points of origin and are great circles. Those running east–west along parallels other than the equator also close on themselves but are small circles, so they are not great circle routes.

Rarely, however, are rhumb lines either great or small circles. More commonly they cross meridians at an oblique angle, tracing out an odd looking path. To cross each meridian at the same angle, the oblique rhumb line has to keep curving on the spherical earth. Its path on the earth forms a spiral, known as a **loxodromic curve.** Figure 12.27 shows the loxodromic curve that results when a constant 70-degree direction line (one which intersects every meridian at an angle of 70 degrees) starting at the equator converges on the north pole.

To show all rhumb lines on maps as straight lines, a special projection is required. Mercator solved this problem in 1569, as figure 12.28 shows, by pulling the meridians and parallels apart at higher latitudes to create a conformal world projection that preserves directions at any point on the projection surface (see chapter 3 for more on the Mercator projection). On a Mercator projection, then, loxodromic curves have been straightened, and any straight line is a rhumb line. This valuable property makes navigation by compass and straightedge straightforward. There are no repeated computations to make, as there are with great circle routes. The rhumb line route, however, is the shortest distance path only along a great circle, which as we pointed out, is limited to meridians and the equator. The 87-degree rhumb line route from Seattle to London is much longer than the great circle route, although it appears shorter on the Mercator projection. By replacing true azimuths with rhumb lines, navigators make their job easier but lengthen their route.

The navigator's dilemma

If you're navigating a long distance, you face a dilemma. Should you use true azimuths along great circle routes to save travel time, while doing a lot more work to determine each azimuth? Or should you use rhumb lines, simplifying navigation planning but lengthening the trip?

There is a solution to the dilemma: use both. First, plot the great circle route as a straight line between your origin and destination on a gnomonic projection (figure 12.26). Next, transfer the great circle route to a Mercator projection as a curve (figure 12.28). Then approximate this curve, which is always concave toward the equator, with a series of straight-line segments called *legs* or *tracks*. In this example, each leg starts at 20-degree-longitude intervals—the interval you use will vary based on how closely you want to follow the great circle route. Now you can measure the true azimuth at the beginning of each leg with a protractor by finding the azimuth angle between the vertical meridian and the direction line for each leg.

Computing directions from geographic coordinates

You can also compute the true azimuth (Az) at the starting point of the great circle path. The azimuth computation uses a combination of the **Law of Sines** and the **Law of Cosines** from spherical trigonometry in an equation for the tangent of the true azimuth, where a and b are the latitudes of the starting and ending points and $\delta\lambda$ is the positive difference in longitude between the two

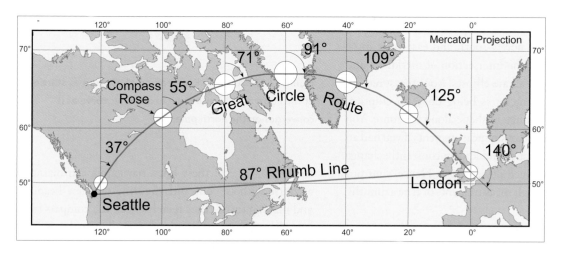

Figure 12.28 **Portion of a world map made with the Mercator projection showing the rhumb line and great circle route from Seattle to London. The great circle route, obtained from a gnomonic projection, has been divided into 500-nautical-mile legs. Since the Mercator projection is conformal, each simplified compass rose is correctly projected as a circle.**

points. Here is an example of this computation to find the starting azimuth for a flight from *a* in Seattle (47.5°N, 122.33°W) to *b* in London (51.5°N, 0°).

$$\tan(Az) = \frac{\cos(b) \times \sin(\delta\lambda)}{(\cos(a) \times \sin(b) - \sin(a) \times \cos(b) \times \cos(\delta\lambda))}$$

$$\tan(Az) = \frac{\cos(51.5°) \times \sin(122.33°)}{(\cos(47.5°) \times \sin(51.5°) - \sin(47.5°) \times \cos(51.5°) \times \cos(122.33°))}$$

$$\tan(Az) = \frac{0.6225 \times 0.8450}{(0.6756 \times 0.7826 - 0.7373 \times 0.6225 \times -0.5348)}$$

$$\tan(Az) = \frac{0.5260}{(0.5287 + 0.2455)}$$

$$\tan(Az) = \frac{0.5260}{0.7742}$$

$$\tan(Az) = 0.6794$$

$$Az = 34.2°$$

You can write functions for your GIS that use these trigonometric calculations from geographic coordinates to compute azimuths between pairs of distant points on the earth and for line segments along the great circle route between the points.

SELECTED READINGS

Blandford, P. 1984. *Maps & compasses, A user's handbook.* Blue Ridge Summit, Penn.: TAB Books Inc.

Jonkers, A. 2003. *Earth's magnetism in the age of sail.* Baltimore, Md.: Johns Hopkins University Press.

Kjellstrom, B. 1994. *Be expert with map and compass.* New York: John Wiley & Sons, Inc.

Maloney, E. 2003. *Chapman piloting.* 64th ed. New York: Hearst Marine Books.

———.1988. *Dutton's navigation & piloting.* 14th ed. Annapolis, Md.: Naval Institute Press.

Selwyn, V. 1987. *Plan your route: The new approach to map reading.* London: David & Charles.

United States Army. 1969. Directions. In *Map reading,* Field Manual (FM 21-26) , 5-1–5-27. Washington, D.C.: Superintendent of Documents.

chapter
thirteen   **POSITION FINDING AND NAVIGATION**

13

Position finding and navigation

There's no feeling as chilling as realizing you don't know where you are or how to get where you want to go. The fear of becoming lost can be so overwhelming that you hesitate to leave your known environment. But you needn't give in to fears of disorientation. If you know how to compare your surroundings with a map, you don't need to worry about losing your way.

Being oriented is knowing your position on the ground and where distant features are located in relation to you. This information lets you navigate (plan and follow a specific route) from place to place. For a long time, people relied solely on their powers of observation for position finding and route planning. Over the centuries we have enhanced our orientation and navigation capabilities by inventing a variety of technical aids. Clocks, compasses, optical sighting devices, electronic direction and distance finders, inertial navigation systems, and global positioning satellites all fall in this category. No matter how technically sophisticated our position and path finding aids, however, they share much in common with long-standing "eyeball" methods. They all use distance or direction information and rely on a few basic geometrical concepts. When our modern technical gadgets fail or aren't at hand, we must fall back on methods that are based on observation.

We'll start this chapter by discussing traditional observation and compass techniques that have proven useful for centuries and underlie modern satellite-based methods of position finding and route planning. We'll begin with orienting the map, then learn about ways to find and map ground positions, and conclude with navigation planning and route following methods. We'll bring these methods up to date with discussions of computerized route planning and GPS-assisted land, water, and air navigation. In chapter 14 ("GPS and maps") we examine in detail the capabilities of handheld GPS units used for land navigation.

ORIENTING THE MAP

To **orient** a map means to determine how directions on the map align to directions on the ground. The term comes from medieval Europe where church scholars drew maps of the known world with the Orient (China) at the top, but today most maps are drawn with true north at the top.

When traveling, people tend to orient maps in one of two ways. Many people keep the map topside up, no matter which way they're facing. This procedure has the advantage that place names, symbols, and features are easy to read. But it has the disadvantage that map directions aren't usually aligned with ground directions. When you're heading south, right on the map is left—in reality. Such reverse thinking can make the mind reel. Imagine trying to make a split second decision in heavy traffic about which way to turn! Since this may not be much of a problem in familiar surroundings, it's easy to forget that there is an alternative.

If you're in an unfamiliar setting or confusing surroundings, it's usually easier to find your way if you first orient the map to your direction of travel. Turning the map until ground and map features are aligned has the advantage that you can always determine directions directly. Although you may have to read place names and symbols upside down or sideways, this is easier than trying to unscramble backward directions. You can finesse the problem of markings and labels being upside down by thoroughly familiarizing yourself with them before your trip and by writing down the key data you need in margins or on a separate piece of paper. A disadvantage of this method is that when following a sinuous route you will have to constantly turn the map to stay aligned with the direction of travel.

Inspection method

One of the easiest ways to orient a map in the field is by the **inspection method.** With this technique, you don't need to know which way north is. Nor do you need any special tools. Two conditions must be met, however. First, you must be able to see one or more linear features or prominent objects in your vicinity. Second, you must be able to identify these same features on the map. Various landmarks satisfy both conditions, and many are shown directly on maps.

One linear feature

There are three forms of alignment with a single linear feature. The most straightforward method is to position yourself on a linear feature shown on the map, such

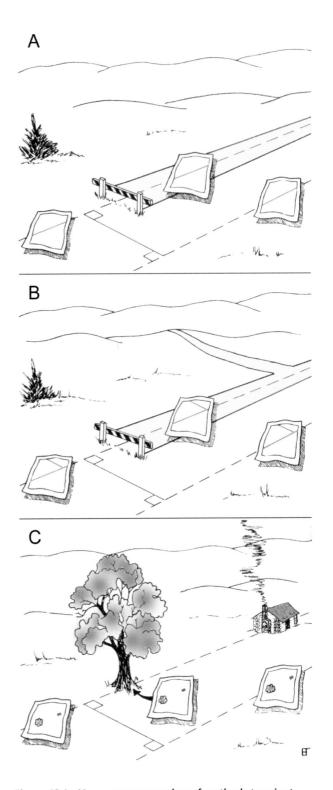

Figure 13.1 You can use a number of methods to orient your map by inspection of a single linear feature (A), two linear features (B), or prominent features (C).

as a road. Then you need only turn the map until the mapped feature lines up with the real one in front of you (figure 13.1A, middle map). The line you see between you and the feature is called a **sight line.** A second method is to assume a position that lies on a straight-line extension of the ground feature (figure 13.1A, left map). Your third option is to take up a position to either side of the linear ground feature and then to align the feature shown on the map so that it is parallel with the ground feature (figure 13.1A, right map). This last form of map orientation is the least accurate of the three methods, but it will usually suffice when you can't line yourself up directly with the linear feature.

The first and third options share one serious problem if you happen to orient yourself where you can't see the end of the feature. When you line up the map feature with the ground feature, you don't know which direction is which, and you could end up going in the opposite direction from what you intend. Because of this potential reversal of orientation, you shouldn't rely on a single linear feature if you can avoid it.

Two linear features

A better practice is to rely on two (or more) linear features when orienting the map by inspection. As with a single feature, you have three options. You can move to a position on either of the two linear features (figure 13.1B, middle map), move to a point which lies on a straight-line extension of either feature (figure 13.1B, left map), or assume a position off to one side of both features (figure 13.1B, right map). Then you simply turn the map until the two features on the map are aligned with the same features on the ground. Again, the third approach is the least accurate, and map reversal leads you to orient yourself in the opposite direction.

Prominent features

Reversal in map orientation can be avoided by using two or more prominent features on the ground that can be found on the map, such as a house or tree. You first move to one of the features (figure 13.1C, middle map) or to a position on a line extended through the two features (figure 13.1C, left map). Next, place a ruler or other straightedge between the features on the map. (If a straightedge isn't available, mentally sighting between the two features will do.) Then turn the map until you can sight along the straightedge or sight line to the other visible point (when you have moved to one feature) or to both features (when you have moved to an extension-line position). If you are located off to the side of the features (figure 13.1C, right map), you will have to estimate your

position relative to the straightedge line between the prominent features.

The success of map orientation by the inspection method depends primarily on the ability to identify ground features that can be found on the map. When prominent features are unfamiliar or obscured (by vegetation, fog, or terrain, for example) or when they are locally nonexistent (as on an open plain or ocean surface), the method of inspection is of little value. In these special situations, a better method of map orientation is to use a magnetic compass.

Compass method

To use the **compass method** of map orientation, first be sure you have a compass and a map that shows the direction of magnetic north with a declination diagram such as that shown at the bottom of figure 13.2 (see chapter 12 for more on declination diagrams). Then do the following:

1. Find the magnetic north indicator on the map. This is usually a barb-shaped arrow placed in the margin of the map.

2. Holding the map under the compass, turn the map until the compass needle lines up with magnetic north on the map.

Since this procedure accounts for magnetic declination (see chapter 12 for more on magnetic declination), your map will now be properly oriented with true north as well.

Figure 13.2 shows why it's so essential to take magnetic declination into account when orienting a map with the compass method. In this example, the magnetic declination is 20° East. When the compass needle is oriented with the magnetic north reference line (A), true north is located 20° west of the compass needle. Conversely, when the compass needle is oriented with true north (B), the map is oriented to magnetic north (20° east of true north).

Unfortunately, magnetic north isn't always shown on maps. Often a topographic map sheet will have been clipped to the area of interest or have its margins trimmed off to make its size more manageable, with the result that the magnetic north indicator is lost. For these reasons, it makes sense to commit the basic declination pattern to memory. For the contiguous United States, declination varies from approximately 25° West (in the Northeast) to 25° East (in the Pacific Northwest) (see figure 12.9 in chapter 12).

If you keep in mind the magnetic declination for the area in which you're traveling, you need never be too

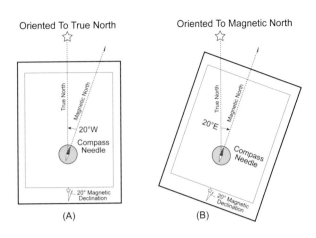

Figure 13.2 When orienting a map to true north (A) or magnetic north (B) with a compass, there must be information about the magnetic declination for the map you are using.

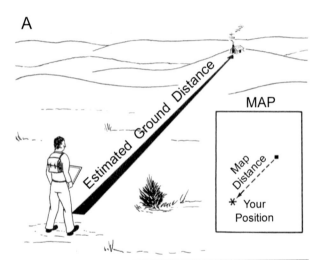

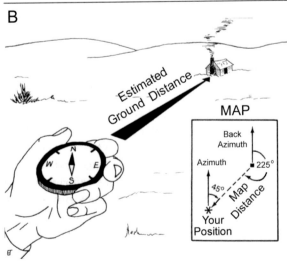

Figure 13.3 One way to find your position is to use distance and direction information gained by inspecting your surroundings (A) or using a magnetic compass (B).

far off in estimating the direction of magnetic north. For large portions of the Midwest, the declination is close enough to 0 degrees that true north (indicated by longitude lines) provides a reasonable substitute for magnetic north. If you don't know what the declination is, then the compass method just won't work for you and you will have to use one of the inspection methods described previously.

FINDING YOUR GROUND POSITION

At times you'll orient your map on the basis of a known ground position or you'll establish your position in the process of map orientation. At other times, you won't know your ground location even after you've oriented your map. But once your map is oriented, you can then figure out where you are on the map.

Distance estimation

The most common technique for determining your ground position is **distance estimation** (or **range estimation**) in which you use your distance and direction to other objects to location your position on the map. There are two useful methods: by inspection and using a compass.

Inspection method

When you use the **inspection method** for distance estimation, the idea is the same as when you used inspection to orient your map. You are simply looking at (or "inspecting") ground features without any special aids. With the inspection method, you find ground features on a map you've oriented and then estimate their distance from you.

To use this method, first orient your map and select a feature on the ground that you can also identify on the map. Next, estimate the distance from your ground position to the distant feature, and convert this figure to map distance units by using the map scale. Now, mark out the computed map distance along the proper direction line and you have established your position (figure 13.3A). You can double-check your work by repeating the procedure with several features that are along different direction lines, but don't be too discouraged if your results don't agree. Distances are hard to judge accurately, although there are several things you can do to improve your estimates.

One trick is to use multiples of familiar distance units, such as a football field in the United States (100 yards) or a football pitch in Britain (100–110 meters). But most people have trouble visualizing these units. A further complication is that farther distances are easier to estimate than nearer distances. As your vision approaches the **vanishing point** (the point on the horizon where parallel lines appear to converge), it becomes harder to judge distance.

Another trick is to memorize what familiar objects look like at different distances. Then you can simply compare the size and general appearance of an object, such as a house or a tree, with the "template" that you hold in your mind's eye, and you will have a good idea of the range. There are problems with this procedure, however. The object may not be of standard size. Even when it is the correct height or width, troubles may arise. Weather, atmospheric conditions, the relationship of the sun's rays to the object, and the intervening terrain may all compound the difficulties of judging distance on the basis of the size of objects. A feature will appear to be a different distance when seen from a low position over a flat surface than when viewed from your position on one hillside to its location on another. If the object is seen from above or below, it will appear smaller and farther away than it actually is. And object lit from behind will seem farther away than one lit from the front. A brightly colored feature will appear closer than a dull feature, and both objects will look closer on a bright, dry day than a humid or foggy one.

Compass method

The second form of distance estimation involves the use of a map and a **magnetic compass** (figure 13.3B), which indicates directions magnetically by the alignment of a magnetic needle or floating disk to the earth's magnetic field. With the **compass method** of distance estimation, you plot lines that intersect at your position. An advantage of this compass technique is that you don't need to orient your map before finding your position. Use your compass to sight on a distant feature and note the reading (45° in this example). This is the **azimuth,** or the angle measured clockwise in degrees between the direction line and the reference point (see chapter 12 for more on azimuths). Using this, you can figure out the **back azimuth.** Back azimuths are calculated by adding 180 degrees to azimuths less than 180 degrees, or subtracting 180 degrees from azimuths greater than 180 degrees (see figure 12.13). Now draw a line from the object on the map along the back azimuth. You can determine the ground distance from your ground position to that feature using one of

the inspection methods discussed above. Using the map scale, convert this figure to map distance units. Finally, mark off the estimated distance on the line that you drew, and that is your estimated ground position.

Resection method

If accurate position determination is crucial, the **resection method** is a better technique for locating position than the distance estimation techniques described previously. Instead of estimating the distance from some object on the ground to your position, you can determine this accurately and plot it on the map to find your ground position. This involves plotting lines that cross, or resect, at your position. The position found using this method is sometimes called a **cross fix.** You can make this plot with two or three lines (figure 13.4A), although it is most accurate to use three lines. You can use either the inspection method or the compass method to determine your position using resection.

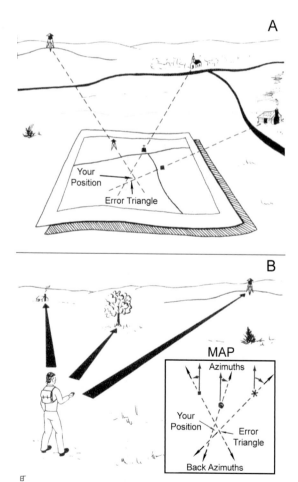

Figure 13.4 Position finding can be carried out by resection based on map inspection (A) or by magnetic compass readings (B).

Inspection method

First, you must be able to find three point features on both the ground and the map. Then, with the map properly oriented, determine the direction line between you and one of the prominent ground features. Then find the same feature on the map. Use a straightedge to draw the direction line on the map. This line is really a **backsight** (or back azimuth), since it's drawn from the known position of a distant feature back to your position. Next, draw backsights for the other two features. You're located where the three lines cross on the map.

You'll get the most reliable position or **fix** (a position determined using a navigational method) when at least two of your sight lines cross at angles close to 90 degrees. If possible, avoid using sight lines which cross at angles smaller than 45 degrees. With small angles, even a slight error in drawing the lines may add considerable error to the location at which the lines should intersect.

If you look carefully at the intersection of the three lines, you will see that they didn't exactly intersect at a point, but rather formed a small triangle called an **error triangle** (figure 13.4). Your location is somewhere in the error triangle; for convenience, you can assume that it is at the center of the error triangle. If this triangle is small, you can feel confident in your results. If the error triangle is large, however, it's advisable to repeat the procedure, taking greater care.

There are several reasons why the line did not intersect at a point on the map. You may have made an error in either in positioning the known locations correctly on the map or in plotting the sight lines. But there's also the chance that you didn't sight to the correct feature or that the mapmaker misplaced one or more of the features on the map in the first place.

More important is the fact that the size and shape of the error triangle in large part depends on the geometrical arrangement of the three known locations. The ideal arrangement would be known locations that form an equilateral triangle with the unknown location at its center. The worst arrangement is when two or all three points are close to each other and are nearly in a straight line with your unknown location. The lines drawn from the distant features with this arrangement typically intersect at a low angle, so that small measurement and plotting errors may significantly shift the points at which the lines intersect.

If only two outstanding point features are identifiable on both ground and map, you can still use the resection method, proceeding as you did when three features were available. Because only two lines are constructed, they will cross at a point rather than form a triangle. Regardless of drawing errors on your part, this will give the illusion of accuracy. Any error you made in either of your sightings won't be evident. Whenever two rather than three lines are used, therefore, you must watch out for this potential hidden error.

Compass method

The **compass method of resection** makes use of back azimuths as resection lines (figure 13.4B). You merely use a compass, rather than a straightedge or visual sighting, to determine the resection lines, the same as you did when you used a compass for distance estimation (figure 13.3). Again, you can make your resection plot with two or three lines. To use the three-point resection method, select three prominent features on the ground that you can also identify on the map. Next, sight on the three distant points with a compass from your ground position and note the azimuths. Plot the back azimuths for the three sightings through their respective map points using a protractor. The three back azimuth lines will form an error triangle, just as your resection lines did when made with a straightedge.

Measurement instruments

As we have seen, determining your ground position requires knowing your distance from some object and translating that to the map. Instruments also can be used to help you determine your position based on this information. Let's take a look at some of them.

Stadia

Many gadgets to help estimate distance are on the market. Sophisticated and accurate surveying instruments and hunting scopes that measure distance use the **stadia principle**, which is based on using a constant angle to determine a distance between a marker and a sighting device. These devices have a built-in pair of horizontal wires, called **stadia,** that are spaced to subtend a small angular distance (that is, they are opposite to and delimit the extent of an angle). The angle subtended is usually on the order of **minutes of angle (MOA).** One minute of angle (one-sixtieth of a degree) will subtend approximately 1 inch (actually 1.023 inches) at 300 feet, 2 inches at 600 feet, and so forth. Surveyors use stadia-equipped telescopes to sight at a graduated **stadia rod** and then compute the intervening distance by using simple trigonometry (figure 13.5).

Laser rangefinder

Newer technology for distance estimation includes the **laser rangefinder** (figure 13.6). Although somewhat expensive, these "point and read" instruments are much more accurate and handy. Infrared laser beams built into binoculars or a monocular (one eyepiece) measure distance up to 1,000 yards (over 900 meters) with an accuracy of one yard (almost a meter). These laser rangefinders have become indispensable around construction sites and realtor's offices, where they are known as **electronic rulers.**

Laser rangefinders can be a lot of fun to use, especially in checking the results of your mental distance estimates. Their cost and weight are directly related to their range and accuracy. Varying from 9 ounces (255 grams) to 3 pounds (1.4 kilos), they're all portable enough to carry.

Altimeter

In mountainous regions, particularly those that are heavily forested or cloud-shrouded much of the time,

you may be unable to locate enough features on both the ground and map to use the inspection or compass methods of position finding accurately. Or you may not have your distance finding instruments with you. In these cases, you might think the GPS receiver would be your savior. But you may be wrong. Handheld GPS receivers have limited capability in heavily forested or dissected terrain because it is not possible for the GPS receiver to get a good reading from the satellites (see chapter 14 for more on GPS).

A pocket altimeter may be your best position-finding aid in such circumstances (figure 13.7). An **altimeter** is an instrument used to measure the height above fixed level, usually sea level. With a topographic map and an altimeter, you can fix positions with the help of only a single feature common to your map and your surroundings. An altimeter position fix is easiest if you are located on or near a linear feature such as a trail, stream, or ridgeline. But it's also possible to determine your position with an altimeter if you can make a sighting on a distant feature such as a building, mountain peak, or lake.

Before you hike into the wilderness with an altimeter in your pack, there's something you should know. An altimeter's primary drawback in position finding is that an instrument of portable size and weight is difficult to keep calibrated. These calibration problems can be traced to the fact that an altimeter is a form of barometer that determines elevation by measuring air pressure.

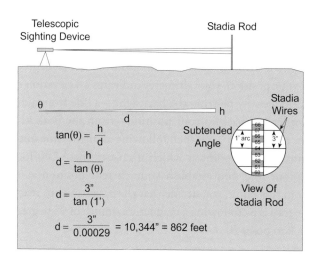

Figure 13.5 Distance estimation instruments include stadia rods and telescopic sighting devices.

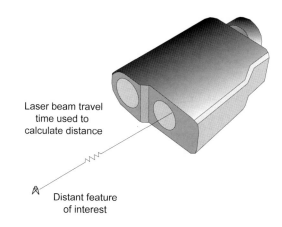

Figure 13.6 Laser rangefinders time laser beam pulses to find the distance to a distant object.

Figure 13.7 In heavily forested or cloud-shrouded mountains, a pocket altimeter can be an invaluable position-finding aid.

Unfortunately, the relation between elevation and air pressure isn't a simple one. Since under ideal conditions air pressure decreases systematically with altitude, the altimeter's scale can in theory be calibrated in meters or feet to reflect changes in elevation. In practice, though, air temperature and humidity also influence atmospheric pressure, and these often change rapidly and unpredictably. All things considered, it is best to view pocket altimeter readings as approximate.

You can do several things to increase the accuracy of your altimeter readings, however. One is to pay a little more and obtain a **temperature compensated altimeter,** which means that the measuring device has been designed to compensate for the effect of temperature changes. Also, always set your altimeter before you start your trip using your topographic map to obtain accurate elevation data. Finally, reset your altimeter as often as possible during the trip. You can do so whenever you come to a place where you can clearly discern the elevation. A hilltop, a saddle between two peaks, and the point at which a trail branches or crosses a stream are good spots to calibrate the altimeter. The shorter the time between your altimeter settings, the less chance that changing weather conditions will cause incorrect elevation readings.

Linear feature method

To establish your position with an altimeter when you're located on a path or other prominent linear feature, there are two steps. First, read the elevation from your altimeter, taking care to tap the device several times to be certain that the pointer hasn't stuck in a wrong position. Next, follow along the appropriate linear feature on your map until you come to the elevation indicated by the altimeter. This will be your approximate ground position. This method works if the linear feature has a steadily changing elevation. Otherwise, it might be hard to pinpoint your position on a meandering path or other lengthy linear feature all at the same elevation.

An example of position fixing with an altimeter and a linear feature is provided in figure 13.8. Let's assume that you have parked your car at the parking lot, with the intent of hiking south along the trail. You note that the parking lot is bisected by the 5,800-foot contour line. Before starting out from the parking lot, therefore, you set your altimeter to this elevation. When you get to the stream, you check the altimeter and note that it is reading about 5,400 feet. When you cross the second stream, you note a reading of about 5,800 feet. Later, after hiking several hours through dense fog, you stop and read your altimeter again. The device indicates that you have climbed to an elevation of 7,000 feet. By tracing your progress up the trail, you decide that you must be located near point B, where the 7,000-foot contour crosses the trail.

Backsight line method

If you aren't located on a linear feature, you can still use your altimeter to fix your position if you can sight on a distant object. Three steps are required. First, plot the backsight line from your position to the visible feature on your map. To do so, you can use one of the inspection methods discussed previously. Next, read the elevation from your altimeter. Finally, visually follow along the sight line until you come to this elevation value. This should be your approximate position.

An example of this sight line method of position finding is given in figure 13.8. In this case, suppose that you lost the trail in the fog shortly after making the position fix for point B. After an hour or so, the fog clears enough so that you can see a mountain peak (D) across a valley to the south of your position. Let's say you have your compass so you can use the compass method of distance estimation. Using this, you determine that the magnetic azimuth to the mountain peak is 140 degrees. At your position, your altimeter indicates an elevation of 6,200 feet. By plotting the backsight—a line from a distant feature back to your position—from the peak as a line on your map, and then following along the line until you get to the point at which it crosses the 6,200-foot contour line, you establish your current position at C. From this point, you decide that the easiest way to reach the trail is by following a magnetic azimuth of 95 degrees back to point B.

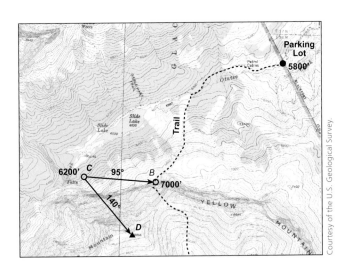

Figure 13.8 Using an altimeter to find your position along a trail and when lost.

LOCATING A DISTANT POINT

It isn't always enough to know how to locate your position on the map. What if you want to determine the position of something else? A pond, house, or tree, for instance, may be clearly visible but not shown on your map. How do you plot its correct map location?

Intersection method

The technique to use here is just the opposite of the method of resection. You use the **method of intersection** to find the position of something to plot on the map. As we have seen, resection lines are formed by backsights. Intersection lines, on the other hand, are made by **foresights**—lines from your position to a distant feature.

Using the intersection method, you sight on the unknown spot from two (or more) ground positions whose map locations are known or can be computed. The point at which these foresights intersect establishes the feature's location. As with resection, the accuracy of the method of intersection is improved by using three rather than two known sighting points. Again, the error triangle created by using three sight lines will give you a measure of confidence in how accurately you have determined and plotted the foresights.

To locate a distant point by intersection, you can use techniques similar to those you used to establish your position. The two main techniques are the inspection method and the compass method.

Inspection method

With the **inspection method of intersection,** you first orient the map and determine your position on the map. Next, lay a straightedge (or visually sight) from your map position toward the distant ground feature. Plot this foresight line on your map (figure 13.9A). Now move to a second position that you can find on your map, sight to the same feature, and plot this second foresight line. The intersecting foresight lines will establish the location of the feature.

Two fixed positions can also be used to determine location of a distant point. Fire towers serve this purpose for firefighters. In this case, one fire station communicates the azimuth of a foresight line (as measured from the map with a protractor) to the other station, which, combined with its own sight-line information, is enough to find the location of the fire.

If only a single identifiable ground point is available, you can still use the intersection method if the location of a second point can be computed. Simply follow the above procedure in making a sighting from the known

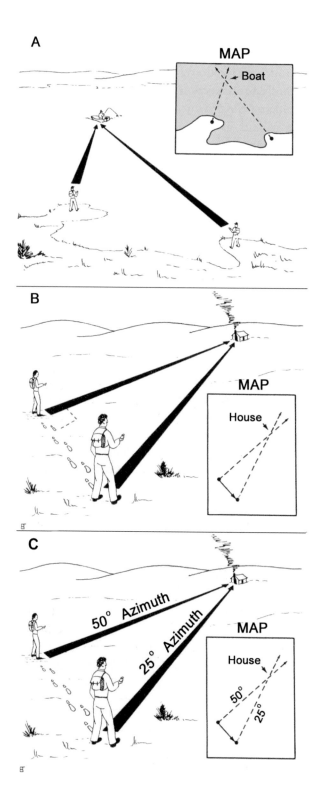

Figure 13.9 The intersection method is commonly used to determine the location of a distant point.

point, and then repeat the procedure from a second point that is a measured distance and direction from the first (figure 13.9B). As before, the sight lines will intersect on the map at the desired ground position.

Compass method

If a map and compass are both available, you may determine the location of a distant feature from **compass method of intersection** by locating your map position, sighting on the distant feature, and recording the foresight azimuth as noted on the compass (figure 13.9C). Next, move to a second known position, sight to the feature, and record the reading of this second foresight. If you plot the two foresight azimuths from the known positions, the point at which they intersect is the map location of the distant feature. The compass method of intersection can also be used when you're working from two fixed positions or from a fixed position and a computed position. The procedures are similar to the inspection methods discussed above in our fire tower example.

Trilateration method

Trilateration is a method for finding the location of a distant feature using only the distances from known locations to the feature. You probably first became familiar with trilateration when you used a drafting compass (the type you draw circles with) to draw intersecting arcs of circles from two points on a sheet of paper. Let's extend this idea to a large-scale map by drawing arcs of circles with radii equal to the ground distance from each point to a distant feature (figure 13.10).

Assume that the distance to the distant feature from a radio tower *(A)* is 6,700 feet (4,165 meters), from a hilltop *(B)* is 3,800 feet (2,360 meters), and from a hilltop *(C)* is 7,000 feet (4,350 meters). If you use the map's scale bar to draw circular arcs of radii 6,700 feet and 3,800 feet outward from points *A* and *B,* you will see that the arcs intersect at two points, each a possible position of the distant feature. Drawing the 7,000-foot radius circle from point *C* eliminates the ambiguity—the three circles intersect at a point on the map next to the creek. As with the resection method, the location is really within an error triangle.

The size of the error triangle can give you an indication of errors you may have made when using the trilateration method. As with the resection method, you may have made an error by not positioning the known locations correctly or because the mapmaker misplaced one or more of the features on the map in the first place. Or you may have sighted to the wrong feature or calculated the wrong distance from one or more of the features. You may also have measured one or more ground distance along a sloping line-of-sight, but then forgot to correct the slope distance to the horizontal distance plotted on the map. Fortunately, we saw in chapter 11 that the difference between the slope and horizontal distance is negligible for the slightly sloping line of sight distances that you normally measure, and if it is more, you can easily correct the distance measurement.

MAPS AND NAVIGATION

Once you've learned to determine your position and that of distant features, you're ready to use maps to navigate from one known location to another. **Navigation** is a two-part activity—planning your route and then following the route you have planned. Planning a route is more than drawing lines on maps, for you need to find out what features or amenities are at different locations so you will know which places you want to visit. In order to discuss route planning and following, it is important to consider the types of special maps and charts available to assist you in navigation. (By convention, maps used for a special navigational purpose, especially by ship and airplane pilots, are called **charts.**) How you plan your trip and travel through the environment will vary depending on the type of map you have at hand and what information it gives you. We'll focus on maps and charts used for land, water, and air travel, and how you can use them for navigation.

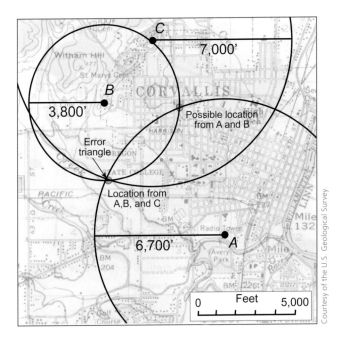

Courtesy of the U.S. Geological Survey.

Figure 13.10 **You can determine the location of a distant feature from distance information alone using the method of trilateration.**

Land navigation

Land navigation is often restricted to specific routes such as trails, roads, subways, or railroads. As a result, travelers are concerned primarily with their route network; they don't much care about the environment as a whole, or about exact compass direction. Maps for such travelers can be greatly simplified, especially if an obvious origin, route, or destination exists. Let's look first at some of the maps and atlases used for land travel. Then we'll turn to computerized methods and GPS-linked systems.

Maps and atlases

Making a simple, schematic map that focuses on a particular route isn't a new idea. The Romans simplified their road maps, such as the medieval **Peutinger table** (figure 13.11, top), by ignoring exact directions and distances, concentrating instead on showing the road network and cities along the routes.

The distortion of geographic space in schematic mapping that began with the Romans has been carried a step further in recent times. Often there has been a concerted attempt to mislead the map user. For example, the scale and direction relationships on nineteenth century **railroad maps** were often carefully arranged to create a favorable impression of the company's service in a region (figure 13.11, bottom). The practice of deliberate map distortion continues today, especially with subway and airline route maps.

The modern hiking or canoeing guide (figure 13.12) is a similar type of simplified route map. Again, in an effort to make the maps as simple and uncluttered as possible, only the route network and prominent surrounding landmarks are depicted. On these maps, however, an attempt is made to retain some indication of correct directions and distances so the hiker or boater can find his or her way along the route. Such maps are well designed for the amateur outdoorsperson but are generally inadequate for sophisticated map users. Furthermore, there is a danger that some unplanned event such as a washed out trail or damaged canoe will force the user to alter course or engage in another mode of transportation for which the map is totally unsuited. The main problem with these simplified route maps, then, is that they aren't flexible. They keep you tied to your route.

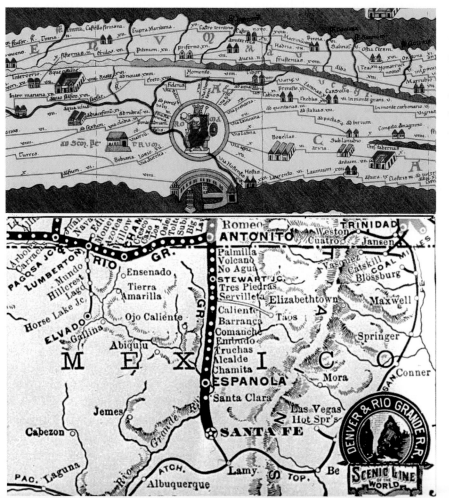

Figure 13.11 A section of the medieval Roman Peutinger tables (top) and a nineteenth century railroad map (bottom) are examples of land travel maps that are more schematic than planimetrically accurate.

PENOBSCOT RIVER
14 MILES

Ripogenus Dam

McKay Station
Put-In

Ripogenus Gorge
Class V

Troublemaker
Class III

Exterminator
Staircase
The Heaters

The Cribworks
Class V

Little Ambejackwockamus Falls
Class III

Big Ambejack
Class IV

Horserace Rapids
Class III

Nesowadnehunk Falls
Class IV

Beautiful Views
Of Katahdin

Big Pockwockamus
Class IV

Abol Falls
Class IV

Little Pockwockamus
Class III

Take-Out

Courtesy of Raft Maine.

Figure 13.12 Canoeing and rafting map for the Penobscot River, Maine.

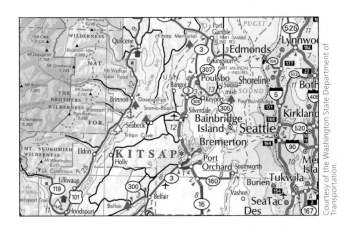

Courtesy of the Washington State Department of Transportation.

Figure 13.13 State road maps usually show a variety of information in addition to highways, making them useful for general reference purposes as well as for navigational assistance. Note the options for taking ferries across Puget Sound.

The **standard road map** provides a general picture for route finding, since its coverage is defined by state, region, or country (figure 13.13). A variety of information is available to the traveler, including possible alternate routes such as railroads and ferry routes, as well as landmarks such as parks, airports, and physical features such as mountain peaks. Map information selected for the purposes of road travel doesn't paint a complete picture of the environment, of course. The traveler concerned with details falling much beyond the shoulder of the road will likely find road maps inadequate.

You can also plan your route by using general purpose maps on which an optimal route has been marked. One such map, put out by the American Automobile Association, is called a **TripTik.** The TripTik contains a planned route between your departure point and arrival destination, but in the event of an emergency it also includes a basemap of the region neighboring the route. A major deviation from your planned route is required to thrust you into "unmapped territory." The paper version of this is a form of a **strip map** printed on multiple pages showing sections of the route in the order that they will be traversed. A strip map may or may not be drawn to scale, but it should show identifying landmarks. It is a quick source of reference that shows essential details of the environment and amenities important to the traveler, such as rest areas, toll booths, and more. The pages are collated into a small book that the navigator can flip through to view the entire route.

The latest version is the **TripTik Travel Planner,** an online mapping application that highlights on a basemap the best route for you to follow between two locations (figure 13.14). To the left of the map is an itinerary box that shows the total distance of your trip and the estimated travel time for the route highlighted on the map at the right. Below are detailed driving directions along the route. A slider bar at the right of the map allows you to zoom in on segments of the route to see more detailed maps. At the highest zoom level, named residential streets are shown along with other details of the local area.

Private companies publish a variety of **road and recreation atlases,** which are books for travel and trip planning that contain a series of maps covering large regions. Many travelers carry national road atlases produced by Rand McNally, National Geographic Maps, Michelin, or other private firms. These atlases generally consist of maps for states, provinces or other regions, showing different classes of roads, mileage along road segments, cities and towns, counties, rivers, and other features of interest. Recently, larger-scale state road and recreation atlases have been produced by companies

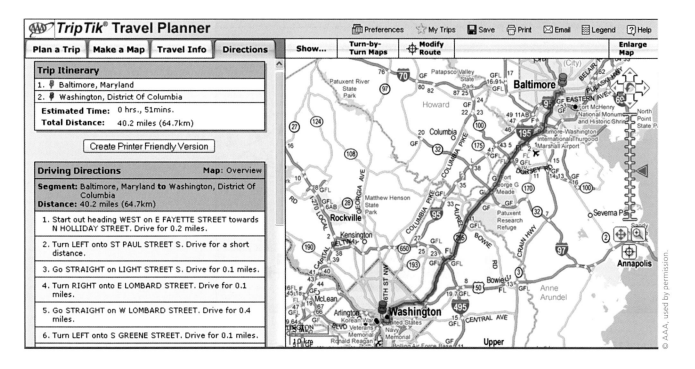

Figure 13.14 The American Automobile Association has popularized a type of route map called a TripTik, shown here in its Travel Planner online version.

such as Benchmark Maps. Their maps show in greater detail the same types of information as national road atlases, plus relief-shaded or layer-tinted topography and detailed landownership information (figure 13.15). Graticule lines at small increments are printed over the map to assist GPS users. Recreation features and other potential **points of interest (POIs)** to travelers are also added to the map.

Large-scale topographic maps are also used for detailed route finding. A 1:24,000-scale U.S. Geological Survey (USGS) topographic quadrangle, for example, shows topography, roads, railroads, streams, buildings, forested areas, and other features that are often important to planning and following a route either by car or on foot. On British Ordnance Survey maps features such as footpaths, bridleways, national trails, and youth hostels are included for ramblers. The graticule and grid ticks along the map's margins are aids to position finding.

Digital route planning

As we mentioned, navigation includes not only following your route, it also includes planning your route. **Route planning** involves determining where you will go and how you will get there. Route planning software such as ArcLogistics lets you view a map your planned destination (figure 13.16). You can target the location by entering a telephone area code, postal ZIP Code, street name,

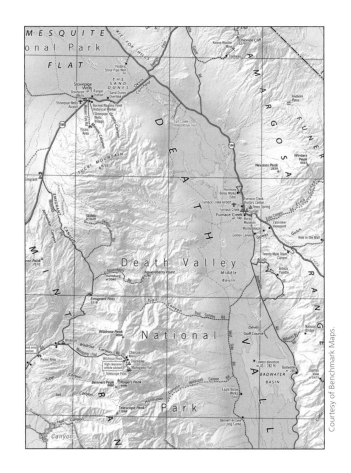

Figure 13.15 Portion of a page from the Benchmark Road & Recreation Atlas for California.

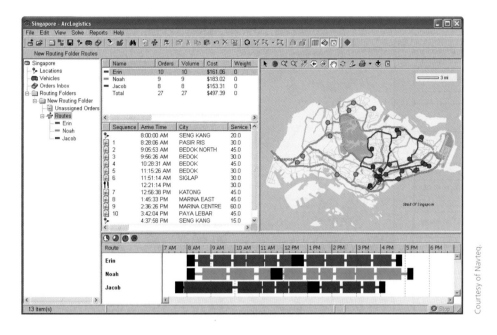

Figure 13.16 Using ArcLogistics software from ESRI, fleet managers can create dynamic routes and schedules that cut fuel costs and improve customer service, while still considering the drivers' needs to take breaks and maintain their vehicles.

street intersection, building address, or geographic place name (city, national park, and so on). Often programs such as this are linked with electronic telephone directories so that you can find the location of anything listed in the phone book. Alternatively, you can find locations of businesses or POIs in the vicinity of a targeted location. The software will also give you the nearest address if you click on a specific map location.

You can view the same location at a wide range of map scales, depending on the amount of geographic detail and extent of geographic coverage you desire. It is now possible to toggle between traditional line maps and air photos of your route at different map scales. You can choose to annotate these maps with road names, route numbers, and other map symbols. You can customize software settings with a profile of your travel preferences so that your personal favorites are automatically displayed. Internet links to vendors' Web pages, and other sites provide access to volumes of additional travel-related information, including road conditions and weather.

This software allows local governments to streamline operations by demonstrating more efficient ways to route vehicles. Applications include housing and building inspections, property appraisals, public works, solid waste pickup, work crews, special needs transportation, and many more. Routing can be fully dynamic, static, or a hybrid, where the software can optimize the route-finding solution that is then manually fine-tuned for your specific requirements.

Personal navigation systems

You may own or have used a pocket-sized, portable handheld **personal navigation system.** These devices combine cell phone and global positioning system (GPS) technology (see chapter 14 for more on GPS) into a navigation system often referred to as off-road, on-the-water, or all-terrain. Such systems are advertised as being ideal navigation devices for the hiker, camper, bird watcher, angler, hunter, or other outdoor enthusiast.

Recent personal navigation systems, such as the Apple iPhone 3G cell phone, combine cell phone and GPS with digital maps displayed on the cell phone screen. Your position is constantly computed from the unit's built-in GPS receiver and also by using a form of resection from cell phone towers. The computed position is displayed on a conventional map, a satellite image, or a hybrid photomap made by combining the map and satellite image. You can pan and zoom on any of these cartographic displays when you are standing still or moving across the landscape. You can also view a list of travel directions and follow a travel route highlighted on the map or satellite image. In urban areas, you can even download current vehicle traffic information and display current traffic flow conditions as green, yellow, and road highlights similar to those in figure 8.17.

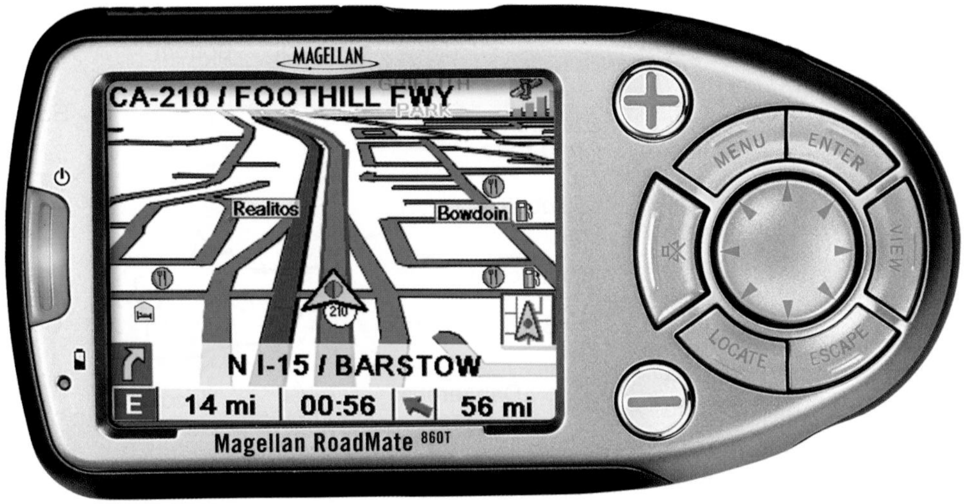

Figure 13.17 The Magellan Roadmate 860 vehicle navigation system displays your map position as you travel through the environment.

Courtesy of Magellan Navigation, Inc.

Courtesy of TravRoute Australia Pty. Ltd.

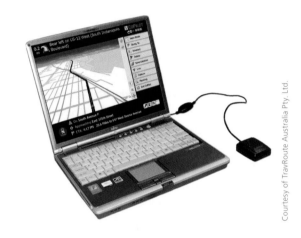

Figure 13.18 Plugging a GPS receiver into a laptop computer creates a powerful unit for analysis and display of positioning data, yet it is portable enough to use in moving vehicles.

Vehicle navigation systems

Potential applications of GPS technology in route finding stagger the imagination. Route finding will soon be transformed into something hardly recognizable today. Passengers cars, trucks, buses, delivery vehicles, and rental cars can be equipped with GPS-enabled (or radio-enabled) **vehicle navigation systems.** The cost of these systems continues to decrease at the same time that the functionality is improving. There are also multiple platforms for these systems. GPS-based vehicle navigation systems can either be built into the car or they can be extensions of the capabilities of a laptop computer you take along in a car.

In-car systems

In-car systems, including the Magellan Roadmate 860 (figure 13.17), are now available for less than $200. If you provide an origin and destination city, the vehicle navigation system will tell you the shortest distance, least time, or most scenic route. If you specify stopover points en route, the system will direct you to these locations as well. You have the choice of seeing your route marked on a map, or seeing an itinerary giving all pertinent identification and distance data for each route segment as with the TripTik travel planner we saw earlier.

In more sophisticated systems, the map display is linked by cellular phone or radio signals to local databases so that you can receive traffic updates and other useful information as you travel. Vast databases of lodging, restaurant, and other travel information, as well as trip planning software, are also available to help you navigate.

Laptop computer systems

It is possible to connect a handheld GPS receiver to a laptop computer with software such as TravRoute Co-Pilot Laptop 10 (figure 13.18). Since computer memory is vast, the unit can download data from databases worldwide so that all types of data can be fed into the system. You can also take advantage of the computer display screen, which is bigger, multicolored, and higher in spatial resolution than the screen on most in-car navigation units. This all adds up to a high-quality map display backed up by powerful analytical capabilities.

A laptop computer system has the advantage of portability and versatility–because the GPS unit isn't permanently mounted in the vehicle, it can be removed when not needed. This may be the most cost-effective way to bring automated positioning technology into your vehicle travels. Furthermore, when you sell your car you don't lose your navigation device as you do when you have an integrated navigation system.

Although GPS technology is the basis of vehicle navigation systems, gaps in coverage can occur when the satellite signals are blocked by steep terrain, large buildings, or heavy forest cover. An **inertial navigation system (INS)** is used in many vehicles to fill these gaps in GPS coverage. An INS uses gyroscope-like devices that measure up-down and sideways changes in the vehicle's angular velocity. This information plus your odometer reading is used to constantly calculate the distance and direction you have traveled from a starting point, and this calculation is used to position a symbol for your vehicle on the map display. When you turn on your car, the INS is initialized at a known starting point through use of the GPS positional reading. Each distance and direction computation involves some error, and since each subsequent computation is based on the previous one, errors accumulate and the estimated position tends to "drift." Luckily, the "drift" is corrected when your vehicle enters good GPS coverage so that the INS can be turned off and reinitialized when needed again.

Marine navigation

To understand marine navigation, it is useful to know about the types of maps used by mariners and how they are used for position finding and navigation.

Nautical charts

The primary tool for marine navigation is the **nautical chart,** a map specifically designed to meet the needs of navigators. These type of maps are designed to be drawn on—**courses** (routes) can be plotted, directions or **bearings** can be marked, distances and times can be noted, and more. In contrast to maps and atlases used for land travel that are produced commercially, nautical charts are most often produced by government agencies. Different types of charts are produced but all have information about the subsurface water region and areas along land that can be seen from water.

They are designed to give all available information necessary for safe marine navigation, including **soundings** (depth measures), **isobaths** showing **bathymetric contours** (lines connecting points of equal depth below the hydrographic datum), obstructions and hazards, bottom types, currents, prominent landmarks near shore, and navigational aids such as buoys, beacons, and lighthouses (figure 13.19).

The scales of nautical charts published for the coastal waters of the United States by the National Oceanic

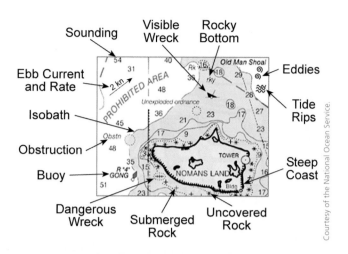

Figure 13.19 Nautical charts contain information specifically for the marine navigator.

and Atmospheric Administration (NOAA) range from 1:2,500 to about 1:500,000. Coverage includes the Atlantic, Pacific, and Gulf coasts, and coastal areas of Alaska, Hawaii, and the U.S. possessions (Virgin Islands, Guam, Samoa, and Puerto Rico). Oceanic mapping agencies, typically referred to as **hydrographic organizations or agencies,** as well as commercial companies produce charts for other areas of the world. For example, the U.S. National Geospatial Intelligence Agency (NGA) and the United Kingdom Hydrographic Office (UKHO) chart their own waters, as well as the world.

Hydrographic charts are also published for certain inland waters. For example, in the United States, charts for the Great Lakes, the upper Hudson River, Lake Champlain, the New York State Barge Canal, and some of the Minnesota–Ontario border lakes are issued at various scales and are designed primarily for navigational use. Most of these charts show the hydrography of water areas, together with the topography of limited areas of adjacent shores and islands, including docks, structures, and populated places visible by boat from the lakes and channels (figure 13.20A).

Charts are also designed for flood control, recreation, and other uses. For example, charts of navigable rivers and connected waters in the United States are also published at several scales. The U.S. Army Corps of Engineers is the primary source of these useful charts (figure 13.20B). In addition, state agencies such as the Department of Natural Resources usually produce hydrographic maps for inland lakes and rivers of significant size or recreational potential.

If you plan to navigate on the water, a good rule is to obtain the largest scale chart available. This will help ensure that you have the most comprehensive graphic summary of information pertinent to making navigational decisions. Before you use the chart, study the notes, scale, symbols, and abbreviations placed near the title or next to the legend, for they are essential for effective use of the chart. *Coast and Geodetic Survey Chart No. 1* shows the symbols and abbreviations used on nautical charts of the Unites States and is recommended to the mariner for study.

The date of the chart is of such vital importance to the marine navigator that it warrants special attention. When charted information becomes obsolete, further use of the chart for navigation may be extremely dangerous. Natural and human-induced changes, many of them critical, are occurring constantly. It is essential, therefore, that navigators obtain up-to-date charts at regular intervals or hand correct their copies with changes. In the United States, changes are announced in a weekly publication entitled *Notice to Mariners.*

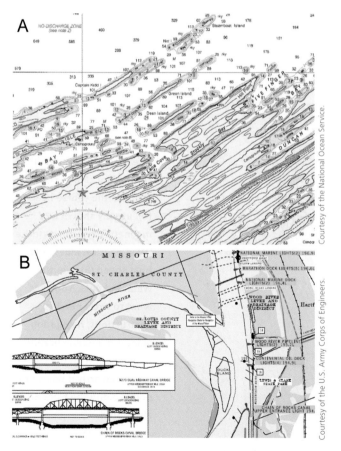

Figure 13.20 Portions of the Isle Royale, Minnesota, hydrographic chart (A) and U.S. Army Corps of Engineers navigation chart for the confluence of the Missouri and Mississippi Rivers (B).

Marine navigation methods

In addition to nautical charts, you also need two special skills for successful and safe marine navigation: piloting and dead reckoning. With **piloting,** you plan your intended route and then find your position and direction of movement by paying attention to landmarks, aids to navigation (like buoys and beacons), and water depth. When **dead reckoning,** you are estimating your present position relative to your last accurately determined location by using direction, speed, distance, and time information. We will discuss these two navigational skills separately, realizing that dead reckoning is a key part of piloting.

Dead reckoning

The term dead reckoning (DR) is a shortened form of the phrase "deduced (ded.) reckoning" used in the era of sailing ships. The basic idea is to deduce your current position from the vessel's direction of travel and speed. Dead reckoning is used both when planning a route and steering a boat along the planned route.

To plan your route, you use large-scale nautical charts to create a **dead reckoning plot.** You map your route on the charts as a series of **intended dead reckoning tracks,** or the routes you intend to follow between selected points. Two track lines are plotted in figure 13.21 for the last part of a voyage to Flounder Bay on Fidalgo Island, Washington. Assume that your boat has a maximum **cruising speed** (the speed at which the boat travels) of 10 **knots** (nautical miles/hour) and a **draft** (depth from the water line to the bottom of the keel) of 3 meters.

Before plotting intended dead reckoning tracks, navigators highlight obstructions, hazards to navigation, and navigational aids along the route (the circled features in figure 13.21). Tracks must be plotted to avoid obstructions and minimize hazards. Track 2, for instance, is drawn perpendicular to the international shipping lane (shown by the purple dashed lines) in order to minimize the time spent crossing it. Track 3 is the shortest straight line to Flounder Bay that gives safe passage distance around Williamson Rocks, Allan Island, and Young Island while having water depths greater than the boat's 3-meter draft. Dead reckoning does not take into account the effects of currents, winds, vessel traffic, or steering errors on tracks, although the navigator must correct for all these things during the actual voyage.

The **course heading** (direction) and distance for each intended track are calculated on the chart with a parallel rule and dividers, using the methods described in chapters 11 and 12. Notice that the nautical convention is to write the course heading (C) above and boat speed (S) below the beginning of each track line, and to write the distance (D) below the center of the line under the track number. Track 2, for example, has course C051M (magnetic azimuth of 51 degrees), distance D3.0 (3 nautical miles), and speed S10.0 (10 knots over the water). In strict nautical terminology, the azimuth is called a **bearing.**

This distance and speed information allows you to calculate the time (T) required to complete each track from the equation: T = D × 60/S. From this equation, completing track 2 should take

$$\frac{3 \text{ nautical miles} \times 60 \text{ minutes / hour}}{10 \text{ nautical miles / hour}} = 18 \text{ minutes}$$

A similar computation gives a crossing time of 26 minutes for track 3. This means that if you plan to begin track 2 at 8:00 AM, your **estimated time of arrival (ETA)** at the start of track 3 is 8:18 AM and at Flounder Bay it is 8:44 AM. These times are written on the chart at the start and end of the two tracks as four-digit numbers 0800, 0818, and 0844 using **military time** notation. The difference between regular and military time is how hours are expressed. **Regular time** uses numbers to identify the 24 hours in a day as 1 to 12 from both midnight to noon and from noon to midnight. In military time, the hours are numbered from 00 to 23, so that midnight is 00, 1 AM is 01, 1 PM is 13, and so on. Minutes and seconds are expressed the same as in regular time.

Piloting

Piloting is following your route using landmarks, water depth and aids to navigation. The first step in piloting is to acquire the charts and other navigational publications that cover the area in which you will travel. In marine navigation, you will obtain small- and large-scale nautical charts, tide tables, and current information for the area. You must also learn about certain characteristics of your boat, particularly its maximum cruising speed and draft.

Using the dead reckoning plot in figure 13.21 as an example, imagine that you're on a boat trip trying to hold the course on the plot. You'll notice that this plot doesn't take water currents into account. Currents can be strong in this locale, so it's best to look at a map or atlas showing predicted currents at the time of your voyage.

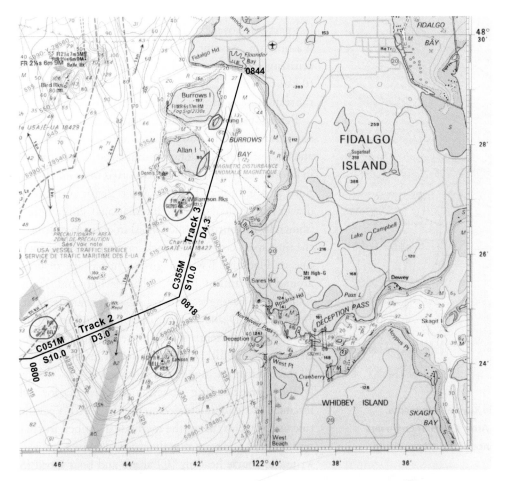

Figure 13.21 Route planning by dead reckoning consists of plotting intended tracks on nautical charts. Track information includes the course heading (C), usually shown as a magnetic azimuth; distance (D) in nautical miles; and boat speed (S) over the water in knots. You may want to circle nearby obstructions, navigation aids, and possible hazards to navigation on the chart. Estimated arrival times at points along the track are also noted.

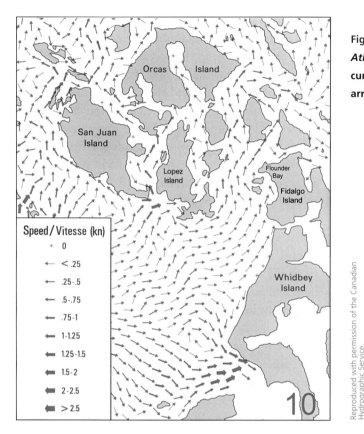

Figure 13.22 Southeast corner of page 10 from the *Current Atlas* covering the San Juan Islands of Washington. Predicted current directions and speeds are shown with current vector arrows for a one-hour period.

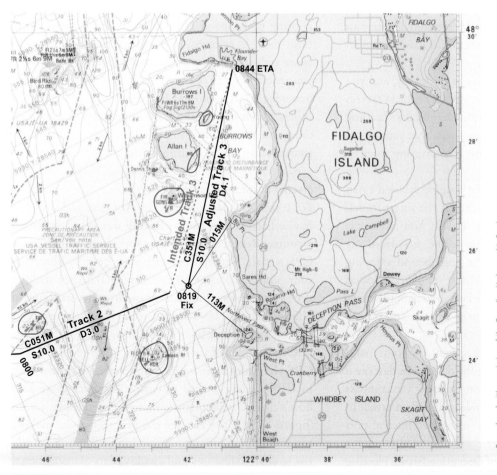

Figure 13.23 Dead reckoning plot showing intended tracks and adjusted track 3 necessitated by a position fix.

Assume that page 10 in the *Current Atlas* (produced by the Canadian Hydrographic Service) for the Strait of Juan de Fuca to the Strait of Georgia (figure 13.22) shows the currents predicted for this area at 0800 on the day of your voyage. The arrows on the map show that the current is predicted to be light (<0.25 knots) and at almost right angles to track 2, so you can safely ignore that current. However, the arrows in the vicinity of track 3 indicate a slightly stronger current (0.25 to 0.5 knots) in the direction of travel. The *Current Atlas* shows that, on average, the current will be around 0.4 knots. Adding this current speed to the 10-knot boat speed, you obtain an estimated **speed over the ground (SOG)** of 10.4 knots. At this speed, you calculate that it should take 25 minutes to complete track 3 using the $T = D \times 60/S$ equation: 4.3 nautical miles × 60 minutes/10.4 nautical miles/hour.

The current adjusted ETA at Flounder Bay is 0843, one minute earlier than the ETA would be if not adjusted for currents. Because it is so minor, you can ignore this one-minute difference caused by currents. However, if at another location your boat had a 10-knot maximum speed and the *Current Atlas* showed a predicted current of 6 knots, you would need to adjust your ETA to take these stronger currents into account. What if the current flowed in the opposite direction to the one you are taking and the current is stronger than your boat speed? Then you would not be able to pass, and you would have to wait until the current changed or you would have to find an alternate route.

Now, let's assume your voyage is underway, and you have correctly reached the starting point for dead reckoning track 2. Steering the boat, you set a compass heading of 051° magnetic. You note on the chart that you started the track at 0800, and you adjust your cruising speed to 10 knots (figure 13.23).

At 0806, you estimate that the boat has traveled 1 nautical mile and that you have entered the international shipping lane for large tonnage vessels. You must now look and listen intently for ship traffic, first from the southbound and then from the northbound shipping lane, during the seven minutes that it should take you to cross the 1.2-nautical-mile-wide lane drawn on the chart. Let's assume that no ships have to be avoided, and at 0818 you complete track 2.

Before changing your bearing, you use your hand bearing compass (a special handheld compass described in chapter 12) to take a position fix using the compass resection method of position finding described earlier in the chapter (mariners call this "using two lines of position"). You obtain a magnetic azimuth of 015° to Biz Point and 113° to the center of Deception Island. Plotting the corresponding 195° and 293° magnetic back azimuth lines from these points, you note their intersection as a 0819 fix. Evidently, in this scenario, currents, winds, or steering errors have caused your actual position to differ considerably from that predicted by dead reckoning.

Your course must be adjusted, and you quickly plot a new dead reckoning track 3 from the position fix to Flounder Bay. Using your parallel rule and dividers, you find the new track to have a course of 351° magnetic and a distance of 4.1 nautical miles. At 10 knots, you should cover this track in 25 minutes, so your ETA at Flounder Bay is still 0844.

Setting your compass to 351° magnetic, you motor toward Flounder Bay, visible in the far distance, at 10 knots. At 0827 you see to your left the flashing red light and hear the gong shown on the chart at Williamson Rocks. In another six minutes you have the east edge of Allan Island safely to your left. You have noted and decided to ignore the slight movement of the compass dial caused by the local magnetic disturbance shown on the chart. In another five minutes you have traveled nearly a nautical mile and you have Young Island at a safe distance to your left. The entrance to Flounder Bay is now clearly in sight and you steer to the entrance, arriving at 0845, almost perfectly on time.

Your imaginary voyage is a classic example of daytime piloting under good weather and water current conditions using a compass, dividers, parallel rule, nautical chart and, above all, your eyes for guidance. But what if you are crossing the open ocean or traveling at night or in heavy fog? Dead reckoning has been the traditional mode of navigation even under these conditions, but modern electronic devices have simplified the task and made navigating under these conditions safer for mariners.

Electronic piloting systems

For several decades now, navigators of ships and airplanes have relied on **electronic piloting devices** to maintain a course. Recent developments link maps to radio and GPS signals. You can navigate your boat by observing your progress along a path on a map and then making decisions en route. You can choose from several types of piloting systems. In each case, your changing position is plotted using the radio or GPS signals.

One example of these electronic piloting systems is a **GPS chartplotter** in which positional data from a built-in GPS receiver is used to correctly superimpose your boat's current position on a nautical chart display screen (figure 13.24). You can see the progress you're making toward your destination, the position of your vessel relative to charted features, and a wide variety of navigational data. The display is a moving chart covering several hundred miles of coastline and dozens of conventional printed charts at different scales. Simply pointing to locations on the display screen creates and stores **waypoints**, or specified geographical location, spots, or destinations defined by longitude and latitude that are used with a GPS for navigational purposes (see chapter 14 for more on waypoints). The detailed chart information displayed makes it much easier to understand your position and heading relative to navigation hazards. The continual electronic plotting of your position eliminates the time-consuming hand plotting of position fixes on paper charts.

The electronic chart can be entered into the GPS receiver in several ways. You can read the chart of the area you will be in into the unit's memory from a digital chart database on a CD-ROM before taking the unit into the field. You select the geographic area, and in many cases the features you want to load, such as marine aids to navigation or an orthophoto background for land areas, until the area you are traveling in is covered or the memory in the receiver is filled.

More commonly, a data cartridge holding the desired chart is plugged into the unit. After traveling some distance, you may have to insert a new cartridge. These cartridges are often sold by the manufacturers of cartographic GPS units and are much more expensive though—at the same time—more extensive than paper maps.

Both inland and offshore **chart cartridges** with a series of electronic charts are available from several companies in different formats. **Inland chart cartridges** (examples include Garmin Inland G-chart, Lowrance IMS Smart-Map, and Inland C-MAP *NT*) contain electronic charts that cover all or part of a state. The chart data are in vector format, allowing you to change from very small to very large scales and to pan seamlessly across the map area. As you zoom in, freeways, national and state highways, state boundaries, and large water bodies appear. Zooming in further, minor roads and smaller lakes appear, along with dams, boat ramps, marinas, and public facilities. Chart information is organized in layers, so that you can selectively add or remove different classes of features from the display. **Offshore chart cartridges** (examples include Offshore C-MAP *NT,* Garmin Offshore G-chart, and Navionics Offshore Cartography) operate like inland charts, but contain sounding, navigational aid, bottom condition, and harbor information.

Air navigation

In the United States, the National Aeronautical Charting Office (NACO), administered by the Federal Aviation

Photograph courtesy of Garmin. Copyright © 2007 Garmin Ltd. or its subsidiaries.

Figure 13.24 Electronic piloting devices such as the Garmin GPSmap 4008 chartplotter include a screen of sufficient size and resolution to display your position on a readable chart of the area in your vicinity. Notice the orthophoto map that is used for the land areas.

Administration (FAA), publishes and distributes U.S. government **civil aeronautical charts** and flight information publications. Aeronautical charts published in the United States are specially designed for use in air navigation, and emphasis is given to features of greatest aeronautical importance (figure 13.25). Five-chart series are produced; we'll look at each, starting with the smallest-scale chart.

Aeronautical charts

World aeronautical charts at 1:1,000,000 scale cover land areas at a standard size and scale for navigation by moderate speed aircraft and aircraft operating at high altitudes (figure 13.25A). Basemap information includes cities (shown by color tints), principal roads, railroads, distinctive landmarks, drainage patterns, and relief. Aeronautical information includes visual and radio aids to navigation, airports, airways, restricted areas, and obstructions.

Sectional aeronautical charts at 1:500,000 scale are designed for visual navigation of slow moving to medium speed aircraft (figure 13.25B). The topographic information consists of the relief and a selection of visual checkpoints used under visual flight rules. The checkpoints include populated places, drainage patterns, roads, railroads, and other distinctive landmarks. The aeronautical information on sectional charts includes visual and radio aids to navigation, airports, controlled airspace, restricted areas, and obstructions.

Aeronautical Chart Series

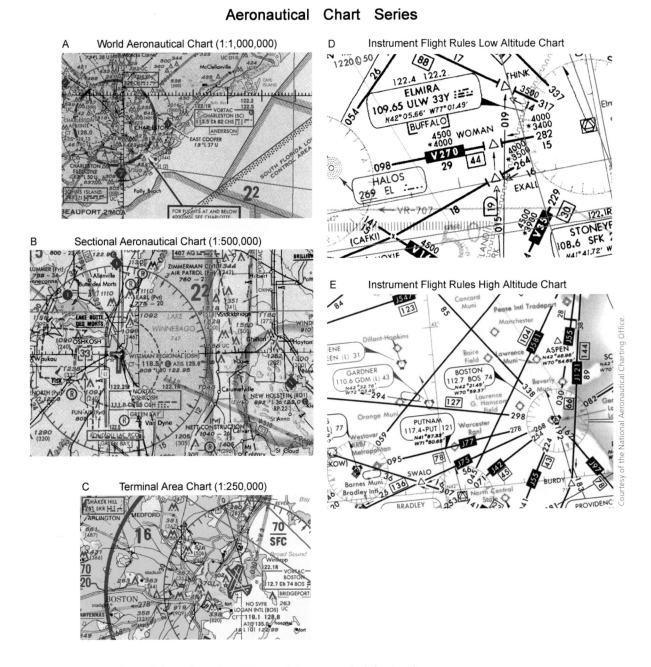

Figure 13.25 **Portions of charts from five different U.S. aeronautical chart series.**

Courtesy of the National Aeronautical Charting Office.

Terminal area charts at 1:250,000 scale depict the airspace around major airports (figure 13.25C). The information found on these charts is similar to that found on sectional charts, but it's shown in much more detail because of the larger scale.

Instrument flight rules, low-altitude charts provide aeronautical information for navigation below 18,000 feet above mean sea level (MSL) (figure 13.25D). This chart series includes airways, limits of controlled airspace, VHF radio aids to navigation, airports that have an instrument approach procedure or minimum

3,000-foot hard-surface runway, off-route obstruction clearance altitudes, airway distances, reporting points, special-use airspace areas, and military training routes.

Instrument flight rules, high-altitude charts are designed for navigation at or above 18,000 feet MSL (figure 13.25E). Included on these charts are jet route structure, VHF radio aids to navigation, selected airports, and reporting points.

As with nautical charts, once information on an aeronautical chart has become obsolete its further use for navigational purposes becomes a definite hazard. One of the most important things on any chart is its date. On aeronautical charts, the date is usually given in large, red type in the lower right margin of the sheet or on the outer legend page of the folded chart. Other information for the chart includes its name, type, scale, date effective, and edition of the chart. A note will say in red that the chart will become obsolete for use in navigation upon publication of the next edition, which is scheduled for a stated date.

Air navigation methods

Aircraft navigation, like marine route planning and following, is a combination of dead reckoning and piloting with the aid of charts and position-finding devices. The details of flight planning and electronic route following are learned in flight school, but we can look at the basics of using a sectional aeronautical chart to plan and follow a flight path between two airports. Let's use a portion of the Seattle 1:500,000-scale sectional chart (figure 13.26) to plan a round trip flight from Pasco, Washington, to Pendleton, Oregon. Our first step is to find on the aeronautical chart the municipal airports for the two cities, and then plot a straight-line course from the Pasco airport (A) to the Pendleton airport (B). Finding the two airports on the chart is a simple task because each airport is shown with a symbols that includes the orientating of the runways, as well as their respective lengths.

Also notice that the airports on this chart are near the center of the compass roses, but this isn't always the case. Magnetic compass roses generally are centered on **VOR (VHF omnidirectional range) navigational beacons,** which may be located at airports, but are not always. Medium-to-large airports may have VOR beacons, but sometimes they are located near well-frequented air traffic lanes that are miles from the nearest airport (you can check this out on sectional charts). The compass rose allows pilots to track their routes inbound and outbound on magnetic headings and also to determine their exact location using the resection method we discussed earlier with signals from two or more VOR beacons.

We explained in chapter 3 that aeronautical charts in the United States and other countries are made on the Lambert conformal conic map projection because directions are correct in local areas and straight lines drawn on the chart are very close to great circle routes. Hence, you can assume that a straight-line dead reckoning course between the two airports is the most direct route and that the angle this line makes with north as the reference line is the correct heading to follow.

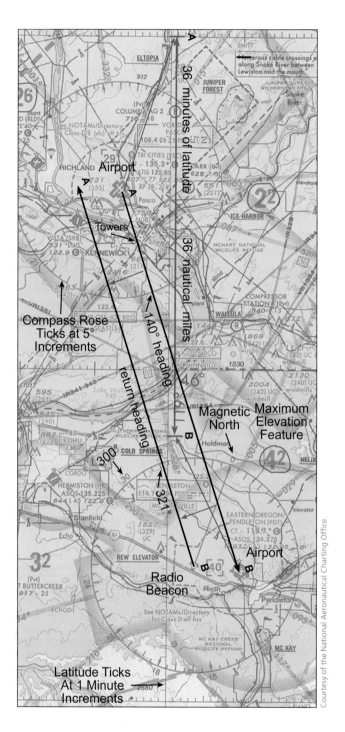

Figure 13.26   Air navigation using a 1:500,000-scale sectional aeronautical chart involves planning dead reckoning flight paths and then piloting along these paths.

The heading from the Pasco to Pendleton airports is read from the compass rose that is centered on the Pasco airport. Notice on the compass rose centered on a point about 4nm west of the Pendleton airport that a line from its center points to magnetic north, which is labeled as 0 (degrees). Azimuth ticks on aeronautical chart compass roses are at 5-degree intervals, with arrowhead ticks at every 30 degrees. The trailing 0 is left off the azimuth number so that, for example, 6 means 60 degrees and 30 means 300 degrees. The ticks on the Pasco compass rose show a magnetic azimuth of 140 degrees to the Pendleton airport.

You would next find the distance in nautical miles between the two airports. You can measure this distance from the scale bar printed on the chart, or by using dividers to measure the length of the flight line using the scale indicators along one of the graduated meridians printed on the map at 30-minute-longitude increments (the vertical red line in figure 13.26). The latitude ticks are spaced at one-minute intervals along the meridian line, so you can determine that the flight line measures 36 minutes of latitude. Knowing that one minute of latitude represents one nautical mile on the ground, you can determine that the distance between the two airports is 36 nautical miles.

The heading from Pendleton to Pasco airports should be the back azimuth of 140 degrees, or 320 degrees (see chapter 12 for more on azimuths and back azimuths). But you may want to check for local magnetic variation by drawing the return flight line from the compass rose for the Pendleton airport. The problem is that the compass rose is not centered on the airport symbol, but rather on the VOR navigation beacon a few miles to the west. For dead reckoning from the airport, you must use a parallel rule (see chapter 12) or similar device

to shift the flight line from the airport to the radio beacon and then find the return heading from the compass rose. We measured an azimuth of 321 degrees, one degree more than the back azimuth based on the Pasco compass rose. There must be a slight change in magnetic declination or a local magnetic disturbance causing this small change in heading between the two airports.

Piloting between the two airports during the day is a matter of holding the aircraft on the planned magnetic heading, estimating the distance traveled, identifying charted features on the ground, and flying at a sufficiently high altitude to avoid ground obstructions. Several types of safe altitude information are found on the visual flight rules (VFR) aeronautical chart. Prominent features on the chart are the large blue numbers (we have circled several in red) in the center of the 30-minute quadrangles bounded by latitude and longitude lines. These are the **maximum elevation figures** for the quadrangle in thousands and hundreds of feet. For example, the number 42 to the north of the Pendleton airport means that maintaining an altitude of 4,200 feet above mean sea level within the quadrangle will clear all terrain features and obstructions such as towers and antennas—but only by a very small margin of as little as 10 feet. That's why sound flight planning (and federal regulations) call for adding 1,000 feet to this figure (2,000 feet in mountainous terrain).

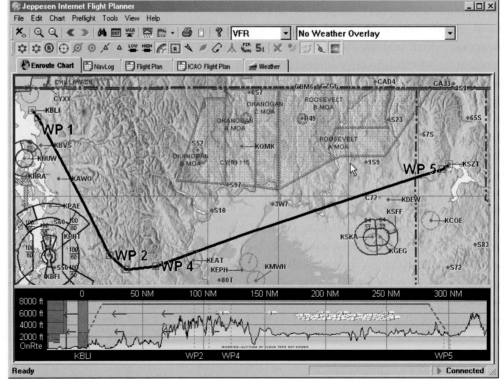

Figure 13.27   Flight routes can be planned interactively on personal computers with the Jeppesen Internet Flight Planner.

The upside down V- and W-shaped symbols with two dots at their base near the Pasco airport are towers that you must avoid on take-off and landing. Below some of the symbols, like the one southeast of the Pasco airport, are labels such as 666 and (311) that tell you the elevation of the tower and its height in feet, respectively.

Interactive route planning and piloting

Modern air navigation relies on digital computers, digitized aeronautical charts, and aviation databases, along with GPS receivers linked to computers through interactive navigation software. Interactive route planning and piloting technology ranges from very expensive systems in commercial and military aircraft to low-cost hardware and software designed for a private airplane. Let's first look at the latter starting with an example of interactive route planning software for laptop computers, and then turn to stand-alone GPS aviation units with built-in navigation software and chart pages.

The route planning tool in the Jeppesen Internet Flight Planner allows you to use your personal computer to interactively plan a route using current flight information. You begin by entering your departure and arrival times, the name and model of your aircraft, and the names of the departure and destination airports. You then plan your route on a digital **en route aeronautical chart** for the area between the two airports, such as the chart for a flight between Bellingham, Washington, and Sand Point, Idaho, shown in figure 13.27. You can pan and zoom on this chart as well as toggle on and off geographic information such as elevation and terrain features, hydrography, political boundaries, transportation features, and the latitude-longitude graticule. You can also turn on and off navigation information like approach and departure flight altitudes for major airports, and areas with flight restrictions. Selecting the weather information menu box brings up a terrain profile below the chart that shows you predicted weather conditions between the two airports, including cloud type, height and thickness, wind speed, and wind direction.

You can now plan your flight on the en route chart. The default is the great circle route between the airports, but you may need to plan a more complex route with several legs in order to avoid areas with flight restrictions. You create these legs by adding waypoints to the initial great circle route using a "rubberband" tool similar to that found in graphics software. For example, in figure 13.27, four straight-line legs between five waypoints were added to the initial great circle route between Bellingham and Sand Point, creating a route through the Cascade Range that avoided the airspace control areas for Boeing Field

Figure 13.28 GPS aviation units like the Garmin GPSmap 496 link satellite tracking technology to digital aeronautical charts and navigation displays.

in Seattle and the flight restrictions across much of northeastern Washington. The terrain profile below the chart (note the highly exaggerated vertical scale) is for this modified route, showing that an altitude of 7,800 feet should be maintained to safely clear the terrain along the route and to stay above the cloud layer in eastern Washington.

A strong point of the Jeppesen system is that it is Internet-based. Basic program information and current flight planning data are loaded into your computer when you first log on to the system. When you later use the system, the program in your computer accesses the Internet to obtain updates on airport information, flight restrictions, surface obstructions, and weather conditions. Hence, your planned route is based on the most up-to-date information.

Light aircraft navigation and route planning is now easier because of the linking of GPS to digital aeronautical charts through **GPS aviation units.** A GPS aviation unit such as the Garmin GPSmap 496 (figure 13.28) consists of a high-accuracy receiver and a user interface similar in appearance to a vehicle navigation unit. The display is structured around a map page that allows you to selectively show aeronautical information from an internally stored database, such as Jeppesen databases that include position and facility data for thousands of airports, as well as navigational aids, airspace boundaries, coastlines, lakes, rivers, cities, and highways. An aircraft icon shows your present position on the continuously moving map, and your ground track and route are recorded along a line of dots that mark the track. It is also possible to upload constantly updated, high-resolution weather data and images for the United States and other regions that can be overlaid on the map page (see chapter 9 for more on these types of image maps).

The display usually is configured as a split screen with a rotating compass card to the right of the map page that indicates your current heading. The compass arrow and adjacent course deviation needle point to your desired heading, showing your position relative to the desired course. A "bug" indicator shows you how to steer back to the desired course should you deviate. Boxes to the right of the compass card can be configured to display flight information such as the altitude of your aircraft, the azimuth of your current heading, course to steer, distance to the next waypoint or final destination, speed, estimated time of arrival, and other useful information.

Whether you are navigating or planning to travel by land, water, or air, you now have a better understanding of the aids to navigation available to you and the techniques for using them. We started out the chapter with an observation about the desperate feeling we have when we are lost. With the information in this chapter, you can avoid those feelings and be assured of successful position finding and route planning whether on land, by water, or in the air.

SELECTED READINGS

Ames, G. P. 2003. Forgetting St. Louis and other map mischief. *Railroad History* 188 (Spring–Summer): 28–41.

Baker, R. R. 1981. *Human navigation and the sixth sense.* New York: Simon & Schuster.

Blandford, P. W. 1992. *Maps & compasses.* 2d ed. Blue Ridge Summit, Penn.: TAB Books.

Canadian Hydrographic Service. 1999. *Current atlas: Juan de Fuca Strait to Strait of Georgia.* Sidney, B.C.: Canadian Hydrographic Service.

Dilke, O. A. W. 1998. *Greek and Roman maps.* Reprint ed. Baltimore, Md.: Johns Hopkins University Press.

Geary, D. 1995. *Using a map & compass.* Mechanicsburg, Penn.: Stackpole Books.

Hodgson, M. 1997. *Compass & map navigator.* Riverton, Wyo.: The Brunton Company.

Jacobson, C. 1997. *The basic essentials of map & compass.* 2d ed. Merrillville, Ind.: ICS Books, Inc.

Kjellstrom, B. 1994. *Be expert with map & compass: The complete orienteering handbook.* Revised ed. New York: Macmillan General Reference.

Langley, R. B. 1992. The federal radionavigation plan. *GPS World* 3 (2) : 50–3.

Lankford, Terry. 1996. *Understanding aeronautical charts.* Maidenhead, U.K.: McGraw-Hill Education Europe.

Larkin, F. J. 1993. *Basic coastal navigation: An introduction to piloting.* Dobbs Ferry, N.Y.: Sheridan House.

Maloney, E. S. 2003. *Chapman piloting.* 64th ed. New York: Hearst Books.

Randall, G. 1989. *The outward bound map & compass handbook.* New York: Lyons & Burford, Publishers.

Seidman, D. 1995. *The essential wilderness navigator.* Camden, Maine: Ragged Mountain Press.

United States Army. 1993. *Map reading.* Field Manual, FM 21-26. Washington, D.C.: Department of the Army, Headquarters.

Wilkes, K. 1994. *Ocean navigation.* Revised by P. Langley-Price and P. Ouvry. London: Adlard Coles Nautical.

chapter fourteen GPS AND MAPS

14

GPS and maps

The **Global Positioning System (GPS)** has found wide use in a variety of navigation, data collection, and mapping applications because of its many advantages: ease of use, low cost, coverage over large areas, relatively high-speed data collection, ability to collect either two- or three-dimensional data, ability to provide velocity data, and ability to provide accurate time information.

GPS receivers give **absolute positions**—a specific position defined by coordinates such as latitude and longitude or UTM or state plane grid coordinates (see chapters 1 and 4 for more on these coordinate systems). Absolute position coordinates are often used in navigation, especially sea and air travel, but also when travelling by car or on foot. If you know your absolute position, you can understand your location relative to other features using ancillary data, such as from maps or geographic information systems (GIS).

GPS data are often used in geographic information systems. GIS data have two components, a spatial component, called a **feature;** and a component that describes some characteristic or characteristics of the feature, called an **attribute**. GPS receivers determine positions in all three dimensions. These positions can be stored either as **waypoints** (geographic locations, spots, or destinations), or as point features in a GIS. Strings of points can be connected into line or area features.

But positions by themselves are difficult to interpret, so attribute information that is tagged to the position is also often collected. It's possible to key in attributes as labels or identifiers with standard receivers, but the task quickly becomes laborious, and the length of the attribute fields and the number of attributes is limited. If you're going to tag attributes for more than a few locations, you'll want to use specially designed GPS receivers that incorporate dedicated software, making it convenient to collect location data with attributes. Once collected, GPS data can be uploaded to a GIS for use in inventory, analysis, mapping, and other applications.

A BRIEF OVERVIEW OF GPS

Various technologies have been used to determine position and help with navigation. For example, since World War II **LORAN** (long-range navigation) has been used all over the world for marine navigation. LORAN is based on radio signals from multiple transmitters that are used to determine the location and speed of the receiver. But LORAN has shortcomings, such as limited accuracy, radio interference, and susceptibility of the signals to be affected by the geography of the area. Because of its need for a technology free from these limitations, in the early 1990s the United States **Department of Defense (DOD)** spent 12 billion dollars to devise and implement the **NAVSTAR** (Navigation System with Time and Ranging) Global Positioning System, and GPS was born.

Although you will often see the acronym GPS used in reference to worldwide radio navigation systems that provide **autonomous** (independent) geographic positioning, the generic name is really **Global Navigation Satellite System (GNSS).** The acronym GNSS will likely become more popular as other countries create their own systems.

The NAVSTAR system became fully operational in 1994. In 1983, President Ronald Reagan issued a directive guaranteeing that GPS signals would be available at no charge to the world when the system became operational. GPS users continue to enjoy this freedom from cost today.

Using radio signals transmitted along a line of sight from satellites, GPS receivers determine their horizontal and vertical location to within a few meters (and down to the submeter level, depending on the unit). Because of the highly precise atomic clocks on the satellites, receivers at a position on the ground can be used to calculate the exact time it takes to receive the GPS signals. The signal travel time is used along with the velocity of the signals to determine the distance from the satellite to the ground position, and multiple distances are used to determine location through a process called **space trilateration,** which we will discuss in detail later in the chapter.

GPS components

The GPS consists of a space segment, a control segment, and a user segment that are organized and administered separately but are integrated together to constitute the complete system (figure 14.1).

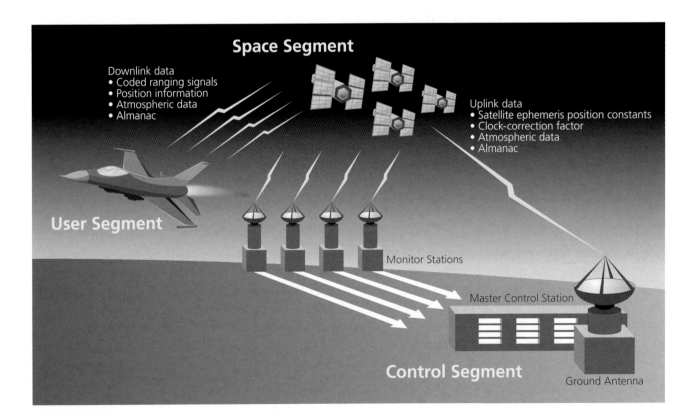

Figure 14.1 The three segments of the GPS are used together to provide a worldwide radio navigation system for autonomous navigation and geographic positioning.

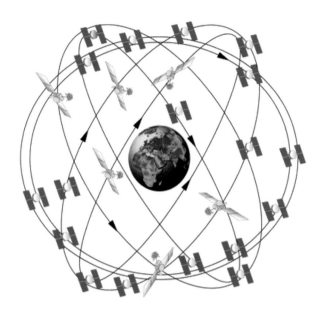

Figure 14.2 **The original GPS constellation consisted of 24 satellites in six orbital planes.**

Space segment

The **space segment (SS)** is a **constellation** of GPS satellites (figure 14.2), positioned in **medium earth orbit** at an altitude of 20,200 kilometers (12,552 miles) above the earth's surface, below the geostationary orbit (35,786 kilometers or 22,236 miles) used for weather satellites, as we saw in chapter 9. The satellites are placed in six orbital planes evenly spaced 60 degrees apart so that at any time five to eight satellites are visible from any point on earth, except in the north and south polar regions.

The satellites are placed in **posigrade orbit** (one that moves in the same direction as the earth's rotation), and since each satellite's altitude is nearly three times the earth's radius, it takes just about 24 hours to pass over the same place on earth. Actually, the satellites lose about 4 minutes a day so that each satellite will appear in the same place in the sky every day, just 4 minutes earlier—a fact that you can take advantage of to find a good satellite configuration using your GPS.

Originally, the GPS constellation consisted of 24 satellites (21 plus 3 operating spares), but as of February 26, 2008, there are now 32. The additional satellites improve the precision of GPS positioning calculations by providing redundant measurements. Although a few older GPS devices might not recognize the new satellites, most should be able to pick them up.

Control segment

To use GPS satellites as references for distance measurements, we need to know exactly where they are. Because of their high altitude, the satellites are clear of the atmosphere and they orbit according to very simple mathematics, so their positions are highly predictable. Nevertheless, the DOD continuously monitors the satellites to determine their exact positions and to detect and report minor fluctuations in their operating conditions.

The **control segment (CS)** of the GPS is the set of ground monitoring stations that track the location and **health** of each satellite, which includes conditions such as clock error and satellite malfunctions. Satellites are tracked by their own signals and an extrapolated **ephemeris** (which describes the positions of each of the satellites at all times) is uploaded periodically by the CS for transmission with the timing signals from the satellites so the receivers will know where to look for the satellites.

These updates from the ground stations also synchronize the highly precise atomic clocks on board the satellites to within a few nanoseconds of each other, and a satellite clock drift model is uploaded from the CS to again be transmitted to the GPS receiver. For the NAVSTAR-GPS, the CS originally consisted of a set of five ground stations in Colorado Springs, Colorado (where the **master control station** is located); Kwajalein; Diego Garcia; Ascension Island; and Hawaii. During 2005, six additional stations were set up in Washington, D.C. ; Ecuador; Argentina; England; Bahrain; and Australia. With this CS configuration, every satellite can be seen by at least two ground stations at any time.

User segment

The **user segment (US)** of the GPS is the receiver, which has an antenna tuned to the radio frequencies transmitted by the satellites. The receiver also has a clock and a processor to "decode" the satellite's signal. The receiver clock is not nearly as accurate as the satellite clocks, so this is taken into account when refining the GPS position. Depending on the unit, the receiver may also have a graphical display to show location and speed information to the user. For these displays, the position of the GPS receiver (actually of the receiver's antenna) onboard ships, airplanes, and ground vehicles is shown on a basemap displayed on a computer monitor. The basemap is retrieved from disk storage, and the extent of the area shown on the monitor is determined using locational information from the GPS receiver.

A GPS receiver is often described by the number of channels it has, which tells you how many satellites it can monitor at one time. Originally, GPS receivers could only monitor four or five satellites simultaneously, but most receivers today have between 12 and 20 channels—more than the number of satellites typically in view.

Many GPS receivers can relay their information via radio signals to other electronic equipment; they can also be connected to devices, such as computers, using a serial port, **USB** (a plug-and-play interface between a computer and add-on devices), or **Bluetooth** (short-range wireless networking) technology.

Selective availability

Since GPS was originally developed for military purposes, the DOD intended to prevent hostile forces from fully utilizing the capabilities of GPS by introducing random errors that would cause positions to be off by up to 100 meters (about 300 feet), though it typically added horizontal errors of up to about 10 meters (about 30 feet) and vertical errors of about 30 meters (about 90 feet) vertically (figure 14.3). This intentional degradation of the GPS signals is called **selective availability (SA).**

In 2000, the U.S. government, recognizing the increasing importance of GPS to commercial and civil users, decided to discontinue the deliberate degradation of accuracy of the GPS signals. Its decision to turn off

the SA feature was influenced by a number of factors. In the 1990s, the U.S. **Federal Aviation Administration (FAA),** which is responsible for regulating and overseeing all aspects of civil aviation in the United States, began pressuring the military to turn off SA permanently so that the FAA could save millions of dollars annually by replacing their own radio navigation systems (which required continual monitoring and maintenance) with GPS. In addition, private companies devised ways to circumvent the effects of SA for civilian use. Improved accuracy was attained through post-processing of the data. Eventually real-time data correction became possible using **GPS augmentation,** in which additional information is used along with the satellite signals to provide more accurate positioning. Many users were willing to pay the increased cost associated with circumventing SA, and certainly most of the hostile forces it was meant to thwart could do the same.

The future of GPS

The SA capability of GPS could theoretically be turned back on at any time. However, various government agencies have stated there is no intention to have it reintroduced. The suspension of SA has allowed GPS to be widely used in air, marine, and ground navigation worldwide. Other uses have also been realized, such as surveying, land-resource management, precision farming,

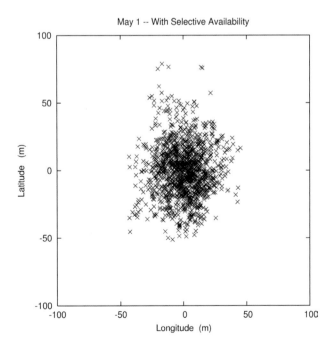

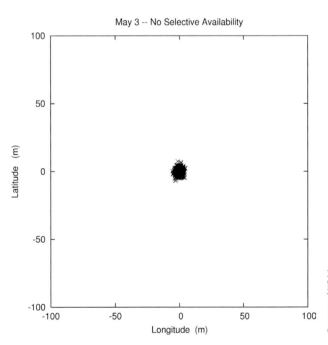

Figure 14.3 **These plots of positions collected over a 24-hour period can be used to compare the accuracy of GPS positioning with and without SA. The plots show that with SA, 95 percent of the points are within a radius of 45.0 meters (147.6 feet). Without SA, 95 percent of the points are within a radius of 6.3 meters (20.7 feet).**

scientific analysis, and, of course, mapmaking. More uses are constantly emerging—recently **geocaching** (outdoor treasure hunting using GPS and other navigational aids, such as maps) has become a popular use of handheld GPS units.

Today, research is underway in the United States, Australia, France, the United Kingdom, and Japan into what is called **ubiquitous positioning.** This next generation GPS would work everywhere, be available at all times, and provide high-level precision, all at an acceptable cost. But ubiquitous positioning is still a long way from reality.

In the meantime, other countries are developing their own international satellite navigation systems. The Russian **GLONASS** (Global Navigation Satellite System), originally completed in 1995 with 24 satellites (21 operating and 3 in-orbit spares, like the NAVSTAR-GPS), had slipped to only 14 operational satellites by 2006. In 2005, the Indian government agreed to share development costs with Russia to update GLONASS. It is now being restored, with 18 satellites due to be operational by 2008, and 24 by 2010.

The European Union is developing a 30-satellite system called **Galileo** that is intended to provide greater precision than is currently available and improve coverage at higher latitudes. Only one of Galileo's four navigation services will be available free of charge, however. China, Israel, Ukraine, India, Morocco, Saudi Arabia, and South Korea have joined this project, which is in the initial deployment phase and should be operational by 2013.

In addition, China has proposed expanding its regional **Beidou** navigation system into a global system called **Compass,** and India's **IRNSS (Indian Regional Navigational Satellite System),** now in the developmental phase, is scheduled to be operational around 2012. Japan's proposed regional system, the **Quasi-Zenith Satellite System (QZSS)** will provide better coverage for the Japanese Islands—its first satellite is scheduled to be launched in 2009.

In September 2007, the DOD announced its plans for the next generation of the NAVSTAR-GPS—**GPS III.** This system of satellites will not be capable of implementing SA, so the policy to turn off SA will ultimately be made permanent. It was originally envisioned that the initial launches of the GPS III satellites would begin in 2005 with complete replacement by 2011. The first launch has now slipped to 2009, so it is unclear when this new generation of satellites will be fully operational.

HOW GPS WORKS—THE BIG PICTURE

GPS satellites transmit signals to receivers on the ground that decode the satellite's message into precise time and location information. GPS receivers need a clear view of the sky because the radio signals from the satellites are transmitted along a line of sight. So, GPS receivers are only useful outside and do not work well in mountainous areas, under forest canopies, next to tall buildings, or in other locations that block or deflect the radio signals (we will learn later in this chapter how these limitations can be overcome to some extent through GPS augmentation).

Each GPS satellite transmits information about its location and the current time. All the GPS satellites are synchronized so the satellite signals are transmitted at the same instant. The CS ground stations precisely collect this information as the satellites orbit the earth.

The distance from the receiver to the GPS satellites can be determined by calculating the amount of time it takes for the satellite signals to reach the receiver (since they travel at the speed of light). When the receiver estimates the distance to at least four GPS satellites, it can accurately calculate its position in three dimensions through space trilateration, as we will explain later in the chapter.

The accuracy of a GPS-determined position depends on the receiver and the **field conditions** (location characteristics, length of time at the location, and arrangement of the satellites in the sky). Most handheld GPS units have 30- to 45-foot (10- to 15-meter) accuracy. Other receivers use a method called **differential GPS (DGPS)** for much higher accuracy, as we will see later in the chapter. DGPS requires one receiver fixed at a known ground location and one **mobile receiver** that is used in the field. The mobile receiver uses the information from the fixed receiver to correct its GPS positions to obtain accuracy to less than 1 meter.

HOW GPS WORKS—THE DETAILS

Now you have the big picture idea of GPS, but how *exactly* does it work? For much of your GPS work with maps, having the big picture is sufficient. For special applications, such as surveying, resource management, and land-use planning, you will likely need more detail. Much of your need to know the details will depend on how you will be using GPS and what level of accuracy you require. However, all GPS units work off the same basic principles.

Satellite position

One thing you need to know to determine GPS position is the exact location of the satellites in space. Each satellite makes two revolutions in one **sidereal day** (the period it takes for the earth to complete one rotation about its axis with respect to the stars). Each satellite's orbital period is approximately 11 hours and 58 minutes, so that at the end of a sidereal day (approximately 23 hours and 56 minutes) the satellite is again over the same position on earth. In terms of a **solar day** (24 hours), the satellites are in the same position in the sky about four minutes earlier each day. The orbit for any one satellite traces the same ground track each day, except that there is a small drift to the west (−0.03 degrees per day).

For most casual users, whatever satellite configuration is available at the time usually will suffice. But for times when you desire the highest possible accuracy, the ability to predict the satellites' locations can help you in **mission planning.** This is done in the office prior to going out in the field so that you can identify the optimal times for recording GPS data. Since the quality of GPS data is dependent on the geometry between the receiver and satellites, you can determine ahead of time when there will be enough satellites and when they will be in an optimal geometric arrangement.

Space trilateration

The basis of GPS positioning is three-dimensional trilateration from satellites, called **space trilateration. Trilateration** is a method of determining relative positions using the geometry of triangles. For GPS, trilateration uses the known locations of two or more satellites, and the measured distance between the GPS receiver and each satellite to find an absolute position, called a **fix.**

We'll explain how space trilateration works through an example (figure 14.4). Let's say we measure our distance from a satellite and find that it's 11,000 miles. Knowing we're 11,000 miles from a particular satellite narrows down all the possible locations we could be to

SPACE TRILATERATION

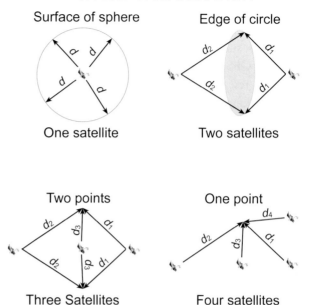

Figure 14.4 **A signal at distance d away from a single GPS satellite could be at a location anywhere on the surface of a sphere of radius d. Signals from two satellites at distances d1 and d2 intersect and define a position anywhere along the edge of a circle at the intersection of the two spheres. With three satellites, the signals with distances d1, d2, and d3 intersect at two points in space. You could now throw one of the points out if you could tell that it is ridiculous. So now you have one point that has an x,y,z coordinate. With four satellites, there is no ambiguity, since the four distances mitigate any error due to the imprecision of the less accurate receiver clock.**

the surface of a sphere that is centered on the satellite and has a radius of 11,000 miles. Next, we measure our distance to a second satellite and find that it's 12,000 miles. That tells us that we're not only somewhere on the surface of the first sphere but we're also on the surface of a second sphere that's 12,000 miles from us. So we're somewhere on the circle where these two spheres intersect.

If we make a measurement from a third satellite and find that we're 13,000 miles from that one, our position is narrowed down even further, to the two points where the 13,000 mile sphere cuts through the circle at the intersection of the first two spheres. So by **ranging** (finding distance from one location to another) from three satellites, we can narrow our position to only two points. One of the two points is a position too far from earth, so we can decide to immediately reject it, leaving us the one point that is our GPS fix. Using the distance from a fourth satellite, we can eliminate the ambiguity, if we

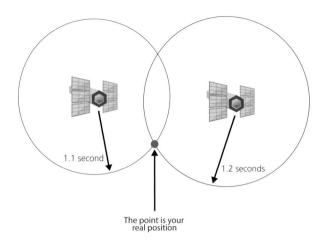

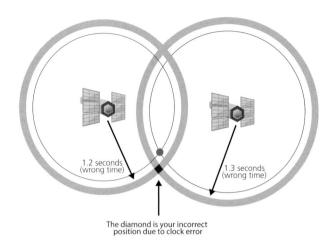

Figure 14.5 In two dimensions, you could determine your position using two ranges that would intersect at two points, one of which can be thrown out because it is not near the earth's surface.

Figure 14.6 If clock error is introduced, your position will change because the two ranges will intersect at a different point.

have accurate receiver clocks synchronized to **GPS time** (the time recorded by the GPS satellites).

We have seen that three satellite ranges can be used to locate a point on the earth, assuming the distance calculations are accurately determined and that one of the points can be rejected. Accurate distance calculation requires that both the GPS receiver and satellite clocks are precisely and continually synchronized, but this is impossible because the clocks in GPS receivers are not as accurate as the atomic clocks in the satellites. The CS ground station corrections that synchronize all the GPS satellite clocks do not apply to the clocks in the receivers. So to compensate for the lack of perfect timing, a fourth satellite range is used.

To illustrate how the fourth measurement eliminates the clock errors, it might be easier to think about the problem in two dimensions, because this is easier to visualize. However, the same principle applies in three dimensions. And because satellite range is calculated from time, we can illustrate the measurement solution using your distance in time from the satellites.

Imagine that you are located 1.1 seconds from one satellite and 1.2 seconds from another as shown in figure 14.5. Two ranges will intersect at two points, but one will be so far from the earth's surface that you can throw it out. So you have now narrowed your position down to a single point.

If your GPS receiver's clock were 0.1 second fast, then your position not would be the same because the ranges would intersect at a different point (figure 14.6).

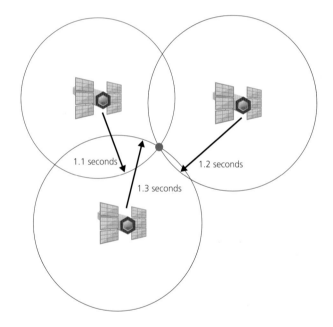

Figure 14.7 If there were no clock error, ranges from additional satellites would intersect at a single point, which would be your position.

With no receiver clock error, adding a third satellite will result in three ranges that all intersect at a single point, which is where you would be located (figure 14.7).

Taking into account your receiver's clock error for the added range, there would be three intersections, and none of them would be your real position. Using the example in figure 14.8, this tells you that no single point exists that can be 1.2 seconds from satellite *A,* 1.3 seconds from satellite *B,* and 1.4 seconds from satellite *C.* This impossible solution signals the receiver to either increase or decrease the time used in its equations so that the three ranges will provide a solution that intersects at a single point. This is done algebraically—by solving for the unknowns, the receiver can ultimately determine the single point that is your real position.

In the example in figure 14.8, there are three unknowns, but recall that we are illustrating the problem in two dimensions—in three dimensions, there are three position unknowns and one time unknown, so four satellites is the minimum required for a unique solution. In other words, a fourth satellite range helps to correct for receiver clock error to more accurately determine the GPS position relative to ellipsoidal height as well as two-dimensional latitude and longitude.

Elevation determination

Since four satellite ranges narrow down our position accurately to a single point in three-dimensional space, that point has a location that we can describe using absolute coordinates. And since we reference a position on the earth's surface relative to accepted coordinate systems on the WGS 84 ellipsoid (see chapter 1 for a discussion of ellipsoids), we use these systems with the GPS point as well. Geodetic latitude and longitude is normally used for the horizontal position, and height above or below the ellipsoid surface is used for the vertical position.

There's a complication, though. GPS position is based on the WGS 84 ellipsoid, which is actually a mathematical

model representing the earth's nearly spherical surface. GPS receivers give us a height relative to the surface of the ellipsoid based on the distance from the center of the earth to the point. But is this the same as elevation above sea level? No. As we saw in chapter 1, mean sea level elevation is based on the **geoid,** the gravity surface equal to mean sea level. And since GPS heights are based on an ellipsoidal approximation of the earth's shape, they do not have the same underlying geometrical model of the earth as the elevation we really want to know—height relative to mean sea level. Luckily, GPS heights on the WGS 84 ellipsoid can be mathematically converted to elevation above or below sea level.

GPS ACCURACY

To understand the potential for using GPS with maps, it is important to understand the limitations of the system. GPS has been designed to be as accurate as possible, but there are still errors. Added together, these errors can cause a deviation of anywhere from a few centimeters to more than a few meters from the actual GPS receiver position. Sources of error may be either random or systematic. The good thing about the systematic errors is that they can usually be corrected, at least to some degree. The random errors are harder to deal with, but even for these there are some possible corrections.

GPS error, called **bias,** can be attributed to three primary sources: the satellites, which can be affected by clock and orbital errors related to the ephemeris; the GPS signal, which can be affected by ionospheric and tropospheric refraction; and the receiver, which can be affected by the calibration of GPS antenna, GPS signal reflection or deflection, and the receiver's own clock bias.

Random error is called **range noise** or noise in the distance measurement. This is primarily attributed to **observation noise** (random "white noise" in the GPS signal partially associated with ionospheric influences) plus the random parts of **multipath error** from the reflection of the satellite signal off other objects before it reaches the GPS receiver. Systematic error can be attributed to atmospheric effects, receiver bias, orbital bias, clock bias, multipath error, and satellite. All these and other error sources leave residual random error in GPS measurements. Since all of these can be corrected to some extent, let's take a closer look at each.

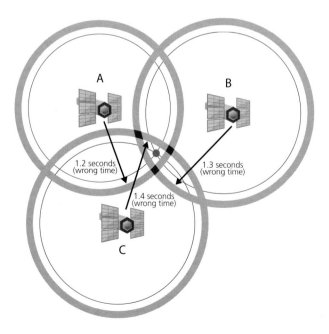

Figure 14.8 **From the intersection of three ranges with clock error, your position can be determined algebraically by solving for the unknowns in the equation.**

The error budget

The **error budget** describes the contribution of each source of error to the overall error of the satellite range measurement (table 14.1). The systematic error is coded in the signal from the GPS satellite—this includes information about its own error (clock, atmosphere, and position) and information about the other satellites from the ephemeris that is uploaded to the satellites from the CS ground stations.

Source of error	Uncorrected error
Ionosphere	Up to 30 meters (100 feet)
Troposphere	Up to 30 meters
Signal noise	Up to 10 meters (30 feet)
Ephemeris data	1-5 meters (3-15 feet)
Clock drift	Up to 1.5 meters (4.5 feet)
Multipath	Up to 1 meter (3 feet)

Table 14.1 Example of an error budget.

A large part of the error can be attributed to *satellite clock bias,* but this can be accounted for if clock corrections are used to synchronize the satellite's clock time with the GPS receiver's clock (figure 14.9A).

Error also relates to *ionospheric effects,* which can alter the speed, and to a lesser extent, the direction of the GPS signal (figure 14.9B). Ionospheric correction is also carried in the satellite's signal, but this can only correct about three-fourths of the ionospheric effect.

In addition, error is related to *receiver clock bias* (figure 14.9C). Most often, quartz crystal clocks are used in GPS receivers because they are compact and inexpensive, have a long life span, and require little power to operate. However, they are not as accurate as the atomic clocks on the GPS satellites, and they are more sensitive to temperature, vibration, and sudden impacts, so they introduce some error in the GPS fix.

Another source of error is *orbital bias,* which is addressed in the ephemeris generated by the CS ground stations rather than by the satellite itself (figure 14.9D). This **ephemeris bias** relates to error in the position of

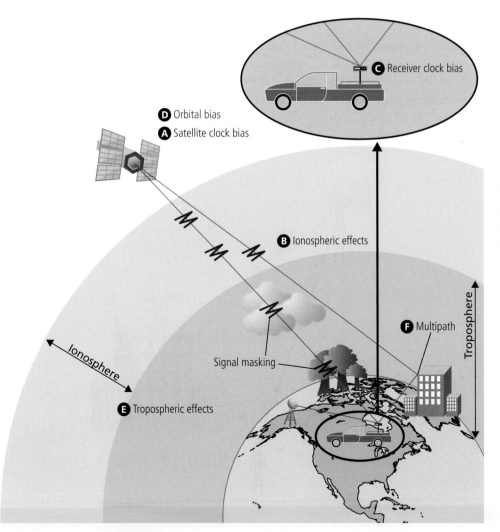

Figure 14.9 A number of sources of error contribute to the overall bias of the range measurement.

the satellite due to the earth's gravitational pull, the nonspherical nature of the earth's gravitational field, attractions of the moon and sun, and solar radiation pressure.

Range bias is also caused by *tropospheric effects*, which add a slight distance to the range that the receiver measures between itself and the satellite because the troposphere affects the speed of light (figure 14.9E). Since this is much more random than the ionospheric effect, most of this error cannot be corrected.

The remaining bias is the amount of error that is attributed to multipath as well as receiver error due to rounding off the number of decimal places or calculation errors in such functions as the addition of fractions (figure 14.9F).

Multipath error

Multipath error occurs when the GPS signal bounces before it reaches the antenna on the GPS receiver (figure 14.10). When the satellite signal arrives at the receiver via more than one path, the result is an inaccurate position reading. Because it is difficult to completely correct, multipath error is a serious concern to the GPS user even with high-precision GPS units.

Causes of multipath error include topography, tall buildings, vehicles, rock cliffs, tree canopies, and other obstructions to your line of sight to the satellites. Multipathing is a major problem in environments such as deep natural and **urban canyons** (locations in an urban landscape where you are surrounded by very tall buildings).

To reduce the error effects of multipathing, avoid sites where it could be a problem, if you can. You can also consider placing the GPS antenna on the ground or moving away from the obstructions to the GPS signal. If you have a laser range finder, you can record the fix off a reflecting disc at the point you want to locate. Signals from low-elevation satellites are more susceptible to multipathing, so another solution is to wait if possible until the satellites are in a better configuration.

Dilution of precision

The effect of the alignment, or **geometry,** of the satellite constellation on the accuracy of a GPS position is called **dilution of precision,** or **DOP.** DOP is also sometimes referred to as **geometric dilution of precision** or **GDOP.** The components of DOP include **position dilution of precision (PDOP),** which relates to three-dimensional *(x, y, z)* determination of the x,y,z position and **time dilution of precision (TDOP),** which relates to timing of the GPS signals. PDOP is made up of the **horizontal dilution of precision (HDOP),** which relates to the determination of geodetic latitude and longitude, and the **vertical dilution of precision (VDOP),** which relates to the determination of height.

You can think of DOP as a numerical indicator of the precision obtainable with a given satellite geometry. A value of 1 means that the highest confidence can be given to the GPS determination, but this is rare. Nonetheless, there are military and other applications that demand this highest possible precision at all times. A range of 2

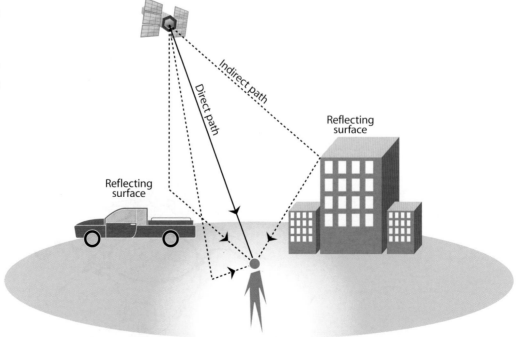

Figure 14.10 Multipath errors occur when GPS satellite signals reflect off surfaces before they reach the GPS receiver's antenna.

Indirect path

Direct path

Reflecting surface

Reflecting surface

to 3 is excellent, and position determinations can be considered as meeting the nominal accuracy specifications of the receiver in use. A range of 4 to 6 is good and is often used for vehicle navigation or resource management data collection. DOP values of 7 or higher are generally not recommended for positioning unless there is no alternative (for example, you are under dense canopy or rugged terrain). Values closer to 20 indicate that the GPS positions should only be used for rough estimates, and values up to 50 tell you that the GPS positions are too inaccurate to use at all.

Depending on the type of DOP, the satellite constellation will yield different values. For example, HDOP values become worse (higher) if the satellites seen by the receiver are high in the sky. To illustrate this, first look again at figure 14.6. Note that the ranges of two satellites intersect to form a small diamond-shaped area—the real position will be somewhere in that small area rather than at a point. When the geometry of the satellites is "poor," the area within which the real position lies is larger, so the accuracy is reduced (figure 14.11). This happens when the satellites are spaced close together overhead.

The greater the angle between the satellites, the lower the HDOP, and the better the measurement.

However, when the satellites are positioned overhead, the VDOP values will improve (figure 14.12).

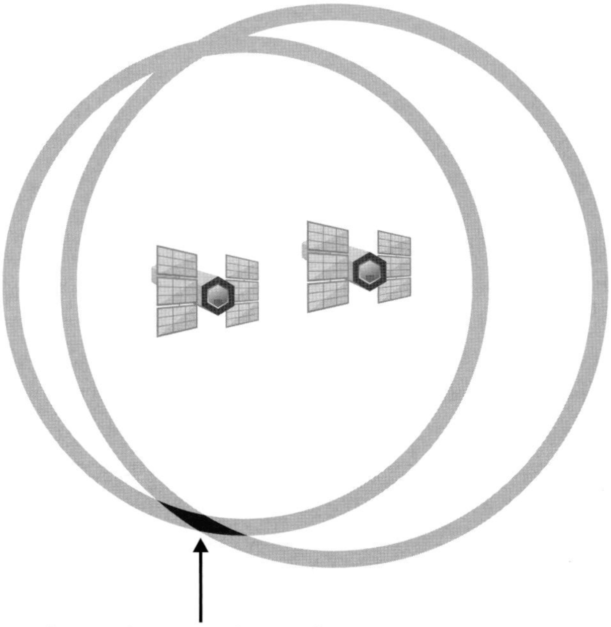

In contrast, VDOP values get worse when the satellites are closer to the horizon (figure 14.13). When this happens, the angles between the satellites are larger and the region within which the real 3D position lies is greater, so there is more margin for error.

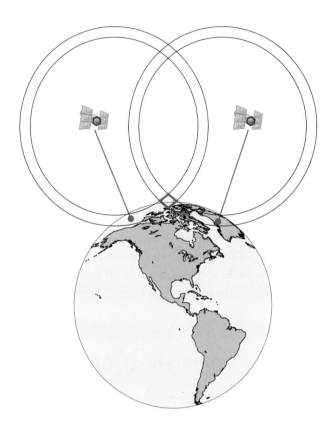

Figure 14.12   Good VDOP values require that all the satellites are positioned overhead.

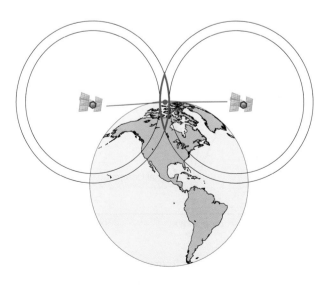

The area of uncertainty becomes larger as satellites get closer together

Figure 14.11   Poor geometry of the satellites increases the area within which the real position lies.

Figure 14.13   Poor VDOP values are a result of all the satellites being positioned near the horizon.

Since PDOP is a combination of HDOP and VDOP, PDOP values are best if at least one satellite is positioned vertically above the receiver and at least three others are evenly distributed closer to the horizon in different sectors of the sky. DOP can be degraded (increased) by signals that are blocked by the terrain, vegetation, buildings, vehicle structure, and other obstacles.

Elevation accuracy

GPS receivers give us an elevation relative to the surface of the ellipsoid based on the distance from the center of the earth to the point. Since this not the same as elevation above mean sea level, which is what we really want to know, GPS elevations are converted to elevation above or below mean sea level. That means that the elevation value of a GPS position is first determined relative to the center of the earth, and then it is mathematically adjusted to mean sea level.

This mathematical transformation coupled with VDOP introduces some error into the elevation of a GPS position. Usually this error is about double the horizontal error, so a position accurate to within 10 meters (30 feet) horizontally will be accurate to within 20 meters (60 feet) vertically.

Differential GPS

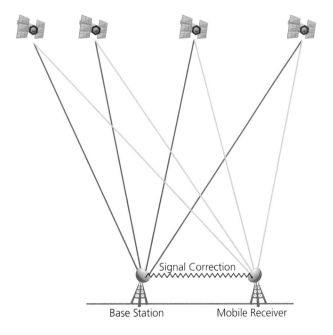

Signal Correction

Base Station   Mobile Receiver

Figure 14.14   Using signals from several satellites, a mobile GPS receiver can use signal correction data from a base station receiver at a surveyed control point to greatly improve the accuracy of GPS positioning.

HIGH-ACCURACY POSITIONING

To obtain the highest accuracy using GPS, you have to have a clear view of the sky at all times, and the GPS signal has to be able to avoid trees, buildings, and other obstructions that could cause the GPS signal to be blocked or reflected and cause multipath errors. You also want to have the maximum number of satellites with the best viewing geometry possible. In addition to meeting these criteria, there are other ways you can improve the accuracy of GPS positioning.

As we have seen, a number of factors contribute to the accuracy of a GPS fix. Some errors can be factored out using mathematics and modeling. Others can be reduced if you take the field conditions into account. But even when these errors have been reduced or eliminated, the accuracy of the position still may not be great enough. There are two other ways you can increase the accuracy—the first involves more expensive GPS units that are able to use higher-frequency **carrier phase signals.** The other is GPS augmentation, which is becoming more widely available to all GPS users at lower costs.

Carrier phase enhancement

Some higher cost GPS receivers are designed to take advantage of higher-frequency GPS signals. This is called **carrier phase enhancement** or **real time kinematic (RTK) navigation.** The idea is to refine the receive measurements using higher-frequency carrier phase signals, which have shorter cycles that are much closer together, and more accurate. This is a more complicated subject that relates to special GPS receivers, so if you find yourself in need of this information, you should refer to more detailed sources. We will continue with methods that are readily available and can be used by a wider range of GPS users.

GPS augmentation

It is also possible to gain more precision by augmenting GPS with navigation signals other than the ones sent by the satellites. **GPS augmentation** involves using external information, often integrated into the calculation process, to improve the accuracy, availability, or reliability of the satellite signal. There are a number of GPS augmentation systems in place, and they vary based on how the GPS sensor receives the information. Some systems transmit additional information about sources of error (such as clock bias, ephemeris, or ionospheric delay), others provide direct measurements of how much the signal was off in the past, while still others provide additional navigational or vehicle information to be integrated in the position calculation process.

Three widely used augmentation systems are differential GPS, inertial navigation systems, and the Wide Area Augmentation System.

Differential GPS

A technology called **differential GPS (DGPS)** can be used to achieve horizontal accuracy of 1 to 5 meters, or even better. Differential correction requires a second GPS receiver called a **base station.** The base station receiver collects its GPS data at a stationary position on a precisely known point, typically a surveyed control point (figure 14.14). Since the fixed receiver knows its location precisely from the surveyed measurement, it can tell to what extent the GPS fix it computes from the satellite signal is off from its actual position. A correction factor can then be computed by comparing the known location with the GPS location determined by using the satellites—this is called **differential correction.**

This correction is then communicated in real time by cellular phone or radio to a mobile receiver in the field, sometimes called a **rover,** which is assumed to be working under similar atmospheric conditions. The mobile receiver corrects its GPS positions using the base station information so that its fixes will be accurate to within less than a meter (3 feet), depending on the quality of the receiver.

Differential correction eliminates most of the errors in the GPS error budget described earlier. After differential correction, the GPS error budget changes as shown in table 14.2.

Source of error	Uncorrected error	With differential correction
Ionosphere	0–30 meters (0–100 feet)	Mostly removed
Troposphere	0–30 meters (0–100 feet)	Small residual error
Signal noise	0–10 meters (0–30 feet)	Small residual error
Ephemeris data	1–5 meters (3–15 feet)	Small residual error
Clock drift	0–1.5 meters (0–4.5 feet)	Small residual error
Multipath	0–1 meters (0–3 feet)	Not removed

Table 14.2 GPS error budget after differential correction.

By reducing many of these errors, differential correction allows GPS fixes to be computed at a much higher level of accuracy.

In the above scenario, the ground station uses a radio link to instantaneously send corrections to the mobile receiver. In order for you to take advantage of this **real-time correction,** your receiver would have to have special capabilities and a special antenna. It would also have to be able to receive a signal to correct its GPS locations.

This differential correction signal can come from a variety of sources, such as a GPS receiver over a known location, as we described above. Your receiver could also pick up a free signal from beacons or transmitters that are administered by reputable agencies such as the Coast Guard, Department of Transportation, and the Army Corps of Engineers, in the United States. Alternatively, you could subscribe to a service for FM radio signals in range of your GPS receiver from commercial companies like DCI and Accupoint. Or for few hundred dollars you can subscribe to receive satellite signals from companies like Raycal and Omnistar (academic institutions usually get discounted or free service for a limited time).

But this real-time correction is not needed with all DGPS applications. **Post-processing** is an attractive alternative, attractive because of its lower cost and because it does not require a sophisticated receiver.

Post-processing can be used for any GPS data that are collected with time tags to tell precisely when each fix was recorded. Post-processing requires base station data for a location near where the mobile receiver collected its positions. The files from the base station and the mobile receiver are downloaded to a computer and special processing software computes corrected positions for the mobile receiver's GPS data. These corrected data can then be viewed in or exported to a GIS. You can even send your data over the Internet to a service that will process your data for you if you do not have the time, software, or expertise to do it yourself.

The source you choose to provide the base station data will depend on how close the base station is to where you collected your GPS data, as well as the reputability of the source, access to the source, and to a lesser extent, cost. For mariners in the United States, Coast Guard beacons provide correction information for coastal areas as well as harbors and waterways. Other government agencies also collect base data, as do commercial firms, and some universities. And, of course you could also set up your own base station—but this is the most costly solution, especially when maintenance of the station is taken into account.

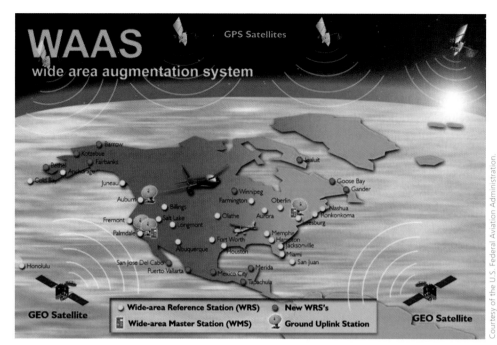

Figure 14.15 **WAAS consists of a control segment of WRFs and a space segment of geostationary WAAS satellites.**

Inertial navigation systems

So far, we have suggested that you need line of sight to the GPS satellite in order to receive its signal. What if you want to use GPS inside or underground (like in a cave, a tunnel, or a mine)? Or what if you are in a deep natural or urban canyon where multipathing causes significant errors in the GPS signal? In these cases, **inertial navigation systems (INS)** are invaluable. An INS is a navigation aid that uses a computer and motion sensors to continuously track the position, orientation, and velocity of the GPS receiver without the need for external information.

The INS is first **initialized** (provided with its position and velocity from another source, such as the user or a GPS receiver). Then, using accelerometers or gyroscopes, which are motion-sensing devices, the INS can detect a change in its geographic position (a move east or north, for example), a change in its velocity (speed and direction of movement), and a change in its orientation (rotation about an axis). It continually computes its own position and velocity by integrating information received from the motion sensors and reporting these updates back to the receiver. It does this by measuring the linear and angular **accelerations** (rate of change of velocity) to the system. Since it requires no external information (after initialization), it is immune to jamming and deception.

As you can imagine, the advantage of an INS is that it requires no external data to determine its position, orientation, or velocity, once it has been initialized. But INS can usually only provide an accurate solution for a short period of time. The benefits of using GPS with INS

are that if GPS loses its signal, the INS can continue to compute the position during the period of lost signal. Additionally, the INS can be initialized using the GPS signals, and the INS can provide updates between the 1-second GPS updates.

Because the two systems are complementary, they are often used together. On submarines, for example, they are used to provide non–line of sight navigation, while on aircraft and spacecraft they provide information that augments the GPS signals for higher data transmission rates. These combined systems are becoming more popular for use in car navigation systems so that travel though tunnels and on streets lined with tall buildings does not interrupt GPS positioning.

Wide Area Augmentation System

Even without selective availability, GPS by itself does not have an adequate level of reliability, integrity, accuracy, and availability to support a nationwide autonomous aviation navigation system. The **Wide Area Augmentation System (WAAS)** was created by the FAA to make GPS reliable enough for commercial aviation in all weather conditions without the need for other navigation equipment. Additional signals increase the reliability, integrity, accuracy, and availability of GPS for aviation in the United States. Other GPS users can also take advantage of this free system for their GPS uses.

WAAS consisted originally of a network of 25 **wide-area reference stations (WRS)** (ground-based facilities that gather GPS data used for WAAS corrections) and a number of geostationary satellites broadcasting a radio signal in the same band as the GPS satellite signals (figure 14.15). The number of ground stations has since increased, so that as of October 2007, there were 38 WRS, 20 in the conterminous United States, seven in Alaska, one in Hawaii, one in Puerto Rico, five in Mexico, and four in Canada.

The WRFs make measurements of small variations in the GPS satellites' signals. These measurements are sent to **wide-area master stations (WMS)**, which calculate corrections for WAAS using the data they receive from the WRSs. The corrections are compiled and processed into messages that contain corrections for the satellite positions and information about the satellite health status. These messages are sent to geostationary WAAS satellites every 5 seconds or less. The WAAS satellites broadcast the correction messages back to earth, where WAAS-enabled GPS receivers use the corrections to improve accuracy while computing a GPS fix.

Augmented GPS is of great value to surveyors and has become essential to commercial air and sea navigators. Before long, inexpensive augmented handheld GPS receivers will be available for recreational users as well, so you can receive correction data from one of a rapidly growing number of GPS augmentation systems.

EXPRESSING GPS INFORMATION

Now that we have seen how GPS can be used to give us accurate geographic coordinates and highly accurate time, let's turn to how this information is expressed. In order to convert between the different time and horizontal and vertical position systems, the permanent memory in a GPS receiver holds equations and tables of data. The receiver's computer uses these equations and tables to process the satellite signal information. These data are also used in making computations when determining direction (true or magnetic), distance (in English or metric units), speed (in miles or kilometers per hour, or knots), and position with respect to different grid coordinate systems (see chapter 4 for more on grid coordinate systems). These data in the GPS memory bank can also be used to process information that you enter through the keypad on the GPS unit or through a connection to a computer.

Time

Usually we think of GPS in terms of position or navigation, but because of the precision that its atomic clocks afford, GPS has become a principal foundation for determining accurate time. It is used extensively both as a source of time and as a means of transferring time from one location to another. GPS time is often used in scientific applications that require careful time measurement, such as monitoring earthquakes, calibrating vehicle navigation systems, and synchronizing computer networks, telecommunications networks, and electronic equipment. GPS positioning might also be used in these applications, but precise time is the more important aspect of the information that GPS can provide.

Occasionally there is a need for you to know the details of how a GPS receiver expresses time, although this generally is not a concern for the casual user. When it is a concern, it is useful to know that there are three main ways time is expressed with GPS: GPS time, Universal Time Coordinated (UTC), and the time from each satellite's atomic clock, which is referenced to International Atomic Time.

GPS time

GPS time began at midnight, Saturday, January 5th, 1980, and is counted in weeks and seconds of a week from that instant. A new **GPS week** begins at every Saturday/Sunday midnight transition. The days of the week are numbered, so that Sunday is 0, Monday is 1, and so on. Within each week, the **second of the week (SOW),** noted as a number between 0 and 604,800 (seconds + 60 seconds per minute × minutes + 3,600 seconds per hour × hours + 86,400 seconds per day × days), further refines the time. For example, in GPS time, Thursday, July 23, 2009, 23:05:53:00 is week 1,541 428,753 seconds (53 + 60 × 5 + 3,600 × 23 + 86,400 × 4).

Universal Time and UTC

Universal Time (UT) is the same as **Greenwich mean time (GMT)**. UT is a time scale that is tied to the rotation of the earth—it is based on the apparent motion of the **mean sun.** The mean sun is a hypothetical sun that would appear at the equator to move from east to west at the mean (average) speed that the real sun moves. **UT1** is UT corrected for a phenomenon known as polar motion. Since people usually mean UT1 when they say UT, we will use UT1 here.

The mean sun is based on an earth that rotates uniformly, which the real earth does not. To compensate for the minor variations in the earth's rotational speed, GPS uses **Universal Time Coordinated (UTC).** It is a time frame based on atomic clocks and aligned with the 0° longitude time zone. UTC, the "coordinated" version of multiple UT1s (all of which fall within 1 second of each other), is estimated and produced by the United States Naval Observatory.

International Atomic Time

International Atomic Time (Temps Atomique International, or TAI) is defined as the weighted average of the time kept by about 200 atomic clocks in over 50 national laboratories worldwide. Since GPS time is kept by atomic clocks, it uses TAI. The UTC and TAI timescales were identical on January 5, 1980. It is possible to relate UTC to TAI by using a correction defined by the **International Earth Rotation Service (IERS)**. The correction is given in multiples or increments of whole seconds. As of January 1, 2006, TAI was ahead of UTC by 14 seconds.

Datums

In chapter 1, you learned about parallels and meridians on the ellipsoid and how they can be defined mathematically by their geodetic latitude and longitude. This is the coordinate system of choice for GPS because it is a worldwide standard and it works for the entire earth. But there is a problem—even though we have a clear definition of latitude and longitude and we have a precise method of measurement (using GPS), there can still be ambiguity in the coordinates that the GPS gives us.

Why is it so hard to define geographic coordinates? The difficulty lies in defining the shape of the earth. Since the earth is not a sphere, in order to establish latitude and longitude it is necessary to define a reference ellipsoid. In chapter 1, you learned that for a long time the Clarke 1866 Ellipsoid, the reference ellipsoid for the North American Datum of 1927 (NAD27), provided a best fit for horizontal positions for North America. The Geodetic Reference System of 1980 (GRS80), the reference ellipsoid for NAD83, is now the datum of choice for the continental United States because of the increased quality in measurements and technologies used to determine the ellipsoidal approximation to the earth's shape and size.

Also in chapter 1 you learned that there are vertical datums. The North American Vertical Datum of 1929 (NAVD29) was calculated from 26 tide-gauging stations across the United States and Canada. The updated North American Vertical Datum of 1988 (NAVD88) accounts for crustal movement and subsidence as well.

With these and many other datum choices, you can see that the determination of horizontal and vertical GPS coordinates is influenced by the horizontal and vertical datum that you use.

The GPS receiver memory holds the data and equations needed to make both horizontal and vertical datum calculations. Parameters for more than 100 datums used around the world are stored in many GPS receivers. These parameters are used when the computer in the receiver makes conversions between the datums.

GPS receivers default to the WGS84 ellipsoid. Unfortunately, this datum isn't used on all large-scale maps, particularly older editions. If you're not using a map based on WGS84, the horizontal discrepancy between your map datum and WGS84 will create positioning errors beyond those we have already discussed. This additional locational error can be as much as 100 meters (300 feet) in the United States and 600 meters (1,800 feet) worldwide.

To minimize datum errors, you have two choices. You can adjust computed fixes to your map manually, which is a laborious task, or you can enter the datum that was used to make your map and let the receiver make the datum conversion before displaying position coordinates. The second choice best uses the power of the receiver and is preferred.

Coordinate systems

The **geographic coordinate system** (geodetic latitude and longitude) is usually the default for expressing horizontal positions. But most GPS receivers can convert latitude and longitude to several types of grid coordinates (see chapters 1 and 4 for more information on geographic and grid coordinates). Conversion to UTM grid coordinates is available on all receivers. In addition, you might find state plane grid coordinates and a variety of regional and proprietary grids. Some receivers even let you define your own grid coordinate system.

Elevation

The elevation most often found by GPS receivers is **height above the ellipsoid (HAE)** which is elevation relative to the surface of the ellipsoid. But since HAE has little practical meaning for most people, the height *displayed* on most handheld GPS receivers is **orthometric height** (height above the geoid), which is familiar to us as height above mean sea level (MSL), which most of us are familiar with. Orthometric heights for the continental United States are referenced to NAVD 29 or more recently to NAVD 88, although other vertical datums could be used.

The GPS receiver converts ellipsoidal height to orthometric height by interpolating the **geoid model** (a lookup table stored in the receiver) and using the simple calculation that we explained in chapter 1:

$$h = H + N,$$

where h is the ellipsoidal height, H is the orthometric elevation, and N is the **geoidal undulation,** or the difference between height of the geoid and the ellipsoid.

Be sure to check how your GPS is displaying elevation as the differences can be significant and the consequences great if you are not paying attention.

Units of measurement

The choice of grid coordinate system usually determines the units of measurement. The UTM grid coordinate system, for example, uses metric units, while the state plane grid coordinate system uses imperial units (feet). Geographic coordinates are usually displayed in degrees, minutes, seconds (DMS), but also can be shown in decimal degrees (DD) (see chapter 1 for more on these systems). There normally is a menu option to change to the units of your choice.

Compass declination

GPS receivers report direction with respect to both true and magnetic north—you can choose which to display. To make these conversions, the receiver must hold in its permanent memory magnetic declination data for the entire earth. Computations done with these data let the receiver serve as a form of electronic compass. But the standard handheld GPS receiver is no substitute for a conventional magnetic compass, as we'll see later in this chapter.

Databases

In addition to specific information needed to perform corrections and conversions, information pertinent to the use of the GPS receiver can either be held in memory or connected to when needed. For example, a database of airport facilities is commonly found in GPS units used in planes, while a database of port facilities and other navigational information is found in units used in boats. A database for highway travel contains information on local fuel, food, and lodging services, as well as entertainment, recreation, and other points of interest. These are used in vehicle navigation systems that are GPS-enabled, as described in chapter 13. Even handheld GPS units have the capability, albeit limited, to store waypoint information from a database.

The community of handheld and car navigation system users is growing. To help you use these systems to their full advantage and to share interesting places, online databases are available from which you can download waypoints that others have entered and to which you can upload waypoints you may have collected. To provide the widest use, all waypoints are recorded in geographic coordinates in which the latitude, longitude format is NDDMM.MMM where N is north, DD is degrees in 2 or 3 digits, and M is minutes in thousandths. For example, here is the waypoint record from the International

GPS Global Positioning System Waypoint Registry for the Washington Monument in Washington, D.C.:

Washington Monument
A popular tourist spot in Washington, D.C.
Category=Monu
Waypoint: N3853.516 W7702472
http://www.graftacs.com/horton/gpsgis.html
Submitted by: horton##graftacs.com
Submitted on: 12 Feb 2001

You can find waypoints of interest to you by searching either by area or by category and then area. Categories include restaurants, museums, stores and shopping places, hotels and other lodging, emergency services such as hospitals and police stations, and even the homes and graves of famous people!

Positional accuracy

Taking all the error factors we discussed above into consideration, what is the positional accuracy of GPS? **Positional accuracy** is determined by comparing the true measurements of both the horizontal and vertical measurements of the GPS position to the actual ground coordinates on the earth's surface. Because there are several ways to define accuracy, you must first ask, "What type of accuracy am I interested in?"

One way to state accuracy is with respect to the lowest achieved distance error or the probability of being less than or equal to a specified linear range 50, 68, 95, or 99 percent of the time. Accuracy is sometimes stated with respect to a single fix and sometimes with respect to averaging fixes over a period of time. Unfortunately, the proper clarification is often missing when you see a statement of the accuracy your GPS unit.

If we adopt a 95 percent rule using what is called the **circular error probable (CEP)** method, we can describe accuracy by giving the radius of a circle containing 95 percent of all possible fixes (figure 14.16). Using this method, manufacturers of GPS receivers can state the horizontal accuracies of their units—for example, they might say that 95 percent of the fixes can be expected to fall within 30–45 feet (10 to 15 meters) horizontally of their true position on earth. Don't forget that the vertical accuracy is about two times the horizontal accuracy.

Beware of vendor hype. You'll see a wide range of numbers given for GPS receiver accuracy. These figures commonly represent the most optimistic case under the best of all conditions. Such luck is rare. The numbers also vary with the definition of accuracy used. CEPs of 50, 68, 95, and 99 percent are commonly reported. If vendors

1st ring (inner) is CEP at 50% of the fixes
2nd ring is RMS at 63% of the fixes
3rd ring is 2RDRMS at 95% of the fixes
4th ring (outer) is 3DRMS at 99.999% of the fixes

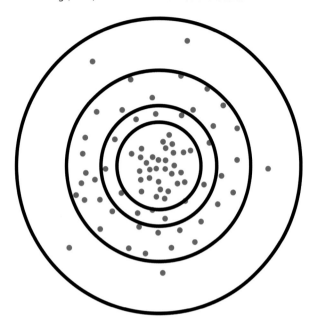

Figure 14.16 The circular error probable (CEP) can be used to assess and express GPS accuracy—it is based on the radius of a circle inside which there is a stated probability (in the example, 50 percent) of the fixes being located.

adopt a 50-percent rule, of course, their products may seem to the unwary to be more accurate. You must take care to judge the competition against the same standard.

If you're willing to average signals over a longer period of time, you can use your handheld receiver to achieve differential GPS horizontal accuracy as good as 10 feet (3 meters). A handheld receiver using corrected coordinates from WAAS should have a positional accuracy of under 30 feet (10 meters) without averaging signals. If you are moving, your accuracy is lower. Conveniently, many GPS receivers will tell you the **estimated position error (EPE),** an estimate of how good the horizontal position is, when it displays your coordinates (see appendix B for more on this and other GPS acronyms). Clearly, augmented GPS holds great promise. But its added expense and inconvenience may not be justified if you can live with up to 30- to 45-foot (10- to 15-meter) positional accuracy.

On the other hand, surveyors and other mapping professionals need accuracy in the submeter range. Accuracy in the submeter range is now quite routine, and accuracy in the centimeter range is possible under the best of conditions. Such accuracy requires sophisticated equipment

and data processing. This is costly in terms of time, equipment, and the training of skilled professionals. But these GPS users are usually willing to put up with the extra effort and cost to obtain this increased accuracy.

In the future it's likely that all GPS receivers will be equipped for augmentation and that the required correction data will be communicated from permanent ground stations via satellite to your receiver using a free global system similar to WAAS or a local system such as INS. You will then enjoy 10- to 20-foot (3-to 6-meter) horizontal accuracy worldwide with an inexpensive, handheld receiver.

The accuracy of your vertical position is another matter. Handheld receivers typically have 95 percent of readings falling within 92 feet (28 meters) for a given fix. This error can be greater than your local relief in much of the world! With augmentation, however, you should be able to obtain a vertical accuracy of 8 meters (26 feet) or better 95 percent of the time with a handheld receiver.

TYPES OF GPS RECEIVERS

GPS receivers can be categorized into three **grades** based on the level of accuracy they provide and special functions they offer.

Recreation grade units are used primarily for outdoor sports or recreational activities. These units have a relatively low level of horizontal accuracy (between 15 to 45 feet [5 and 15 meters] under favorable conditions), but they have the advantage of being inexpensive, commonly available, and user-friendly. Most of these units can store the locations of waypoints and some offer the ability to download the points to your computer. These units usually cost less than $500.

Mapping grade (also called **resource grade**) GPS units offer a higher level of accuracy and are designed to facilitate GIS data collection. Features captured as points, lines, or polygons can be automatically transferred to a GIS. Many mapping grade GPS receivers also provide the capability to record attribute data for the features. Some have the ability to connect with other devices such as laser range finders so that remote fixes can be recorded (see chapter 13 for more on these devices). Units that provide accuracy of around 1 meter cost about $5,000; submeter-accuracy units cost closer to $10,000.

Surveying-grade GPS units provide the highest horizontal accuracy (within 1 centimeter) and are used for precision surveying applications such as road, bridge, or other construction projects. These units cost closer to $20,000.

Multisensor receivers

A few manufacturers have recently addressed the stationary compass and altitude inaccuracy problems inherent in GPS technology by marketing **multisensor receivers.** In these units, the GPS receiver integrates information from multiple sensors; such as a wireless signal detector or an **altimeter** (an instrument to measure the height above sea level). The Garmin GPSMAP 76CSx, for instance, includes a WAAS-equipped differential GPS receiver capable of 10-foot (3-meter) horizontal accuracy. In addition, the unit has a barometric altimeter and electronic compass, so you can obtain horizontal positions, elevations, and heading (direction) azimuths when you're standing still. The altimeter still must be calibrated frequently for elevations to be accurate (see chapter 13 for more on altimeters), but even without constant calibration the altimeter values probably will be more reliable than elevations computed by the GPS receiver because of the error issues we discussed earlier.

At this point, let's turn to how a handheld GPS receiver is used, since many of us will have the opportunity to work with these at some time.

Figure 14.17 **The Garmin GPSMAP 76CSx is a typical handheld recreation-grade GPS receiver. Notice the dedicated function keys above the display screen.**

HANDHELD GPS RECEIVERS

You can use a recreation-grade **handheld GPS receiver** for personal use, such as hiking, boating, or traveling by car. Although GPS receivers share many traits, we advise you to consult the user's manual that comes with your GPS unit because vendors use slightly different designs and terminology. Your manual will explain how to use labeled **hard keys** or **function keys** (buttons for particular functions) on the face of your GPS unit, how to use different key combinations and sequences to activate **soft keys** (items) in the display menus, and what the menu jargon and cryptic acronyms mean.

The appearance of a handheld GPS receiver is usually rather stark (figure 14.17). For example, the Garmin GPSMAP 76CSx receiver has a small (approximately 2 inch by 3 inch or 5 centimeter by 7.5 centimeter) color or black-and-white display screen. This screen has rather poor spatial resolution, which explains why a typical screen displays only a simple graphic. The use of symbols, abbreviations, and acronyms increases the amount of information that can be shown on the small screen display.

On the receiver, you also find a few dedicated function keys, including a *rocker* key, the large center button used to control the horizontal and movement of the cursor

on the display pages. Several additional hard keys, such as *power, menu, zoom in, zoom out, quit* and *enter* have somewhat self-explanatory functions.

On each screen page you can scroll up or down to a soft key (menu item) and then activate that function by pressing *enter*. To enter an alphanumeric character, use the *rocker* key to scroll through the alphabet (A to Z) or digits (0 to 9) until you reach the desired character, and then press *enter*. You have to repeat this process for each letter in a word and each digit in a number. In addition to selecting menu times, the *enter/mark* key is used to mark your current location as a waypoint.

The *find* function allows you to find the locations of waypoints and mapped features such as cities, interstate exits, and rest areas. You can select features by their name or the map-panning feature can be used to find features nearest to a location.

Pressing and holding the *find* key brings you to the *man overboard (MOB)* function, which marks an immediate fix so you can get back to it. You might use MOB, for example, if you accidentally drop your fishing rod (or your partner!) in the water while crossing a lake. A couple of quick keystrokes activate the MOB function on most receivers, marking the coordinates of the spot. You can then navigate back to this fix as you would to any destination. If you've ever tried to recover something dropped in a lake while the boat was moving or you were far from shore, you will appreciate the potential of MOB.

The handheld GPS receiver also includes an antenna. Internal antennas make for a more compact unit. External antennas may be attached to the unit's side and will need to be extended vertically for use. External antennas may also be attached to the unit with a cable so that they can be remotely mounted, say on a vehicle's roof or the top of a backpack.

The *page* key lets you browse through a half dozen or so screen pages or menus that give you access to most of the receiver's functions. The Garmin GPSMAP 76CSx, for example, has five primary pages:

1. A *GPS satellite page* shows the positions of the satellites within range of your receiver and the strength of signals received from each satellite (figure 14.18). Satellites just outside of range are shown in gray so you can see if you will soon gain another signal (or if you will lose one) and from which location. Satellite positions are displayed on a north-oriented diagram consisting of two circles. The outer circle represents the horizon, and the inner circle represents 45 degrees above the horizon. Your current location is at the center of the circles. The coordinates of your position are given at the upper right of the screen and a "heading bug" shows your direction of movement if you happen to be moving at the time.

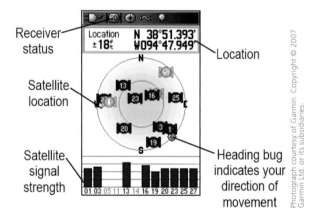

Figure 14.18   The GPS satellite page on the Garmin GPSMAP 76CSx receiver shows the positions of satellites in view, the strengths of their signals, your location, and your direction of movement.

2. A **map page** shows a crude basemap of your position relative to features such as political boundaries, streets and highways, cities, railway lines, lakes and rivers, and navigational aids (figure 14.19). Information such as your current heading and speed of travel are displayed in data field boxes above the map. You can use the *zoom in* and *zoom out* keys to change the map scale to view a larger or smaller area.

Figure 14.19   The map page on the Garmin GPSMAP 76CSx receiver shows your current speed and compass heading, as well as a basemap with features such as cities, roads, political boundaries, and rivers.

3. The **compass page** indicates directions and headings in degrees. The compass card on the display rotates as you change direction (see chapter 12 for more on compasses). The **true azimuth** (the angle from the direction line to the true north reference line) for your current heading is shown by a "bearing pointer" on top of the compass card (figure 14.20). This pointer always points toward your destination waypoint, so you can see how your current heading deviates from the straight-line path to your destination. Note that the GPS receiver cannot give you this information if you are not moving because it cannot tell which way you are pointed if you are standing still (unless your receiver is INS enabled). Navigation information such as your speed, distance, and estimated time of arrival to the waypoint is displayed above the compass card.

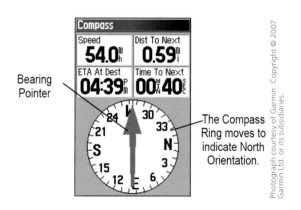

Figure 14.20   The compass page on the Garmin GPSMAP 76CSx receiver guides your steering when following a course.

4. A **trip computer page** gives you detailed information about the course you are following. You can customize the page to show, for example, an odometer indicating how far you have traveled, the length of time you were moving and stopped, your maximum speed, and your current elevation (figure 14.21).

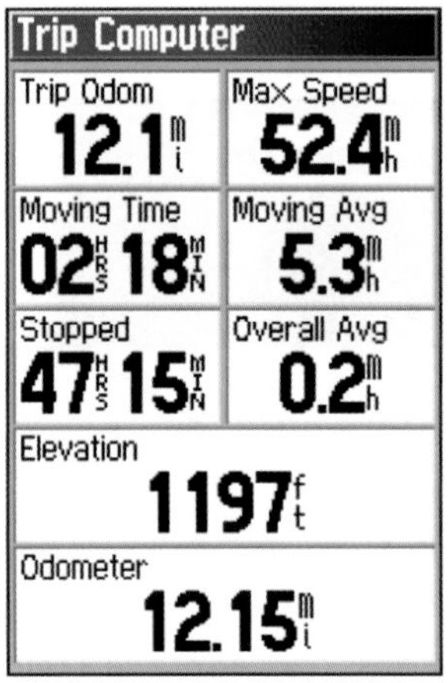

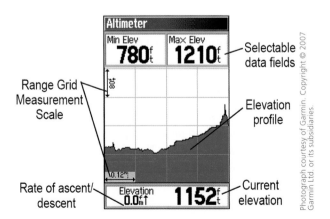

Figure 14.21 The trip computer page on the Garmin GPSMAP 76CSx receiver gives you detailed information about the course you are following.

5. An **altimeter page** shows your current elevation, the maximum and minimum elevation along the route you have followed, an elevation profile for your trip, and your **rate of ascent or descent** , which is the how many feet per minute you are gaining or losing in elevation (figure 14.22). You can also select from a menu the different types of data or **data fields** you want to display in the field window.

Figure 14.22 The altimeter page on the Garmin GPSMAP 76CSx receiver gives you detailed information about elevation along your course and your rate of ascent or descent.

USING YOUR HANDHELD GPS RECEIVER

You can best judge the potential of GPS technology by reviewing what it can do. We've said GPS is a revolutionary advancement in positioning and navigation technology. So how does it compare with traditional methods of position and route finding? In this section, as we describe what handheld receivers such as the Garmin GPSMAP 76CSx can do, we invite you to make a direct comparison with the map and compass methods discussed in the previous chapter.

Finding your position
Position determination (finding your location) with GPS couldn't be simpler. Merely turn the power on, and in a few seconds the receiver displays your two- or three-dimensional position on one of several pages. If you need more precise measurements, augmented GPS and higher-cost GPS units can be used.

Orienting your map
There are two ways to orient your map on the map page. The *North Up* option orients the map like a standard north-oriented paper map. If you select *Track Up,* the map is oriented with your straight-ahead direction of travel. A north arrow on the map shows you the direction to true north.

Entering waypoints
You can use your GPS receiver to enter a waypoint by navigating to the location of interest and marking its position. The receiver will record the coordinates of the position and allow you to enter a short description. If you know the coordinates of a distant point of interest, you can enter it as a waypoint as well. Use the *rocker* key to move the pointer on the map page to the distant point, then *save* the coordinates of the point. You can also enter the coordinates manually by pressing the *find* key, selecting the waypoints page, and entering the name and coordinates of the point as a new waypoint. Waypoints that you enter manually can also come from maps, gazetteers, Web sites, and other geographical sources.

Reading from tables
Alternatively, you can access tables with waypoint coordinates derived from various geographical sources, such as maps and gazetteers. Entire books and Web sites are devoted to providing this information. Using these sources, you can manually enter the waypoints as

described in the previous section, or you can upload them if they are in digital format and your GPS receiver is capable of receiving such information.

Navigating to a waypoint

Once you have selected a waypoint as your destination, you can use the course pointer on the compass page on the main menu to navigate to the distant position.

Planning a route

A GPS receiver can be used to plan a route, but the receiver requires you to enter route data. Coordinates for landmarks or destinations stored as waypoints in your receiver's memory can be organized into a **route sequence.** You create routes with the Garmin GPSMAP 76CSx using the routes page from the main menu and adding waypoints from the *find* menu. It's wise to save coordinate information for routes that you might want to repeat by writing them in a notebook or entering them into a computer file because the coordinates held in a GPS receiver can be lost if the unit is damaged.

The problem with using coordinates from past trips or other sources is that handheld receivers have limited temporary storage capacity, so unless they all fit into your receiver's memory, you may find yourself frequently uploading and downloading the waypoints you want to use (if your unit has this capability). Today's receivers let you store anywhere from 250 to 3,000 waypoints organized into 20 to 200 routes. Once you exceed this capacity, you must clear some coordinates before you can enter new ones. The coordinates of these purged points are lost unless you download them onto some other storage device. Some GPS units let you save to a memory card, a CD or DVD, or directly onto your computer.

Following a route

The awesome power of GPS is revealed when following routes. Your GPS unit tells you which way to go, where you've been, how you're doing, and how to get back on track if you're off course. Let's look at each of these functions.

Where am I going?

As we discussed earlier, to plan your route, you can enter your destination's coordinates into your receiver. Once you've done so, you merely press the *menu* key, choose the route page, highlight the navigate soft key, and select your destination from the waypoint list. Your receiver will display your route on the map page, and the direction and distance to your destination on the compass page.

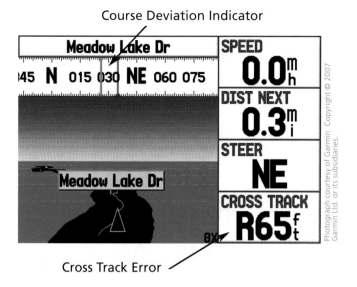

Course Deviation Indicator

Cross Track Error

Figure 14.23   The highway page on a Garmin GPS receiver shows your speed and distance to the next waypoint, in addition to your cross track error and a course deviation indicator showing the direction to steer to correct your navigation error.

If your route has several legs, your receiver will automatically shift to the next leg when you approach an upcoming waypoint. The new direction and distance data will be displayed. This process will be repeated until you reach your destination.

Where have I been?

The map page displays your progress along a route. It shows your starting point and each waypoint you've visited. It also shows the path you took between waypoints to get to your current position. Some units display landmarks, city locations, or navigation aids stored in the receiver's memory. The map is crude but sufficient enough to give you the information you need in the field.

You may want to save information about your route. This might be the case, for example, if you're following a trail or road not found on your paper map. Later, when you get back to your computer, you can download the route coordinates and plot a computer map that you can overlay on your paper map. Then you can use the overlay to update your printed map.

How am I doing?

Your GPS receiver also provides graphic steering guidance and navigation information en route. This information is displayed on another page on the GPS receiver called the highway page. The highway page commonly includes a **course deviation indicator (CDI)**—a graphic display showing how far you've strayed left or right of your desired path. This information is called your **cross track error (XTE or XTR)**. Such a portrayal of navigation error can be very useful in guiding you back on course (see figure 14.23).

Compass and highway pages also display a variety of speed and time information: *speed (SPD), speed over ground (SOG), velocity made good (VMG), estimated time en route (ETE),* and *estimated time of arrival (ETA)* are some of these functions. These acronyms, and others relating to GPS, are explained in detail in appendix B. This information may be useful if you're traveling in a vehicle along a fairly straight course at a constant rate of speed. But at slow speed over an irregular course, such as walking along a winding trail, the data are unreliable. Also remember that the GPS uses straight "as the crow flies" distances in making these calculations.

When you near your destination, your receiver may beep or display a message that you have arrived. At this point, you should be within viewing distance of your ground destination, and the GPS unit has done its job. Even experienced navigators complete the final few meters of their journey by eye, not by instrument or map!

How do I get back?

Most GPS receivers have a **backtrack function** that helps you retrace a route back to its starting point. Pressing a few menu keys reverses your route and displays it as a graphic plot. As you follow the route back toward your starting point, the navigation display works the same as when you were outbound on the route.

HANDHELD GPS LIMITATIONS

GPS technology is revolutionary, but it is not magical. A receiver knows coordinates, not geography. It uses these coordinates to compute distance, direction, and speed. It assumes a straight-line path between waypoints when making these computations.

But a GPS receiver lacks the sense of place that makes locations special to us. Since it can't read your mind, its computations aren't tempered by such human factors as needs and desires. It can tell you the shortest distance as measured in miles, but it can't tell you which route is the quickest, safest, or most interesting. You glean such functional information from maps, either those in your head (mental) or those in your hand (cartographic). Maps help you to see spatial relations at a glance, to put the environmental pieces together to form a bigger picture. The map suggests alternative places to visit and routes to take. You can then program these places and routes into your GPS receiver.

Keeping a notebook

We've already discussed how the small size and poor resolution of display screens on handheld GPS receivers can create problems. One difficulty is that the displayed terms, graphics, and messages must be abbreviated. Most GPS units will, by default, code waypoints sequentially by number (001, 002, 003, and so on). Having hundreds of numerical identifiers in your receiver's memory is a mental challenge when you want to select a destination or create a route.

Fortunately, receivers have a provision for manually entering a short waypoint label in place of the default numerical identifier. A label such as "TRAILHD," "CAMP," or "PARK" is easier to use than numbered identifiers, but you soon run out of meaningful short words and recognizable abbreviations. There is no space to say anything significant about a waypoint that you want to recall later, and keying in multiple characters is a laborious process.

You can get around the waypoint entry and recall problems by annotating your map with the proper numerical identifier each time you enter one into your receiver's memory. This can be quite a chore, especially in the field, and it makes a mess of your map if you use it for many trips. Furthermore, you would have to add still more annotations to your map to capture the character of the landscape along each leg of your trip.

A better way to augment waypoint identifiers is to keep a notebook or log. In a notebook, you can expand on each short waypoint identifier to whatever degree you want. You can also note what terrain features to watch for as you approach waypoints. If nothing else, your notebook makes good reading later when you want to daydream about past trips.

Having your compass with you

We have stressed how maps can help overcome some of the limitations of GPS technology. The same can be said of the magnetic compass. Your GPS receiver cannot replace your compass. You need both.

True, a GPS receiver can do some of the things compasses do. For instance, your GPS receiver can give you the direction to a distant point for which it has coordinates. This function mimics compass use. If you're moving along at a brisk walk or faster, it can give you your direction of travel. Again, this mirrors using a compass.

But if you aren't moving, your GPS receiver's likeness to a compass breaks down if the receiver does not have a built-in electronic compass or INS capability. Under these conditions, it doesn't know which way is north or which way you are facing. And remember, it can't indicate direction if you are not moving. This is a concern because it makes it difficult to keep your map oriented if you're traveling slowly. It can also be a problem with faster movement over a path that keeps changing direction.

An even greater difference between GPS and compass technology is that GPS uses satellite signals that can be easily blocked. Terrain, buildings, vegetation, and even a boat sail may block line-of-sight satellite signals that are crucial in determining a fix. Multipathed signals that reflect off nearby objects can confuse the GPS receiver and lead to degraded position calculations.

Other things can happen to the satellite signals as well. Problems can occur with signal transmission at the satellite, resulting in a weak signal or none at all. Satellites must be a certain distance above the horizon or their signals will be weak or blocked. The geometrical configuration of the satellites has an impact on the reliability of position computations. Sometimes you can do little but wait for the satellites to move when you find yourself in situations in which they are all positioned near each other, nearly all are directly overhead, some are low on the horizon and could be blocked by trees or terrain, or critical ones are behind obstructions such as trees or buildings.

Problems will more often occur with your receiver. A GPS receiver needs a power supply to run its internal computer. For GPS units used in planes, boats, and automobiles, the vehicle's power source will supply the GPS unit. For most other units, the power supply is usually a battery. We all know that batteries lose power with time, use, and cold temperatures. Since a loss of power will cause the receiver to fail, the batteries must be replaced or recharged after use. The best way to conserve batteries is to rely primarily on your map and compass and use your GPS receiver only intermittently to check your progress and make navigation adjustments.

In contrast, a magnetic compass uses the earth's magnetic field as its energy source and signal. No electrical power is needed, and magnetic energy is neither blocked nor reflected by landscape features.

GPS and magnetic compass technologies are complementary—one enhances the other. When your GPS receiver fails, is lost, or is damaged, your compass could be life-saving. Carrying both devices is a form of insurance and risk management.

A warning to GPS novices

You may know exactly where you are and where you want to go but still be lost. Your GPS receiver will tell you your location and the direction and distance to your desired destination. It will do this every few seconds until it loses power. But this information is of little use if you can't move (because of an injury, bad weather, or other calamity) and have no way to communicate your location to a rescuer. Before you leave home, it's a good idea to arm yourself with a backup device that can let someone know where to find you.

In many parts of the world, a battery-powered cell phone may get your message out. Using this ground-based phone system means you must be located within a region that is served by a transmitter or receiver. The system has the advantage that you can call an existing telephone number, and someone can call you back if you leave a message. The disadvantage of cellular technology is that transmitters are found only in densely populated areas and along major travel routes. The prudent GPS user will allow for these contingencies and always have a backup plan in place.

SELECTED READINGS

Ackroyd, N., and R. Lorimer. 1994. *Global navigation: A GPS user's guide*. 2d ed. London: Lloyd's of London Press, Ltd.

Clarke, B. 1998. *Aviators' guide to GPS*. 3rd ed. New York: McGraw-Hill Professional.

Grubbs, B. 2005. *Basic essentials using GPS*. 2d ed. Guilford, Conn: Falcon.

Hinch, S. W. 2004. *Outdoor navigation with GPS*. Santa Rosa, Calif: Annadel Press.

Hotchkiss, N. J. 1999. *A comprehensive guide to land navigation with GPS*. 3d ed. Herndon, Va.: Alexis Publishing.

Hurn, Jeff. 1989. *GPS: A Guide to the next utility*. Sunnyvale, Calif.: Trimble Navigation.

——. 1993. *Differential GPS explained*. Sunnyvale, Calif.: Trimble.

Kaplan, E., and C. Hegarty. 2006. *Understanding GPS: Principles and applications*. 2d ed. Norwood, Mass: Artech House.

Kennedy, M. 2002. *The global positioning system and GIS*. 2d ed. Boca Raton, Fla.: CRC Press,

Leick, A. 2004. *GPS satellite surveying*. 3d ed. New York: John Wiley & Sons, Inc.

Letham, L. 2003. *GPS made easy: Using global positioning systems in the outdoors*. 4th ed. Seattle, Wash.: The Mountaineers.

Sweet, R. J. 2003. *GPS for mariners*. New York: McGraw-Hill.

Van Sickle, J. 2001. *GPS for land surveyors*. 2d ed. Ann Arbor, Mich.: Ann Arbor Press, Inc.

chapter
fifteen   **AREA AND VOLUME MEASURES**

15

Area and volume measures

Estimating the sizes of features in our environment is a basic mental activity important to our survival. We subconsciously estimate the relative sizes of objects as part of our visual perception of the world, but we can also numerically determine the *areas* of two-dimensional features. Knowing the area of a land parcel, for example, is critical in a real estate transaction and in assessing property taxes. The areas of land parcels can be computed directly from the land survey description for the parcel, but it is often easier to measure the area from a large-scale map or compute the area from grid coordinates read from the map.

The same can be said for determining the *volume* of a feature that has depth or height in addition to area. The volume of earth to remove in a construction project, for example, is an important piece of information for civil engineers. Again, although volumes can be computed directly from surveyed depths and heights, it is usually easier to determine volumes from the elevation and water depth information shown on large-scale topographic maps and nautical charts.

Another characteristic of a feature is its center of area, or *centroid.* This is more often determined for two-dimensional features, although it could also be determined for three-dimensional features. A lot of information about a feature can be obtained from knowing its centroid. For example, the centroid of a feature is the purest example of that feature. The trees at the centroid of a forest stand, for instance, may best represent the entire stand. Cultural features are located at or near centroids. For example, a marker is located at the center of the 48 conterminous United States (figure 15.1). Certain phenomena tend to disperse outward from a feature's centroid, such as disease epidemics.

Measuring the *shape* of a feature may at first seem difficult to do and of little practical value. But shape descriptions are important in a variety of activities. Legislative redistricting is a classic example; in political theory, a prime characteristic of districts is their spatial compactness. The original "gerrymander" was created in 1812 by Massachusetts governor Elbridge Gerry, who for political purposes created a long, narrow, salamander-shaped district.

Let's look at the methods that we can use to measure areas, volumes, and shapes of features and to determine their centroids from maps. We'll begin with the measurement of area.

AREA

The size of a two-dimensional figure or region is called its **area.** When you want to know the area that lies within a region's boundaries, you can measure the area in several ways. Visual area estimation will suffice if you need only a rough approximation. But if you need a more precise numerical value, you should rely on more rigorous measurement methods. Both visual methods and more precise calculation methods are covered here.

Counting methods

A number of methods can be used to determine a feature's area. Because of its simplicity, the most common technique for computing area is a visual method that involves **grid cell counting.** A variation of this method involves counting dots, although the idea is similar in both cases. A third method uses strips instead to determine the areal extent of features.

Grid cell counting method

One procedure is to superimpose a grid of small evenly spaced squares on the map (figure 15.2). To determine the area of a feature such as a lake, you tally the grid cells that fall fully within its boundary, and then total the cells that fall partially within. You then divide the number of partial cells by two and add this value to the number of whole cells. Finally, you multiply the resulting sum by the ground area covered by each cell.

Finding the ground area covered by a cell is easy if you know the cell dimensions and the map scale. Say you have 0.1-inch-by-0.1-inch cells on a 1:24,000-scale topographic map. Since this scale also can be stated as 1 inch to 2,000 feet (see chapter 2 for more on map scale), a grid cell must be 200 feet by 200 feet on the ground, or 40,000 square feet. You may want to convert square feet to acres, which is easy since an acre[1] is defined as a 208.7-foot-by-208.7-foot square, or 43,560 square feet. Each of your grid cells is thus 40,000 square feet divided by 43,560 square feet/acre, or 0.918 acres on the ground.

Another calculation method is to find the map area of the feature, then use the map's *representative fraction (RF)* to convert to ground area (see chapter 2 for more on representative fractions). In the previous example, each grid square has an area of 0.1 inch by 0.1 inch, or 0.01 square inches. If a feature measured 86.5 grid cells, its map area must be 86.5 cells × 0.01 square inches per cell, or 0.865 square inches. To convert map area to ground area, you multiply by the square of the RF. Therefore, 0.865 square inches on a 1:24,000-scale map converts to 0.865 square inches × 24,000[2] = 0.865 ×

Figure 15.1 A stone monument with a brass plaque in central Kansas installed in 1940 before Alaska and Hawaii joined the union claims this site as "the Geographic Center of the United States."

576,000,000, which equals 498,240,000 square inches on the ground. You will probably convert square inches to square feet, knowing that there are 12 inches by 12 inches or 144 square inches per square foot. Your ground area in square feet is thus 498,240,000 square inches ÷ 144 square inches/square foot, or 3,460,000 square feet, which further converts to 79.3 acres (dividing by 43,560 square feet/acre).

You can take advantage of electronic image scanning and image processing technology to increase the speed and accuracy of grid cell counting. Such technology is especially helpful if you're measuring features with intricate boundaries, or if you have a large number of features or a large area to measure, especially if you are using a fine-resolution grid. Some high-end geographic information systems (GIS) also include image analysis functions, so you could use software such as ArcGIS for this. The data used in digital image processing are stored as a grid of square cells called **pixels.** Area measurement is simply a matter of pixel counting, and most GIS or digital image processing systems can tell you the number of pixels on the image that are within a certain area.

To use the grid cell counting method, the first step is to electronically scan the map containing the features to be measured into a grid so that the pixels can be counted. You may be able to acquire from a Web site or CD-ROM previously scanned maps, such as U.S. Geological Survey (USGS) **Digital Raster Graphics (DRGs)**, which are scanned 1:24,000-scale topographic maps (see appendix A for details). Otherwise, you can scan the portion of the map containing the features of interest using your desktop scanner. Current desktop scanners can scan up to 8.5-inch-by-14-inch color images at spatial resolutions ranging anywhere from 72 to 4,800 dots (pixels) per inch (dpi). Lower-resolution scans are sufficient for measuring the area of all but the smallest features.

You must be able to identify on your computer monitor the features on the scanned image that you want to measure. You can do this electronically by identifying the features of interest visually and digitizing them by hand using your computer's mouse as the tracing tool. Next, you assign the area you digitized a unique color that doesn't appear anywhere else on the image. When this is done, area computation is easy—just query the software for the number of pixels of this color (30,220 in the 1:24,000-scale topographic map example in figure 15.3), then multiply by the ground area covered by a pixel. The easiest way to find the ground area of a pixel is by counting the number of pixels along a vertical or horizontal line of known ground length. In figure 15.3, for instance, 84 pixels were found to span the lake where it was 300 meters wide. Then you simply divide the length of the horizontal line by the number of cells to get the size of one side of a pixel. The **unit area value** (the ground area per cell) is equal to that value squared (assuming the pixels are squares and not rectangles).

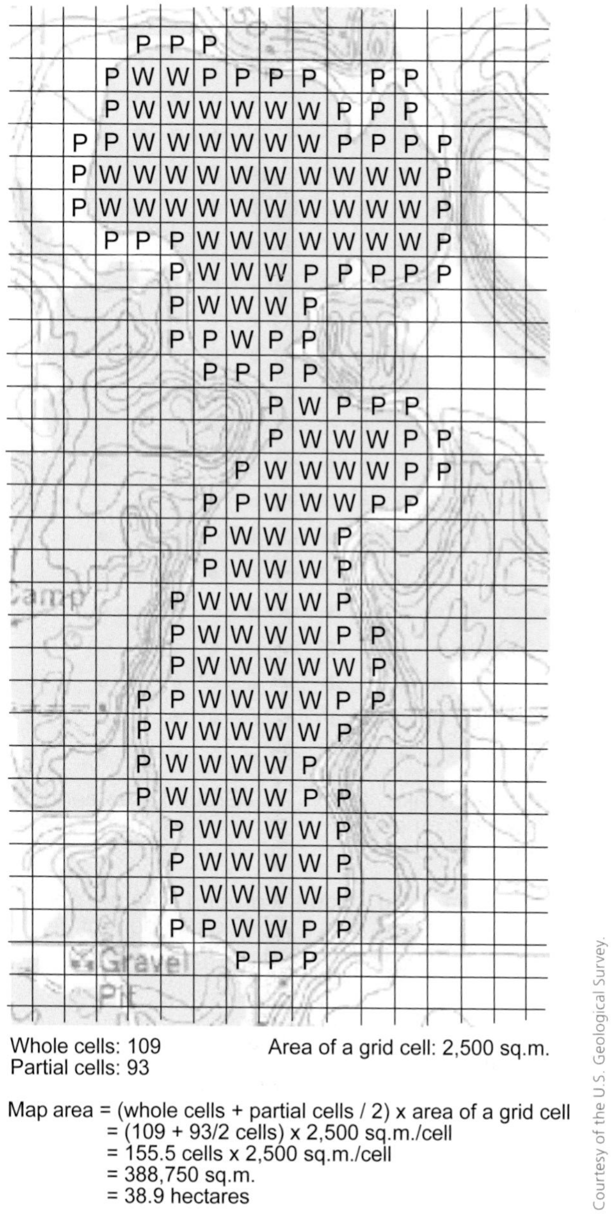

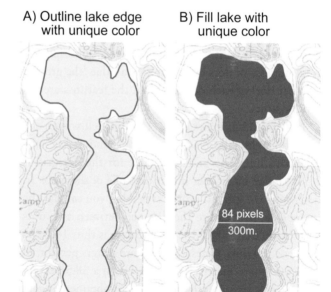

A) Outline lake edge with unique color

B) Fill lake with unique color

84 pixels
300m.

C) Use software to find number of pixels with unique color: 30,220 pixels

D) Calculate ground area pixels
84 pixels = 300 meters, 1 pixel = 3.57 meters on a side, and 3.57^2 = 12.75 square meters in area

E) Multiply number of pixels by ground area/pixel
30,220 pixels x 12.75 sq.m. per pixel = 385,500 sq.m. or 3.85 hectares

Courtesy of the U.S. Geological Survey.

Whole cells: 109
Partial cells: 93

Area of a grid cell: 2,500 sq.m.

Map area = (whole cells + partial cells / 2) x area of a grid cell
= (109 + 93/2 cells) x 2,500 sq.m./cell
= 155.5 cells x 2,500 sq.m./cell
= 388,750 sq.m.
= 38.9 hectares

Courtesy of the U.S. Geological Survey.

Figure 15.2 The surface area of this lake in ground units can be measured by summing the number of grid cells that fall completely within its boundary and half of the number of grid cells that fall partially within its boundary, then multiplying by the area of the grid cell in ground units (2,500 square meters).

Figure 15.3 Desktop scanners and digital image processing software can be used to generate data for area computation by grid cell counting, since each pixel in the data is a grid cell.

Notice that the digital grid cell counting method using pixels has an all-or-nothing character. You have feature pixels and nonfeature pixels, but no in-between pixels along boundaries as with manual grid cell counting. In theory, the partial grid cells falling on the edges of the feature accounted for in manual counting methods should produce a more accurate result. In practice, however, the coarse grids in manual grid cell counting are actually less accurate than automated higher-resolution pixel counting.

Dot grid counting method

The second visual measurement method is similar to the grid cell counting method, but uses a grid of dots placed over the map instead of a grid of cells. A **dot grid** is nothing more than a square grid of alternating solid and open dots (figure 15.4). To measure an area, you count all the dots that fall entirely within the feature's boundary. You then add the number of either open or solid dots falling on the feature's boundary with the total of those falling entirely within the boundary. Next, multiply the total number of dots by the unit area value (the ground area per dot) of each dot to obtain the feature's area in ground units.

In figure 15.4, for example, 146 dots fell within the lake and eight solid black dots were found on the lake boundary line. The unit area value for the dot grid is 2,500 square meters per dot at the scale of the map. To determine the ground unit area per dot, you can use the map's scale to measure the distance between two dots and square that value. The ground area is thus 146 plus eight, or 154 dots, times 2,500 square meters per dot, or 385,000 square meters. This converts to a 38.5-hectare lake, slightly less than the 38.9 hectares obtained by grid cell counting in figure 15.2.

Measurement accuracy

With either of these visual counting methods, smaller grid cell sizes and more closely spaced dots improve the accuracy of the area computation. But this greater accuracy is attained at the expense of more tedious cell or dot counting. Thus, you should choose a cell size or dot unit value based on how much accuracy you need (see chapter 10 for more on precision).

Exactly what grid cell width or dot grid spacing should you use? There are no strict rules for these choices, but a few researchers have looked at the problem both from a theoretical viewpoint and by experimenting with a variety of feature sizes and shapes. The first thing this research has shown is that repeating the counting procedure with the grid shifted and rotated differently between each

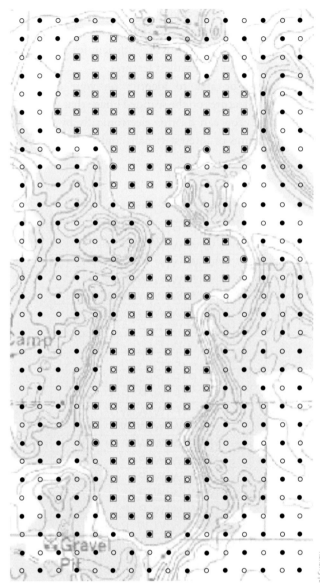

⊡ Dots inside lake: 146
⦿ Black dots on edge: 8
   Total dots: 154

Ground area per dot: 2,500 sq.m.

Area = No. of dots x ground area per dot
= 154 dots x 2,500 sq.m. per dot
= 385,000 sq.m.
= 38.5 hectares

Figure 15.4 **To estimate the area of a feature using a dot grid, count the number of dots within the feature and the number of either solid or open dots on its boundary, sum them and multiply by the ground area per dot.**

repetition, followed by averaging the measured areas, gives a result closer to the true feature area than the area obtained by a single counting. This makes sense, because at certain grid orientation angles there could be rows of dots aligned parallel to one or more edges of the feature. In these situations, slightly shifting the grid could cause a large number of grid cells or dots to shift from inside to outside the feature. If you average the area totals obtained by repeating the cell or dot counting at different grid orientations, you'll minimize this type of counting error.

The relationship between the density of dots on a grid, defined by the spacing between dots, and measurement error has also been studied for features of different shapes and areas varying from 1 to 100 square centimeters. The curves in figure 15.5 give you a feel of what to expect. The graph shows that, regardless of the dot spacing, measurement error decreases as the area of the feature being measured increases. However, the rate of decrease lessens with increasing feature area, so that using a dot grid with more closely spaced dots would minimally improve measurement accuracy for large features.

This graph also shows you the dot spacing needed to measure features of different approximate sizes with a certain allowable measurement error. For instance, to measure an approximately 10-square-centimeter feature with an allowable error of 1.5 percent requires a dot grid with 0.2-centimeter dot spacing. Table 15.1, derived from the curves in figure 15.5, gives the maximum dot

spacing for several feature area size ranges if the allowable measurement error is 1.5 percent.

Feature size range	Maximum dot spacing
< 4 centimeters^2	0.1 centimeter
4 to16 centimeters^2	0.2 centimeter
16 to 36 centimeters^2	0.4 centimeter
36 to 64 centimeters^2	0.6 centimeter
64 to 100 centimeters^2	0.8 centimeter
>= 100 centimeters^2	1.0 centimeter

Table 15.1   Dot spacing required to measure features in different size ranges with an allowable error of 1.5 percent.

The recommended maximum dot spacing values tell you that there should at least 100 dots falling within the feature, regardless of its approximate size, if you want the measured area to be within 1.5 percent of the true area. To have 100 dots within all features, you will have to use several dot grids of varying densities for different-sized features, such as the six grids suggested in table 15.1.

These guidelines also can be extended to grid cell counting, as the appropriate grid cell width and height for each feature size range should be the same as the maximum dot spacing in the table. The general rule, then, is that at least 100 grid cells must fall within each feature. You can also keep these guidelines in mind as we discuss the third counting method, which uses strips rather than cells or dots.

Strip method

If the feature has an irregular or curved border, you can use a modification of the grid cell counting method called the **strip method** (figure 15.6). With this technique, the first step is to position a series of equally spaced parallel lines that define narrow strips spanning the feature. The strips should entirely cover the extent of the feature from top to bottom. The easiest way to begin drawing the strips is to determine the width of a strip using the map's scale and, starting at either the top or the bottom, place it so that the furthest extent of the feature falls midway inside the strip. Then draw the rest of the strips the same width as the first strip. Be sure they extend midway outside the feature on both ends.

Next, draw vertical lines to define the ends of each strip. These vertical lines should be drawn where the boundary crosses the strip so that areas falling inside and outside of the feature are approximately equal (see illustrative diagram in figure 15.6).

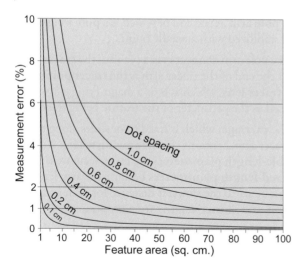

Figure 15.5   Dot grid measurement error (as a percentage of the total feature area) decreases exponentially as the size of the feature increases, regardless of the dot spacing on the grid. For any feature size, a point of diminishing returns is reached when further reducing the dot spacing only marginally improves measurement accuracy.

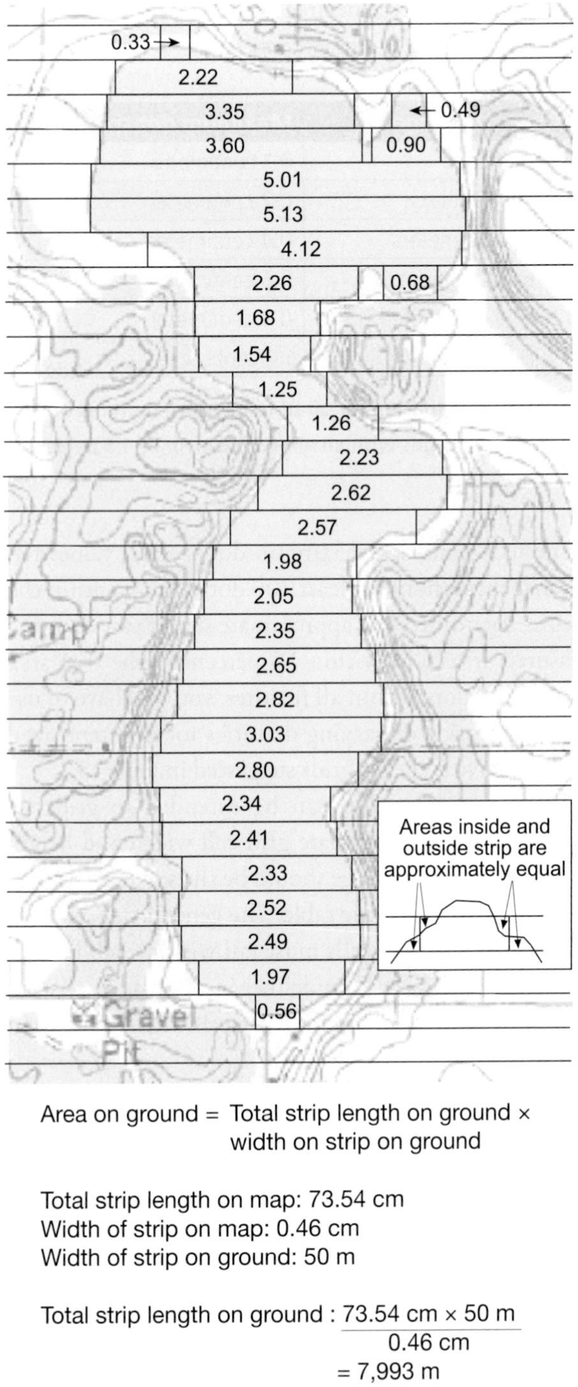

Area on ground = Total strip length on ground ×
width on strip on ground

Total strip length on map: 73.54 cm
Width of strip on map: 0.46 cm
Width of strip on ground: 50 m

Total strip length on ground : $\dfrac{73.54 \text{ cm} \times 50 \text{ m}}{0.46 \text{ cm}}$
= 7,993 m

Lake area = 7,993 m × 50 m
= 399,700 sq.mi.
= 39.97 hectares

Figure 15.6 The strip method is a less time-consuming way to compute the area of irregular features.

Courtesy of the U.S. Geological Survey

For very irregular features, there will probably be more than one set of perpendicular lines for many of the strips as the boundary weaves in and out of the extent of the strip. Measure the length of the each strip between the perpendicular lines, and sum these lengths.

By multiplying the total strip segment length by the strip width, you will have the total strip area of the feature. Using the unit area value determined by the map's scale, you can convert the strip area on the map to ground area. Obviously, the narrower the strips, the greater the accuracy of your results, but also the greater the computational effort.

Polar planimeter

In previous decades, a mechanical instrument called a **polar planimeter** (figure 15.7) was the tool of choice for area measurement on maps and aerial photographs. (The term "polar planimeter" derives from tracing the figure without moving the pole—which is placed outside of the feature—and "planimeter," which relates to measurement of the area of a plane shape). A polar planimeter is a mechanical device that measures area through the use of an integrating wheel. Polar planimeters, though much slower and somewhat less accurate than modern digital measurement instruments, are still widely available and used because they are fast, efficient, and especially useful for measuring small or irregularly shaped features. The accuracy is comparable to visual counting methods (0.5 percent), and digital models are even more accurate.

All planimeters have some basic features in common:

1. A **pole arm** at the end of which is a **pole weight** that is stabilized with a needle point.

2. The pole arm connects via a pivot to the **tracer arm.** At the end of the tracer arm is the **tracer point** with a **tracer lens**, which acts as a magnifying glass that is used as the perimeter of the feature being traced.

3. The **carriage,** which is fixed to or slides along the tracer arm, contains the **measuring wheel. Adjustable-length planimeters** are more accurate than **fixed-length planimeters** because the arm length can be set to the shortest distance for each application, and shorter lengths are more accurate.

4. The results are displayed on **counting dials** (also called **recording dials**). One counting dial records the number of whole and partial revolutions that the measuring wheel makes down to tens of square inches. The other is the **Vernier scale**, which allows the partial revolutions to be determined with greater precision down to $1/100$ th of a square inch. The Vernier scale (also called a **Vernier unit**) is a graduated scale calibrated to indicate fractional parts of the subdivisions of the larger scale.

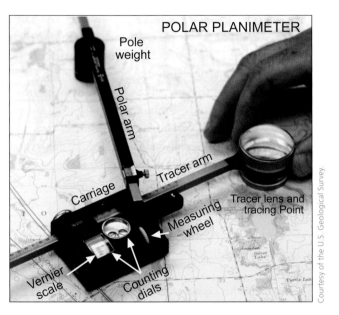

Figure 15.7   **Use of a polar planimeter for computing the area of a mapped region requires tracing the region's outline with the instrument's cursor and converting the resulting map area reading to ground units.**

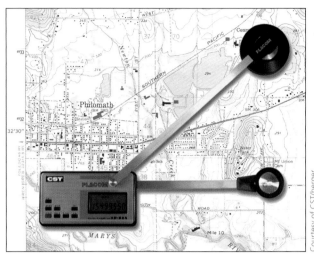

Figure 15.8   **A digital polar planimeter automatically converts areas to ground units, which are displayed on a small screen.**

You begin by setting up the polar planimeter with the pole weight at the end of the polar arm anchored securely off to the side of the area feature to be measured. You should be able to move the tracing point in the center of the tracer lens attached to the end of the tracer arm freely around the feature boundary. If you have an adjustable-length planimeter, set the tracer arm length to the shortest possible length that covers the area you want to measure. Set the counting dial and Vernier scale to zero, mark a starting point on the feature boundary, then trace the outline of the feature in a clockwise direction. When you come back to the starting point, the counting dial and Vernier scale will tell you the area you've traced. It's a good idea to trace the outline several times to obtain an average reading.

Most polar planimeters give you the area in square centimeters or square inches. This means your final step when using these instruments will be to convert this value into ground units, using the square of the map scale as explained previously.

Digital polar planimeters (figure 15.8) are computerized and require initial settings for units and scale. They show on a digital display the area traced in square inches or centimeters, and on some models you can key in the ground units that you want the results to be converted to automatically (acres, square meters, square kilometers, etc.). Most digital polar planimeters have various memory functions that allow you to add

areas, accumulate measurements, and average multiple measurements. With the most advanced units, you can download the data into a personal computer. The accuracy of these modern digital polar planimeter models is higher than the manual ones, about 0.2 percent.

Coordinate methods

There are two coordinate methods you can use to calculate area—the one you use will depends on whether you are measuring area with planar or spherical coordinates. Boundary points for smaller features, such as tracts of land that have small differences in latitude and longitude, should be calculated using the planar grid coordinate method. The spherical coordinate method should be used for larger area features spanning more than one grid coordinate zone. A good example is calculating the areas of countries or large states.

Planimetric coordinate methods

If you have the *x,y* grid coordinates for the points defining the boundary of a tract of land, you can use this information to calculate the area of the tract. Area computation as performed using ArcGIS software and GIS is based on using grid coordinates of area feature boundaries (see chapter 4 for more on grid coordinate systems).

The **planimetric coordinate method** of area measurement is based on finding the areas of trapezoids formed by drawing horizontal lines from the boundary points to the vertical y-axis. In figure 15.9, for example, the area of tract ABCDEF is determined from the grid coordinates $(x_1,y_1)...(x_5,y_5)$ for points P_1 through P_5. To compute the area of the tract, five trapezoids must be formed. Trapezoid 1 begins at the intersection of the horizontal line from P_1 and the y-axis. Its sides are the segment of the y-axis from this point to the horizontal line from point P_2, then the horizontal line to P_2, the line segment from P_2 to P_1, and finally the horizontal line from P_1 to the y-axis. The other four trapezoids are constructed in the same manner from the remaining points that define the perimeter of the tract.

The area of each trapezoid is computed by multiplying the difference in the y-coordinates for the first and second point by the average of the x-coordinates for these two points:

$$\text{area}_{\text{trapezoid}} = \frac{(x_i + x_{i+1})}{2} \times (y_i - y_{i+1})$$

Notice that the trapezoid area will be negative when the second y-coordinate is greater than the first. Having positive and negative trapezoid areas is crucial to computing the total tract area. Figure 15.9 shows why. The negative area trapezoids are those between the y-axis and the left side of the tract (3, 4, and 5 in this example), whereas the positive area trapezoids (1 and 2) extend from the y-axis to the right side of the tract. Thus, summing the positive and negative trapezoids eliminates the outside area from the y-axis to the tract, leaving the area inside the tract.

By combining the individual trapezoid area equations, the following equation is obtained for the area of the tract:

$$Tract\ area = 0.5 \times ((x_1+x_2)\times(y_1-y_2) + (x_2+x_3)\times(y_2-y_3) + (x_3+x_4)\times(y_3-y_4) + (x_4+x_5)\times(y_4-y_5) + (x_5+x_1)\times(y_5-y_1))$$

The terms in this equation can be rearranged so that the equation can be extended to any number of boundary points:

$$Tract\ area = 0.5 \times (y_1(x_n-x_2) + y_2(x_1-x_3) + ... + y_n(x_{n-1}-x_1)),$$

where n is the number of boundary points. This equation is stated in words as the following:

To determine the area of a tract of land when the coordinates of its corners are known, multiply the ordinate (y-coordinate) of each corner by the difference between the abscissa (x-coordinate) of the following and preceding corners, always subtracting the following corner from the preceding corner. The area of the tract is one half of the sum of the resulting products.

Spherical coordinate methods

The **spherical coordinate method** for area measurement is similar to the planimetric method but it is based on the use of geographic coordinates (longitude, latitude) in area calculations of spherical triangles instead of trapezoids. In figure 15.10, you can see that the north (or south) pole and two adjacent points along the feature's boundary comprise each spherical triangle. Two of the edges are used to determine the angular distances from the pole to the boundary point along meridians (called the **colatitudes** of the points). The angle between these edges is the difference in longitude between the two boundary points. Triangle areas are positive or negative depending on whether the longitude of the second boundary point is greater or less than the longitude of the first point.

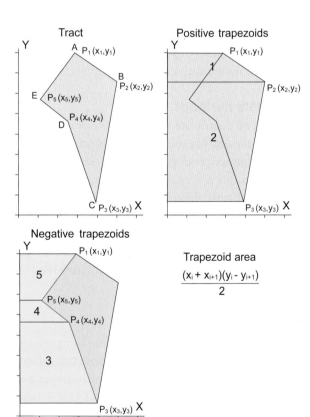

Figure 15.9 The coordinate method to determine the area of a tract of land or any other area feature uses the boundary coordinates.

Geographical region

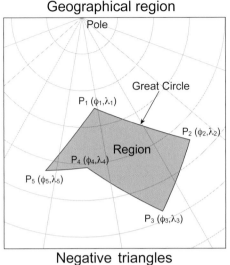

Positive triangles

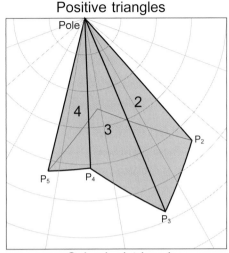

Figure 15.10 Geographic coordinates (latitude and longitude) can be used in area calculations of spherical triangles instead of trapezoids.

Negative triangles

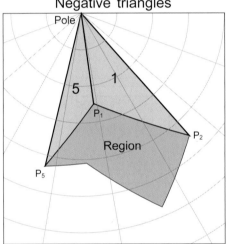

Spherical triangle

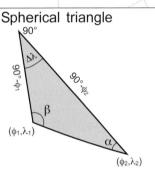

The most commonly used spherical triangle area equation used is the following:

$$Area = (\alpha + \beta + \Delta\lambda - 180°) \times R^2,$$

where R is the earth's radius (6,371 kilometers) and $\Delta\lambda$ is the absolute value of the difference in longitude between the two points. The other two interior angles, α and β, are computed from a spherical trigonometry equation that is one of what are called **Napier's analogies** (a set of trigonometric formulas useful in the solution of oblique spherical triangles):

$$\tan\left(\frac{\alpha + \beta}{2}\right) = \frac{\cos\left(0.5 \times \left((90° - \phi_1) - (90° - \phi_2)\right)\right)}{\cos\left(0.5 \times \left((90° - \phi_1) + (90° - \phi_2)\right)\right) \times \tan(\Delta\lambda/2)}$$

$$\alpha + \beta = 2 \times \tan^{-1}\left(\frac{\cos\left(0.5 \times (\phi_2 - \phi_1)\right)}{\cos\left(0.5 \times (180° - (\phi_1 + \phi_2))\right) \times \tan(\Delta\lambda/2)}\right)$$

Although computing trapezoid or spherical triangle areas is very tedious if done by hand, it is a simple matter for a computer program to calculate areas using either coordinate method. Such a program will quickly, accurately, and effortlessly compute the area of any feature. The computer-based coordinate approach is particularly appropriate when the tract's boundary is complex or involves a large number of boundary points. Increasing the number of boundary points along complex boundaries improves the accuracy of area estimation with a minimal increase in computation time—this is a function that can be found in many GIS. However, increasing the density of boundary points along nearly straight edges does little to improve the accuracy of area computation. By establishing boundary points only at those positions on the boundary where major angular changes occur, you can approximate the figure's area most efficiently.

Irregular surface area

Area measurements and computations so far have assumed that the earth's surface is flat. These **planimetric surface areas** are the values found in most area calculation reports and GIS databases. But there are times when you may want to take the slope of the landscape into account and compute the **irregular surface area** of a feature. Wildlife biologists, for example, may use the irregular surface area as a better estimate of an animal's habitat, and physical geographers use the ratio of irregular surface to planimetric area as a measure of landscape roughness.

Let's begin by seeing how differences in slope increase the surface area of an initially flat 100-meter-by-100-square-meter land parcel with a planimetric area of 1 hectare (figure 15.11). Notice that as the parcel's slope increases from 10 to 100 percent (see chapter 16 for more on slope percentage), the ground area of the sloping surface increases in proportion to the ground distance values (the values in figure 15.11 also relate to the slope error examples that we illustrated in figure 11.9). You can see that for slopes of 10 percent or less the difference between the planimetric and actual ground area is very small—for a 10 percent slope, the parcel is actually 1.005 hectares, not 1.0 as measured on the map. Only for steeper slopes of around 30 percent or greater is the actual ground area noticeably larger than the planimetric area. Slopes this steep are found in hilly areas of rugged terrain.

The calculation of irregular surface area is commonly done using triangles positioned on maps of the landscape, such as the black triangle placed on the hillside shown by the contours in figure 15.12. You need to find the grid coordinates of the three triangle vertices, and then examine the contours to estimate the elevation of each vertex (converted from feet to meters in this example). We made these measurements manually (see chapter 4 for ways to measure grid coordinates), but you can also use the terrain analysis module in a GIS to interactively select the triangle vertices on a digital map that has a digital elevation model (DEM) as a data layer and have the software compute the horizontal coordinates and elevations of each point.

To compute the irregular surface area of a larger region, you will need to find the grid coordinates and elevations

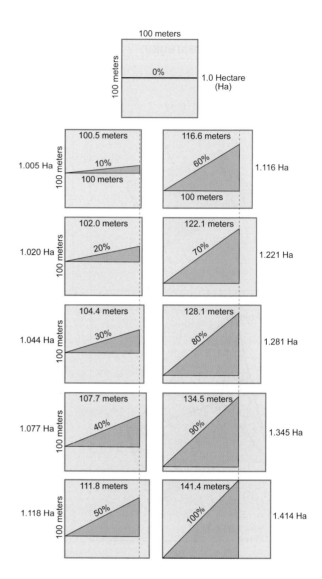

Figure 15.11 The actual ground area of a land parcel with a 1-hectare planimetric area increases in proportion to the ground distance along the sloping surface.

for hundreds of triangles connected together to form what is called a **triangulated irregular network (TIN)**. A TIN is a representation of the land surface made up of irregularly distributed points with three-dimensional coordinates (x, y, and z) and lines that connect them to create a network of nonoverlapping triangles. You can create a TIN, such as that shown by the gray triangle edges in figure 15.12, in a GIS from the digitized contours or the DEM for the region. Or you can download TINs from a variety of sources on the Internet or purchase them for many locales from a number of commercial vendors.

An easy way to compute the area of a triangle on an irregular surface is from the lengths of its sides (a, b, and c in figure 15.12). The length computation is an extension

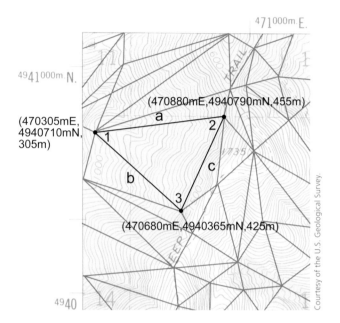

Figure 15.12   Facets or triangles that form a triangulated irregular network (TIN) can be used to compute the ground area of an irregular surface if the grid coordinates and elevations of their vertices are known.

to three dimensions of the distance equation for grid coordinates that we explained in chapter 4. The straight-line distance d between the two vertices (1 and 2) of the triangle edge is found from the eastings, northings, and elevations of the vertices from the equation (see chapter 4 for more on eastings and northings):

$$d_{1-2} = \sqrt{(\text{Easting}_2 - \text{Easting}_1)^2 + (\text{Northing}_2 - \text{Northing}_1)^2 + (\text{elevation}_2 - \text{elevation}_1)^2}$$

Let's apply this equation to the UTM coordinates and elevations in meters for sides a, b, and c (vertices 1, 2, and 3) of our triangle:

$$d_{1-2} = \sqrt{(470{,}680 - 470{,}305)^2 + (4{,}940{,}710)^2 + (455 - 305)^2}$$
$$= \sqrt{(575)^2 + (80)^2 + (150)^2}$$
$$= \sqrt{330{,}625 + 6{,}400 + 22{,}500}$$
$$= \sqrt{359{,}525} = 599.6 \text{ meters}$$

$$d_{1-3} = \sqrt{(470{,}680 - 470{,}305)^2 + (4{,}940{,}365 - 4{,}940{,}710)^2 + (455 - 305)^2}$$
$$= \sqrt{(375)^2 + (-345)^2 + (120)^2}$$
$$= \sqrt{140{,}625 + 119{,}025 + 14{,}400}$$
$$= \sqrt{274{,}050} = 523.5 \text{ meters}$$

$$d_{2-3} = \sqrt{(470{,}680 - 470{,}880)^2 + (4{,}940{,}365 - 4{,}940{,}790)^2 + (425 - 455)^2}$$
$$= \sqrt{(-200)^2 + (-425)^2 + (-30)^2}$$
$$= \sqrt{40{,}000 + 180{,}625 + 900}$$
$$= \sqrt{221{,}525} = 470.7 \text{ meters}$$

Having computed the lengths of triangle sides a, b, and c, we can find the triangle area from **Heron's formula** (attributed to Heron of Alexandria, ca. 60 AD): $Area = \sqrt{s(s-a)(s-b)(s-c)}$ where s is the **semiperimeter** of the triangle, or half the perimeter of the triangle, found by $(a + b + c)/2$.

The semiperimeter of our triangle is (599.6 + 523.5 + 470.7)/2, or 796.9 meters, so its area can be calculated like this:

$$\sqrt{796.9(796.9 - 599.6)(796.9 - 523.5)(796.9 - 470.7)}$$
$$= \sqrt{796.9 \times 197.3 \times 273.4 \times 326.2} = 118{,}415 \text{ square meters.}$$

How does this value compare with the planimetric area we would compute using the distance equation in chapter 4 to find the side lengths from the eastings and northings alone? You can verify that the distance equation gives lengths of 580.5, 509.6, and 469.7 meters for triangle sides a, b, and c, and that Heron's formula will give an area of 114,189 square meters for the triangle. The triangle appears to be on the side of a moderately steep hill, yet its ground area is only 3.7 percent larger than its planimetric area.

Computing the area of a topographically irregular surface from coordinates stored in a computer database using the three-dimensional form of the distance equation and Heron's formula is a straightforward procedure applied widely in GIS terrain analysis software. The accuracy of the areas computed depends on how well each vertex is chosen to capture places where the surface flexes (markedly changes direction) and how uniform the sloping surface is within each triangle. The contours within our triangle example portray a uniformly sloping hillside, with the exception of a small ridge near vertex 1. Other triangles may cross ridgelines or valleys if the TIN is constructed poorly. These topographic variations within the triangles of a TIN cause the computed area to be less than the actual ground area that we would find if we divided the triangle into a number of smaller triangles that more accurately reflected the surface of the terrain.

Area measurements on small-scale maps

Counting methods or the use of a polar planimeter to measure the areas of large geographic features on small-scale maps is problematic due to the geometric distortion of the earth's spherical geometry introduced by the map projection. Counting or polar planimeter methods will give accurate areas only if the features are drawn on an equal area map projection (see chapter 3 for more on these map projections). An equal area map projection correctly portrays the relative areas of features on the earth's surface, but how do we convert map area to ground area? On large-scale maps, you only need to multiply by the square of the map's RF, since the scale is essentially constant across the map. But on small-scale maps, the RF printed in the legend is only true at the centerpoint or along the standard parallel or parallels. The Albers conic equal area projection covering the lower 48 U.S. states, for example, has standard parallels at 29.5°N and 45.5°N. The scale at the U.S.–Canadian border (49°N) and the southern tip of Florida is 1.25 percent larger in the north–south direction and 1.25 percent smaller in the east–west direction. This variation in scale is large enough that we must find another way to accurately convert map area to ground area.

The way to obtain accurate ground areas in this case is to measure the map area of a feature of known ground area, such as a quadrilateral bounded by parallels and meridians (quadrilateral areas are listed in table C.5 in appendix C. You can then find the ground area of the feature from the following proportion:

Feature ground area = quadrilateral ground area × feature map area ÷ quadrilateral map area.

An example is a feature map area of 4.53 square inches on a map where a 1-degree-by-1-degree quadrilateral from 44°N to 45°N measures 7.26 square inches. From table C.5 the ground area of this quadrilateral is 3,412.26 square miles. Hence,

Area = 3,412.26 miles^2 × 4.54 inches^2 ÷ 7.26 inches^2 = 2,134 miles^2.

CENTROID

Another measure for area features is the **centroid.** You may want to find the centroid for an area feature such as a lake. You can think of the centroid as the lake's balance point or center of gravity. To determine the centroid, assume that the lake's surface is completely flat like the outline of a lake printed on a paper map or displayed on a computer monitor.

If you cut the lake out of the map, and find the point where it balances on the tip of a pencil, you will have found the lake's center of gravity. Mathematically, the centroid of a complex-shaped feature like a lake is determined by first subdividing the complex shape into basic shapes. Centroids of simple geometric shapes are easy to find. The centroid of a square or rectangle lies at the midpoint of its height and width (figure 15.13). The centroid of a right triangle is not as obvious, but the rule is that the centroid is one-third of the way along each of its two perpendicular edges. Lines drawn perpendicular to each edge at this location intersect at the centroid.

What about finding the centroid of a more complex shape, such as the lake used earlier in the chapter to illustrate several area computation methods? A basic property of centroids for complex shapes is that they are weighted averages of the centroids for the basic shapes that comprise the complex shape. The weights are simply the areas of the basic shapes.

Centroid locations are defined by x,y coordinates. The following equations are used to find the centroid of a complex shape:

$$x = \frac{\sum_{i=1}^{n} x_i \times area_i}{area_{total}}$$

and

$$y = \frac{\sum_{i=1}^{n} y_i \times area_i}{area_{total}}$$

where x_i and y_i are the coordinates for one of the n basic shapes that make up the complex shape.

These equations are easily applied to a feature defined by contiguous grid cells. In the example in figure 15.14, we have simplified the grid by placing its origin at the lower left corner of a rectangle that bounds the feature. By making each cell 1-by-1 square unit of measurement, $area_i$ always is 1 and hence can be eliminated from the computations. The x and y centroid coordinates are thus simply the sum of the x- and y-coordinates of all grid cell centerpoints, divided by the total number of cells (both whole and partial) that cover the feature n. Figure 15.14 shows an easy computation procedure based on totaling the number of whole and partial grid cells in each row and column.

Determining the centroid of an irregularly shaped area feature defined by a closed string of x,y boundary coordinates is more complex. The computation procedure is closely linked to the coordinate method of area calculation described earlier in this chapter. Adding and subtracting trapezoids formed by each line segment and the y-axis can be extended to the area weights used in centroid computation.

The first step is to find the area and centroid of each trapezoid by dividing the trapezoid into a rectangle and right triangle (figure 15.15.) The centroids of these two basic shapes are found using the formulas given for figure 15.14. Each centroid location is then converted to an x,y coordinate in the grid coordinate system used for the trapezoid. The x-coordinate for the centroid of the right triangle in figure 15.15, for instance, is 2 centimeters (left edge of triangle) plus one-third of the 2-centimeter triangle width, which equals 2.67 centimeters.

The x and y centroid coordinates for each of the two basic shapes are then weighted by their respective areas (height × width, and 0.5 × base height × width). The computation is illustrated in figure 15.15.

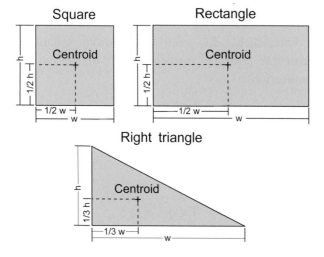

Figure 15.13 Centroids for the basic shapes are used to calculate the centroids of complex area features.

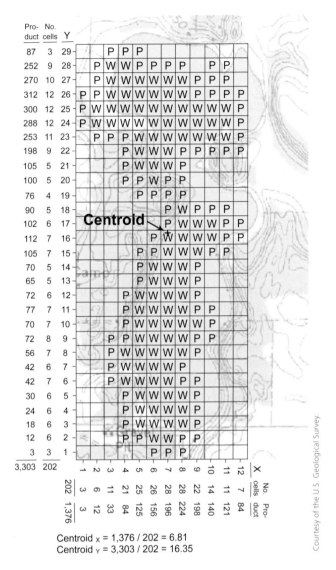

Centroid $_X$ = 1,376 / 202 = 6.81
Centroid $_Y$ = 3,303 / 202 = 16.35

Figure 15.14 The centroid of an area feature such as this lake can be determined by counting the number of square grid cells both completely and partially within the feature along each row and column. The product of the number of cells and the column and row number (x- and y-coordinates) is then found and summed. The x and y products are then divided by the number of cells to find the x- and y-coordinates for the centroid of the feature.

Courtesy of the U.S. Geological Survey.

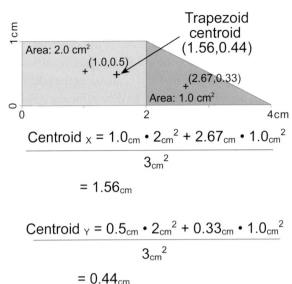

Figure 15.15 The centroid of a trapezoid is computed by dividing it into a rectangle and right triangle. The centroid and area are found for these two basic shapes, then the summation equations described in the text are used to find the x- and y-coordinates of the centroid.

Fortunately, this multistep centroid calculation procedure reduces mathematically to the simple summations:

$$x = \frac{1}{6A} \sum_{i=1}^{n} \left(x_i + x_{i+1}\right)\left(x_i y_{i+1} - x_{i+1} + y_i\right)$$

and

$$y = \frac{1}{6A} \sum_{i=1}^{n} \left(y_i + y_{i+1}\right)\left(x_i y_{i+1} - x_{i+1} y_i\right)$$

where A is the area of the tract computed by the coordinate method, and 1 through n are the tract's x,y boundary coordinates (points 1 and $n+1$ are identical). The only complication is that the summations sometimes give negative instead of positive coordinates, so their absolute value should be used.

Once the centroid coordinates and area of each trapezoid are determined, the centroid coordinate equations are used again to find the x and y centroid location for the irregularly shaped tract such as in figure 15.16. The tract's 12.58-square-centimeter area, computed by subtracting the areas of the three negative trapezoids from the sum of the two positive trapezoids, is the total area used in the equations. Applying the equations as shown in figure 15.16, the tract centroid coordinates for x,y are computed as (3.38, 5.14).

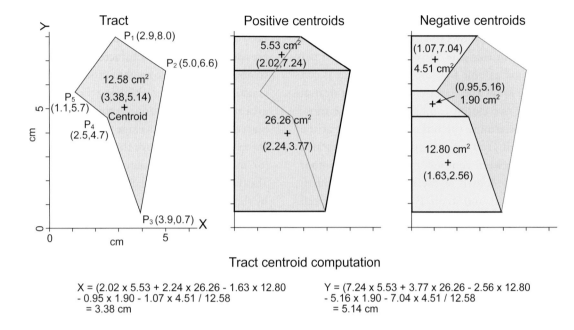

Figure 15.16 The centroid of an irregular tract can be found from its boundary coordinates. Trapezoids are constructed from successive pairs of boundary coordinates. Centroids and areas are computed for each trapezoid, and negative area weights are assigned to trapezoids where the second y-coordinate is larger than the first. The summation equations given in the text are then used to compute x- and y-coordinates of the tract centroid.

VOLUME

An object's **volume** is its area times its height or depth. You can find the volume of a feature outlined on a large-scale map if height or depth information is included on the map as contours, isobaths, soundings, or spot elevations (see chapter 6 for more on these ways of showing heights and depths). Volume computation is simplified if the base or top of the feature is at a constant height or depth, such as mean sea level or the 500-foot contour. Let's look at several ways to compute volumes from maps, beginning with the discrete ordinate method.

Discrete ordinate method
With the **discrete ordinate method,** you determine the average *height* or *depth* of a feature on your map and multiply this value by the feature's area to get the volume. You can combine the processes of determining the area and height or depth by laying a square grid over the feature. This allows you to determine the feature's area using the grid cell counting method described earlier in the chapter. You can also determine the elevation of each cell centroid to find the average elevation and use this value in calculating the volume.

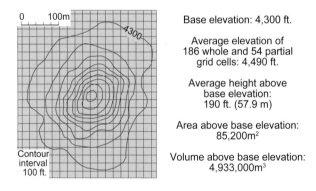

Base elevation: 4,300 ft.

Average elevation of 186 whole and 54 partial grid cells: 4,490 ft.

Average height above base elevation: 190 ft. (57.9 m)

Area above base elevation: 85,200m³

Volume above base elevation: 4,933,000m³

Figure 15.17 You can find the volume of a feature (Devil's Tower, Wyoming, in this illustration) above a base elevation by laying a grid of square cells over your map and using the discrete ordinate method.

Say you want to determine the volume of volcanic rock above a base elevation of 4,300 feet for Devil's Tower, Wyoming, as depicted by contours copied from a USGS topographic map segment (figure 15.17). First, you use the grid cell counting method. You tally the number of full and partial grid cells bounded by the 4,300-foot (1,310-meter) contour to find the total area. You can ignore the other contours for this step, but the height information for the contours is used in the next step.

Then you estimate the elevation for each cell centroid by linear interpolation between neighboring contours. **Linear interpolation** allows you to predict an unknown value (elevation, in this case) if you know any two particular values and assume that the rate of change is constant. So if you know the elevation values of the nearest point along the two closest contours, you can use those two values to determine the unknown elevation at the center of the cell. You would estimate the value at a point midway between contours to be half the contour interval value, one-fourth of the way to be one-fourth the contour interval, and so on. For example, if the cell center appeared to be one-fourth of the way between the 4,400 and 4,500-foot contours, you would estimate the elevation as 4,425 feet.

Next, you use the following equation to compute the average elevation:

$$\text{Average elevation} = \frac{\sum_{1}^{n} elevation._{whole} + \frac{1}{2}\sum_{1}^{m} elevation._{part}}{\left(n + \frac{m}{2}\right)}$$

where n is the number of whole grid cells and m is the number of partial cells. In this example, the average elevation was computed as

$$\text{Average elevation} = \frac{840{,}150 \text{ feet} + \frac{1}{2}(232{,}349 \text{ feet})}{\left(186 + \frac{54}{2}\right)}$$

$$= \frac{840{,}150 \text{ feet} + 116{,}170 \text{ feet}}{213} = 4{,}490 \text{ feet}$$

for the 186 whole and 54 partial cells bounded by the 4,300-foot contour.

The next step is to subtract the 4,300-foot base elevation from the average elevation, giving an average height of 190 feet (57.9 meters) for the area. Finally, you multiply the average height by the area to obtain the volume. The area bounded by the 4,300-foot contour was measured as 85,200 square meters, so the volume is 57.9 meters × 85,200 square meters, or 4,933,000 cubic meters.

Using the coordinates of a random sample of points usually simplifies the volume computation. For the previous volume problem, you would measure the area bounded by the feature outline (in this case, the 4,300-foot contour) using any of the area measurement methods described earlier in the chapter. You then randomly place

points within this contour (figure 15.18) and estimate the elevation at each random point by linear interpolation between neighboring contours. The estimated elevations are then averaged and subtracted from the base elevation to obtain the average height within the area. The volume is again obtained by multiplying the average height by the area at the base elevation. In this example, 100 randomly placed dots had an average elevation of 4,504 feet, so the average height above the base elevation was 204 feet (62 meters), giving a volume of 5,282,000 cubic meters. This value is 7 percent larger than the volume estimated by the grid cell method. This large difference may be due to an inadequate number of random sample points for the locations in which they fell.

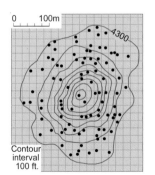

Base elevation: 4,300 ft.

Average elevation of 100 100 random dots: 4,504 ft.

Average height above Base elevation: 204 ft. (62.0 m)

Area above base elevation: 85,200m²

Volume above base elevation: 5,282,000m³

Figure 15.18   Using randomly placed sample points with the discrete ordinate method can make estimating the average elevation within an area easier, but the results may not be as accurate.

How do you determine the number of random sample points to use? Statisticians would say that if the land surface above the base elevation undulated in a random fashion, at least 30 sample points are needed to give a good result. However, land surfaces usually don't vary randomly. An extreme example would be the top of a flat, horizontal mesa in Arizona or New Mexico, where only one sample point is required to accurately determine its height, and many points are required along the steep slopes of the sides.

You can see the effect of sample-point spacing on the accuracy of the average elevation by recomputing the average elevation as additional points are added to the sample. For instance, in figure 15.19, a graph of average elevation versus number of sample points for Devil's Tower, Wyoming, shows a rapid decrease in the variation of computed average elevations with increasing numbers of randomly placed dots. At 100 sample points there is less variation, but the trend in average elevations is still decreasing slightly. This indicates that additional sample points are required to reach a point of minimal variation along the horizontal trend line.

You may be faced with determining the volume of a water body from soundings placed on a nautical chart. These depth points may at first appear to be randomly arranged, but they are probably points taken along transect lines and hence are a **systematic sample** (a sample obtained by randomly selecting a point to start, and then selected additional points by a systematic method; for example, at equally spaced intervals). You can obtain accurate volumes from such systematic height or depth samples if you select at least 30 that are evenly spaced across the water feature. Remember that in the United States, soundings are relative to mean lower low water (see chapter 1 for further information on chart datums), so you must add or subtract the difference between mean sea level if you want a volume relative to the datum on topographic maps. If you are using maps for other areas, be sure to determine what constitutes the datum.

The method of discrete ordinates, using either grid cells or random sample points, can also be applied to an undulating base surface such as the bottom of an irregular ore body. For example, the mine site in figure 15.20 has ore extending from the land surface (shown by brown contours) downward to the bottom elevations shown by blue contours. For each random sample point, the top and bottom elevations are found by linear interpolation between contours. Sample point *A,* for instance, has a surface elevation of 351 feet and a bottom elevation of 305 feet. This gives an ore body depth of 351 minus 305, or 46 feet at this point. The ore body depths then are averaged and multiplied by the ground area of the mine site to obtain the volume of ore in the site.

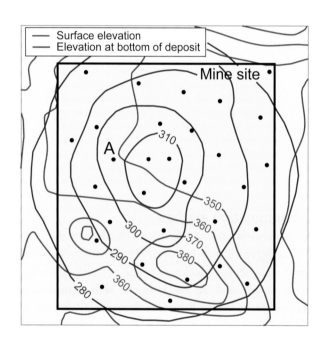

Figure 15.20 **The volume of ore within a mine site can be estimated from random sample points if the top and bottom of the deposit can be defined by contours.**

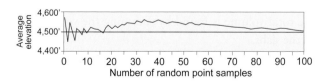

Figure 15.19 **Variations in the computed average elevation for Devil's Tower, Wyoming, rapidly decrease with increasing numbers of randomly placed sample points.**

Figure 15.21 **The volume of an irregular feature can be determined by slicing the feature into horizontal slabs defined by contours and summing the volume of all slabs. The topmost slab is calculated using the right circular cone formula.**

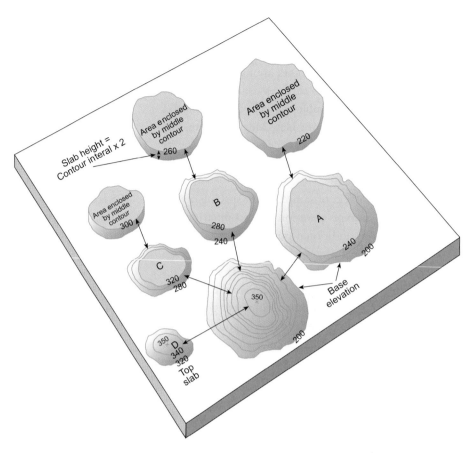

Slab summation method

You can also determine the volume of a feature outlined on a topographic map by summing the volumes of horizontal slabs whose thickness is the vertical difference between contours. To see how the **slab summation method** works for the feature in figure 15.21, start with slab *A* at the bottom of the feature. This slab is defined by three nested closed contours at elevations of 200, 220, and 240 feet. You estimate the volume of earth material between the 200- and 240-foot contours by measuring the ground area bounded by the 220-foot contour, then multiplying this area by the 40-foot elevation difference between the top and bottom contours. The area within the middle contour is used because we assume that the larger volume below balances the smaller volume above the middle contour.

Then you repeat this procedure for slabs *B* (240 to 280 feet) and *C* (280 to 320 feet). The only question is how to estimate the volume of the topmost slab, since we usually assume that the hilltop is not a flat surface but has a pointed summit. We further assume that the top of the feature extends one-half of the contour interval above the last contour. For slab *D* in figure 15.21, the last contour is at 340 feet, so the top of the feature is assumed to be at 350 feet, one-half contour interval higher than the last contour. Sometimes this elevation is noted on the map, in which case you can use that value instead of the one you would guess at in its absence.

You next assume that the three-dimensional form of the top slab from its base contour (320-foot contour in this example) to the top point can be approximated by a **right circular cone** whose volume is $1/3\pi r^2 h$. In this equation, *h* is the cone height and πr^2 is the area of its circular base. Notice that one-third of the cone height (10 feet in this example) is one-half the contour interval. We can thus restate the equation as the following:

The top slab volume equals one-half the contour interval multiplied by the ground area enclosed by the base contour.

The total volume is found by summing the volumes for all the slabs.

Triangulated irregular network volume computation

Volumes are calculated in three-dimensional surface analysis GIS software, such as the ArcGIS 3D Analyst extension, using a TIN to determine the volume of each triangle contained within the limits of particular elevations. The sum of these triangles is used to calculate the total volume, which then can be converted to the desired units (for example, cubic meters or acre-feet). Volumes under individual triangles in the TIN are computed by finding the average elevation of the three triangle corners and multiplying this value by the area of the triangle when projected onto a flat surface of zero elevation, such as mean sea level (figure 15.22).

This method works well because the UTM or state plane grid coordinates of the triangle corners in the TIN are horizontal positions on the earth projected vertically downward or upward to mean sea level. Heron's formula presented earlier in the chapter is often used to compute triangle areas from the UTM or state plane grid coordinates of their three corners. In this example, the three triangle edges were found from the difference in the UTM easting and northing coordinates of their end points to have lengths of 971, 618, and 1,131 meters. We used Heron's formula to compute an area of 299,820 square meters for the triangle. We then found an average elevation of 56.7 meters for the three triangle corners, which—when multiplied by the triangle area—gave a volume under the triangle of 16,999,900 cubic meters.

You can also calculate the volumetric difference between two TINs by subtracting the second TIN from the first. The calculation is performed using vectors, and the result is an output polygon divided into areas in which the first TIN surface is above the second, below the second, or the same as the second. The GIS calculates the difference in height between triangle corner points in the two TINs. The heights for the vertices of the triangle in the first TIN are compared to the interpolated values at the same locations on the second TIN. The procedure is then repeated in inverse, with the measurements of the second TIN being compared against the surface of the first. The differences in height for all triangle corner points from both TINs are used to create the **difference TIN.** Zero values on this difference surface represent the intersection between the two surfaces.

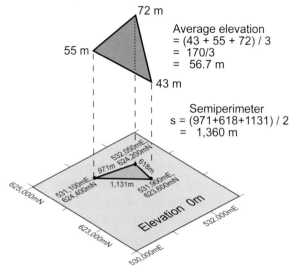

TIN triangle volume above 0m elevation

Average elevation
= (43 + 55 + 72) / 3
= 170/3
= 56.7 m

Semiperimeter
s = (971+618+1131) / 2
= 1,360 m

Triangle area from Heron's Formula
Area = sqrt(1,360 (1,360-971)(1,360-618)(1,360-1,131))
= sqrt(1,360 x 389 x 742 x 229)
= 299,820m²

Volume = Triangle area x average elevation
Volume = 299.820m² x 56.7m

= 16,999,900m³

Figure 15.22 The volume under a triangle in a TIN is computed from the area of its projection on a flat surface at zero meters or feet elevation and the average elevation of its three corners.

SHAPE

An area feature's two-dimensional form is called its **shape.** Shape is one thing that gives each area feature its distinctive geographic character. We use terms such as "compact," "elongated," or "irregular" to describe the shapes of area features. The shapes of features are important to map users because it helps us identify features as well as describe them.

As important as identifying the shapes of area features is for map users, shape is a difficult thing to communicate in words. The outlines or external surfaces of environmental features take so many forms that we can't conveniently attach labels to them all. But we have named a number of simple geometric figures, and we frequently refer to these through comparison. We say that features are somewhat circular, roughly triangular, or approximately rectangular in shape. But the degree of resemblance between the area feature shape (the shape seen on a map) and a standard geometric shape may be difficult to state in words.

Shape indexes

While verbal descriptions of different shapes are sufficient for many purposes, the terms used are often so subjective that disagreements arise among map users. Thus, more objective shape indexes have been devised. A **shape index** is a method for describing the shape of an area feature in numerical terms.

Area correspondence

One method for describing the shape of an area feature is to state in numerical terms the degree of similarity between a standard two-dimensional shape and the feature. If you want an idea of the "squareness" of an irregularly shaped area feature, for example, you can superimpose the feature on a square of approximately the *same map area* and compute the **area correspondence** between the two shapes mathematically (figure 15.23). This can either be done using the intersection of the areas or the union of the areas. If you repeat the procedure using other standard shapes such as circles, you can then describe the feature's shape relative to these select standards in an objective way.

Although an unlimited number of mathematical procedures could be devised, one example of a shape index should make the procedure clear:

$$\text{Shape} = 1.0 - \left(\frac{F \cap S}{F \cup S} \right)$$

where $F \cap S$ is the **intersection** (overlap) area of the feature and standard shape, and $F \cup S$ is the **union** (area in one or the other) of the two shapes.

To compute this shape index, first find the area of intersection where the two shapes overlap. Do this by finding the centroid of the irregular shape, then aligning the standard shape so that its centroid location is the same as the irregular shape. You can measure the areas of intersection or union with a polar planimeter, or you can solve this as a polygon overlay problem in a GIS. You can then find the ratio between this intersection area and the total union area covered by the two superimposed shapes. Finally, subtract this value from 1 so that the shape index will range from 0 when there is perfect coincidence between the two shapes to 1 when there is no overlap.

In figure 15.23, the irregularly shaped feature, the circle, and the square all have a map area of 0.634 square inches. For the square, the intersection area is 0.501 square inches and the union area is 0.785 square inches, so the shape index is 1.0 − 0.501 ÷ 0.785 = 0.362. For the circle, the intersection area is 0.542 square inches, while the union area is 0.750 square inches. Thus, the shape index is 1.0 − 0.542 ÷ 0.750 = 0.272. These shape index values suggest that the irregular shape is slightly closer to a circle than a square. If the index were 0 using the intersection and union values for the square, we would know that the feature is perfectly square. If the same results were found using the values for the circle, the feature would be perfectly circular.

It's not uncommon to find indexes close to 0, especially when determining the shapes of **cultural features** (those produced by humans). The upper index value of 1, however, provides an unreachable limit, for it would mean that the feature and standard shape did not overlap, which is impossible.

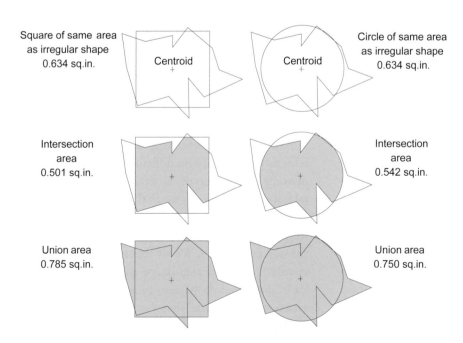

Square of same area as irregular shape 0.634 sq.in.

Centroid

Circle of same area as irregular shape 0.634 sq.in.

Centroid

Intersection area 0.501 sq.in.

Intersection area 0.542 sq.in.

Union area 0.785 sq.in.

Union area 0.750 sq.in.

Figure 15.23 A shape index such as the ratio of the intersection or union of an irregularly shaped feature and standard shapes, such as squares and circles, makes it possible to objectively describe the shapes of different features.

Compactness

In some situations, you might prefer to measure the compactness of an area feature rather than comparing it with standard shapes. **Compactness** of a shape means that an area feature occupies little space in relation to its area. This property is often considered desirable, in part because compact regions are most efficiently serviced and defended. Partitioning space into compact units for administrative and political purposes also conveys a sense of fairness, as suggested by the gerrymandering example at the beginning of the chapter.

A **compact shape** is one in which all points on the boundary are as close as possible to the center. The circle is the most compact two-dimensional shape, because its boundary is everywhere equidistant from its center point. (The same is true of a sphere in three-dimensional space.) The length of its boundary (perimeter) relative to its area is also minimal. For these reasons, **compactness indexes** all use the circle in some way as a standard reference shape.

Compactness indexes are derived by forming ratios between such basic figure attributes as area, perimeter, length of longest axis, and the **inradius,** (the radius of the **inscribed circle**, which is the largest circle contained in a given triangle), or **circumradius** (the radius of the **circumscribing circle**, which is the smallest circle that contains the triangle). For example, the ratio of a feature's area *(A)* to its perimeter *(P)* provides a useful compactness index. This ratio usually is modified by multiplying the area by 4π and dividing by P^2, so that a circle has a compactness of 1. The compactness index is thus:

$$\text{Compactness} = \frac{A \times 4\pi}{P^2}$$

With this index, the more elongated and irregular the feature, the closer to 0 its index value will be. As an example, refer to figure 15.24. Since the feature illustrated in this figure has a perimeter of 4.48 inches and an area of 0.634 square inches, its compactness index is $0.634 \times 4\pi/4.48^2$, or 0.40. This could be considered as a medium compactness index value.

The examples of standard shape and shape compactness we've discussed in this section are fairly primitive mathematically, but they can all be done with a map in hand. And they illustrate several attributes of all numerical indexes you need to consider before putting too much faith in their values. Based on the decisions you must make when determining compactness or comparing a feature with a standard shape, a feature can have several different index values. Shapes that look quite different

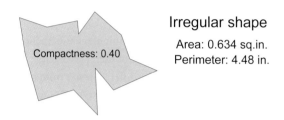

Irregular shape
Area: 0.634 sq.in.
Perimeter: 4.48 in.

Compactness: 0.40

Figure 15.24 Compactness indexes are often based upon the ratio of the area and perimeter of an irregularly shaped feature.

can also have the same index value. The indexes we've described don't have unique numerical values for different shapes. Despite these drawbacks, however, shape indexes are still useful in many map analysis situations, especially if they're simple, unit-free, independent of the feature's size, and intuitively appealing.

We have seen that there are some fairly straightforward methods useful for determining the area, volume, and shape of a feature and for discovering its centroid. Many of these draw on concepts described in earlier chapters such as map scale, grid coordinates, and map projections. Together, these concepts can be used to help you tackle some of the more complex challenges of map use.

NOTES

1. The acre was originally an English unit of measurement that described the area that a yoke of oxen (two oxen) could plow in a day. It originally differed in size from one area to the next, but was ultimately fixed at 4,840 square yards, or 43,560 square feet.

SELECTED READINGS

Barrett, J. P., and J. S. Philbrook. 1970. Dot grid area estimation: Precision by repeated trials. *Journal of Forestry* 68 (3): 149–51.

Boyce, R. R. and W. A. V. Clark. 1964. The concept of shape in geography. *Geographical Review* 54: 561–72.

Dickinson, G. C. 1969. Measurement of area. In *Maps and air photographs,* 132–41. London: Edward Arnold Publishers, Ltd.

Dury, G. H. 1972. Geometric analysis. In *Map interpretation.* 4th ed. London: Pitman & Sons, Ltd, 163–77.

Earickson, R. J., and J. M. Harlin. 1994. *Geographical measurement and quantitative analysis.* New York: Macmillan College Publishing Co.

Frolov, Y. S., and D. H. Maling. 1969. The accuracy of area measurement by point counting techniques. *The Cartographic Journal* 6 (1): 21–35.

Gierhart, J. W. 1954. Evaluation of methods of area measurement. *Surveying and Mapping* 14:460–69.

Kimerling, A. J. 1984. Area computation from geodetic coordinates on the spheroid. *Surveying and Mapping* 44 (4): 343–51.

Lawrence, G. R. P. 1979. Measurements from maps. In *Cartographic Methods.* 2d ed., 82–104. London: Methuen & Co., Ltd.

———. 1979. Map analysis. In *Cartographic Methods.* 2d ed., 82–104. London: Methuen & Co., Ltd.

Lee, D. R., and G. T. Sallee. 1970. A method of measuring shape. *Geographical Review* 60 (4): 555–63.

Maling, D. H. 1988. *Measurement from maps: The principles & methods of cartometry.* New York: Pergamon Press, Inc.

Proudfoot, M. 1946. *The measurement of geographic area.* Washington, D.C.: U.S. Bureau of the Census.

Whyte, W, and R. E. Paul. 1997. *Basic surveying.* Amsterdam: Elsevier.

chapter
sixteen **SURFACE ANALYSIS**

16

Surface analysis

Rises and falls in the ground surface have a major influence on human behavior. We speak of the ease of hiking on a gentle slope, or the effort involved in traversing a steep road. A lake bottom is said to drop off rapidly from shore, and a steep downgrade is a danger to truck drivers. The concept can be applied to any surface representing a geographic phenomenon that has continuously changing values, not just terrain. Undulations in surfaces that represent physical phenomena also have important meaning for us. For example, **temperature gradients,** the differences in air temperature between different locations, are critical in weather forecasting. In all cases, we are experiencing something about vertical variations on the earth's surface.

Qualitative terms describing vertical landform change are of limited use because they can take on quite diverse meanings for people under different conditions of age, fitness, stress, or experience. So we turn to maps to provide quantitative information from which we can create mathematical measures and graphical illustrations of the amount of elevation change along lines or across areas on surfaces.

In this chapter you will see that the amount of *slope* in a particular direction can be computed from the contours on a topographic map or the isobaths on a nautical chart. This slope information allows you to create a *slope map* for an area. *Slope direction* is used to create *aspect maps.* Combining aspect information with sun altitude and angle allows us to calculate the *illumination* of a surface. *Gradient* helps us determine the steepest or gentlest path up or down a slope. *Curvature* tells us whether the surface is upwardly convex or upwardly concave. Changes in elevation along a line on the map can be shown graphically by drawing a *profile* of the terrain. You can also use terrain information to create *visibility maps* that show areas visible and hidden from view at a particular location on the earth. All of these are important forms of surface analysis that allow us to answer important questions about our environment. Let's begin by looking at slope.

SLOPE

The vertical change in the elevation of the land surface, when determined over a given horizontal distance—along a road or stream, for instance—is known as its **slope.**

Determining slope

Slope can only be computed between two points. If you want to know the slope at a specific location, therefore, you must use the elevation at the location and another point in computing the slope. You determine the elevation difference (**rise**) from the contours or isobaths on the map, then measure the map distance between the two points and convert to ground distance (**run**) using the map scale. Slope is computed in the example map in figure 16.1 as rise over run.

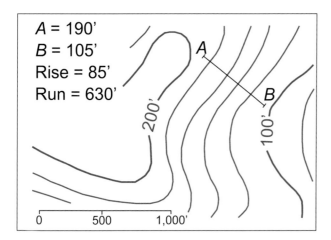

$A = 190'$
$B = 105'$
Rise $= 85'$
Run $= 630'$

0   500   1,000'

Figure 16.1 Slope is computed from the elevation difference (rise) between two points *A* and *B* on a topographic map and the horizontal distance (run) between the two points.

Slope computations are based on a surface assumed to increase or decrease in height uniformly between two points. Such surfaces are called **linear surfaces.** This assumption is important because linear interpolation that relies on this assumption is used for these computations (see chapter 15 for more on linear interpolation). Deviation from a linear surface within the distance over which the slope is computed will result in **slope measurement error.** While the actual slope of the curved surface varies continually in steepness between the two points, the slope computation will give a constant slope value. The greater the distance over which slope is computed and the more curved the surface, the greater the potential slope measurement error. For this reason the two points used in the slope calculation are usually nearby points.

Although we generally use the term "slope" to refer to the amount of rise or fall of the ground surface, the concept is equally applicable to any phenomenon with a magnitude that changes as a function of changing distance, as we mentioned. For example, a surface magnitude of population density often slopes steeply downward away from the center of a city.

Expressing slope

There are three primary ways to quantitatively express the slope between two points. In each, the lower the slope value, the flatter the terrain, and the higher the slope value, the steeper the terrain. The slope values may be expressed as a ratio, as a percentage, or as an angle.

Slope ratio

The simplest way to express slope is to describe it as the **slope ratio** between the elevation difference (rise), and the ground distance (run) between the two points (figure 16.2A). Mathematically, this is written as y/x. In figure 16.1, a rise of 85 feet over a run of 630 feet gives a slope ratio of 85/630, which is 0.13. Slope ratios also can be negative—a fall of 30 feet over a run of 150 feet gives a slope ratio of −30/150, which is −0.20. Notice that the vertical and horizontal distances must always be in the same units of measurement.

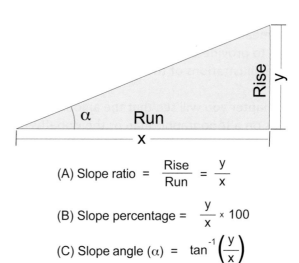

(A) Slope ratio $= \dfrac{\text{Rise}}{\text{Run}} = \dfrac{y}{x}$

(B) Slope percentage $= \dfrac{y}{x} \times 100$

(C) Slope angle (α) $= \tan^{-1}\left(\dfrac{y}{x}\right)$

Figure 16.2 The slope of a surface between two points can be expressed as a ratio (A), as a percentage (B), or as an angle (C), where tan⁻¹ is read as "angle whose tangent is..."

Slope percentage

You can also express the ratio as a **slope percentage** (also called **percent rise**) (figure 16.2B). To do so, simply multiply the slope ratio by 100. In the example above, the slope ratio 0.13 is also 0.13 × 100, or 13 percent. Slope percentages range from 0 to near infinity. A flat surface is 0 percent, and as the surface becomes more vertical, the slope percent becomes increasingly larger. In figure 16.3, you can see that when the angle is 45 degrees, as in triangle *B,* the rise is equal to the run, and the slope percent is 100 percent. As the slope angle approaches vertical (90 degrees), as in triangle *C,* the slope percent approaches infinity.

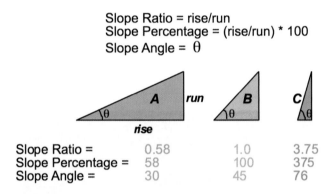

Slope Ratio = rise/run
Slope Percentage = (rise/run) * 100
Slope Angle = θ

	A	B	C
Slope Ratio =	0.58	1.0	3.75
Slope Percentage =	58	100	375
Slope Angle =	30	45	76

Figure 16.3 These triangles help demonstrate the comparison of the values for expressing slope as a ratio, as a percentage, and in degrees.

Slope angle

A third way to express slope is in degrees that relate to the **slope angle** (figure 16.2C). Slope angle values range in degrees from 0 to 90. This way of specifying slope is based on the fact that the slope ratio *(y/x)* is the trigonometric tangent of the slope angle. Consequently, the slope angle is the inverse tangent (*tan⁻¹,* sometimes called the **arctan**) of the slope ratio (the angle whose tangent is the slope ratio). A slope angle of 45 degrees is a 100 percent slope, since this is the angle whose tangent and slope ratio is 1.

Slope maps

Slope maps are useful tools for land-use planners, resource managers, and others who need to know the steepness of the terrain. They have been used successfully in runoff modeling, urban development (some cities have a slope-zone ordinance), logging, farming, and other activities. Slope maps help to identify constraints related to landforms, such as grades that are too steep to support

buildings. They also help you to evaluate potential environmental impacts due to alterations, such as erosion and slope failure. When used with aerial photographs, slope maps are excellent tools to look for potential landslide or erosion areas, identify drainage features, evaluate landform patterns, explore land-use suitability, and more.

Slope zone maps can be generated as isoline maps of specific slope categories. To construct a slope map you must first select the slope categories you want to map. This should be a relatively small set (0–6 percent, 7–11 percent, 12–19 percent, 20–34 percent, 35–59 percent, and >60 percent, for example) since these maps are somewhat tedious to produce unless you are using a geographic information system (GIS). You also have to know the scale of the topographic map and the contour interval.

One way to map slope by hand and eye is with a **slope template** (also called a **tick sheet**), which is a scaled template of selected slope isolines drawn so you can use them to easily identify your slope categories. Using the map's scale and contour interval, you can make a clear plastic template of straight-line ticks spaced so that they cover the range of slopes to be mapped. The template in figure 16.4, for example, has ticks spaced increasingly closer so that there is a 1 percent change in slope for each succeeding pair of lines. The easiest way to map slope is to measure the slope between two contours; these measurements are always perpendicular to the **axis** (direction) of the two contours.

The formula for measuring the distance between contour lines to determine the isolines for a slope map is

$$D = \frac{100 \times CI}{P \times S}$$

where D is the distance in inches between contour lines, CI is the contour interval in feet, P is the percent slope isoline, and S is the scale of map (in the equivalent of feet, such as 1 inch equals 400 feet). The contour interval and the map scale could also be in other equivalent units, such as meters.

As mentioned above, when creating a slope map, the easiest approach is to use slope measurements between two contour lines; making slope measurements across several contour lines is more difficult and requires interpolation. At hilltops and ridgelines as well as valley bottoms, assume a constant or lessening slope (but be conservative). Along streams, assume a constant slope.

Next, compare the spacing of lines on the slope template with the spacing of contours across the map. Draw outlines around small areas on the map that have slopes falling within the same slope zone, so that the entire map ends up divided into different slope zone regions. The hillside in the upper left corner of figure 16.4, for instance, has slopes of around 23 percent that fall within the 20–34 percent slope zone for the selected slope categories we identified earlier. Figure 16.5 is an example of a slope zone map manually produced using a slope template.

Because the method above can be time-consuming and tedious, you will often find **raster slope maps** that have been generated using a GIS and raster elevation data (figure 16.6). For these, it is useful to know how the maps were produced so that you will know how to use them appropriately. Usually these maps are generated from digital elevation models (DEMs) or other surface data that have been converted to a raster (gridded) format. On these maps, the slope is shown as pixels that are colored to indicate the slope, expressed in either percentages or degrees, as explained above. The colors identify the rate of maximum change in elevation value from each pixel.

Most often, these kinds of maps show slopes for terrain surfaces; however, you will sometimes find that the slope of other types of continuous surface data, such as population or rainfall, has been mapped to identify sharp changes in values on the surface.

Constant slope path

Sometimes you may want to determine a **constant slope path,** or a path that maintains a constant steepness. Suppose, for example, that a road or trail must have a constant slope of five degrees. If you have a contour map available, one way to define this path is to set a map divider at a ground distance *(GD)* equal to the contour interval *(CI)* divided by the tangent of the slope angle (α). A **map divider** is an instrument that allows you to set a specified distance so you can "walk" it along a path—the spacing of divider points and number of steps give an estimate of the length of the path. This procedure is illustrated in figure 16.7.

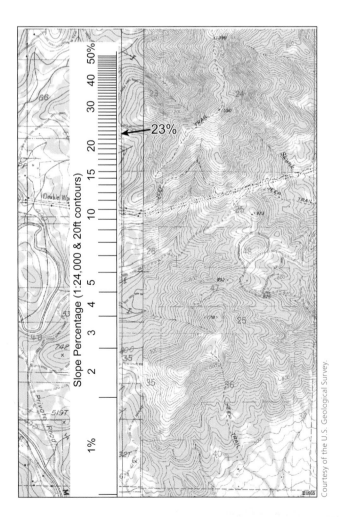

Figure 16.4 Slope templates make it convenient to convert the contours on a topographic map into a slope zone map.

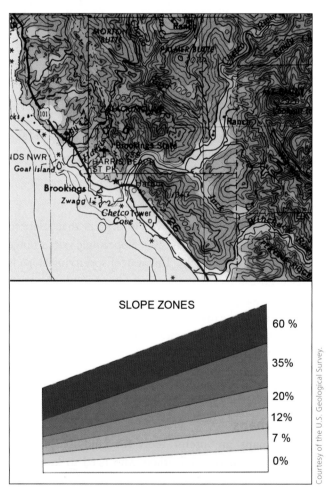

Figure 16.5 Slope templates can be used to create slope zone maps, which are useful land-use planning and management tools.

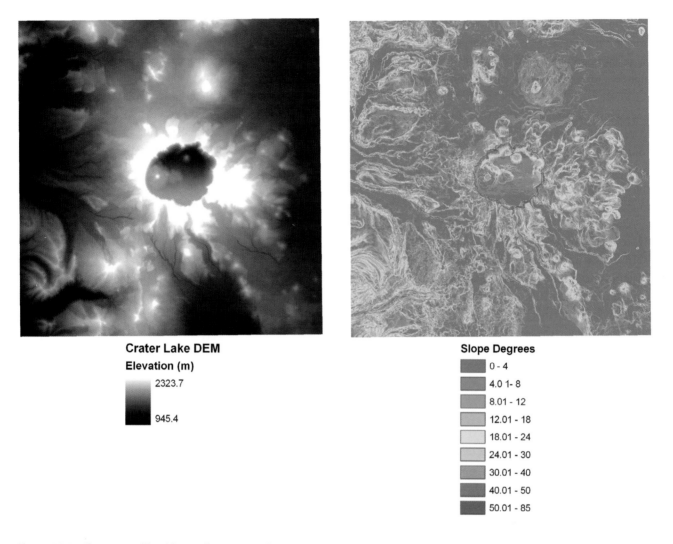

Crater Lake DEM

Elevation (m)

2323.7

945.4

Slope Degrees

0 - 4

4.0 1- 8

8.01 - 12

12.01 - 18

18.01 - 24

24.01 - 30

30.01 - 40

40.01 - 50

50.01 - 85

Figure 16.6 Slope maps like this one for Crater Lake, Oregon, can easily be created using GIS and DEMs.

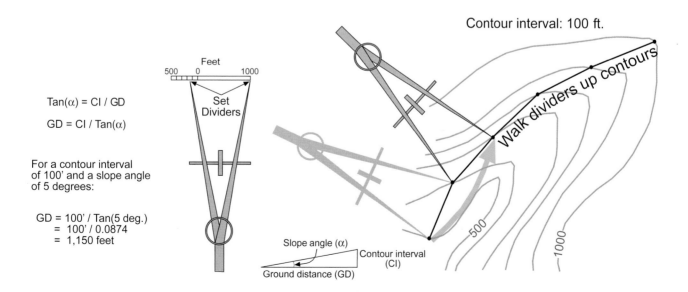

Contour interval: 100 ft.

Tan(α) = CI / GD

GD = CI / Tan(α)

For a contour interval
of 100' and a slope angle
of 5 degrees:

GD = 100' / Tan(5 deg.)
   = 100' / 0.0874
   = 1,150 feet

Feet
500   0   1000

Set
Dividers

Slope angle (α)

Contour interval
(CI)

Ground distance (GD)

Walk dividers up contours

500

1000

Figure 16.7 You can determine a constant slope path by "walking" preset map dividers up the slope.

Assuming your beginning point is located on a contour line, the first step is to place one foot of the dividers at the beginning point and the other foot on the next higher contour line. Mark this second point; then rotate the lower foot of the dividers to the next higher contour line, while keeping the other foot stationary. Mark this third point. Continue the process until you reach the contour line that lies at or just below the elevation of your destination point. Finally, connect the points, including the beginning and destination points, with line segments to map the constant slope path.

Maximum slope path

Rather than find a constant slope path, you may want to determine a **maximum slope path,** or a path that does not exceed a specified slope angle. An example is a railroad track grade that must stay below a certain slope angle. Although the grade is determined by available locomotive horsepower and traction efforts required to pull a specified number of railroad cars up a grade, as well as the braking effort required to control the movement when going down the grade, the path can still be determined using the methods described here. Similarly, highways have maximum grades allowable depending on the types of vehicles that will be using them and their maximum speeds. Government regulations often specify the maximum slope permissible for such routes.

The same procedure for finding a constant slope path is used to find a maximum slope path, but three things can happen. First, the maximum allowable slope path may intersect the destination contour between the destination point and the steepest possible route from the origin (figure 16.8A). This means that a straight-line course from origin to destination will fall within acceptable slope limits. When this happens, you may have a great deal of flexibility in choosing the actual route to take.

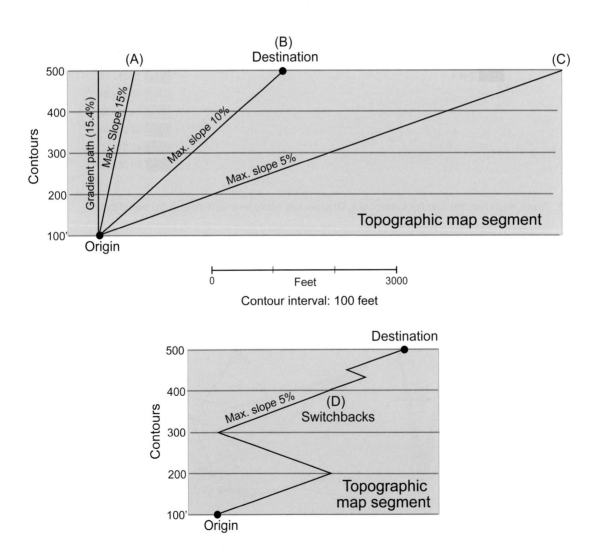

Figure 16.8 Routes laid out with a maximum slope restriction may end up in a series of switchbacks on steep slopes. Idealized topographic map segments with horizontal contours are used to show you constant slope paths at their simplest.

The second possibility is that the maximum allowable slope line will connect directly with the destination point (figure 16.8B). There will be no flexibility in route location in these rare cases.

The most complex situation is when the maximum permissible slope line intersects the destination contour beyond the destination point (figure 16.8C). This means that a relatively indirect course must be taken between origin and destination to meet the maximum slope restriction. This last case explains the prevalence of "switchbacks" on steep mountain trails and roads (figure 16.8D).

GRADIENT

Gradient is another way to think about the form of a sloping surface. **Gradient** is a vector that describes both magnitude and direction, the maximum amount of vertical change on a surface, and the direction in which that change occurs. Using gradient, you can find the steepest route down a hill or the steepest grade up a mountain. This steepest route is called the **gradient path.** Each location on a hill or other terrain feature has different slope angles in different compass directions, but from any location there is only one gradient path where every point along the path follows the steepest possible route (figure 16.9).

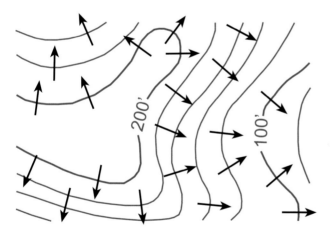

Figure 16.9 These gradient vectors show the steepest downhill (or uphill) direction and magnitude at selected points. As with aspect, they are always perpendicular to the contour lines.

Gradient path

To visualize the gradient path, it's helpful to look at a three-dimensional terrain model. You could place a drop of water at the highest point and observe the path the water drop takes moving downward under gravity. If contour traces were drawn on the terrain model surface, you would see the water move downhill along a least-resistance path at right angles to the contours (figure 16.10). This is to be expected, since the maximum gradient always is at right angles to the contour lines. The water merely traces the gradient path (or steepest slope route) from high to low points on the landform.

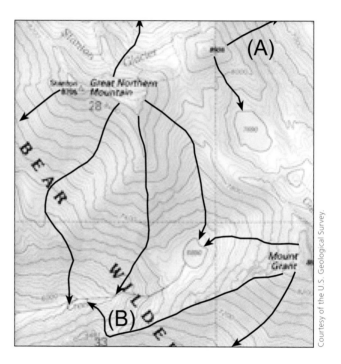

Courtesy of the U.S. Geological Survey.

Figure 16.10 The gradient path traces out the least-resistant path on a contoured terrain surface. If contour lines are parallel, the gradient path is a straight line perpendicular to the contours (A). Otherwise the gradient path traces out a curved route (B) perpendicular to all contours.

If contour lines in an area of linear terrain (constant slope) are parallel, the gradient path will be a straight line perpendicular to the contours (figure 16.10A). Terrain surfaces aren't usually linear, of course. This means that long, straight gradient paths aren't all that common. When water flows down a non-linear terrain surface along a gradient path, its course direction must be altered so that all contours are crossed at right angles (figure 16.10B).

Determining gradient

Finding the gradient for points on a surface is a bit more complicated than computing slope ratios, percentages, or degrees because not one but two calculations are necessary. In addition to the slope magnitude, you also have to compute the gradient direction, usually a true azimuth (see chapter 12 for more on azimuths). To make these computations, you determine the slopes in two perpendicular directions, usually north–south and east–west.

To understand how the gradient is computed, imagine laying a grid of square cells over the map so that the cell boundary lines are equally spaced along the x- and y-coordinate axes. In figure 16.11, for instance, 50-meter-by-50-meter grid cell boundary lines are defined by UTM grid coordinate system easting (x) and northing (y) coordinates (see chapter 4 for more on the UTM grid coordinate system). You can find the elevation at the cell center by interpolating between the contour lines in the vicinity of the cell.

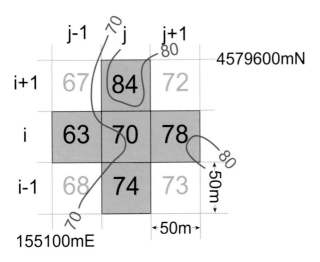

Figure 16.11 **The gradient can be computed by first finding the elevations for a grid of square cells laid over a contour map. Here elevations at cell centers are given in meters for 50-meter-by-50-meter cells. These values are used in the computations explained below.**

The first step in calculating the gradient for each cell is to find the slope in the x (easting) direction—this is called the **x-slope**—as well as the slope in the y (northing) direction—the **y-slope.** There are several ways to compute the x-slope and y-slope, but a common procedure is to use the elevations of the grid cell and its immediate east–west and north–south neighbors. For example,

the x-slope can be calculated as the average slope between the center cell at row i and column j $(Cell_{i,j})$, and the cells immediately to the left $(Cell_{i,j-1})$, and right $(Cell_{i,j+1})$. The y-slope is calculated as the average slope between the center cell $(Cell_{i,j})$, and the cells immediately above $(Cell_{i-1,j})$ and below $(Cell_{i+1,j})$. For a grid cell spacing of d meters, the equations used to find the x-slope and y-slope are the following:

$$slope_x = \frac{1}{2} \times \left(\frac{Cell_{i,j} - Cell_{i,j-1}}{d} + \frac{Cell_{i,j+1} - Cell_{i,j}}{d} \right)$$

$$slope_x = \frac{Cell_{i,j+1} - Cell_{i,j-1}}{2d}$$

$$slope_y = \frac{1}{2} \times \left(\frac{Cell_{i,j} - Cell_{i+1,j}}{d} + \frac{Cell_{i-1,j} - Cell_{i,j}}{d} \right)$$

$$slope_y = \frac{Cell_{i-1,j} - Cell_{i+1,j}}{2d}$$

Notice that although the elevation of $Cell_{i,j}$ is a key part of the initial equations, it cancels out and does not appear in the final equations—only the elevations of the immediate neighbor cells are used in computing the x-slope and y-slope. In figure 16.11, for instance, $Cell_{i,j} = 70$ meters, $Cell_{i,j+1} = 78$ meters, $Cell_{i,j-1} = 63$ meters, $Cell_{i+1,j} = 74$ meters, and $Cell_{i-1,j} = 84$ meters. For the 50-meter square cells ($d = 50$ meters), the slope equations are the following:

$$slope_x = \frac{78m - 63m}{100m} = 0.15$$

and

$$slope_y = \frac{84m - 74m}{100m} = 0.10$$

Once you find the x-slope and y-slope, you can compute the maximum slope ratio, called the **gradient magnitude** (G), using the following equation:

$$G = \sqrt{(slope_x)^2 + (slope_y)^2}$$

In our example, the gradient magnitude at $Cell_{i,j}$ is

$$G = \sqrt{(0.15)^2 + (0.10)^2} = \sqrt{0.0225 + 0.01} = \sqrt{0.0325} = 0.18$$

expressed as a slope ratio, or an 18 percent upward slope expressed as a slope percent.

You now have the slope magnitude. The next step is to compute the gradient direction—this is expressed as a grid azimuth. You find the gradient azimuth by using the following equation:

$$\theta = tan^{-1}\left(\frac{slope_x}{slope_y}\right)$$

where θ is the azimuth angle measured from the grid north reference line (see chapter 12 for more on grid north), and tan^{-1} (sometimes called **arctan**) is shorthand for "angle whose trigonometric tangent is..."

In our example

$$\theta = tan^{-1}\left(\frac{0.15}{0.10}\right) = tan^{-1}(1.5) = 56°$$

so the uphill gradient is 18 percent at a 56° azimuth from grid north. The downhill gradient (indicated by a negative value) is just the opposite: –18 percent at a 56° + 180°, or 236°, grid back azimuth (see chapter 12 for more on back azimuths).

Gradient maps

Like gradient magnitude computations, gradient azimuth computations are tedious if not done by computer software using DEM data. Once such computations have been made, whether by hand or computer, it is possible to create a **gradient vector map** (figure 16.12) showing with vector lines the maximum slope magnitude and downward direction. When working with terrain data, such a map is useful in visualizing the flow of water or material such as soil over the surface. With nonterrain data, surface gradient vectors may be suggestive of movement or flow of forces, goods, or ideas. Thus, a gradient vector map of barometric pressure may help to explain the pattern of winds, and a gradient vector map of commodities can help explain the flow of goods among regions.

Courtesy of Golden Software, Inc.

Figure 16.12 Gradient vector maps suggest patterns of flow associated with surfaces.

ASPECT

Another important characteristic of surfaces is aspect. **Aspect** identifies the downslope direction of the maximum vertical change in the surface determined over a given horizontal distance (that is, the slope). Aspect is the same as gradient direction. It can be thought of as the compass direction a hill faces and, in fact, aspect is usually expressed in terms of the compass direction of the slope (figure 16.13). Aspect is measured clockwise in degrees from 0 (due north) to 360. The aspect value indicates the direction the slope faces. Flat areas with no downslope direction are given a value outside the 0-to-360 range, for example, –1. Aspect is computed in the same manner as gradient direction, as explained above and illustrated in figure 16.13.

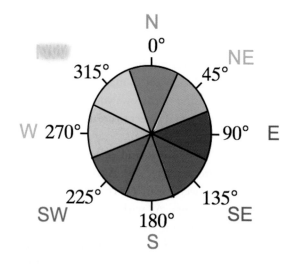

Figure 16.13 Aspect is expressed in degrees from 0 at due north, to 90 at due east, 180 at due south, and so on.

Aspect can be used to find all north-facing slopes on a mountain as part of an analysis to determine where snow will melt later in the season. You can also use aspect to find all southerly slopes in a region to identify locations where the sun will illuminate properties you may be thinking of buying. Aspect would also be important in identifying flat areas that may be suitable for a plane to land in an emergency.

Aspect maps

Aspect maps are dasymetric maps (see chapter 8 for more on dasymetric maps) of selected aspect categories, for example, north, northeast, east, southeast, and so on. The aspect categories are usually selected as a function of the intended use of the map. For example, if the map is to be used for site planning, the aspect categories shown would relate to factors that can affect microclimatic conditions such as shadows and frost. In this case, the map may portray the aspect categories shown in figure 16.13.

Aspect zone maps are constructed from contour line data, not slope data, to show specific aspect categories. Aspect is determined relative to the direction of contour lines, and it is always perpendicular to the contours. To create an aspect zone map, first orient your map so that you can clearly determine the directions for the aspect categories. For example, if you orient your map so that north is aligned to the top, the north–south aspect isoline will be perpendicular to the top of the map, and the east–west aspect isoline will be parallel to it. Begin to mark points of tangency to the contour line direction—using arrows may help. Follow landforms from top to bottom or bottom to top along the points of tangency. You will find that the aspect is easier to determine and map in steeper slopes because of the increased information provided by the greater number and the closer spacing of the contours. Use a minimum mapping unit to set the minimum width and size of areas you will map (see chapter 10 for more on minimum mapping units). For example, you may decide not to map areas smaller than one acre. Finally, using the arrows you have drawn, you can connect lines of equal aspect to create the final aspect zone map.

As with slope, you will also find aspect maps generated using GIS software (figure 16.14). Again, it is important to know how these were created so that you can use them to their best advantage. As with slope maps, they are most often generated using DEMs or other elevation data that have been converted to raster format. Using these data, aspect is identified as the downslope direction of the maximum rate of change in value from each grid cell to its neighbors. It is computed in the same manner as

gradient direction from the elevation value for each cell in the raster. Recall that aspect is expressed in positive degrees from 0 to 359.9, measured clockwise from north. Cells in the input raster that have zero slope (that is, they are flat) are assigned an aspect value of –1. If the center cell contains no data, the output will have no data. As with slope maps, if any neighborhood cells contain no data, they are assigned the value of the center cell, and aspect is then computed.

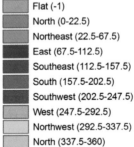

	Flat (-1)
	North (0-22.5)
	Northeast (22.5-67.5)
	East (67.5-112.5)
	Southeast (112.5-157.5)
	South (157.5-202.5)
	Southwest (202.5-247.5)
	West (247.5-292.5)
	Northwest (292.5-337.5)
	North (337.5-360)

Figure 16.14   An aspect map for Crater Lake, Oregon, generated from DEM data using a GIS. The coloring is based on Moellering and Kimerling's MKS-ASPECT scheme (see Selected readings, page 365), which is a method of coloring the aspect categories for terrain or other surfaces.

Aspect-slope maps

Sometimes aspect and slope are mapped together because they are so integrally related. These maps often use a color scheme that combines hue (what we normally think of as color–red, green, blue, for example) for the aspect azimuth and saturation, or brilliance of a color for the slope value (see chapter 7 for more on color hue and saturation). A portion of such a map is shown in figure 16.15 below. The hues correspond to the aspect; the more brilliant colors are higher slope, and the less saturated ones are lower.

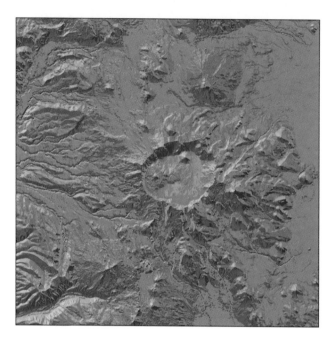

Figure 16.15   An aspect-slope map shows both aspect, through variation in color hue, and slope through variation in color saturation.

ILLUMINATION

Illumination is also an important factor to consider when analyzing the terrain surface. **Illumination** is the amount of light that hits a surface. It can be used to determine whether or not certain regions of the landscape fall in shadow. This information can help land managers to know how much sunlight is illuminating the land, which relates to growing season, snowmelt, soil moisture, and more. A stream biologist, for example, might want to know how much shade is cast upon a stream since that will relate to the warmth of the water and subsequently the biotic communities in the stream.

Determining illumination

The process used to make these maps is nearly identical to relief shading (see chapter 6 for more on relief shading). For these maps, the cartographer has to define the **azimuth,** which is the angular direction of the sun, measured from north in clockwise degrees from 0 to 360, as well as the **altitude,** which is the angle of the sun above the horizon (figure 16.16). Altitude units are in degrees, from 0 on the horizon to 90 directly overhead.

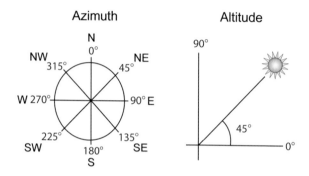

Figure 16.16   Illumination is determined using two parameters that identify the location of the sun—azimuth and altitude.

Illumination is calculated using a GIS and DEM data by first setting a position for the sun using azimuth and altitude. The illumination value of each cell relative to its neighbors can then be calculated. In order to calculate the illumination of a surface for a particular time and day, you need to know where the sun was positioned relative to the surface. You can use a solar ephemeris to get this information. An **ephemeris** charts the movement of celestial bodies (planets, moons, etc.) and predicts their positions at a given time; a **solar ephemeris** gives us information about the sun's position. The solar ephemeris is used to set the local sun angle and azimuth for the day and time that is being considered when the illumination map is made.

Illumination maps

Illumination maps take into account the location of the sun and any objects that impede the sun's illumination such as hills, trees, or buildings. In figure 16.17 below, the darker areas are in shadow, and the lighter areas are illuminated. Note that illumination maps such as this one often display relief inversion (see chapter 6 for a discussion of relief inversion) because the azimuth of the illumination is not from the northwest.

CURVATURE

So far we have made slope computations based on a linear surface, or one that is assumed to increase or decrease in height linearly between two points. In actuality, the slope is likely a curved surface curving continually between the two points. The greater the distance over which slope is computed, the greater the potential curvature between the two points. **Curvature** is the amount that a surface deviates from being flat (linear). Curvature can be used to determine many important aspects of a surface. For example, it can describe the physical characteristics of a drainage basin in an effort to understand erosion and runoff processes.

Scientists in a number of fields are interested in describing the **slope curvature** of hillsides that are altered by natural or human causes into **concave** or **convex** forms. For example, soil scientists and geomorphologists divide hillsides into different **hillslope elements** (figure 16.18) because the amount of soil erosion and genesis differs on each part of the hill. Hillslopes are classified into summit, shoulder, backslope, footslope, and toeslope elements. The **summit** is the level area at the top of the slope. Below that is the **shoulder,** which is convex. This descends to the steepest part of the slope, the **backslope,** which is a linear portion of the hillslope. The backslope descends to a concave portion of the hillslope called the **footslope,** which then merges with the **toeslope,** which either has a linear slope or is slightly concave. These landscape positions can greatly influence hydrological and other processes such as soil genesis.

Let's look at how we can distinguish linear, convex, and concave slopes, both on contour maps and by numerical slope curvature analysis that you can perform with GIS software.

Linear slopes (figure 16.19B) that increase or decrease in height uniformly are shown on a topographic map by evenly spaced contours. The separation between contours tells you the relative steepness of hillsides, assuming a constant contour interval on the topographic map showing the features. Widely separated contours equally spaced on the map tell you that the hillside has a uniform gentle slope, whereas closely spaced contours show a uniform steep slope.

The elevation values in a DEM can be used to illustrate mathematically small linear parts of hillsides. Notice in figure 16.19 that the elevations from a 10-meter resolution DEM are 20, 15, and 10 meters from the top to the bottom of the hillslope, so that the average of the top and bottom elevations is the same as the middle elevation value of 15 (figure 16.19E).

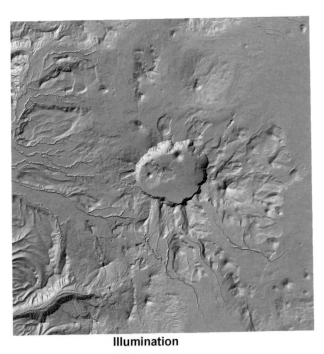

Illumination Value

High : 254

Low : 0

Figure 16.17 An illumination map for Crater Lake looks similar to a shaded relief map, but the azimuth of the sun in this example is from the southeast at 135 degrees and the sun angle is 35 degrees. For the proper relief effect, turn this page upside down.

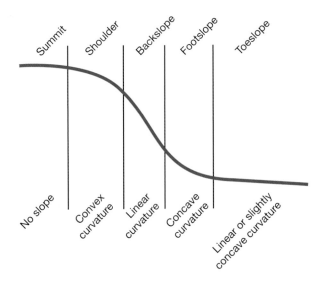

Figure 16.18 There are five hillslope elements, defined by their position and slope; these can have considerable influence on geomorphic and hydrologic processes.

Figure 16.19 Hillsides can have linear, convex, or concave sections that you can see from the spacing of contours on a topographic map or from the elevation values in a DEM.

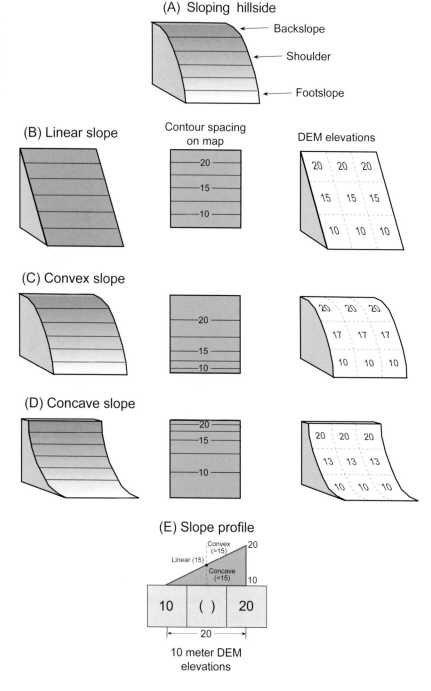

(A) Sloping hillside

Backslope

Shoulder

Footslope

(B) Linear slope

Contour spacing on map

DEM elevations

(C) Convex slope

(D) Concave slope

(E) Slope profile

Convex (>15)

Linear (15)

Concave (<15)

10 meter DEM elevations

An outward curving **convex slope** (figure 16.19C) will have contours more widely spaced at the top of the hill, with a progressive increase in spacing between contours downslope. Standing at the top of the convex hill, you will be able to see only the top part of the convex hillside. Notice that the DEM elevations from top to bottom are 20, 17, and 10 meters, so that the middle elevation of 17 is greater than the average of the top and bottom elevations. This gives you a general rule—the middle elevation on a convex slope is greater than the average of the top and bottom elevation values.

An inward curving **concave slope** (figure 16.19D) has the opposite downslope contour arrangement—contours are more closely spaced at the top of the hill and more widely spaced downslope. You would be able to see the entire concave hillside if your viewpoint from the hilltop was not obstructed by vegetation or other objects. The DEM elevations from top to bottom are 20, 13, and 10 meters, so that the middle elevation of 13 is less than the average of the top and bottom elevations. This gives you another general rule—on a concave slope, the middle elevation is less than the average of the top and bottom elevation values.

Determining curvature

Curvature is the second derivative of a surface, or the slope of the slope. To fully understand the influence of curvature of a surface, it is necessary to understand the two types of curvature—profile curvature and planform curvature.

Profile curvature

Profile curvature is parallel to the direction of the maximum slope. A negative value (figure 16.20A) indicates that the surface is vertically convex at that cell. A positive profile (figure 16.20B) indicates that the surface is vertically concave at that cell. A value of zero indicates that the surface is linear (figure 16.20C). Profile curvature affects the acceleration or deceleration of flow across the surface. Note that this is the same as the linear, convex, and concave slopes shown in figure 16.22.

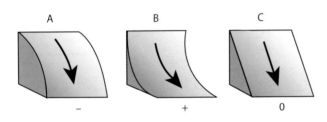

Figure 16.20   Profile curvature is parallel to the slope and indicates the direction of maximum slope. It affects the acceleration and deceleration of flow across the surface.

Planform curvature

Planform curvature (commonly called **plan curvature**) is perpendicular to the direction of the maximum slope. A positive value (figure 16.21A) indicates the surface is horizontally convex at that cell. A negative plan (figure 16.21B) indicates the surface is horizontally concave at that cell. A value of zero indicates the surface is linear (figure 16.21C). Profile curvature relates to the convergence and divergence of flow across a surface.

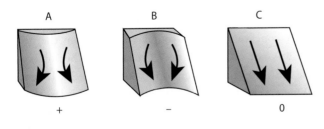

Figure 16.21   Plan curvature is perpendicular to the slope and affects the convergence and divergence of flow across the surface.

Combinations of curvature

Understanding the combinations of plan and profile curvature is important (figure 16.22). The slope affects the overall rate of movement downslope. Aspect defines the direction of flow. The profile curvature affects the acceleration and deceleration of flow and, therefore, influences erosion and deposition. The plan curvature influences convergence and divergence of flow. Considering both plan and profile curvature together allows us to understand more accurately the flow across a surface.

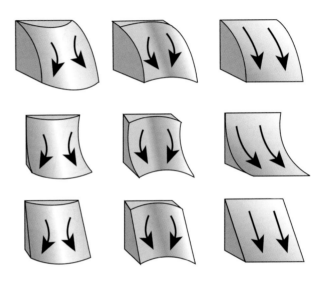

Figure 16.22   Combinations of profile and plan curvatures help us understand flow across a surface.

PROFILES

So far in this chapter we have concentrated on the measurement, calculation, and mapping of slope information. Slope zone and gradient path maps show you changes in these quantities across the map. Knowing the spatial distribution of slopes is important in land management and site planning. But even skilled map users often find it difficult to visualize the three-dimensional nature of changes in the terrain from slope and gradient path maps.

Another way you can visualize changes in slope is to view the change in elevation of a surface along a line. This is called a **profile.** Architects call this the **elevation view,** which is a three-dimensional view of an object from the side. Since you usually can't see a side view of the land surface directly (as you can see the side of a building in an architectural drawing), you must create a profile from elevations along a line on the surface.

Profiles have a number of practical uses beyond providing a dramatic picture of the vertical or magnitude aspects of geographic distributions. For instance, topographic profiles are used to plan the routing of roads, railroads, pipelines, and canals. When it comes time to determine the earth-moving requirements of such projects, profiles are used to compute the volume of material involved in making surface cuts and fills.

Constructing a straight-line profile

You can construct a straight-line profile from contours on topographic maps, or isobath lines on nautical charts, by following these steps:

1. First draw a straight line on the map between the points of interest. This is called the **profile line** (line *AB* on the map at the top of figure 16.23).

2. Determine how many intervals separate the contours with the highest and lowest values that cross or touch the profile line, and add two to this number (totaling seven in this example).

3. On a sheet of paper, draw as many equally spaced horizontal construction lines as the number you determined in the previous step. **Construction lines** are lines you draw simply for the purpose of creating or "constructing" the profile. You can save yourself some work by using graph paper for this purpose.

4. Orient the paper so that the bottom construction line is aligned directly below the profile line.

5. Label the construction lines with contour elevation values, beginning at the bottom line with the lowest value (one interval below the lowest contour crossed by the profile line) and proceeding to one interval above the highest contour value.

6. From every point at which the profile line is crossed or touched by a contour, draw a vertical dashed line downward to the construction line that has the same elevation value.

7. Draw a smooth curve through successive points at which the dashed vertical lines intersect the construction lines. Remember that continuous geographic distributions are usually smooth rather than angular so you want to connect the intersecting points using smooth lines. When constructing a profile of the terrain surface, you should take local landform conditions into account in modifying the profile. If you know there are sharp drops such as cliffs, you can add those into the profile.

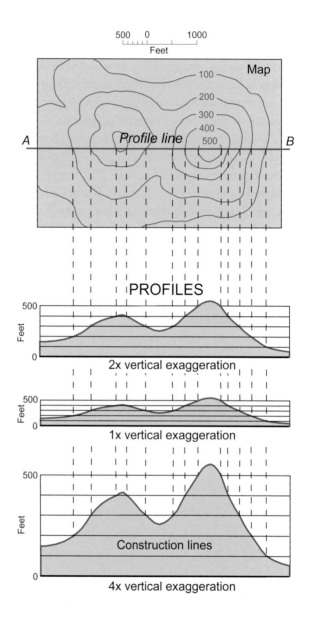

Figure 16.23 A profile can be constructed from a contour map along any desired profile line, such as from *A* to *B*. Use vertical exaggeration if the profile does not give the same impression of landform variation that you would have in the field.

Vertical exaggeration

One arbitrary decision you will have to make is the amount of vertical exaggeration to use. **Vertical exaggeration** is the ratio between the vertical scale and the map's horizontal scale. It can be used to emphasize subtle changes in a surface, but too much exaggeration will give a false impression of the surface variations. (Note, for example, the extreme vertical exaggeration evident in the terrain profile shown in figure 13.27.) You should not use vertical exaggeration unless there is a definite need for it, for example when there are subtle slope features that you have to show. If there is no vertical exaggeration, slopes are shown to scale, and slope angles can be scaled from the profile. Large vertical exaggeration is the primary cause for misinterpreting profiles. If you have to use vertical exaggeration, use the smallest exaggeration to produce the results you desire.

The effects of vertical exaggeration are shown in figure 16.23. The middle profile's vertical scale is the same as the map scale. The top and bottom profiles were drawn by making the vertical scale larger than the horizontal scale. Such vertical exaggeration not only increases the height of the profile, but also steepens and lengthens the hillsides.

The spacing of the horizontal construction lines defines the scale of the vertical axis on the profile and therefore

the vertical exaggeration. To exaggerate minor terrain features, you increase the spacing; to deemphasize features, you decrease the spacing. The unexaggerated middle profile in figure 16.23 is said to have a "1× vertical exaggeration," and it is drawn "to scale." The top and bottom profiles in this figure have been vertically exaggerated by factors of two (2×) and four (4×). If the map scale is 1:24,000 (1 inch to 2,000 feet), for example, the vertical scale for the 2× exaggeration will be 1:12,000 (1 inch to 1,000 feet). The 2× profile may appear more realistic than the overly flat 1× and excessively steep 4× profiles. When dealing with a region of high relief, you may occasionally find it necessary to deemphasize the mountain peaks somewhat (vertical exaggeration <1×) while exaggerating the foothills to make the profile look natural.

Constructing an irregular profile

A profile line doesn't have to be straight, of course. With a little extra work, you can create an irregular profile for a stream bed or hiking trail as well (figure 16.24). To do so, follow these steps:

1. Determine and clearly mark on the map the irregular profile line.

2. Follow step 2 in the previous example ("Constructing a straight-line profile") by determining how many intervals separate the contours with the highest and lowest values that cross or touch the profile line, and add two.

3. Measure the total length of the irregular profile line, and draw a horizontal construction line of this length directly below the map.

4. Beginning at one end of the profile line, measure the length of the first line segment (segment 1–2 in figure 16.24), mark off this distance from the left end of the construction line with a small tick, and write the line segment endpoint number below the tick.

5. Repeat the step above until you reach the end of the profile line—nine line segments were measured, plotted, and labeled in figure 16.24.

6. Above the initial construction line, draw as many equally spaced horizontal construction lines as you determined in step 2. Label these lines from bottom to top as you did in step 5 of the previous example.

7. Find the elevation at the endpoint of each profile line segment by interpolating between contour lines, then plot the vertical position of each elevation point directly above the endpoint tick marked on the horizontal axis.

8. Draw smooth curves through the elevation points plotted in step 7 to finish the profile.

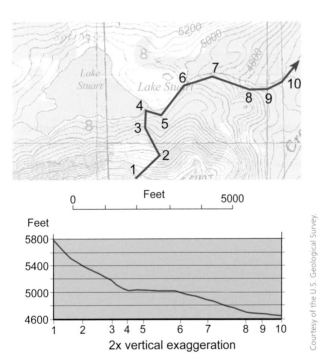

Courtesy of the U.S. Geological Survey.

Figure 16.24 The line along which a profile is constructed needn't be straight. For example, the profile of an irregular hiking trail can be constructed by scaling trail distances along the horizontal axis using the steps outlined in the text.

Topographic profiles

Terrain profiles based on contour lines aren't a perfect reconstruction of the topographic surface, of course. A profile constructed from contour lines gives elevation information only along the profile line, from which you must infer the complete terrain surface. If contour lines are optimally spaced with respect to terrain undulations, your profile will accurately show the surface. If they're poorly situated, however, your portrayal of the surface will be poor. Also be aware that a few topographic maps have a variable contour interval, so you must adjust the spacing of your horizontal construction lines in these cases.

Terrain analysis and GIS software have greatly simplified the construction of topographic profiles. You simply trace the path of your profile on a digital topographic map displayed on your monitor, and the software finds the elevations along the profile line and automatically constructs the profile using the same concepts as those explained above. You still must tell the program the amount of vertical exaggeration to use, but you can change the exaggeration almost instantaneously if you are not pleased with your initial choice.

Nontopographic profiles

Scaling the vertical axis of nontopographic profiles is complicated by the fact that the magnitude units are not comparable to those making up the horizontal dimension. The inch or centimeter units used to measure precipitation, for example, bear no physical relation to the ground units measured in kilometers or miles. This means that with nontopographic profiles there is no such thing as vertical exaggeration. There is, however, such a thing as alternately emphasizing or deemphasizing aspects of the profile by altering the spacing of the horizontal construction lines. You will have to take special care that you don't end up with a false impression of a distribution based solely on the way a profile is constructed.

CROSS SECTIONS

A diagram of the vertical section of the ground surface above or below a profile line is called a **cross section.** We'll use as our example a cross section showing the subsurface orientation of rock layers, which are a basic component of a **surficial geology map,** or a map showing bands of differing rock type exposure. Constructing a cross section from a surficial geology map is an easy matter because these maps are created by drawing contacts between rock layers at the surface on a topographic basemap containing contours (figure 16.25).

Creating a cross section from surficial geology information superimposed on contours involves the following steps:

1. Construct a terrain profile at a given vertical exaggeration from a profile line drawn on the surficial geology map by following the procedure described earlier in the chapter.

2. Mark on the profile line the intersection of contact lines between rock layers, such as layers *A, B,* and *C* in figure 16.25.

3. Find the dip angle for each rock layer from the strike-dip symbols on the surficial geology map. For instance, the symbol |– 27 means that the dip angle for rock layer *B* is 27 degrees downward to the east, assuming the map to be north-oriented.

4. Plot the subsurface rock layer contacts as straight lines drawn at the dip angle downward from horizontal.

5. Color the subsurface beds to match the surficial geology map.

Notice that the bottom cross section in figure 16.25 is drawn with a 2× vertical exaggeration. When the cross

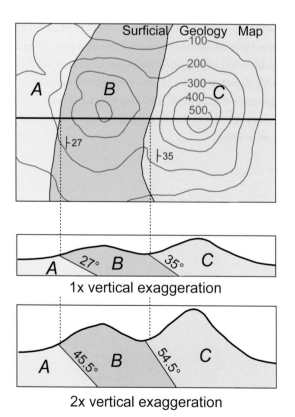

Figure 16.25 Geologic cross sections can be created from surficial geology maps and profiles, as explained in the text.

section is exaggerated vertically, the dip angles must be adjusted to match the amount of exaggeration. There are two ways to do this.

The first method starts with the construction of a cross section without vertical exaggeration (1×). Below this you draw a profile that reflects the amount of vertical exaggeration you want to use. For instance, the 2× cross section in figure 16.25 is drawn twice as high as the 1× cross section. You then drop vertical lines downward from both the top and bottom of the subsurface rock layer contact lines on the unexaggerated cross section. Now find the point at which the line from the top intersects the top of the exaggerated profile, and the point at which the line from the bottom intersects the bottom of the cross section. Those points define the endpoints of the vertically exaggerated contact line.

The second method is to compute the vertically exaggerated dip angles. The following is the equation to do this:

$$Angle = tan^{-1}(tan(dip) \times v.e.)$$

where dip is the angle relative to the horizontal plane and *v.e.* is the vertical exaggeration. The 27- and 35-degree dip angles in figure 16.25, for instance, were found to be 45.5 and 54.5 degrees on a cross section with 2× vertical exaggeration. The computation for the first dip angle (27 degrees) is the following:

$$Angle = tan^{-1}(tan(27°) \times 2) = tan^{-1}(0.509 \times 2)$$
$$= tan^{-1}(1.18) = 45.5°$$

Geologic mapping software often includes a cross section construction module. You only need to trace the profile line on a digital geologic map and give the vertical exaggeration for the cross section. The software finds the elevations along the profile, along with the contact lines between rock layers, as well as the strike and dip of each rock layer in the vicinity of the profile line. The profile is then created with the vertical exaggeration you specify, and the vertically exaggerated apparent dip angle is computed for each contact line. Lines are drawn downward from each contact at the computed angle, and each rock layer is labeled as to its formation name.

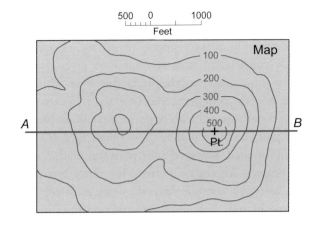

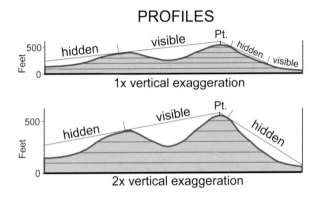

Figure 16.26 Topographic profiles can be used to identify a viewshed by determining the intervisibility of landscape features from a particular viewpoint along a line of sight. The hidden and visible portions of the landscape are correctly depicted on the profile with no (1×) vertical exaggeration (top profile). Visibility may be shown incorrectly when a vertically exaggerated profile is used (bottom profile).

VISIBILITY ANALYSIS

Terrain profiles can also be used in **visibility analysis** to provide a way for you to see what environmental features are visible from a given vantage point (figure 16.26). It is often useful to identify areas on the ground that are hidden from view. For instance, trails and campgrounds in a wilderness region might be built to minimize the visual impact of nonwilderness activities, such as clear cuts, garbage dumps, and quarries. Viewsheds can also be used in the laboratory to determine whether a certain geographic feature would be visible from a possible building site, thereby saving you a trip into the field. This information, known as **intervisibility,** can be invaluable when orienting yourself in the field.

You can determine intervisibility from a point in the landscape by following these steps:

1. Plot your viewpoint on an unexaggerated terrain profile at the correct height above the surface. The viewpoint *(Pt.)* on the profiles in figure 16.26 is 50 feet above the top of the hill, for example.

2. Draw lines from the viewpoints that are tangent to (just touch) the profile. Continue each line to where it touches the profile again (right side of figure 16.26), or to the edge of the profile diagram (left edge of figure 16.26).

3. Label the visible and hidden portions of the profile.

If profiles are constructed in several directions from a selected viewpoint (or set of viewpoints, such those along a trail), visible or hidden areas on the ground can be plotted on a **viewshed map** (also called an **intervisibility map**). GIS and DEM data can be used to mathematically create closely spaced profile lines (figure 16.27, left), determine the visible and hidden portions of each line, then connect the visible-hidden boundary points on adjacent profile lines into visible and hidden areas (figure 16.27, right).

In this chapter, we've looked at ways to analyze surfaces, but map analysis includes much more than just surface studies. In the next two chapters, we shift our focus to analysis of the abundance and arrangement of distributions and spatial associations among different patterns.

SELECTED READINGS

De Floriani, L., P. Marzano., and E. Puppo. 1994. Line-of-sight communication on terrain models. *International Journal of Geographical Information Systems* 8: 329–42.

Evans, I. S. 1972. General geomorphometry, derivatives of altitude and descriptive statistics. In *Spatial analysis in geomorphology,* ed. Chorley, R. J., 17–90. London: Methuen.

Greenlee, D. D. 1987. Raster and vector processing for scanned linework. *Photogrammetric Engineering and Remote Sensing* 53 (10): 1383–87.

Jenson S. K., and J. O. Domingue. 1988. Extracting topographic structure from digital elevation data for geographic information system analysis. *Photogrammetric Engineering and Remote Sensing* 54 (11): 1593–1600.

Lee, J. 1991. Analyses of visibility sites on topographic surfaces. *International Journal of Geographical Information Systems* 5: 413-29.

Moellering, H. J., and A. J. Kimerling. 1990. A new digital slope aspect display process. *Cartography and Geographic Information Systems* 17 (2): 151–59.

Moore, I.D., R. B. Grayson, A. R. and Ladson. 1991. Digital terrain modelling: A review of hydrological, geomorphological, and biological applications. *Hydrological Processes* 5: 3–30.

Ruhe, R. V. 1975. *Geomorphology.* Boston, Mass.: Houghton Mifflin.

Skidmore, A. K. 1989. A comparison of techniques for calculating gradient and aspect from a gridded digital elevation model. *International Journal of Geographical Information Systems* 4: 323–34.

Spencer, E. W. 2000. *Geologic maps.* Upper Saddle River, N.J.: Prentice Hall.

Wilson, J. P., and J. C. Gallant, eds. 2000. *Terrain analysis: Principles and applications.* New York: John Wiley & Sons, Inc.

Zevenbergen, L. W., and C. R. Thorne. 1987. Quantitative analysis of land surface topography. *Earth Surface Processes and Landforms* 12: 47-56.

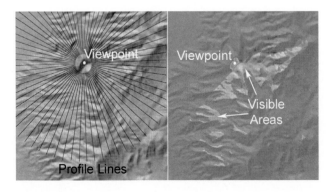

Figure 16.27 An intervisibility (viewshed) map is created from a large number of profile lines radiating away from a viewpoint. Visible and hidden areas are inferred from the visible and hidden portions of each line.

chapter
seventeen **SPATIAL PATTERN ANALYSIS**

17

Spatial pattern analysis

Looking at the landscape from your window seat in an aircraft on a clear day, you may have noticed that features on the ground were arranged in different patterns. Flying over western Oregon, for example, you will see irregularly shaped patches of forest that appear to be randomly placed on hills and valleys, but in other areas they form a regular, checkerboard-like arrangement of forest stands at different stages of tree growth. You may recognize towns from the air by the clustered arrangement of residential buildings around their commercial centers.

It is often the *arrangement* of the features that allows us to identify them. When we look at the landscape as represented on a map, our attention is also drawn to the way features are organized spatially. For example, vegetation patches shown with a green tint will have a pattern on the map that is similar to what you would see from the air. Towns on the map will appear clustered by the arrangement of streets and buildings. We recognize the **spatial pattern** by the arrangement of the features on the landscape.

Seeing how features are arranged in different patterns as represented on maps is a key part of understanding their spatial distribution. For example, biogeographers who study the spatial pattern of trees depicted on image and cartographic maps want to know the density of trees in different areas (for example, the number of trees per stand) and whether the trees are spaced in a random, regular, or clustered arrangement. This information helps them understand the habitat requirements for animal species in their region. A forest manager may use the same tree density and pattern information to determine how a forest should be replanted or thinned to achieve optimal growth and the healthiest stands of trees.

Different densities and arrangements of features may be easy for you to see on a map, but another person may not see the features in the same way. To lessen disagreements arising from different visual impressions of the same distribution on the map, mathematical measures of patterns have been devised that you can use to describe these densities and arrangements in a more consistent and repeatable manner. **Feature counts** and **measures of spatial arrangement** are two fundamental aspects of spatial pattern analysis. Let's begin our discussion of spatial arrangement by looking at counts of point features.

FEATURE COUNTS

When making feature counts, you divide the landscape into a number of **data collection units,** and then make **counts** of the number of point, line, or area features within each unit. Data collection units, such as counties, census tracts, or city lots often are irregular in size and shape. They can also be identical in size and shape, such as the 24 square **quadrats** (rectangular data collection units) in figure 17.1. Counts of point, line, and area features are easily made within each unit. Foresters, for instance, make point counts of the number of harvestable trees within stands. Hydrologists may make line counts of the streams or gullies by type in different watersheds. Urban planners often make area counts such as the areas of buildings within city blocks.

These point or line counts are often standardized (made to conform to a predefined expected format) by dividing the count value by the area of its data collection unit to obtain a **density value,** such as the number of trees per square kilometer in a stand or the number of streams per square mile of watershed. Dividing an area count by the data collection unit area gives you an **area proportion,** such as the portion of a city block covered by buildings.

Each feature you count can then be weighted by an attribute (descriptor) of the feature, such as the diameter of the tree at breast height (used to estimate the amount of timber volume in a single tree or stand of trees) or the length of the stream or the number of people in a building. These weights give you a numerical *value* for each count that you can sum to get a *total value* for each data collection unit, such as the total timber volume in a stand or the total length of streams in each watershed or the total number of people in a city block. Maps of counts, total values, or densities show you the spatial pattern of feature abundance across the landscape.

There are different mathematical methods for analyzing feature counts. The method you use depends on whether you are analyzing point, line, or area features. Counting the number of features within small data collection units on the earth or a map is the beginning point for spatial pattern analysis.

Point counts

Dividing the landscape into collection units is the first step in spatial pattern analysis. The area of interest in figure 17.1, for example, as been divided into 24 quadrats of the same size. You can then count the number of features in each data collection unit.

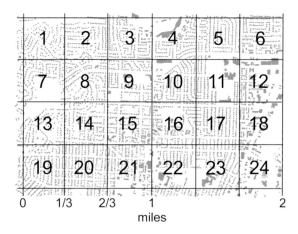

Figure 17.1 Twenty-four data collection units, each one-ninth of a square mile in area, have been placed over a large-scale engineering plan showing building footprints for a portion of Corvallis, Oregon.

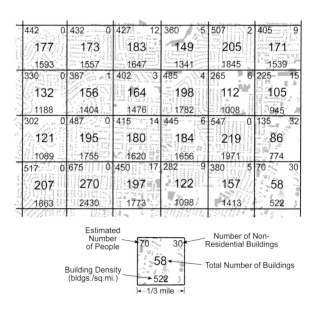

Figure 17.2 Point counts of buildings within data collection units can help you understand the spatial pattern of buildings in a city (see text for explanation).

Point counts give you information about the *abundance* of point features in each data collection unit. When counting point features, you disregard the actual area covered by each. In figure 17.2, for example, the point at the center of each building is counted even though the buildings vary considerably in size. The number printed in the center of each unit is the total number of buildings counted. These counts tell you that there is a wide range (58 to 270) in the abundance of buildings among units.

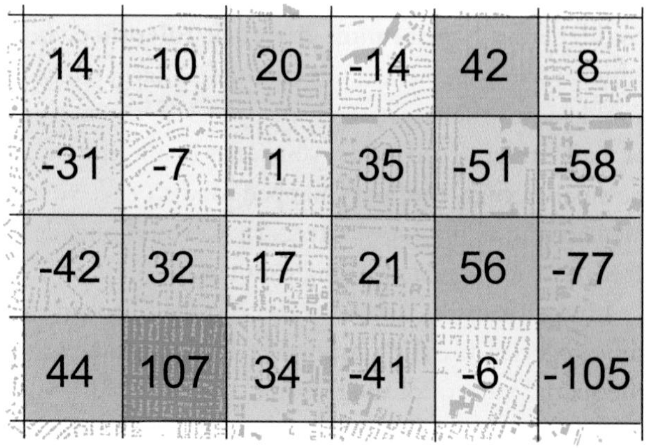

14	10	20	-14	42	8
-31	-7	1	35	-51	-58
-42	32	17	21	56	-77
44	107	34	-41	-6	-105

Figure 17.3   A map of above- and below-average numbers of buildings in the data collection units shows you the relative abundance of buildings. The number in the center of each quadrat is the positive or negative difference from the average of 163 buildings per unit.

Your search for spatial patterns in the point features on the map may be aided by finding the **relative abundance** of features in each data collection unit. Relative abundance is determined by first finding the average number of features in a unit (163 buildings in this example). You can then identify the above- or below-average units on the map, such as in figure 17.3. You can now examine the map of above- and below-average abundance to see if data collection units with low and high building abundance appear to be clustered together or randomly located on the map. From the map in figure 17.3, you can see that there appears to be little spatial clustering of below- and above-average relative building abundance on the map.

You may be able to gain additional information about the features by making point counts of feature subcategories, or a further subdivision of qualitative or quantitative categories. In our building count example, the subcategories of residential and nonresidential buildings have been counted separately in order to better understand the land-use patterns in the city. In figure 17.2, the number of nonresidential buildings is printed in the upper right corner of each data collection unit. The subcategory counts show you that the largest number of nonresidential buildings occurs in the two data collection units at the lower right corner of the map. These units also have the lowest total number of buildings. You can also see that the units with the largest total number of buildings have zero nonresidential buildings. These observations can lead you to conclude that for this area of interest there appears to be an inverse relationship between the total number of buildings and the number of nonresidential buildings in a data collection unit.

The relationship between residential and nonresidential buildings can also be studied by computing the percentage of nonresidential buildings in each data collection unit. Dividing the number of nonresidential buildings by the total number of buildings in a data collection unit, and then multiplying this proportion by 100, you will find that most of the units are between 0 and 10 percent nonresidential. The exceptions are the two units with the smallest total number of buildings, which are 37 and 52 percent nonresidential.

Another thing you can do is weight each point count by the value of some attribute of the feature. For example, on average 2.5 people live in each residential building. You could weight each count by 2.5 (each single count now has a value of 2.5), then total the weighted counts to get an estimate of the number of people residing in each data collection unit. In figure 17.2, these population estimates are printed in the upper left corner of each unit. This is an example of equal weighting of counts, but for most attributes the weights will be different for each count. For instance, the number of people per apartment building will be different than the number of people per single family residence.

When analyzing feature count data, you normally will adjust for differences in the sizes of data collection units. The square units in figure 17.1 require no adjustment because each is $1/3$ mile by $1/3$ mile, or $1/9$ of a square mile in area. Data collection units such as tree stands, watersheds, counties, census tracts, and block groups (figure 17.4) could vary widely in shape and area, however. In these cases, you should divide the number of features counted by the area of the data collection unit to get a feature density value for each unit. In figure 17.2, you would divide the number of houses in each unit by one-ninth of a square mile to obtain the building density per square mile. These density values are printed at the bottom center of each zone. In this example, the building densities are directly proportional to the building count because the collection units are equal in size, but density values will be noticeably different for data collection units varying widely in size.

Line counts

Linear features are counted within data collection units in the same manner as point features. You first disregard the width of each feature and treat it as a one-dimensional straight or curved line. This simplification of each street from an area to a line is appropriate since the width of each street on the map is often exaggerated to make streets easily visible on the map, especially when using a cased line symbol. The centerline in this case is a better representation of the actual location of the street on the ground.

Counting the number of streets in each data collection unit seems a simple task, but there is a problem. A street may cross several data collection units, so you must choose whether to count the street in each unit or only in the unit where most of the street lies. You can get around this problem by breaking each street into segments. A **street segment** is the portion of the street between two adjacent intersections, or between the intersection of a dead-end street and its endpoint. You can then identify each segment as we did with the orange point symbols on the map in figure 17.5. Then you can count the number of streets in each unit by tallying the street segment points. This is the number printed in the center of each data collection unit.

You would expect that the middle left data collection unit with 19 street segments would have a much lower abundance of streets than the middle unit at the bottom with 39 street segments. The count of 39 is misleading, however, since it includes numerous small pieces of streets on the boundary of the unit. To lessen this problem, you

could define a **minimum distance tolerance,** below which a street segment is not counted. Alternatively, you could locate the street segment points at the midpoint of each street. To do this, your area of interest should be smaller than the total mapped area so that you can identify the midpoint of streets that extend outside the area in which you are making your counts.

Linear features like streets can be broken into subcategories for the purpose of learning more about the spatial pattern of the features. For example, you can gain an understanding of the historical development of subdivisions in a city by counting the number of dead-end streets in each data collection unit, particularly streets ending in cul-de-sacs. Data collection units with an abundance of dead-end streets were probably constructed from the 1970s to the present, whereas a grid pattern of intersecting streets indicates an earlier approach to subdivision design. In figure 17.5, counts of dead-end streets are printed in the upper right corner of each data collection unit. Notice the high counts in the

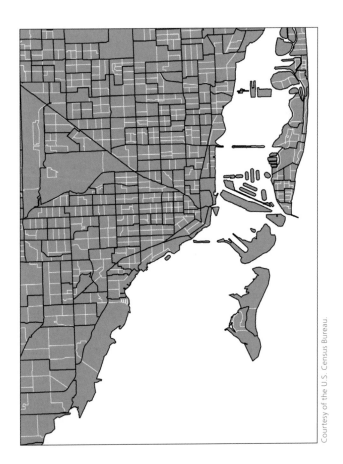

Figure 17.4 United States census data collection units range from coarser-level tracts (shown with black outlines) to finer-resolution block groups (shown with white outlines). As in this example of the Miami metropolitan area, they are usually irregular in shape and vary widely in area.

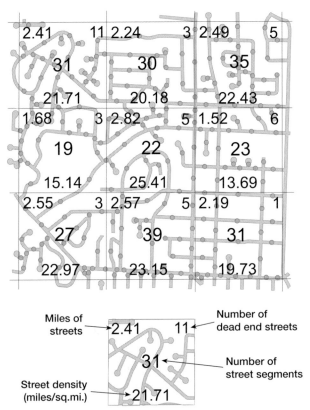

Courtesy of the U.S. Census Bureau.

Figure 17.5 Line counts within data collection units include the number of street segments and the total length of streets in miles.

upper left data collection unit. This, indeed, is an area of recent subdivision construction.

Weighting each street segment by its length on the ground may give you more accurate information about the abundance of streets in each data collection unit. Once you have delineated the street centerlines in a unit, you can use any of the distance measurement methods described in chapter 11 to determine the length of each street segment in the unit. In figure 17.5, the total length of street segments is printed in the upper left corner of each unit. Notice that the ratio of the difference in total street length between the units with the lowest (1.68) and highest (2.57) values for the length of streets (1.68/2.57) is considerably larger than the ratio of total street segments between the units with 19 and 39 street segments (19/39). Weighting the street counts by length appears to have adjusted for the large number of small street segments counted in the upper right unit.

You may also want to standardize the street length values to adjust for differences in the sizes of data collection units. To do this, use the total street length values to compute a street density value for each data collection unit. Dividing the street lengths by the area of each unit, you will obtain a street density expressed as miles of street per square mile. In this example, street densities shown in figure 17.5 at the bottom of each unit are directly proportional to street length because the collection units are equal in size, but values will be different for data collection units that vary in size.

Area counts

Recall that we said when counting point features, you disregard the actual area covered by each. You might be tempted to confuse point counts of areal feature with area counts. **Area counts** are tallies of the total area occupied by a certain type of feature within a data collection unit. You may be able to determine the total surface area of features having two-dimensional extent if they are portrayed on the map without size exaggeration. The building footprints shown in figure 17.6 were drawn without exaggeration, for example (see chapter 10 for more on cartographic exaggeration). Using the area computation methods described in chapter 15, you can compute the map area of each building, and then convert the map area to ground area if you know the map scale. Making these measurements and doing the map-to-ground area conversion for the buildings in each data collection unit, we obtained the total ground area covered by buildings in each unit. These area counts are shown in the center of each unit. The percent coverage value at the bottom of each data collection unit is the total area

of buildings divided by the one-ninth of a square mile covered by each unit and then multiplied by 100.

You should remember that this method of computing area feature sizes works only if the features are portrayed faithfully on the map. Faithful size portrayal can be expected on very large-scale maps such as engineers' plans in the 1:1,000 to 1:10,000 scale range. Buildings and other features drawn larger than their actual size should be expected at smaller map scales due to cartographic generalization.

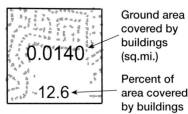

Ground area covered by buildings (sq.mi.)

Percent of area covered by buildings

Figure 17.6   Area counts of buildings in data collection units give the total and percent of ground area covered by buildings.

SPATIAL ARRANGEMENT

The **spatial arrangement** of features is the way things are placed geographically. To clarify what we mean by spatial arrangement, it is convenient to look first at ways to analyze the arrangement of point features. Then we will look at ways to analyze the arrangement of line and area features. Many of the measures we will look at can be used for points, lines, and areas. When used with lines or areas, the centroids of the features are used; when used with points, the features themselves are used in the calculations. We will be examining a variety of techniques, starting with their use with point features.

Point feature arrangement

Point features are in theory dimensionless. If we use them to demonstrate what is meant by spatial arrangement, we can focus solely on *direction* and *distance* relationships between points and not have to deal with the added dimensions of length and area. There are three basic arrangements of point features that relate to distance—regular, clustered, and random. We will discuss these first and then look more closely at direction relationships.

The arrangement of a set of point features is said to be **regular** when there is equal spacing between points. Square and equilateral triangular grids are the most regular arrangement of features on a flat surface. If, for example, the subsurface structure in an oil field were homogeneous, oil wells would likely be regularly arranged to minimize drilling costs yet maximize the area drilled. You would expect the mapped locations of wells to be arranged in a square or equilateral triangular pattern.

When point features tend to be grouped into one or a few small areas, a **clustered** arrangement results. For instance, if an oil company can lease surface rights only to a few small parcels of land over an oil field, it may decide to drill a number of wells clustered together in each parcel and angle the pipes out in different directions under the surrounding area. The wells will be mapped as clusters of closely spaced point features.

When there is no apparent order in the spatial arrangement, it is said to be **random.** There is likely to be some clustering and some regularity in random patterns, but not enough so that the pattern as a whole looks either clustered or regular.

Several quantitative methods have been devised to define randomness, clustering, and regularity in spatial distributions. **Spatial autocorrelation** among features is one of the most widely used measures of the degree to which point features tend to be clustered together

(positive spatial autocorrelation) or evenly dispersed (negative spatial autocorrelation). We'll look at three popular measures: variance–mean ratio, (global) Moran's *I* autocorrelation index, and the nearest neighbor statistic.

Variance–mean ratio

The **variance–mean ratio** (VMR) is used to examine the spatial arrangement of point patterns. It does not take into account any attributes of the features. These measures assess whether the points have a pattern that is significantly different from what would be expected given a random process. These measures can help us answer the question, "Do the points deviate from spatial randomness?"

The VMR is based on superimposing a grid of regularly shaped data collection units called **quadrats** over the area of interest on your map. We'll use the square grid of data collection units for counts of buildings shown in figure 17.1 as our quadrats. You count the number of features *(x)* falling within each quadrat, such as the total number of buildings counted in each data collection unit in figure 17.2.

You next calculate two basic statistics. The first is the **arithmetic mean** $(\bar{x})$ of the quadrat values or counts. The mean is defined as the sum of the quadrat values $(\sum x)$ divided by the number of quadrats *(n)*, and is written mathematically as the following:

$$\bar{x} = \frac{\sum_{i=1}^{n} x_i}{n}$$

The second statistic required is the **variance,** which is defined as the average of the squared deviations of the individual quadrat values from their means. The equation for the variance is written as the following:

$$var. = \frac{\sum_{i=1}^{n} (x_i - \bar{x})^2}{n}$$

The mean and variance for the building count data in figure 17.2 are 163.4 and 2,116.0, respectively.

Through experimentation, researchers have found that, for a perfectly random arrangement of points, the variance of quadrat values about their arithmetic mean is equal to the mean itself. Therefore, the ratio of the variance and mean for a random distribution is equal to 1. For the VMR to be close to 1, the pattern must meet three conditions:

1. Only a few quadrats will have no population occurrences.

2. A large number of quadrats will have an intermediate number of occurrences.

3. A few quadrats will have a relatively large number of occurrences.

To interpret the VMR, use the following guides:

- VMR < 1 implies a regular arrangement
- VMR = 1 indicates a random arrangement
- VMR > 1 implies a clustered arrangement

The upper right map segment in figure 17.7 appears to be a random arrangement of point features. The counts within each quadrat total 81, so the arithmetic mean is 81/9, or 9. The sum of squared deviations is

$$(7-9)^2 + (6-9)^2 + (13-9)^2 + (10-9)^2 + (11-9)^2$$

$$+ (12-9)^2 + (6-9)^2 + (9-9)^2 + (7-9)^2 = 56$$

This gives a variance of 56/9, or 6.2. The VMR, then, is 6.2/9 or 0.7, so the point arrangement is slightly more regular than random, which would be 1.0.

At the opposite extreme from randomness, point features may be arranged in a more regular than random manner. With regular arrangements, the vast majority of quadrats will have similar counts. The variance between quadrat values will therefore be less than the mean of quadrat totals, and the VMR will be less than 1. Therefore, small VMRs can be taken as an indication of spatial pattern regularity.

The top left map in figure 17.7 has the 81 points arranged in a square grid so that each quadrat contains 9 points. The arithmetic mean is again 9, but the variance is 0 because there are no deviations from the mean in the quadrats. So the VMR is 0/9 or zero, the value expected for a perfectly regular arrangement of point features.

If the point arrangement is characterized by large numbers of both empty quadrats and quadrats containing many individuals, we can say that the distribution is clustered. Because such a pattern will make the variance between quadrat values greater than the mean of quadrat totals, the VMR will become larger than 1. As a result, large VMRs indicate that individual population occurrences are clumped together in space.

The bottom two maps in figure 17.7 have the same regular and random arrangements as in the maps above them, but the point spacing has been reduced so that all points fall in the center quadrat. The two arrangements both have an arithmetic mean of nine, but the sum of squared deviations is now the following:

$$8 \times (0-9)^2 + (81-9)^2 = 8 \times 81 + 5,184 = 5,832$$

The variance is now computed as 5,832/9 or 648, giving a VMR of 648/9 or 72 for both maps. This high value indicates a tightly clustered arrangement. Notice that this does not indicate how the points are arranged within the one quadrat where they all lie. Rather, it again gives us an indication of the overall pattern in the quadrats.

Let's now compute the VMR for the building counts in figure 17.2, using the data collection units as quadrats. We mentioned above that the arithmetic mean for the 24 quadrats is 163.4 buildings, and the variance is 2,116. This gives a VMR of 13, which indicates a slightly clustered arrangement of buildings in this part of the city. A city planner might use this information to make a map showing different building arrangements within the city. This map could be used by planners studying how residences and businesses are clustered within the areas they help to manage.

The VMR gives you a mathematical way to describe the arrangement of a spatial pattern, but it does have drawbacks. One criticism is that the terms "random," "clustered," and "regular" are somewhat deceptive because they really refer to variations in the quadrat data values. They don't refer to the spatial pattern of features on the ground from one quadrat to the next, as you might expect. Since the relative location of quadrats having different numbers of population occurrences isn't considered in the computations, the resulting measure of arrangement is not geographic in nature.

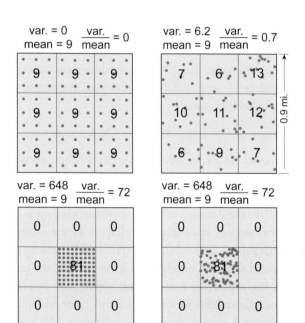

Figure 17.7 **Variance–mean ratios for 4 arrangements of 81 point features within 9 quadrats.**

Another criticism of the VMR is that its ability to detect patterns that are not random depends on the quadrat size used in the analysis. In apparently random or regular arrangements, the quadrat analysis may indicate clustering as the size of the quadrats is steadily increased. The most marked demonstration of this problem occurs when the quadrats have an area about the same size as the area of the spatial grouping of point features, as in the bottom maps in figure 17.7.

This relationship between quadrat size and the ability to detect clustering can be put to good use, however. By computing the VMR several times with quadrats of different sizes, you can determine at what scale clustering occurs. This ability is important because it lets you detect patterns that are not random in a distribution when visual inspection alone may reveal no sign of clustering. In other words, the human eye seems to use a built-in quadrat size in judging pattern arrangement. The scale at which clustering occurs is important in ecological studies. For example, botanists study the spatial arrangement of plants on the landscape and the factors responsible for their arrangement.

Moran's *I* autocorrelation index

There are ways to define uniform, clustered, and random arrangements quantitatively that take the spatial locations of data collection units into account. One commonly used method is to compute the degree of spatial autocorrelation among the features. **Spatial autocorrelation** deals with the degree to which features on the earth's surface are both spatially and numerically similar to other features located nearby. The notion that there is a high degree of spatial autocorrelation in the arrangement of many features is what underlies **Tobler's First Law of Geography:** "All things being equal, closer things are more related than distant things."

The Moran's *I* index is a commonly used measure of spatial autocorrelation. With this measure we ask "Given the spatial pattern, are the *values* different from what we would expect if they came from a random distribution?" Moran's *I* requires an attribute associated with each feature for the analysis. This attribute can be a count, a severity, or a magnitude; in fact, it can be any numeric value, as long as the value isn't the same for every feature. Moran's *I* helps us to determine if the values and their associated features are clustered, randomly distributed, or dispersed. This measure helps us answer the question, "Can we detect any spatial order in the location of points?"

Moran's *I* measures the correlation in terms of proximity between adjacent features of the same phenomenon. As with the VMR, you begin by superimposing a grid of regularly shaped quadrats over the area of interest on your map. As an alternative to quadrats, Moran's *I* can be computed for any contiguous area data collection units such as census tracts, city blocks, or counties. Autocorrelation within any numeric attribute at the ratio level of measurement (see chapter 8 for a discussion of measurement levels) can be measured by the Moran's *I* index, as long the numerical value is a density, percentage, or other value that is standardized by the area of each data collection unit.

To compute Moran's *I,* you begin by counting the number of features *(x)* falling within each data collection unit, such as the total number of buildings in figure 17.3 we counted in each quadrat in figure 17.2. You again determine the arithmetic mean and variance for your counts.

The next step is to compare the values for each data collection unit with values in the other units. If the First Law of Geography is valid, you would expect to see that adjacent units have similar values and that the similarity decreases as the distance between units increases. You can see if this relationship of similarity to distance holds true by measuring what statisticians call the covariance between pairs of data collection units.

Covariance is a measure of the degree to which the values for each data collection unit deviate from the arithmetic mean for all units. The following equation is used to compute covariance for units *i* and *j*:

$$cov_{ij} = (x_i - \bar{x})(x_j - \bar{x})$$

To find the covariance for the building count values in quadrats 1 and 2, for instance, you would use the following calculation:

$$\text{Cov}_{12} = (177 - 163.4)(173 - 163.4) = 130.6$$

Covariance values (rounded to the nearest whole number) for all combinations of the 24 quadrats in figure 17.1 are presented as a table in figure 17.8. This **covariance table** is read like the distance table you may have used on a state highway map or road atlas (see chapter 11 for more on distance tables). To read the covariance table, look up one data collection unit number along the side and the other along the top. Then find the cell where they intersect. The covariance value for quadrats 1 and 2, for example, is found near the top of the leftmost column as 131.

Notice that the covariance values can be either positive or negative numbers. Large negative covariances indicate a large difference between data collection unit values, with one value above and the other below the average. Large

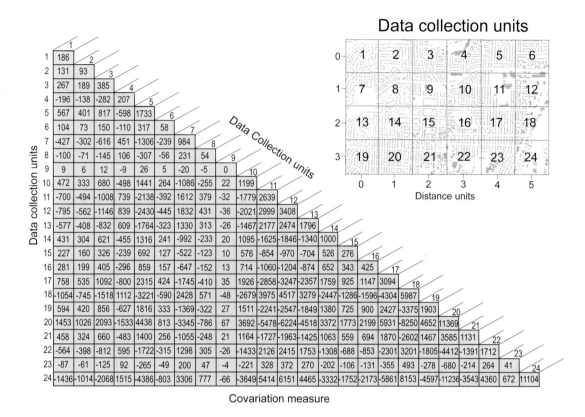

Figure 17.8   **Covariance values for all combinations of the 24 quadrats in the building count example. The covariance table is read like a distance table on a road map, with rows being the first data collection unit and columns the second.**

positive covariances are obtained when both unit values are similar and well above or well below the average. So

- *Large negative covariance* implies a large difference in values
- *Large positive covariance* implies that all values are (1) similar, and (2) well above or below the mean.

The next step in computing the Moran's *I* index is to weight the covariance values by the physical distances between the pairs of data collection units. You do this by first computing the distances *(d)* between unit centroids (see chapter 15 for more on centroids) using the coordinate equation form of the Pythagorean theorem described in chapter 11. For a regular grid of data collection units, it is easiest to use the coordinate system shown for the map in figure 17.8, where the centerpoint of quadrat 1 is at (0, 0), and the distance between adjacent centerpoints is 1. So for our example, the equation is

$$d = \sqrt{(col_2 - col_1)^2 + (row_2 - row_1)^2}$$

where *col* and *row* are the column and row locations of the two quadrats.

The **distance table** in figure 17.9 is read in the same manner as the covariance table in figure 17.8. To find the distances between data collection unit 1 and units 2–6, for example, you read downward from the top of column 1 to see that the distances are 1, 2, 3 ,4, and 5.

Distance weights

You next compute **distance weights** from the distances in the table. The weight can be applied in reference to any magnitude or intensity associated with a feature. The higher the value, the greater the weight for that feature. A feature with a larger weight will have more influence on the pattern than something with a smaller weight. There are a number of ways in which you can define the weights. The type of weighting you use will depend on what you are analyzing, so let's take a moment to look at a few of these.

Inverse distance weighting is based on the concept of impedance, or distance decay (figure 17.10A). All features affect or influence all other features, but the closer two features are in space, the more they interact with and influence each other. Because every feature is a neighbor of every other feature, you will generally want to

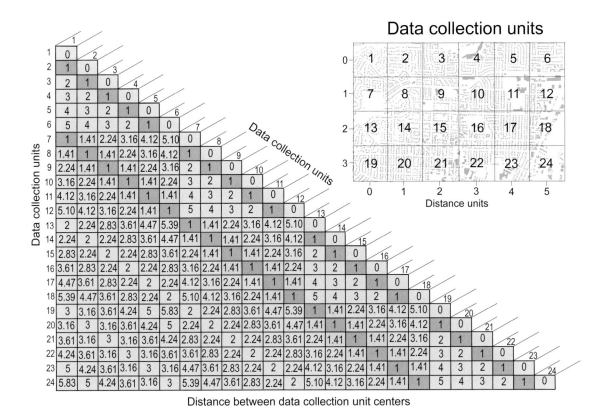

Figure 17.9 Distance table showing the straight-line distance between centerpoints of each pair of data collection units in the building count example. Centerpoints of adjacent units are 1 distance unit apart. Dark red boxes are distances used in the rook's (row-column) case, and light red boxes indicate distances used in the king's (row-column and diagonal) case for distance weighting.

specify a cutoff distance threshold to reduce the number of required computations with large datasets. Inverse distance weighting methods are most appropriate with continuous data, or for modeling geographic processes. For example, if your analysis relates to the arrangement of a particular plant species that propagates by seed dispersal, inverse distance is probably most appropriate weighting system to use.

Inverse Euclidean distance (straight-line distance) is appropriate for modeling continuous data such as temperature variations. **Inverse Manhattan distance** measures distance along line segments arranged at right angles to each other—this method might work best when analyzing locations along road networks. With **inverse distance squared,** the influence of distance drops off more quickly, so closer things have even more influence, and farther things have even less.

With **fixed distance weighting,** you impose a **distance band** or "circle of influence" on the analysis (figure 17.10B). Only features within a specified **critical distance** (threshold) that you define are analyzed. You

should use the fixed distance weighting when you want to evaluate the statistical properties of your data at a particular spatial scale. Fixed distance weighting works well for polygon data where there is a large variation in polygon size.

Zone of indifference weighting combines inverse distance and fixed distance (figure 17.10C). Features within the critical distance of a feature are included in analyses for that feature. Once the critical distance is exceeded, the level of impact quickly drops off. Zone of indifference weighting works well when fixed distance is appropriate, but there are no sharp boundaries on neighborhood relationships. For example, you might not think too much about distance if you are going to a grocery store within 2 miles of your home, but if the grocery store is 10 miles from your home, distance becomes more of an impedance.

Delaunay triangulation is a way to construct neighboring areas for a set of features by creating triangles from point features by using each point as a triangle vertex

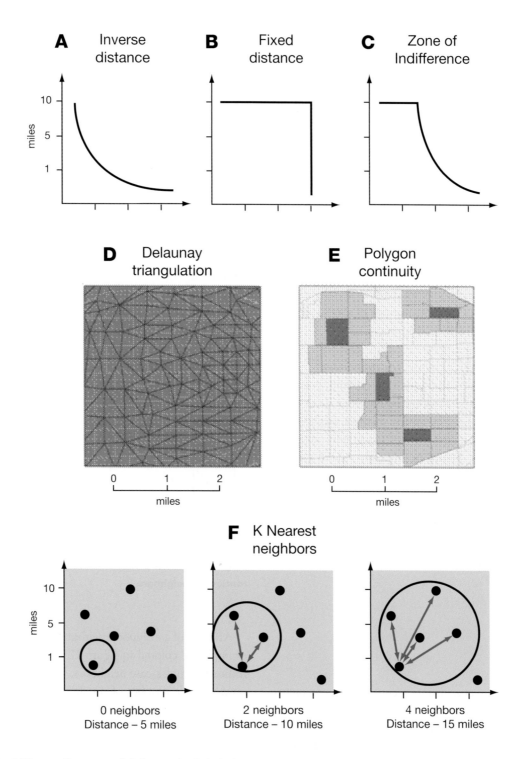

Figure 17.10 Different distance weighting methods include inverse distance (A), fixed distance (B), zone of indifference (C), Delaunay triangulation (D), polygon contiguity (E), and K nearest neighbors (F).

(figure 17.10D). Vertices connected by triangle edges are neighbors. This is a good option when your data include **island polygons** (isolated polygons that do not share any boundaries with other polygons) or in cases where there is a very uneven spatial distribution of features because it ensures that all features have neighbors.

With **polygon contiguity** (also called **first order polygon contiguity**), polygons that share a boundary with a target polygon are considered to influence that polygon and are included in computations for the target polygon (figure 17.10E). Polygons that do not share a boundary do not have an influence and are not included in computations. The polygon contiguity method is effective when polygons are similar in size and distribution, and when spatial relationships are a function of polygon proximity. Use this weighting method with polygons when you are analyzing a contagious process or continuous data are represented as polygons.

With **K nearest neighbors,** relationships are assessed within the spatial context of a fixed number of closest neighbors (figure 17.10F). The number of neighbors specified *(K)* will be included in computations for that feature. The *K* nearest neighbors option ensures that you have a minimum number of neighbors for your analysis. When the distribution of your features varies across the area of interest so that some features are far away from all other features, this method works well. When the spatial scale is less important than the number of neighbors, the *K* nearest neighbors method is appropriate.

Computing the Moran's I autocorrelation index
Once you have chosen the weighting method, you can compute the Moran's *I* autocorrelation index. Let's look at an example that uses the inverse distance method we described above. The First Law of Geography suggests the distance weight should be inversely related to the distance between data collection units—the smaller the distance, the larger the weight. Two inverse distance weighting functions are commonly used:

$$w_{ij} = \frac{1}{d_{ij}}$$

and

$$w_{ij} = \frac{1}{d_{ij}^{2}}$$

Using the first inverse distance equation to complete our example of computing the Moran's *I* index for the building counts, the distance weights between data collection unit 1 and units 2–6 are computed as 1.0, 0.5, 0.33, 0.25, and 0.20.

The second inverse distance equation is an inverse distance squared weighting called the **gravity function.** It is given this name because the denominator is identical to that used by Newton in his equation for the strength of gravity between two objects distance *d* apart. The gravity function rapidly reduces the weight with increasing distance between pairs of data collection units. For instance, the distance weights between unit 1 and units 2–6 are now computed as 1.0, 0.25, 0.11, 0.06, and 0.04.

The inverse distance weighting is often limited to each data collection unit's contiguous neighbors (figure 17.11). Two types of contiguous neighbors are used–the rook's case and the king's case–because of the similarity to the movements of these two chess pieces. The **rook's case** is the most restrictive because contiguity is defined by the four adjacent quadrats sharing a row or column edge with the center quadrat. The distances between the centers of the adjacent quadrats are set arbitrarily at 1 distance unit so that the inverse distance weights are also 1. The **king's case** includes diagonal neighbors, expanding the neighborhood to eight quadrats. All diagonal distances to adjacent neighbors are the square root of 2, shortened to 1.41 in figure 17.11. The inverse distance weights for the diagonal neighbors using the 1/d equation are therefore 1.0/1.41, or 0.707.

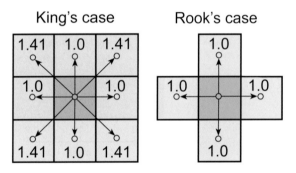

Figure 17.11 Using king's- and rook's-case nearest neighbors restricts the distance weighting to each quadrat's contiguous neighbors.

All pairs of contiguous quadrats in our building count example are colored in figure 17.9, with the darker red indicating rook's-case neighbors. Notice how few of the total possible number of quadrat pairs are used in the Moran's *I* index calculation when the neighborhood is restricted to the king's or rook's case.

The Moran's *I* index is the sum of distance weighted covariance values for all pairs of data collection units, divided by the variance of the values in all units times the sum of distance weights. To compute the Moran's *I* index for *n* data collection units, the equation is the following:

$$I = \frac{\sum_{i=1}^{n} \sum_{j=1}^{n} \left(cov_{ij} \times w_{ij} \right)}{\left(var \times \sum_{i=1}^{n} \sum_{j=1}^{n} w_{ij} \right)}$$

The **expected value** for the index is the value that you would get if you had a random distribution of the values for the features. It turns out that the expected value of Moran's I index for a perfectly random arrangement of n data collection units is obtained from the following equation:

$$I = \frac{-1}{(n-1)}$$

For large numbers of data collection units, the expected value for a random arrangement approaches zero. Values greater than the expected value are characteristic of clustered arrangements, whereas values less than expected are found with regular "checkerboard" arrangements of high and low unit values. So,

- $I < 0$ implies a regular arrangement
- $I = 0$ indicates a random arrangement
- $I > 0$ implies a clustered arrangement

In our building count example, the expected value for the Moran's I index is $-1/23$ or -0.0435. The following values were computed for the four distance weighting options discussed above:

> 1/d weight for all quadrats: 0.0065
> 1/d² weight for all quadrats: 0.0549
> 1/d weight for king's case: 0.0736
> 1/d weight for rook's case: 0.1781

Notice the large range in the values for the Moran's I index computed using the different distance weighting schemes. You can see that the values increase as the number of quadrats used in the weighting is increasingly restricted to contiguous neighbors. This is a drawback of the method—the index values depend on the distance weighting scheme that you use.

Another drawback of the Moran's I autocorrelation index is that, like the VMR, its ability to detect spatial pattern depends on the data collection unit size used in the analysis. In an apparently clustered population, the value of the Moran's I index may indicate randomness, clustering, and regularity as the size of the data collection units is steadily increased. Varying the size of data collection units in the analysis will allow both large clusters and smaller clusters to be identified. This could be useful for health resource allocation, for example. A public health agency could use the Moran's I statistic with house value data and the centroids of increasingly disaggregated census units to determine if there are pockets of low- or high-value housing within a city. Knowing this, a public health agency could review the location of clinics relative to low-income areas. If these areas form large clusters, the agency could make sure

that a clinic is located within the areas. If the areas are smaller and more widely dispersed, then mobile services to provide health care, like visiting nurses or mobile medical clinics, might be a better option.

A third criticism of the Moran's I index is that it measures the spatial arrangement of values within data collection units, not the actual locations of point features. Studying the spatial arrangement of a set of locations requires the use of another measure, the nearest neighbor statistic.

Nearest neighbor statistic

The VMR and the Moran's I autocorrelation index are quantitative measures of the distance component of spatial arrangement, that is, the distance between related data collection units. A second component of spatial arrangement is **spacing,** defined as the locational arrangement of objects with respect to each other rather than relative to data collection units. Spacing is independent of any boundary one might draw around the points. We can measure spacing by finding the distance between each point feature and its **nearest neighbor.**

The average distance between point features and their neighbors is called the **average spacing.** To compute this statistic, you first have to calculate the distance between all pairs of points using the distance equation (see "Determining distance by coordinates" in chapter 11). You can then determine which feature is a point's nearest neighbor.

Nearest neighbors and distances for seven points are shown in figure 17.12. Notice in this example that the distance from two pairs of points to their nearest neighbors is **reflexive.** This means that point A is the nearest neighbor of point $B,$ and point B is also the nearest neighbor of point A. In this example, points 1 and 2 are reflexive, as are points 4 and 5. Usually, however, point B will have some other nearest neighbor, and the relationship between A and B won't be reflexive. For example, the point nearest to 3 is 2, but the point nearest to 2 is 1, so the relationship between points 2 and 3 is not reflexive.

The **nearest neighbor statistic (NNS)** compares the average spacing with the expected mean nearest neighbor distance associated with a random distribution of point features. The average spacing (r) is computed by summing the nearest neighbor distances (δ_i) and dividing the total by the number of points (n). This is written mathematically as the following:

$$r = \frac{\sum_{i=1}^{n} \delta_i}{n}$$

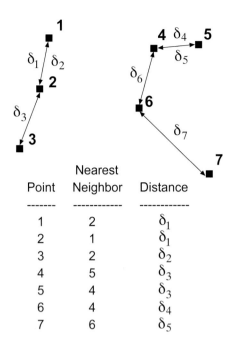

Point	Nearest Neighbor	Distance
1	2	δ_1
2	1	δ_1
3	2	δ_2
4	5	δ_3
5	4	δ_3
6	4	δ_4
7	6	δ_5

Figure 17.12 Nearest neighbors and their distances for seven point features.

The expected mean nearest neighbor distance *(rd)* associated with a random distribution of point features is computed for *n* points within area *A* using this equation, which has been derived by researchers who have studied many random point patterns:

$$rd = 0.5 \times \sqrt{\frac{A}{n}}$$

The NNS equation, or the ratio between actual mean distance and expected mean nearest neighbor distance for a random arrangement, is written mathematically as the following:

$$NNS = \frac{r}{rd}$$

where:

NNS = 0 implies maximum clustering
NNS = 0.25 implies a tightly clustered arrangement
NNS = 0.50 implies a clustered arrangement
NNS = 1.00 indicates a random arrangement
NNS = 2.14 implies maximum regularity (a regular equilateral triangular arrangement)

Applying the nearest neighbor equation to the four point patterns in figure 17.13, we obtain the NNS values shown at the top of each box. Each box is one square inch

in area on the map and contains 100 point features. The expected mean distance for a random arrangement is computed as the following:

$$rd = 0.5 \times \sqrt{\frac{1.0 \text{ in}^2}{100}} = 0.5 \times 0.1 \text{ in} = 0.05 \text{ in}$$

The top left box contains a uniform square grid of points spaced at 0.1-inch increments. The average spacing for these points is the following:

$$r = \frac{\sum_{i=1}^{100} 0.1 \text{ in}}{100} = \frac{10 \text{ in}}{100} = 0.1 \text{ in}$$

These values give an NNS of 0.1 inches/0.05 inches, or 2.0 for this arrangement of point features. Because this is almost 2.14, we can see that this indicates near maximum regularity.

The top right box contains 100 randomly placed points as determined from a random number generation computer program. This program also found the nearest neighbor and distance for each point from the x,y coordinates for the random points. The average spacing was computed as 0.050 inches, giving an NNS of 0.050 inches/0.050 inches, or 1, which we know indicates a random point arrangement.

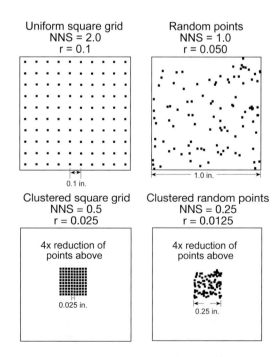

Figure 17.13 Nearest neighbor statistics for four different point patterns, each within a square region one square inch in area.

The bottom two boxes contain the same points as the top boxes, but at a 4 times scale reduction. The average spacing for the lower left box is therefore 0.025 inches, giving an NNS of 0.025 inches/0.05 inches, or 0.5. This value is characteristic of a clustered point arrangement. For the lower right box, the average spacing is 0.0125 inches, giving an NNS of 0.0125 inches/0.05 inches, or 0.25, indicating a tightly clustered arrangement.

The bottom boxes illustrate the importance of understanding how the size of the area of interest relates to the location of all point features. If we reduce the boxes at the bottom of figure 17.13 to one-sixteenth of their original size so as to just enclose each set of points, the NNS will be 2 for the lower left arrangement and 1 for the lower right arrangement. The arrangements within these smaller regions would be correctly characterized as regular and random but they would not be clustered as they are within the larger regions.

Directionality

It is also possible to measure the **directional** trend for a set of points. That is, you can determine whether features are farther from a specified point in one direction than in another direction. Directionality in a spatial distribution is important if you are trying to determine the dispersion around a point is biased in a particular direction. An example is the dispersion of toxins from a pollution source. Knowing the directionality would help risk managers know where to allocate their resources.

A common way that directionality is determined is to calculate the **standard deviational ellipse.** It is called a standard deviational ellipse because the standard deviation of the x- and y-coordinates from their average values, called the **mean center,** is used to define the axes of the ellipse. The distance is measured separately in the x and y directions, then these two measures are used to define the axes of an ellipse encompassing the distribution of features.

The result of the standard deviational ellipse calculation is an ellipse that represents two standard distances (the long and short axes) and the orientation of the ellipse (the rotation of the long axis measured clockwise from 12 o'clock) (figure 17.14). The ellipse allows you to see if the distribution of features is elongated and has a particular orientation. While you can get a sense of the orientation by examining the features on a map, the standard deviational ellipse makes the trend clear.

You can calculate the standard deviational ellipse using either the locations of the features or the locations influenced by a weight that indicates which points

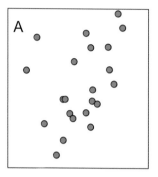
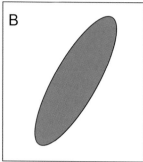

Figure 17.14 the standard deviational ellipse measure can be used to determine if there is directionality in the distribution of features. The ellipse provides a clear indication of elongation in the distribution of features.

are more important than others. The latter is called a **weighted standard deviational ellipse.** Calculations are based on either Euclidean or Manhattan distance. **Euclidean distance** is the shortest distance between two points "as the crow flies"; **Manhattan distance** is measured along a route that has segments oriented only vertically or horizontally, like the rectangular street network found in downtown Manhattan.

If the points are normally distributed about the mean center, the one standard deviation ellipse polygon will cover approximately 68 percent of the features in the cluster. Two standard deviations will contain approximately 95 percent of the features, and three standard deviations will cover approximately 99 percent of the features in the cluster.

Line feature arrangement

All of the measurements above can also be used for line and area features. For example, line counts or values within data collection units can be used to compute the VMR or the Moran's *I* autocorrelation index. In our street segment example, you can see if the count (number of street segments) or value (the total miles of streets in quadrats) are arranged in a random, clustered, or regular manner using the equations explained for point feature arrangements. The number of street segments within each of the nine quadrats in figure 17.5 has an arithmetic mean of 29.2 and a variance of 31.7. The resulting VMR of 1.09 indicates a high degree of randomness in the street segment counts, since 1 indicates complete randomness.

In addition to analyzing how random line features or the values associated with them are, there are other aspects of line feature arrangement that you can analyze quantitatively. Let's look at two of these, connectivity and hierarchy.

Connectivity

Many landscape features found on maps are linked to other features through a variety of **linear connections.** Roads, rivers, bus lines, airline routes, and a host of other connections link one place with another. The set of places and their connections is called a **network.** Although near things are more likely to be better connected than distant things, juxtaposition (contiguity) alone doesn't ensure connectivity. For example, there may be no road or air linkages between adjoining countries despite their common borders. In contrast, places separated by a great distance in space may be closely linked, as people who draw drinking water from a river discover when a toxic chemical spill occurs far upstream. The connectivity of a network helps us understand **accessibility,** which is a measure of how reachable a place is from distant locations.

In network theory terms, networks consist of nodes and links. A **node** is a point at which a line ends or where two or more lines come together or intersect. A **link** is the linear connection between two nodes. The linear connections can be represented by straight-line links or links with vertices that define their curvature, and nodes can be represented by points. The examples in figure 17.15 show street segments represented by straight-line segments (links) and intersections (nodes) as orange points.

Some networks are **totally connected** (figure 17.16A). In other words, all pairs of nodes are connected by the most direct link. These networks are efficient in terms of access because direct movements are possible to the greatest degree. **Circuit networks,** which lack direct connections between all pairs of places, but which still have more than one possible path between some pairs of places, are somewhat less connected (figure 17.16B). Other networks are only **partially connected** (figure 17.16C), with links between some pairs of nodes passing indirectly through at least one additional node. Least connected are **branching networks (trees),** which have only one possible path between pairs of places (figure 17.16D). Networks that are less than totally connected can be relatively inefficient in terms of access because they may require indirect movements between some pairs of places. Airline "hubs" are a familiar example of such reduced connectivity. You often find yourself going out of your way to get to a regional hub in the process of making "connections" to your destination.

When encountering these various networks, your curiosity should be aroused so that you seek explanations for why some networks are more connected than others, and what influence these connections will have on the flow between nodes and, as a result, on spatial patterns and processes. Consider, for example, designing residential street networks to minimize through-traffic and lower the speed of travel or where the water that runs off a streets ends up. Also, remember that there are many ways of measuring network efficiency, including the cost of building and operating the network.

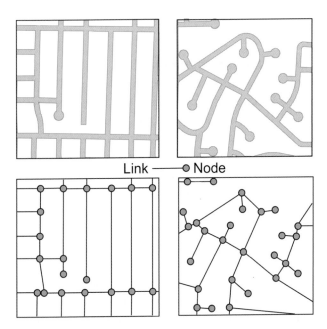

Figure 17.15 **Streets are the links and street intersections along with cul-de-sacs and dead ends are the nodes in the road network above.**

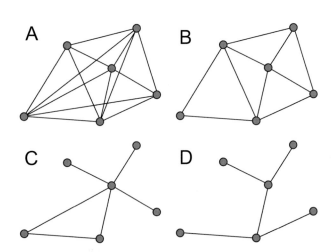

Figure 17.16 **Completely connected networks (A) have connections between all nodes. Circuit networks (B) are generally more connected than partially connected networks (C) or branching networks (D).**

Connectivity measures tell you the degree to which line features are interconnected. Urban planners use these measures to determine how well connected the streets are within different city neighborhoods. There are a number of connectivity measures we can use to objectively compare networks with one another and to relate a network's connectivity to its flow using numerical descriptions. One measure is the connectivity of the network as a whole; another is the connectivity of a selected place with respect to the entire network.

Connectivity of a network
A measure of the **connectivity of a network** is devised by forming a ratio between the actual number of links between nodes in a network and the maximum possible links for a network linking with that number of nodes. Often this ratio is multiplied by 100 to express the **percentage connection** in a network:

$$Connectivity\ (C) = \frac{AL}{PL} \times 100$$

where *AL* is the actual number of links and *PL* is the maximum number possible.

To use this connectivity measure, you need a simple way to determine the maximum possible number of links in a network. You can do so quite readily by counting the number of nodes *(n)* in the network and then using a formula for the maximum possible number of links. The problem is that there are three types of typical networks, so there are three different formulas we could use.

In the first type of network, links are **symmetric** (movement occurs in both directions), and links can't intersect without defining a new node. This form is typical of road or railroad networks that don't have one-way links, overpasses, or underpasses. With such a network, the formula *3(n − 2)* defines the maximum possible number of links between *n* nodes (figure 17.17A).

In a second type of network, the links are still symmetric, but they can cross one another without defining a new node. A network of airline routes is a good example, since routes don't connect in mid-air. Overpasses and underpasses in a road network fall in this category as well. With this sort of network, the maximum number of links between *n* places is given by the expression *n(n−1)/2* (figure 17.17B).

A third kind of network is characterized by **asymmetric links,** such as one-way links that can cross one another without defining a node. With such a network, the formula *n(n−1)* defines the maximum possible number of links between *n* nodes (figure 17.17C).

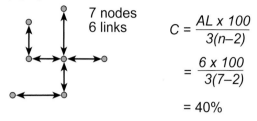

(A) Symmetric with intersection nodes

7 nodes
6 links

$$C = \frac{AL \times 100}{3(n-2)}$$

$$= \frac{6 \times 100}{3(7-2)}$$

$$= 40\%$$

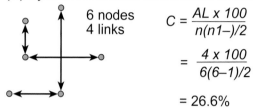

(B) Symmetric without intersection nodes

6 nodes
4 links

$$C = \frac{AL \times 100}{n(n1-)/2}$$

$$= \frac{4 \times 100}{6(6-1)/2}$$

$$= 26.6\%$$

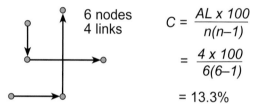

(C) Asymmetric without intersection nodes

6 nodes
4 links

$$C = \frac{AL \times 100}{n(n-1)}$$

$$= \frac{4 \times 100}{6(6-1)}$$

$$= 13.3\%$$

Figure 17.17 **By computing the percentage connection in different types of networks, we can objectively compare their spatial accessibility.**

Connectivity of a place
Rather than study the connectivity of a network as a whole, you might want to change your focus and determine the connectivity of a place with respect to the total network. You may do so, quite simply, by counting the linkages to or from that place. Obviously, in a totally connected network there will be no difference in the connectivity of the various places. In partially connected networks, however, there usually will be variation in the connectivity of individual places. Those with many links will be highly connected to the system, while those with few links will be weakly connected. The next step, of course, is to explain why connectivity varies as it does. Our example of an airline hub fits well here, as you are probably already familiar with airports that serve to link less-accessible destinations.

Hierarchy
The **hierarchy** inherent in a network of line features can also be studied quantitatively. We will discuss network hierarchy, in which nodes can contain more than one source or "parent" (like a family tree). Since environmental

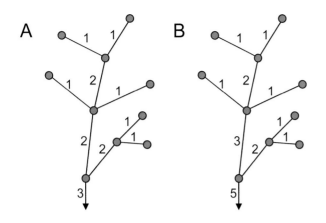

Figure 17.18 Strahler order (A) and Shreve order (B) are commonly used to describe the hierarchical nature of stream networks.

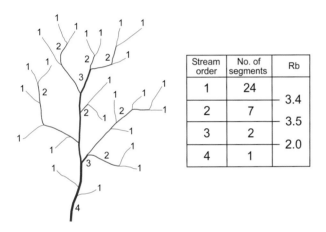

Stream order	No. of segments	Rb
1	24	
		3.4
2	7	
		3.5
3	2	
		2.0
4	1	

Figure 17.19 The bifurcation ratio of a branching network such as a dendritic stream system provides an objective measure of its hierarchical organization.

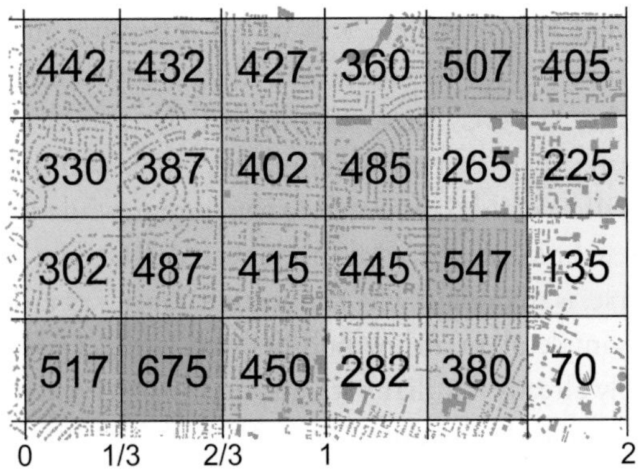

442	432	427	360	507	405
330	387	402	485	265	225
302	487	415	445	547	135
517	675	450	282	380	70

0   1/3   2/3   1   2
miles

Figure 17.20 Autocorrelation measures can be used to analyze the spatial arrangement of quadrat feature density values, such as these population densities per one-ninth of a square mile.

phenomena usually differ in some quantitative characteristic, such as size or importance, we can use this attribute information to arrange the links hierarchically. Sometimes the hierarchical nature of a network is more apparent in the environment than on a map, for maps often abstract information by classifying and simplifying features, which can mask these differences. A map may show all roads as thin black lines, for instance, while in reality the roads form a hierarchy from super highways to gravel lanes. On the other hand, maps can also reflect the hierarchy by grouping similar things and symbolizing the groups differently. For example, you will often find road maps used for navigation with a range of symbol types from interstates to state highways to primary roads to secondary roads and sometimes to gravel lanes.

In some cases maps show hierarchies even more clearly than the environment does. On the ground, railroad tracks might look pretty much the same. A map, however, might show railroads in terms of a hierarchy of high to low traffic. Such a map could quickly tell you which tracks accommodated many trains a day and which were used by only one train a week.

Orders of hierarchy

One way to analyze the hierarchical structure of line networks is to look at the relationship between links at different **orders of hierarchy** in the network. We will use an example of the **bifurcation ratio** (the ratio of the number of stream segments of one order to the number of the next higher) that hydrologists compute when analyzing the number of streams in each level of the stream hierarchy within a watershed.

The method of assigning a numeric value for connectivity of the links in a network is called **ordering,** and because this is often applied to stream systems, it is called **stream ordering.** Knowing the number of tributaries (indicated by order) allows us to infer some basic characteristics of streams. For example, first-order streams have no upstream concentrated flow of water, so we know they are dominated by overland flow. Because of this, they are most susceptible to nonpoint source pollution problems and can derive more benefit from wide riparian buffers than other areas of the watershed. Two common stream ordering methods are Strahler order and Shreve order.

With the **Strahler order** (figure 17.18A), the uppermost stream links are assigned an order of one. Stream order increases when streams of the same order intersect. Therefore, the intersection of two first-order links will create a second-order link, and the intersection of two second-order links will create a third-order link. The intersection of two links of different orders, however, will not result in an increase in order. For example, the intersection of a first-order and second-order link will not create a third-order link; instead, the link is assigned the highest order of the connecting links, in this case, two. The Strahler method is the most common stream-ordering method. However, because this method only increases at intersections of the same order, it does not account for all links and can be sensitive to the addition or removal of links.

Shreve order (figure 17.18B) accounts for all links in the network. As with the Strahler method, the uppermost stream links have an order of one; however, the orders are additive for all links. For example, the intersection of two first-order links creates a second-order link, the intersection of a first-order and second-order link creates a third-order link, and the intersection of a second-order and third-order link creates a fifth-order link.

Because the orders are additive, the numbers from the Shreve method are sometimes referred to as *magnitudes* instead of orders. The magnitude of a link in the Shreve method is simply the number of upstream links.

Bifurcation ratio

Once each link in the network has been classified by its order, your next step is to make a tally of links falling into each category. This tally will permit you to calculate the **bifurcation ratio** (R_b), which is defined as the ratio of links of one order to the number of links of the next highest order. Written symbolically, the bifurcation ratio is the following:

$$R_b = \frac{N_x}{N_{(x+1)}}$$

where N_x is the number of links of any order x and $N_{(x+1)}$ is the number of links of the next highest order. As an example, the bifurcation ratio between first- and second-order stream links is found for the **dendritic** (branching) drainage basin illustrated in figure 17.19 by computing the following ratio:

$$R_b = \frac{N_1}{N_2} = \frac{24}{7} = 3.4$$

The results of these computations and others in the same drainage system are shown in the table in figure 17.19. You could compare these results with statistics from other drainage basins to study the nature of different stream systems. You could also compare the results using different stream-ordering methods. The next step would

be one of interpretation. You might want to explain any differences that emerge in climate, slope, and terms of related factors. Streams, geology, and other environmental factors can help to explain the spatial patterns you find in a stream network.

Area feature arrangement

As with lines, the methods we describe for analysis of the randomness of points or values associated with points can also be used for areas, when the points are substituted with the area centroids. For example, when using the Moran's *I* autocorrelation index to study the spatial arrangement of area features, the trick is to first convert the counts within data collection units into density values or to use some other value attribute for the area features, and then to assign those values to the unit centroids. For example, the population estimates for the 24 quadrats in figure 17.20 can be expressed as number of people per one-ninth of a square mile. (Although population densities are usually expressed in people per square mile or hectare, the quadrat area is more meaningful in this example.) If we assume the density is constant within each quadrat, these values can then be assigned to the centroid of the quadrats.

The resulting population density map (figure 17.20) shows these values with light to dark pink tones representing low to high densities. The king's-case Moran's *I* index you would compute for these population densities is 0.0879, which is greater than the −0.0435 expected value for a random arrangement. You would conclude that the quadrat totals have a slightly clustered arrangement, which should agree with the pattern that you see on the map.

In addition to analyzing how random area features or the values associated with them are, you can examine quantitatively other aspects of area features. What you choose to examine will depend on what aspect of the landscape you are interested in. For example, a biologist might be interested in different kinds of area features in the landscape, such as habitat types. An environmental-ist might be interested in how connected the habitats are because some animals require corridors of movement. A forester might be interested in the configuration or arrangement of forest management units to determine where stands of different growth characteristics are located relative to one another. Geographers, landscape ecologists, and scientists in a number of other fields have developed a variety of methods to examine these and other important aspect of the landscape using quantita-tive measures called landscape metrics.

Landscape metrics

Landscape metrics quantitatively describe patterns in categories on maps that relate to the spatial characteristics of individual areas called **patches.** They can also be used to examine classes of patches or entire landscape **mosaics** made up of the complete set of patches. Although a large number and wide variety of landscape metrics have been developed, it would be safe to group them into two general categories: those that quantitatively describe the composition of the patches on the map without reference to spatial attributes, and those that quantify the spatial configuration of the patches on the map.

Composition relates to the variety and quantity of patch types without consideration of the spatial arrangement of patches within the landscape mosaic. Because we are interested in features that are shown on maps, we will not consider these types of metrics in this book. **Spatial configuration** relates to the spatial character and arrangement, position, or orientation of patches within the patch class or landscape mosaic. These landscape metrics require spatial information to be used in the calculations. There are many configuration metrics, so we will focus on only a few of the most popular ones in this chapter.

Spatial configuration metrics can be can be used to describe the patch itself or the arrangement of patches within the landscape mosaic. Metrics that describe individual patches include average patch *size and shape,* as well as *variance in size and shape.* These patch metrics take into account how a patch is influenced by its neighboring patches. Any patch metric, such as mean or variance, can also be used with patch classes and the landscape mosaic. These metrics at the class or landscape level take into account the relative location of individual patches within the patch mosaic. They can help to describe how ecological processes and features are affected by the overall configuration of patches within the context of the larger landscape mosaic. For example, variation in size could relate to the ability of a habitat unit to support a given biotic population. Or patches with longer perimeters might be better able to support a population that prefers that habitat along the edge of two different kinds of patches.

Popular spatial configuration metrics include patch size, patch density, patch shape complexity, core area, isolation/proximity, contrast, dispersion, contagion, interspersion, subdivision, and connectivity. Often these can be determined for each patch and then averaged for the class or the landscape.

Patch size is a fundamental characteristic of a patch that is easy to calculate. In chapter 15, we describe a number of methods to determine area. **Patch shape complexity** relates to the geometry of patches as simple and compact, or irregular and convoluted. Different ways of measuring shape are also described in chapter 15. The **perimeter-area fractal dimension,** a shape-complexity metric based on the relative amount of perimeter per unit area, can be easily determined from perimeter/area ratios once these values are known. Landscapes with area features for which the perimeter/area ratio is low will have a fractal dimension close to 1 (like one-dimensional linear features), while those containing highly complex convoluted features (the perimeter/area ratio is high) have fractal dimensions approaching 2. This metric therefore describes the relative "edginess" of a patch.

Core area is the internal area of the patches after a user-specified edge **buffer** (a zone of a specified distance around features) is eliminated. This metric combines patch size, shape, and edge-effect distance into a single measure. For example, smaller patches with greater shape complexity tend to have less core area. Although it is possible to calculate this metric using mapped features, it is far easier to compute this using a GIS that contains a patch boundary coordinate dataset. Other metrics that are easily computed with a GIS include the following:

- **Isolation,** which relates to the distance between patches across space.
- **Proximity,** which relates to the nearness of patches across space.
- **Contrast,** which describes the relative differences among patch types with respect to an attribute of the patches.
- **Dispersion,** which describes the regular or clumped arrangement of patches—the nearest-neighbor statistic is an example of this type of metric.
- **Contagion,** or the tendency of patch types to be spatially aggregated.
- **Interspersion,** which describes the intermixing of patches of different types.
- **Subdivision,** or the degree to which a patch type is divided up into separate patches.
- **Connectivity,** which relates to the functional connections among patches and depends on the process or characteristic of interest. We described various connectivity measures previously.

Let's turn now to the demonstration of how you can calculate two popular landscape metrics, fragmentation and diversity, for area features on your maps.

Fragmentation

Fragmentation is a metric often used in environmental and landscape analysis to describe the spatial arrangement of habitat **patches** (ecological data collection units, usually land cover in different classes). It can at times be an important measure of environmental health. When habitats are fragmented, patches can eventually become so small that they are all edge. When this happens, the interior dwellers of the habitats may become extinct.

We'll examine fragmentation using the land-cover map created from remotely sensed imagery in figure 17.21. Each grid cell on the map is placed in one of five land-cover categories. You can see that some large areas appear to be a single category, whereas other areas seem to be a "salt and pepper" mixture of several categories. You can use a **fragmentation index** to quantify your visual impression of the map.

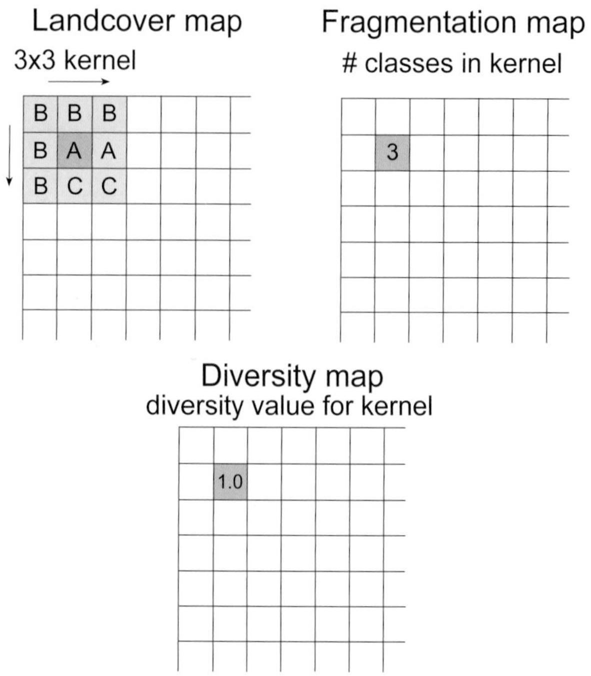

Figure 17.22   A 3-by-3 grid cell kernel can be systematically moved across a gridded land-cover map to obtain the number of classes in the kernel for each cell location on the map. Fragmentation and diversity metrics can be computed and mapped from the kernel data.

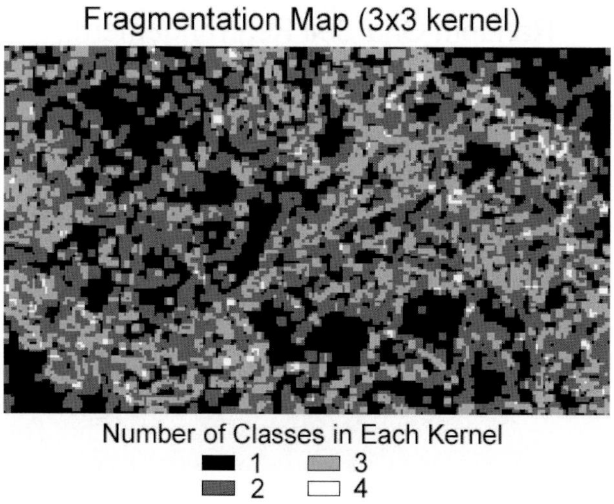

Figure 17.21   Section of a land-cover map for a forested area north of Corvallis, Oregon.

The fragmentation index is based on moving a 3-by-3 grid cell window, or **kernel,** across the gridded map (figure 17.22, top left). At each cell location, the number of different classes found in the neighboring nine-cell kernel is tallied. The total number of categories is then placed in the corresponding center cell to create a **fragmentation map.**

Sweeping the kernel across the land-cover map produced the fragmentation map shown in figure 17.23. Notice that from one to four classes were found in the different kernel positions, but only one or two categories were tallied at most positions.

Figure 17.23   A fragmentation map shows the number of land-cover classes found in the kernel at each cell location as it is moved across the gridded map.

Landcover Map Grid Cells
Favg = 0.118

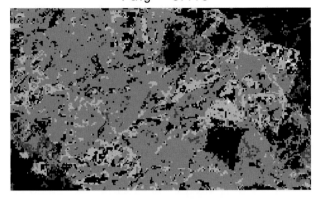

Clustered Grid Cell Rearrangement
Favg = 0.003

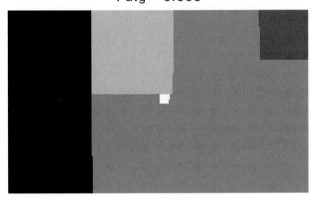

Random Grid Cell Rearrangement
Favg = 0.247

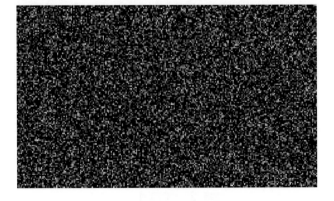

Figure 17.24 An average fragmentation measure value is best interpreted relative to the average values obtained for highly clustered and totally random arrangements of the cells in the original land-cover map.

The values in each cell on the fragmentation map can now be used to compute an **average fragmentation index** for the m grid cells in the map. The following equation is used:

$$F_{avg} = \frac{\sum_{i=1}^{m} (n-1)}{m \times (c-1)}$$

where c is the kernel size (9 in this example) and n is the number of categories found in the kernel at map position i. The equation for our 139-row-by-238-column (33,082 grid cell) land-cover map example is the following:

$$F_{avg} = \frac{\sum_{i=1}^{33082} (n-1)}{33082 \times (8)}$$

This equation yields an average fragmentation of 0.118 for the land-cover map. To interpret this number, you must look at the maximum and minimum values possible. It is easy to see that a minimum value of 0 is computed when the entire mapped area is of a single land-cover category. At the other extreme, a value of 1 is reached when nine or more classes are so fragmented on the map that nine different classes are tallied for the kernel at each cell location.

A more meaningful interpretation of the average fragmentation index is to use the category tally data obtained from the map to compare the average value with the minimum and maximum values possible—this is the **fragmentation index range** for the collection of categories on the map. To find the fragmentation index range, we would use highly clustered and totally random rearrangements of the cells in the original land-cover map (figure 17.24). The minimum value is estimated by rearranging the cells on the fragmentation map into highly compact regions, as seen on the middle map in figure 17.24. The value for randomness can be obtained by rearranging the cells into a totally random arrangement, as on the bottom map in figure 17.24.

The clustered arrangement has an average fragmentation index value of 0.003, and the random arrangement has an average value of 0.247. You can interpret the 0.118 average fragmentation for the map as approximately half way between highly clustered and totally random (exactly half way would be 0.122). Does this interpretation of the average fragmentation measure agree with the degree of fragmentation you see on the map?

Diversity Map (3x3 kernel)

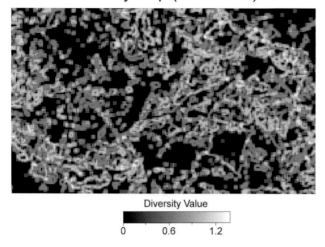

Diversity Value

0   0.6   1.2

Figure 17.25 This diversity map shows Shannon index *(H)* values computed in a 3-by-3 kernel for each grid cell in the land-cover map in figure 17.19.

Diversity

Diversity is another measure of interest to ecologists and others because it describes spatial variation of the various patch types (such as land cover or ecosystems) within the landscape. You have seen that the fragmentation index is determined by counting the number of categories for the kernel at each cell location. A **diversity index** is an even more robust measure of pattern arrangement because the relative abundance of each category is also determined. The widely used **Shannon diversity index (H)** finds the proportion of each category *(p_i)* present for the kernel at each cell location (figure 17.22, bottom). You can then use your computer or hand calculator to multiply each proportion by its natural logarithm *(ln p_i)*, and the products are summed and multiplied by –1, using the equation

$$H = -\sum_{i=1}^{c} p_i \ln p_i$$

for the *c* different categories present for the kernel at each cell location.

Figure 17.25 is a map showing Shannon diversity index *(H)* values for each grid cell in our land-cover map example. Notice how similar this map appears to the fragmentation map in figure 17.23.

The diversity map contains the information needed to compute the **average diversity** for the mapped area containing *m* grid cells from the following formula:

$$H_{avg} = \frac{-\sum_{j=1}^{m}\sum_{i=1}^{c} p_i \ln p_i}{m}$$

The average diversity for our land-cover map example is 0.46. The theoretical range of H_{avg} for a 3-by-3 kernel is from 0 when the map is of a single category, to 2.2 when nine different categories are found in the kernel so that p_i is 1/9 for all cells in the kernel. In the latter case, you would conclude that the map has a more diverse than uniform diversity value arrangement.

As with the fragmentation index, it is more meaningful to interpret the diversity index through a comparison with the **diversity index range** (the minimum and maximum values possible) for the collection of categories on the map. You could again use the highly clustered and totally random rearrangements of the cells in the land-cover map (figure 17.24, center and bottom) to find this range. If you compute the diversity index for these two maps, you will obtain *H* values of 0.014 and 0.90. Given the 0.046 average diversity value, you can conclude that the land-cover map categories are intermediate in arrangement between tightly clustered (not diverse) and completely random (highly diverse).

In this chapter, we've looked at ways to measure the patterns that features make on our maps. But there's still more to map analysis. Next, we must move our focus from the abundance and arrangement of distributions to spatial associations among different patterns. The comparison of patterns is the topic of the next chapter.

SELECTED READINGS

Barber, G. M. 1988. *Elementary statistics for geographers*. New York: John Wiley & Sons, Inc.

Boots, B. N., and A. Getis. 1988. *Point pattern analysis*. Newbury Park, Cal.: Sage Publications, Inc.

Davis, J. C., and M. J. McCullagh, eds. 1975. *Display and analysis of spatial data*. New York: John Wiley & Sons, Inc.

Ebdon, D. 1985. *Statistics in geography*. 2d ed. New York: Blackwell Publishing, Inc.

Goodchild, M. F. 1986. *Spatial autocorrelation*. Norwich, U.K.: Geo Books.

Haggett, P., and R. J. Chorley. 1969. *Network analysis in geography*. London: Edward Arnold Publishers, Ltd.

Lewis, P. 1977. *Maps and statistics*. New York: Halsted Press.

McGarigal, K., S. Cushman, and E. Ene. 2004. *Landscape metrics: A comprehensive guide to their use and interpretation*. Amherst: University of Mass.

McGrew, J. C., and C. B. Monroe. 1993. *An introduction to statistical problem solving in geography*. Dubuque, Iowa: William. C. Brown Publishers.

Mitchell, A. 2005. *The ESRI guide to GIS analysis, volume 2: Spatial measurements & statistics*. Redlands, Calif.: ESRI Press.

Monmonier, M. S. 1974. Measures of pattern complexity for choropleth maps. *The American Cartographer* 1 (2): 159–69.

Rogers, A. 1974. *Statistical analysis of spatial dispersion*. London: Pion.

Taylor, P. J. 1977. *Quantitative methods in geography: An introduction to spatial analysis*. Boston: Houghton Mifflin Co.

Wilson, A. G., and M. J. Kirby. 1980. *Mathematics for geographers and planners*. New York: The Clarendon Press.

chapter
eighteen **SPATIAL ASSOCIATION ANALYSIS**

18

Spatial association analysis

In our analysis of spatial patterns in the previous chapter, the focus was on measuring the abundance and arrangement of a single set of features displayed on a map. We can also compare two or more sets of features found in the same area to determine their degree of **spatial association.** If you can determine where and how environmental features change together across the landscape, you may be better able to explain why this spatial association occurs. When you can describe how environmental features vary together within a geographic area, you can often predict similar associations in other areas.

The suspicion that there is an association between different features located close together is often supported. In some cases the association is **direct.** For instance, fishermen who realize that underwater features such as weed beds, logs, and large rocks provide protection for fish are likely to have the best "luck." But the spatial association also may be **indirect.** Just because two features are found together at the same location doesn't guarantee a meaningful association between them. Both features may be associated with a third phenomenon and not to each other. Continuing our fishing example, really big trophy fish can be caught along the edge of weed beds because this is where small fish (their food) are found, not because large fish prefer to live close to weeds.

It's difficult to imagine that direct or indirect spatial associations among features in our environment will ever be totally predictable. The ability to *predict* implies a deep and complete knowledge of the environment that goes beyond what you can see on maps and images. But it may be possible to increase your understanding of how environmental features interact by quantitatively analyzing their degree of spatial association. In this chapter we describe several measures of spatial association between two sets of static features, and then discuss measures of spatial movement and diffusion for dynamic features mapped at several points in time.

TYPES OF SPATIAL ASSOCIATION

There are two basic forms of spatial association. The first is between **discrete features** such as individual houses or trees. For example, you might look at sets of point, line, or area features in the same geographic area. The closer they are to each other, the greater their spatial association. Or you might detect a spatial association between numerical values computed for features in the same data collection units. When two sets of values increase or decrease similarly within the units, they have a strong **positive spatial association.** A strong **negative spatial association** exists when a high value for the first feature and a low value for the second feature occur in each unit. When there's no systematic spatial relationship between feature values, you can assume there's no spatial association between values for the two features.

The second basic form of spatial association is between **continuous phenomena,** or geographic phenomena that have continuously changing values (see chapter 8 for more on continuous surfaces). For instance, you might look at how two or more continuous physical phenomena such as elevation and temperature, or cultural phenomena like population density and average annual income, change in magnitude at the same time. The highest positive spatial association is when the highest values and the lowest values for the phenomena occur at the same locations, and the strongest negative spatial association occurs when one phenomenon is highest where the other is lowest, or vice versa. No systematic relation between magnitudes at different locations indicates no spatial association.

When analyzing relationships between sets of features, you need to be aware of **lagged spatial association.** This occurs when there is a positive or negative association between two sets of features, but one set is shifted systematically relative to the other. A good example of lagged spatial association is the relationship between streets and houses. Houses are associated with streets, but the two features don't coincide spatially (figure 18.1). The systematic setback from the street for houses is the **lag distance.**

Lagged spatial associations are easiest to analyze when only one direction is involved. In the previous example, houses are set back relative to streets, but no other directions need to be considered. Similarly, paper mills might be found upwind of zones of air pollution, and farmsteads on the Great Plains might be found downwind of shelterbelts (barriers of trees and shrubs that protect against the wind and reduce erosion).

Lagged associations are more complex and harder to analyze when several directions are involved, which is commonly the case. Lake water quality may be associated with many landscape features in the surrounding watershed. Upstream water sources that feed the lake may be polluted, while wetlands along the lake edges may help filter sediments and chemicals in runoff from surrounding areas. Both these positive and negative influences on the overall water quality have lagged associations.

Figure 18.1 Spatial correspondence between two categories of features is often used as an indicator of their degree of association, but it is important to check for lagged spatial association. Notice the systematic shift in the position of houses relative to street centerlines.

JUDGING SPATIAL ASSOCIATION VISUALLY

For centuries, people have used subjective visual judgments to define the spatial association between two sets of point features on maps. The method continues to be popular, largely because it comes so naturally to us. This is because the human visual system has been found to be an amazingly powerful information processing system. We can take advantage of this to visually assess spatial association.

Visual judgments of spatial association are easiest to make when the features being compared are superimposed or overlaid. **Overlay** is the process of superimposing two or more maps to better understand the relationships

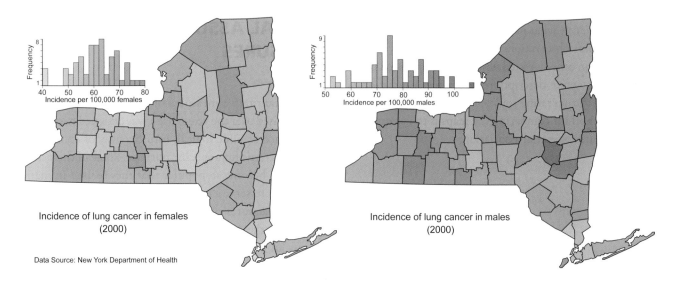

Figure 18.2 Choropleth maps showing the incidence of lung cancer in females (left) and males (right) for New York counties in 2000 can be used to test your ability to visually judge spatial association.

among the geographic features of interest and their attributes (see chapter 5 for more on overlaying). Perhaps the easiest way to see the spatial association is to view the two sets of features printed on the same map, or if one set is plotted on a clear medium that is overlaid on the set. With geographic information systems (GIS), this is easily done; however, it may still be difficult to see the associations if the two patterns are complex, if the features are dense or if there are many types of features displayed on the map as well.

In this case, it might be easier to compare maps in a side-by-side arrangement—a task that requires looking from one map to the other while keeping track of the relative spatial positions of features on the respective maps. **Map comparison** (visually or quantitatively examining resemblances between maps) is simplified, of course, if the maps to be studied are on the same map projection at the same scale and orientation. When this isn't the case, it may be worth your effort to **rescale** (enlarge or reduce), reproject, and reorient one of the maps to match the second.

Another way to compare maps visually is to view them in an alternating time sequence, as when they're toggled on a computer screen. Although the relative effectiveness of this technique isn't known, the method is increasing in popularity as computer map displays have become commonplace.

You can test your ability to judge spatial association visually by comparing the two choropleth maps in figure 18.2 (see chapter 8 for more on choropleth maps). These maps of lung cancer incidence in females and males were created from incidence rates published in 2000 by

the New York State Department of Health. On these maps, the range of county (including New York City borough) lung cancer incidence rates (incidence per 100,000 females or males) was divided into 20 categories for females and 19 categories for males. Notice that on both maps the same light-to-dark red tone progression was used to show the categories. Your task is to describe the spatial associations between female and male lung cancer incidence in New York.

What do these maps tell you? First, you probably see that the incidence of lung cancer is higher overall for males, so you must mentally adjust for this difference between males and females on the two maps. Then you may see a large number of counties with intermediate lung cancer incidence rates for both males and females. You may next see that there are a few areas where the incidence of lung cancer is low for both females and males. But there also are areas with high incidence rates for both sexes. Finally, you probably notice that there are a number of areas where the incidence rate is high for males or females, but not both.

Your list of observations made by comparing the two maps is an important first step in describing spatial associations, but visual observations lack quantitative rigor. You might describe relations between two mapped sets of point features as having a "high correlation," a "poor correspondence," or "a moderate association." Since the meaning of such phrases isn't clear, however, other people won't know exactly what you mean. The problem is compounded by the fact that other people may make different observations when viewing the same maps. Without quantitative measures of spatial association, there's no way to tell exactly what people's descriptions mean, how similar people's observations are or if some observations are better than others. Visual observations also often lack repeatability—your descriptions of spatial associations may change the next time you compare the maps. A more reliable method for comparing the maps is to use quantitative measures of spatial association.

SPATIAL ASSOCIATION MEASURES

Many quantitative measures of spatial association have been devised over the years. Each was designed to capture the essence of some logical, intuitive way of comparing spatial patterns for a specific purpose. Some of these measures have become so popular that they're used in many disciplines to serve a broad range of purposes.

Most measures of spatial association require you to go back to the datasets from which the maps were made. In our lung cancer incidence example, we obtained the county totals for 2000 from the Web site for the New York State Department of Health (table 18.1).

	Female	Male
Albany	61.8	75.4
Allegany	69.9	92.7
Bronx	41.7	62.3
Brooklyn	40.1	58.4
Broome	61.7	74.6
Cattaraugus	72.9	75.7
Cayuga	63.8	67.8
Chautauqua	53.3	68.3
Chemung	67.0	89.3
Chenango	48.9	70.8
Clinton	68.3	78.2
Columbia	58.7	65.0
Cortland	54.5	74.3
Delaware	55.5	71.4
Dutchess	59.3	76.8
Erie	63.1	83.2
Essex	64.7	72.2
Franklin	63.5	87.6
Fulton	62.3	94.5
Genesee	59.7	91.0
Greene	66.9	98.5
Hamilton	79.1	68.3
Herkimer	58.9	73.3
Jefferson	69.4	93.0
Lewis	54.0	70.6
Livingston	53.8	92.1
Madison	69.5	91.2
Manhattan	50.4	53.6
Monroe	55.1	69.2
Montgomery	54.6	70.6
Nassau	57.4	59.5

Niagara	67.1	86.5
Oneida	64.9	77.1
Onondaga	68.3	78.1
Ontario	60.3	85.6
Orange	62.3	75.0
Orleans	61.3	87.0
Oswego	72.1	87.8
Otsego	50.7	82.8
Putnam	72.3	79.7
Queens	41.9	55.9
Rensselaer	71.8	91.0
Richmond	59.2	78.5
Rockland	53.9	53.1
Saratoga	61.5	80.1
Schenectady	62.8	74.5
Schoharie	61.3	107.1
Schuyler	77.6	95.8
Seneca	72.4	85.5
Steuben	62.8	83.9
St. Lawrence	74.5	86.8
Suffolk	63.7	70.3
Sullivan	69.5	74.2
Tioga	59.5	74.8
Tompkins	58.1	61.3
Ulster	66.7	70.2
Warren	57.6	76.8
Washington	62.0	98.8
Wayne	48.7	69.2
Westchester	52.3	58.9
Wyoming	48.4	78.7
Yates	66.1	75.0

Table 18.1 Lung cancer incidence rates in New York counties (incidence per 100,000 people).

To determine spatial association, the first thing you could do with the county data is find the **average** ($\bar{x}$) male and female lung cancer incidence rates for the 62 counties in New York. The average (also called the **arithmetic mean**) is the sum of the rates divided by the number of counties, obtained using the following equation:

$$\bar{x} = \frac{\sum_{i=1}^{62} county_i\, incidence}{62}$$

You will find that the average female incidence rate is 62.1, and the average male incidence rate is 77.6 cases per 100,000 women and men, respectively.

You will also want to know the overall variation in the female and male lung cancer incidence rates for the 62 counties. The **standard deviation** (σ) is one of the most widely used measures of variation within a dataset—it is a statistic used as a measure of the dispersion or variation in a distribution. It is computed by taking the square root of the variance (see chapter 17 for more on computing the variance) for the n (62) data collection units:

$$\sigma = \sqrt{\frac{\sum_{i=1}^{n} (x_i - \bar{x})^2}{n}}$$

Using this equation, you'll find the standard deviation for the female and male data to be 8.42 and 11.77, respectively. A beginning statistics book will tell you that if the data values are normally distributed in the familiar bell-shaped curve, 68 percent of the values will fall within one standard deviation ($\pm 1\sigma$) of the average.

Notice that the legends for both maps in figure 18.2 are **frequency diagrams** (histograms) of the rates of lung cancer incidence. A frequency diagram shows you the number of data collection units within different ranges of data values. The frequency diagrams in the figure 18.2 map legends clearly are bell-shaped in appearance, so the standard deviation is an appropriate way to quantitatively summarize the variation within the data. Look again at table 18.1. You will see that for females, 44 of the 62 counties (71.0 percent) fall within one standard deviation of the average (between 52.8 and 69.6). The male cancer incidence data are even closer to the 68 percent value for a perfect normal distribution of values—42 counties (67.7 percent) fall within one standard deviation (between 65.8 and 89.3).

The next step is to find a quantitative measure of the geographic correspondence between the counties with high and low female and male lung cancer incidence rates, which will tell you whether the locations of units that increase or decrease similarly (or in opposite directions) are coincident. To do this, you can use a statistical mapping program or GIS to make maps showing the percentage that each county is above (>100 percent) or below (<100 percent) the average for the state. The county maps created in this manner for the female and male data are shown in figure 18.3, using the same percentage ranges for each map category to facilitate visual map comparison.

What does comparing the maps in figure 18.3 tell you? You may first notice that the range of percentages is greater on the map for lung cancer in males, with one

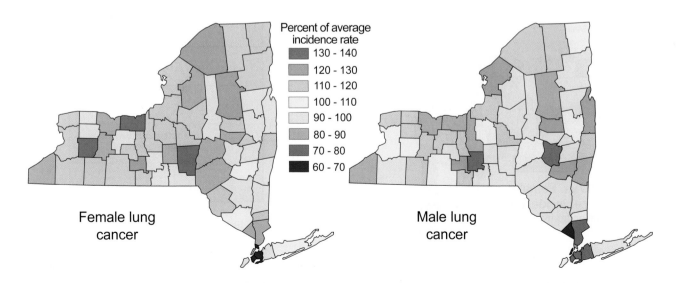

Figure 18.3 Choropleth maps showing the percentage that each New York county is above or below the average female or male lung cancer incidence rate.

county having over 130 percent of the average male lung cancer incidence rate. You will also see that boroughs in the New York City metropolitan area are below average (<100 percent) in lung cancer incidence for both females and males. Several counties in the north central part of the state have high female and male lung cancer incidence rates, and many counties have above average female and below average male cancer incidence, and vice versa. These observations, based on average values for the state, quantitatively reinforce what you concluded by comparing the two choropleth maps in figure 18.2.

You may be able to better see spatial associations between female and male cancer incidence rates if you combine the data on a single map. The female versus male map in figure 18.4 shows a simple way to do this by simplifying the data into four categories. Counties with below average female and male cancer incidence rates are shown in blue. Counties with above average male and female rates are shown in red. Counties with above average female and below average male incidence rates, and vice versa, are shown in light shades of red and blue, respectively. You now clearly see from New York City up the Hudson River Valley (the counties along the southeastern edge) a north–south band of primarily below average counties for both male and female cancer incidence. Counties with above average male and female cancer rates are concentrated in clusters of counties along the northern, southern, and northeastern edges of the state. Between these two bands are many counties that are above average in one and below average in the other. Statisticians would say that the counties with either below average (blue) or above average (red) cancer

incidence rates for both males and females have a positive spatial association. Conversely, lightly shaded blue- and red-toned counties, with large numbers of one and small numbers of the other, have a *negative spatial association*.

The four-category female versus male map highlights positive and negative spatial associations, but it does not tell us if the associations are strong or weak. One way to study **strength of association** is to create a scatterplot of the data used to make the map. A **scatterplot** is a graph on which one set of values is plotted along the x-axis and the other along the y-axis. Figure 18.5 is a scatterplot showing the male and female lung cancer incidence rate data taken from table 18.1. Notice that the values are clustered in a roughly elliptical area in the top half of the scatterplot. If there were a perfect positive association between the incidence of lung cancer among males and females, the percentages would be identical, and all points would fall along the upward-sloping diagonal line defining a perfect positive linear relationship (assuming the axis scales were equal). A perfect negative association would exist if all plotted points fell along the downward-sloping diagonal line defining a negative linear relationship. In contrast, a circular, randomly appearing plot indicates that there is no spatial association. The plotted values between male and female lung cancer incidence rates in New York at the county level form an ellipse, with the long axis roughly aligned with the positive relationship line. On the whole, you can conclude that there's a definite positive association between female and

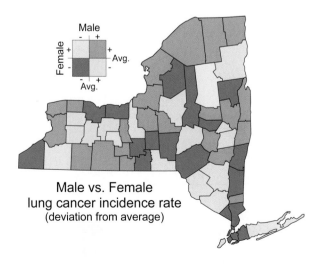

Figure 18.4 This map shows the positive and negative deviation from the average female and male lung cancer incidence rate in each New York county.

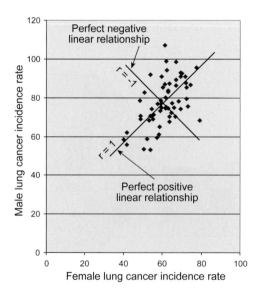

Figure 18.5 A scatterplot of female and male lung cancer incidence data for New York's 62 counties. The correlation coefficient (*r*) indicates the degree of spatial association between the two sets of data.

male lung cancer incidence. This positive association is not surprising, in light of the large number of positively associated counties that you saw in figure 18.4.

You may now wonder if there is a quantitative measure that shows the overall strength of the spatial association between two sets of point features. Several such **parametric measures** have been developed for populations that are assumed to have a normal distribution like the data in our New York example, but spatial cross-correlation is used most widely.

Spatial cross-correlation

Spatial cross-correlation is the similarity between two different datasets for the same data collection units, such as our lung cancer data for New York counties. The first step in determining the **spatial cross-correlation coefficient,** called the **correlation coefficient** for short, is to compute the amount of covariance in the data for each unit. In chapter 17, you saw that **covariance** is the product of the amount that the two values for the unit differ from their respective averages. In mathematical terms, the covariance between two data values x and y for a single data collection unit is the following:

$$cov = (x - \bar{x})(y - \bar{y})$$

The covariance for all data collection units is summed to obtain the **total covariance.** You then divide this value by the number of data collection units (n) to obtain the **average covariance** for the study area. Finally, the **correlation coefficient (r)** is found by dividing the average covariance by the product of the standard deviations for each dataset:

$$r = \frac{\sum_{i=1}^{n}(x_i - \bar{x})(y_i - \bar{y})}{n\sigma_x\sigma_y}$$

The correlation coefficient ranges between −1 to +1, where

$r = +1$ implies a perfect positive linear relationship, and
$r = -1$ implies a perfect negative linear relationship.

Let's do the calculations with the female and male lung cancer incidence data for the 62 New York counties. We have said that the average female and male cancer incidence rates are 61.2 and 77.6 with standard deviations of 8.42 and 11.77.

Putting numbers into the previous equation, we have

$$r = \frac{\sum_{i=1}^{62}(x_i - 61.2)(y_i - 77.6)}{62 \times 8.42 \times 11.77}$$

From the female and male cancer incidence rates in table 18.1, you can compute the covariance values for the first entries, as shown below:

County	Covariance
Albany	(61.8–61.2) × (75.4–77.6) = −1.32
Allegany	(69.9–61.2) × (92.7–77.6) = 131.37
Bronx	(41.7–61.2) × (62.3–77.6) = 298.35

If you compute the covariances for the other counties and sum the numbers, you will find the total covariance for the 62 counties to be 53.25. When you divide the total covariance by the product of the standard deviations (8.42 × 11.77) to compute the correlation coefficient, you obtain an r value of 0.54.

To see what an r value of 0.54 means, let's look at the −1 and +1 limits for the correlation coefficient. For r to be +1 for datasets x and y, all pairs of values must be identical and fall on an upward-sloping diagonal line defining a positive linear relationship between x and y (figure 18.5). Because the equation of this line is $y = x$, the correlation coefficient equation becomes

$$r = \frac{\sum_{i=1}^{n}(x_i - \bar{x})(x_i - \bar{x})}{n\sigma_x\sigma_x} = \frac{\sum_{i=1}^{n}(x_i - \bar{x})^2}{n\sigma_x^2} = \frac{\sigma_x^2}{\sigma_x^2} = 1$$

For r to be −1, each pair of values is greater and less than the average by the same amount, and thus falls on a perpendicular downward-sloping line defining a negative linear relationship. The equation of this line is $y = -x$, so that $-\bar{y} = \bar{x}$ and

$$r = \frac{\sum_{i=1}^{n}(x_i - \bar{x})(-x_i + \bar{x})}{n\sigma_x\sigma_x} = \frac{-1\sum_{i=1}^{n}(x_i - \bar{x})^2}{n\sigma_x^2} = \frac{-\sigma_x^2}{\sigma_x^2} = -1$$

For r to be zero, the total covariance must be zero. A covariance of zero occurs when there is no spatial association between the two sets of values. A scatterplot of values with zero covariance will look random.

Given this understanding of the correlation coefficient, you can conclude that a correlation coefficient $r = 0.54$ indicates a moderate positive spatial relationship between the female and male lung cancer incidence rates for New York counties.

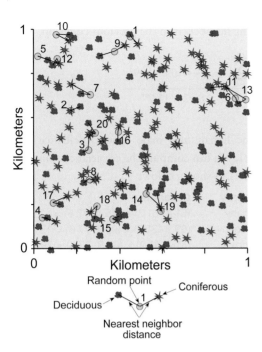

Figure 18.6 Spearman's rank correlation coefficient can be used to measure the spatial correlation between the 100 coniferous and deciduous trees shown on this portion of a forest map (see text for explanation).

Spearman's rank correlation coefficient

We have seen that the cross-correlation coefficient is a measure of spatial association based on computing the covariance for counts of point features within data collection units. You may wonder if there is a measure based on the individual locations of the two types of features, rather than total feature counts within areas. Statisticians have indeed devised measures of the independence between two sets of point features. Two events are **independent** if the occurrence of one of the events provides no information about the occurrence of the other event; that is, the events have no influence on each other.

Let's look at one measure you can use: **Spearman's rank correlation coefficient,** which is based on calculating the distance between nearest neighbors for two sets of point features. Our example (figure 18.6) is a portion of a forest map 1 kilometer by 1 kilometer in area that contains 100 coniferous trees and another 100 deciduous trees mapped as point features. Both categories of trees appear to be randomly arranged on the map, so you might assume that their locations are not correlated with each other. To measure the actual degree of correlation, you can randomly place a set of points on the map. In this example, a random number generator computed 100 random points, and we have shown the first 20 by orange dots on the map.

For each randomly placed point, you must find the nearest neighboring coniferous and deciduous trees, and then measure the distance from the random point to both

NEAREST NEIGHBORS					
	Distance (km.)		Rank		
Rnd. Pt.	Conif.	Decid.	Conif.	Decid.	Rank diff.²
1	0.045	0.010	53	3	$(53–3)^2 = 2,500$
2	0.005	0.050	2	51	$(2–51)^2 = 2,401$
3	0.070	0.027	80	18	$(80–18)^2 = 3,844$
4	0.065	0.025	76	10	$(76–10)^2 = 4,356$
5	0.105	0.046	97	46	$(97–46)^2 = 2,601$
6	0.075	0.030	86	19	$(86–19)^2 = 4,489$
7	0.090	0.040	96	32	$(96–32)^2 = 4,096$
8	0.025	0.035	23	22	$(23–22)^2 = 1$
9	0.055	0.070	68	82	$(68–82)^2 = 196$
10	0.068	0.060	77	73	$(76–73)^2 = 16$
.	.	.	.	.	.
.	.	.	.	.	.
.	.	.	.	.	.
100	8.28	3.59	89	36	$(89–36)^2 = 2,809$

Table 18.2 Nearest neighbors, ranks, and rank differences for computing Spearman's rank correlation coefficient.

nearest neighbors. The straight lines in figure 18.6 show the nearest neighboring trees for the first 20 random points. The nearest neighbors in each tree category are easy to determine visually, and distances can be measured with a ruler, but computer programs that work with digital point location data are normally used to find the neighbors and compute the nearest neighbor distances. Such a program was used to compute nearest neighbor distances in kilometers for the 100 random points. Distances for the first 10 and the 100th random points are listed in table 18.2.

You next sort the distances to find their ranks. In our example, the distances were sorted in ascending order, so that the smallest distance is rank 1 and the largest is rank 100 for both categories of trees. The ranks for the first 10 and last random point are listed in columns 4 and 5 of table 18.2.

For each random point, you next compute the square of the difference between the ranks for the two categories to compute the **squared rank differences.** Column 6 in table 18.2 shows the computation for each value. These values are then summed to give the **total rank difference squared (D²),** which is the following in our example:

$$D^2 = \sum_{i=1}^{100} (rank_{conif.} - rank_{decid.})^2 = 181{,}640$$

This (D^2) value is all you need to compute Spearman's rank correlation coefficient.

Spearman's rank correlation coefficient (ρ) is a nonparametric measure. **Nonparametric** methods are often referred to as "distribution-free" methods because they do not rely on assumptions that the data are fit to a particular probability distribution such as the famous bell-shaped normal curve. Spearman's rank correlation coefficient shows the degree to which the two datasets are spatially independent without having to assume that each dataset has a normal distribution of values. To calculate the ρ coefficient, numerical values must be converted to ranks that are then sorted, just as we have done for the 100 random points in our tree example.

The equation for n ranks is

$$\rho = 1.0 - \frac{6 \times D^2}{(n^3 - n)}$$

which for our example is

$$\rho = 1.0 - \frac{6 \times 181{,}640}{(100^3 - 100)} = 1.0 - \frac{1{,}089{,}140}{999{,}900} = -0.09$$

A ρ value ranges from –1 to +1. The closer ρ is to +1 or –1, the stronger the positive or negative spatial correlation between the two categories of features. To interpret the ρ value use the following guides:

- $\rho = +1$ implies complete **positive spatial dependency** (the two sets of category ranks increase or decrease similarly)
- $\rho = 0$ implies no spatial dependence
- $\rho = -1$ implies complete **negative spatial dependency** (a high rank for one category and a low rank for the second occur for each feature)

The value for our example, –0.09, is very close to the 0 value that would be computed for two datasets that are not correlated—that is, two completely independent datasets. Spatial independence is the expected result, since we must confess that the positions of all coniferous and deciduous trees on the map are not real, but rather were determined with a random number generator so as to be completely independent.

For *complete positive spatial dependency* (ρ is to equal +1), the total squared rank differences must be 0, which happens only when the distance ranks for the two categories are identical. On the hypothetical map in figure 18.7, for example, category 1 and category 2 are identical point features. Because they are identical, each of the 20 randomly placed points on the map has identical nearest neighbors. If the nearest neighbor distances are identical, so are the distance ranks.

Notice that both the regular and random point feature arrangements in the left and right half of the figure have complete positive spatial dependency. Remember that you are measuring the positional correspondence between two datasets, not the spatial arrangement, which was discussed in chapter 17.

A ρ value of –1 implies *complete negative spatial dependency* between the locations of features in the two categories. A negative value is obtained only when the ranks are opposite—ascending for category 1 and descending for category 2, and vice versa. Complete negative spatial dependency is rarely found in two sets of point feature locations, but values of around –0.7 are possible for two regularly arranged, spatially lagged datasets.

In figure 18.8, you can see two hypothetical categories of point features lagged 1 mile to the east and north in a square grid arrangement. Twenty points have been placed randomly on the map so that nearest neighbor distances can be measured and distance ranks can be determined. You can see that random points close to a category 1 nearest neighbor location are farther away from the category 2 nearest neighbor, and vice versa. Table 18.3 lists the ranks for the 20 random points, the squared difference in rank values computed from the ranks, and their summed value of 2,252. Notice that the ranks are roughly opposite, but not completely inverse. You can now compute Spearman's rank correlation coefficient:

$$\rho = 1.0 - \frac{6 \times 2,252}{(20^3 - 20)} = 1.0 - \frac{13,512}{7980} = -0.69$$

This ρ value for two regularly arranged lagged point datasets is probably the greatest negative spatial dependency you can expect to find. The arrangement of real point features on maps is rarely this regular.

We have seen how the spatial association for point features can be measured. Now let's turn to line features.

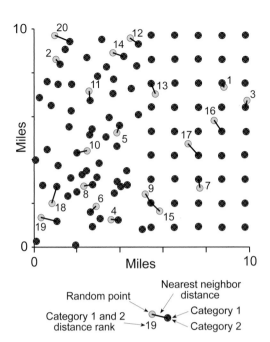

Figure 18.7 When the mapped locations of point features in categories 1 and 2 are identical, the nearest neighbor distances to each randomly placed point are identical, as are the distance ranks. This point arrangement gives a Spearman's rank correlation coefficient ρ value of +1.

Rand. Point	Cat.1 Dist.	Cat.2 Rank	Cat.2 Dist.	Cat.2 Rank	Rank Diff.²
1	2.0	1	11.8	10	81
2	3.1	2	17.3	16	196
3	4.6	3	14.3	13	100
4	5.0	4	20.0	19	225
5	6.8	5	21.0	20	225
6	6.9	6	13.8	12	36
7	8.1	7	11.0	9	4
8	8.2	8	12.4	11	9
9	8.3	9	19.5	18	81
10	9.0	10	15.2	15	25
11	9.1	11	10.0	7	16
12	9.2	12	19.0	17	25
13	11.7	13	15.0	14	1
14	12.1	14	8.1	5	81
15	12.5	15	7.0	3	144
16	14.7	16	10.8	8	64
17	17.0	17	3.7	2	225
18	18.5	18	3.0	1	289
19	19.1	19	9.5	6	169
20	19.8	20	7.7	4	256
				Total =	2,252

Table 18.3 Distance and rank data for the 20 random points in figure 18.8.

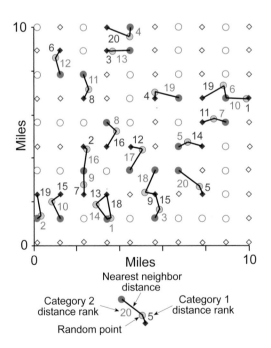

Figure 18.8 On this hypothetical map, two categories of point features are regularly arranged with a 1-mile lag north and east. Notice that when the distance from one of the 20 random points to its category 1 (blue diamond) nearest neighbor is short, the distance to the category 2 (red circle) nearest neighbor is long, and vice versa.

LINE FEATURE SPATIAL ASSOCIATION

You can apply point feature spatial association measures to the analysis of two types of line features shown on separate maps or superimposed on a single map. Visual inspection of the associations between line features displayed on maps is again an important first step. If you look at a topographic map, you may see a high degree of spatial correspondence between a variety of line features. For example, administrative boundaries might be correlated with rivers or ridgelines, since they often follow these kinds of natural features in the landscape. You might also find a lagged association between roads and rivers because roads are often constructed alongside rivers.

Another type of spatial association seen on maps is between different attributes for the same set of line features. For instance, your city's engineering or planning department may have produced a series of street network maps each showing different attributes for the streets. These attributes may be counts or averages for street segments, usually city blocks. You may be able to find block-by-block information for such attributes as speed limits, number of parking places, or traffic counts—information that can be compared quantitatively as well as visually.

Let's compare pedestrian and vehicular traffic counts collected and mapped for city street segments corresponding to blocks in downtown Portland, Oregon (figure 18.9). The actual counts have been generalized into five classes for the range of the pedestrian counts on one map (figure 18.9, top) and vehicle counts on the other (figure 18.9, bottom). For street segments where counts were taken, thick lines ranging from light to dark brown show the five classes. The numerical limits for classes aren't the same on the two maps, and not all blocks with pedestrian counts have vehicle counts. Nevertheless, you can visually compare the two maps, looking for positive or negative correspondence between the brown-toned city street segments. You will most likely conclude that there's no spatial association between the pedestrian and vehicular traffic counts.

How can you quantitatively validate (or invalidate) the visual impression that there is no spatial association between pedestrian and vehicular traffic counts? One idea is to carry out a correlation analysis based on the data used to make the maps. Fortunately, traffic counts and many other attributes for road network datasets are now readily available on the Internet, including the traffic counts for downtown Portland. So, you can download

the actual pedestrian and vehicular traffic counts for the 43 blocks where both were taken and use them to make a scatterplot, like the one shown in figure 18.10.

Looking at the scatterplot, you will see that almost all points fall in a vertical swath on the left half of the pedestrian traffic axis. This arrangement of points indicates minimal correlation between the pedestrian and vehicle counts. If there were a strong positive correlation (and the

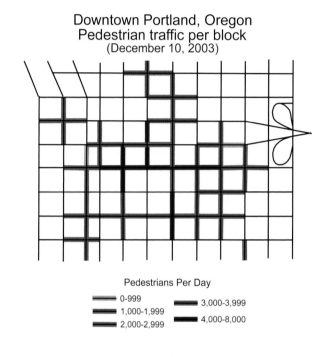

Downtown Portland, Oregon
Pedestrian traffic per block
(December 10, 2003)

Pedestrians Per Day

0-999		3,000-3,999	
1,000-1,999		4,000-8,000	
2,000-2,999			

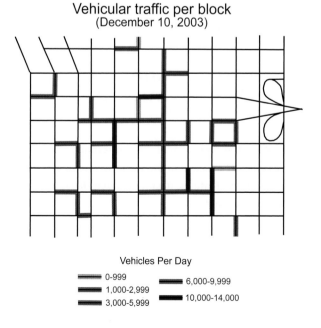

Vehicular traffic per block
(December 10, 2003)

Vehicles Per Day

0-999		6,000-9,999	
1,000-2,999		10,000-14,000	
3,000-5,999			

Figure 18.9 Number of pedestrians and vehicles counted for selected city street segments in downtown Portland, Oregon, on December 10, 2003.

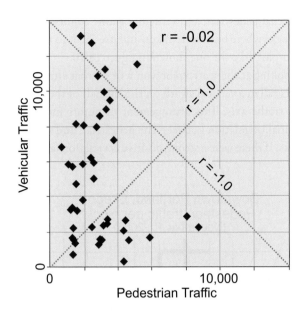

Figure 18.10 Scatterplot of pedestrian and vehicular traffic count data for 43 city blocks of downtown Portland, Oregon.

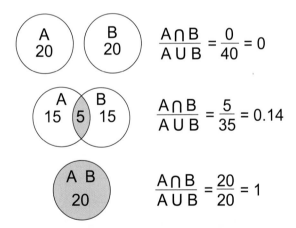

Figure 18.11 The coefficient of areal correspondence (CAC) provides a simple measure of the spatial association of two categories of area features; it varies between 0 (no overlap) and 1 (complete overlap).

scales of the axes were equal), you would see the points aligned along the diagonal from the lower left origin to the upper right. A strong negative correlation would be implied by points aligned along a diagonal line from the upper left to the lower right.

You can then use the spatial cross-correlation measures to determine the similarity between the two datasets for each city block. Using pedestrian and vehicle traffic counts for each of the 43 blocks, you will obtain a correlation coefficient (r) of -0.02. This very low r value is a clear indication that there is no spatial correlation between the two attributes.

AREA FEATURE SPATIAL ASSOCIATION

You may be interested in the spatial association between two types or categories of area features shown on maps or between two attributes for the same set of area features. Let's focus first on the spatial association between two different categories of area features. For instance, you may wish to compare soil types with vegetation zones, or to correlate climate zones with land-use categories. In cases like these, you can think of **spatial association** as the degree to which two categories overlap on the map. The **coefficient of areal correspondence (CAC)** is a commonly used measure of the amount of overlap between two categories of area features.

The coefficient of areal correspondence is the ratio between the area of overlap and the total areal extent of two categories A and B on the map, or in mathematical terms

$$CAC = \frac{A \cap B}{A \cup B}$$

where the symbol $\cap$ means the *intersection* (overlap) and $\cup$ means the *union* (total extent) of the two area feature categories. The CAC coefficient ranges in value from 0 (when there is no overlap) to 1 (when there is perfect areal correspondence) (figure 18.11).

Imagine, for example, that you want to determine the spatial associations between forest land/agricultural fields and gentle-slope/steep-slope zones in a portion of the driftless (nonglaciated) region in southwestern Wisconsin (figure 18.12, top left). If you place a square grid of 2,400 cells (48 rows by 50 columns) over the map, and then create a reclassified grid cell map showing the dominant land use (the land use covering most

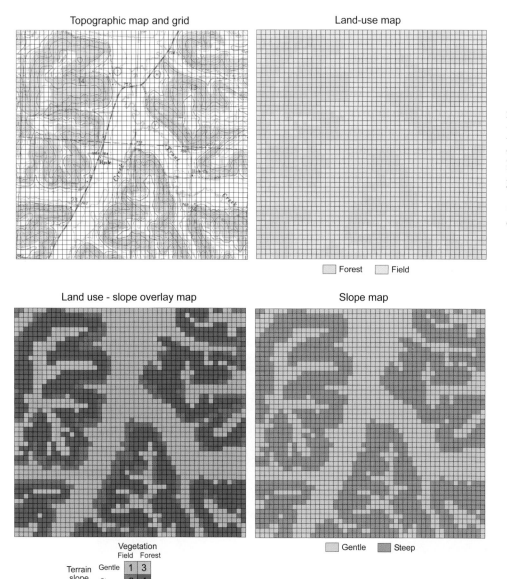

Topographic map and grid

Land-use map

Courtesy of the U.S. Geological Survey.

Forest   Field

Land use - slope overlay map

Slope map

Vegetation
Field Forest

Terrain Gentle **1** **3**
slope   Steep   **2** **4**

Gentle   Steep

Figure 18.12 Notice the areal correspondence between forest cover (shown with a green tint) and steep ground slope (shown by close contour spacing) on this portion of the Barneveld, Wisconsin, 1:24,000-scale USGS topographic quadrangle. For analysis purposes, the study area on the topographic map was divided into 2,400 grid cells. The map at the bottom left shows a four-way classification based on the gridded land-use and slope maps.

of the grid cell) as either forest or agricultural fields (figure 18.12, top right) and another reclassified grid cell map showing the dominant slope zones as either gentle or steep (figure 18.12, bottom right), upon overlaying the two new maps (figure 18.12, bottom left) you'll find the following:

975 cells in gentle-slope agricultural fields (category 1)
343 cells in steep-slope agricultural fields (category 2)
144 cells in gentle-slope forested land (category 3)
938 cells in steep-slope forested land (category 4)

Four coefficients of areal correspondence can be computed from the grid cell totals for the four categories on the land use–slope overlay map:

$$\frac{Gentle \cap Fields}{Gentle \cup Fields} = \frac{category\ 1}{category\ 1 + 2 + 3} = \frac{975}{1,462} = 0.67$$

$$\frac{Steep \cap Fields}{Steep \cup Fields} = \frac{category\ 2}{category\ 1 + 2 + 4} = \frac{343}{2,256} = 0.15$$

$$\frac{Gentle \cap Forest}{Gentle \cup Forest} = \frac{category\ 3}{category\ 1 + 3 + 4} = \frac{144}{2,057} = 0.07$$

$$\frac{Steep \cap Forest}{Steep \cup Forest} = \frac{category\ 4}{category\ 2 + 3 + 4} = \frac{938}{1,425} = 0.66$$

These coefficients tell you that there's a high areal correspondence between gentle slopes and agricultural fields, as well as between steep slopes and forest land because the coefficients for these two combinations are closer to 1 (0.76 and 0.66, respectively). Conversely, there is a low correspondence between agricultural fields and steep slopes and between gentle slopes and forest lands because the coefficients are closer to 0.

Finding the intersection and union areas for two or more categories is a key function of GIS **map overlay analysis tools,** or tools to superimpose maps to better understand the relationships among them using quantitative measures such as intersection and union (see chapter 5 for more on overlay). Areas defined by grid cells, as in our example above, can be overlaid digitally to determine intersection and union areas. Areas defined by *(x, y)* coordinate outlines can also be overlaid and analyzed. Using a GIS, you can determine coefficients of areal correspondence between many area features in seconds.

CONTINUOUS SURFACE SPATIAL ASSOCIATION

Continuous surfaces change smoothly in numerical value across the landscape. Environmental phenomena such as current temperature, precipitation, humidity, or atmospheric pressure are excellent examples of continuous surfaces. Their values change continuously from place to place and also from minute to minute. Ground elevations can be treated as a continuous surface, as happens when contours are created. Surfaces such as these do not vary in time as well as space. Another form of continuous surface is a long-term statistical average of something that isn't continuous at a given point in time. Average monthly precipitation is a good example, since at

Oregon December 2002 precipitation

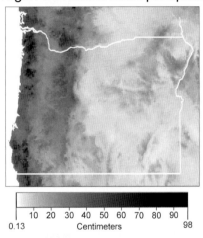

10 20 30 40 50 60 70 80 90
0.13   Centimeters   98

Oregon June 2003 precipitation

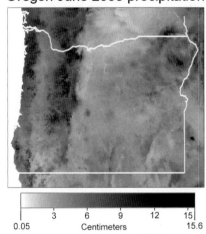

3   6   9   12   15
0.05   Centimeters   15.6

December – June precipitation comparison

1   5   10   15   20
0.37   December – June precipitation   22

Figure 18.13 Continuous surface maps of Oregon precipitation for December 2002, and June 2003 can be compared visually, or a single map showing the ratio of December to June precipitation can be created to quantitatively understand the seasonal variation.

any instant the precipitation can vary from place to place, but you can think of it as a continuous surface of average monthly variation.

There are two basic ways to compare continuous surfaces. First, you can compare different attributes for the same area at the same date. You might compare average temperature and average precipitation data for the same area, for example. Second, you can perform a **temporal change study** for the same attribute at different times. For instance, you might compare monthly temperature in the same area for December and June.

A common method used to study the spatial association between continuous surfaces that have been mapped is to subdivide the surface into a large number of grid cells (data collection units) then find the average value for each cell or the value at its center (you could also use dominance as we saw in our land-use/slope example above.) Fortunately, gridded maps in raster format have already been made for many continuous-surface phenomena, and you can use GIS or digital image processing software to study their degree of spatial association.

Let's explore a temporal change analysis by looking at Oregon precipitation maps for December 2002, and June 2003 (figure 18.13, top left and right). The two monthly precipitation surfaces are subdivided into a 128-row-by-205-column square grid, with an estimated total precipitation value for each grid cell. The range of precipitation for each month is represented by gray tones that vary from white for the lowest value to black for the highest.

Looking at the maps, you can see that there's more precipitation in the western third of Oregon during both months. More detailed visual observations are difficult to make, particularly because the same progression of gray tones on the two maps represent widely differing precipitation ranges (note the legends for each map).

If you have access to the digital data used to make each map, you can use GIS or image processing software to make a new map that is the sum, difference, product, or ratio of the two maps being compared. For instance, you may be able to gain a deeper understanding of winter and summer precipitation differences in Oregon by combining the two monthly maps into a December–June comparison map (figure 18.13, bottom). This map shows the ratio of December to June precipitation for each grid cell. On this map, you can see that the ratio ranges from 22 times wetter in winter in the very southwest corner of Oregon to 0.37 times wetter (or conversely 2.7 times drier) in winter in the high mountains of northeastern Oregon.

The first step in the quantitative comparison of gridded continuous surface maps is to create frequency diagrams

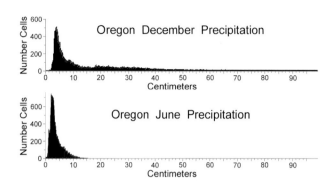

Figure 18.14 Frequency diagrams for the Oregon December and June precipitation data indicate that neither dataset is normally distributed.

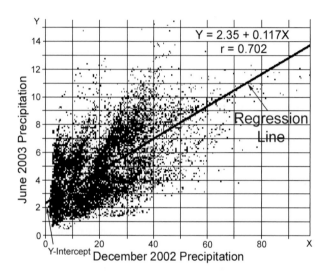

Figure 18.15 Scatterplot of the December and June precipitation data, along with the regression line and equation of best fit. Note that the x- and y-axes are scaled differently on this graph, which has the effect of exaggerating the apparent steepness of the regression line.

(explained earlier in this chapter) for the grid cell values on each map. Frequency diagrams for the December and June precipitation map grid cell values are shown in figure 18.14. You will want to check for normality in the distributions. Notice that neither frequency diagram has the bell-shaped curve that characterizes a normal distribution. Because neither dataset is normally distributed, it is not valid to use summary statistics such as the standard deviation to compare variation in the data.

You can, however, construct a scatterplot for the two datasets to study the strength of their association. The scatterplot for the December and June precipitation data (figure 18.15) shows that the 26,240 pairs of values are dispersed in a roughly oval arrangement, with most pairs in the lower left quarter of the diagram. From this arrangement, you would expect there to be a slight positive linear relationship between the December and June precipitation.

Next you can measure the cross-correlation. You would compute the correlation coefficient (r) using the same equation as for the point-feature comparison described earlier in this chapter. Using a GIS, an r value of 0.702 was computed for the December and June data, a value that indicates a moderate positive correlation between winter and summer precipitation.

You can also use a GIS to perform a regression analysis to analyze the correspondence. **Regression** is a statistical analysis that provides an indication of the degree of linearity in relations between the datasets. The correlation analysis in the GIS gives the regression line for the scatterplot. A **regression line** is the straight line of "best fit" drawn through the paired x, y values. The criterion for "best fit" most commonly used is the **least squares criterion**, which minimizes the squared distances between the y values of the points and the y values on the regression line. As with the scatterplots we mentioned earlier, a strong positive correlation is implied by points aligned along the diagonal from the lower left origin to the upper right, and strong negative correlation is implied by alignment along the diagonal from the upper left to the lower right.

The regression line for the December and June precipitation data is defined by the equation y = 2.35 + 0.117x. The 2.35 centimeter value is the *y-intercept,* the place where the regression line crosses the y-axis of the scatterplot. The y-intercept is also the value we expect for y when $x = 0$ because it falls along the y-axis when $x = 0$. The 0.117 value is the slope of the regression line, close to horizontal for these data (it only appears to be steeper because the two axes are scaled differently). If the data values appear closely clustered along a line so that the correlation coefficient is very close to 1, the regression line equation can be used as a predictor of June precipitation for every value of December precipitation. As you might expect, the regression equation rapidly decreases in utility as a predictor as the correlation coefficient lowers. For our example, we can see that the data are not clustered along the line and the correlation coefficient $r = 0.702$. So we can conclude that December precipitation can only be used as a rough estimator of June precipitation.

The square of the correlation coefficient, called the **coefficient of determination,** provides a measure of the strength of the correlation. Symbolized as r^2, the coefficient of determination can have only positive values ranging from 0 to 1, where

$r^2 = 1$ implies a perfect correlation (positive or negative), and
$r^2 = 0$ implies complete absence of correlation.

Statisticians say that the coefficient of determination gives the percentage of the explained variation (by the linear regression) compared to the total variation in the data. For instance, the r value of 0.702 for the December and June precipitation data gives a coefficient of determination of 0.702^2, or 0.49. You can therefore say that 49 percent of the variability in the December data is explained by the variability in the June data.

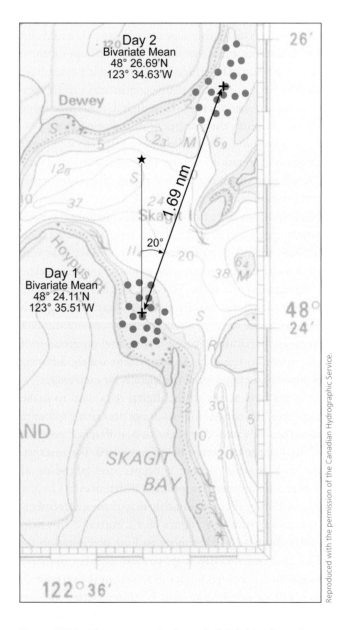

Reproduced with the permission of the Canadian Hydrographic Service.

Figure 18.16 **The movement of a pod of dolphins from day one to day two can be mapped if the position of each animal can be tracked.**

MOVEMENT AND DIFFUSION

So far in this chapter we've treated the features being compared as static phenomena. In most cases, this assumption is justified, and the pattern comparison methods we've discussed will produce meaningful results. But sometimes we're less interested in spatial association between static features than in **movement** (the changes in location) or **diffusion** (spreading or scattering over space) of features over time. Let's look first at how you can analyze the movement of a set of point features, realizing you can also analyze line and area features when line segments or area centroids are used instead. We'll then examine how you can analyze diffusion, again using points but knowing the method can be extended to lines and areas.

Measuring point feature movement

Many physical and cultural features move across the earth as a group—flocks of geese, pods of whales, herds of elk, and companies of soldiers, to name a few. You may want to know basic spatial information about their movement, such as the distance and direction traveled by the group as a whole. In addition, you may be interested in knowing the changes in spatial arrangement of the individuals in the group at the beginning and ending locations.

Our example is a hypothetical pod of 18 dolphins tagged with electronic tracking devices so that the geographic position of each dolphin can be determined daily. The position of each dolphin while stopped at a resting or feeding area on days one and two has been plotted on a large-scale nautical chart (figure 18.16). You can examine the movement from day one to day two of the pod as a whole by measuring the distance and direction between the centers of the pods on the two days.

To do this, you can use a statistic called the bivariate mean to find the center of a set of point features within a group, such as dolphins in a pod in our example. The **bivariate mean,** written as $(\bar{x}, \bar{y})$, is the average of the horizontal and vertical coordinates for n features:

$$(\bar{x}, \bar{y}) = \left(\frac{\sum_{i=1}^{n} x_i}{n}, \frac{\sum_{i=1}^{n} y_i}{n} \right)$$

We computed the bivariate means for the dolphin locations on the two days by first finding the latitude (ϕ) and longitude (λ) of each dolphin from the electronic tracking system. Since the locations are in geographic coordinates, the equation for the bivariate mean becomes the following:

$$(\bar{\phi}, \bar{\lambda}) = \left(\frac{\sum_{i=1}^{n} \phi_i}{n}, \frac{\sum_{i=1}^{n} \lambda_i}{n} \right)$$

The bivariate means for the two days are labeled on the map and plotted with "+" marks on figure 18.16. You can use the two bivariate means as summary statistics for the pod location on each day. Plotting the means as points on the nautical chart also makes it easy for you to measure the average distance the pod moved from day one to day two (you can use the grid coordinate method described in chapter 11). Or you can make a rougher estimate using the 0.1-minute latitude ticks on the right edge of the chart as a ruler (remember that 1 nautical mile equals 1 minute of latitude). We found the distance between the two bivariate means to be 1.69 nautical miles.

In what *direction* is the pod of dolphins moving? You can answer this question by first drawing a straight direction line between the two pod bivariate mean locations (see chapter 12 for more on determining direction). Then use a protractor to measure the angle between the meridian (vertical line) at the first pod location and the direction line to the second pod location. The 20° angle you measure is the true azimuth of the direction between the two pod centers (see chapter 12 for a description of true azimuths).

You may also want to figure out how much the dolphins shifted their locations within the pod from day one to day two—this is called **spatial shift.** To do so, first label each dot on the map with the name of the dolphin at their locations for days one and two. Our dolphins were simply given numbers 1 through 18. The labeled dolphin locations for days one and two (figure 18.17, A and B) can then be superimposed so that the bivariate means coincide exactly (figure 18.17, C). Since each map symbol is labeled by the dolphin's number, you can draw straight lines, called **shift lines,** between pairs of symbols for each dolphin. The length of each line is the amount the dolphin shifted in position from day one to day two relative to the bivariate mean. You can also measure the length of each shift line, add these lengths together, and divide the sum of these lengths by the number of dolphins (18) to find an **average spatial shift** of 0.13 nautical miles from day one to day two.

You can also find **angular shifts** in position from the shift lines if you plot the center of each shift line relative to an x-y axis with compass directions (figure 18.17, D).

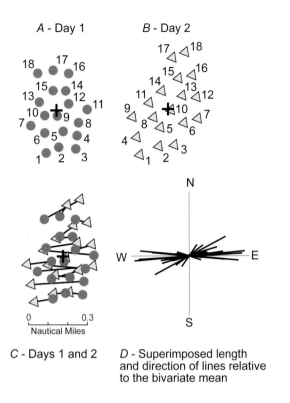

Figure 18.17   You can superimpose the day one *(A)* and day two *(B)* dolphin locations to measure the shifts in position of each animal relative to the bivariate mean *(C)*. Angular shifts can be plotted against compass directions *(D)*.

This diagram clearly shows that most of the angular shifts are close to easterly or westerly, and that none are northerly or southerly. A biologist might try to explain this restricted pattern of angular shifts in the relative positions of dolphins by known facts of dolphin pod behavior or by geographic constraints to movement at or between the two locales.

Measuring point feature diffusion

Spatial diffusion is the process by which features move outward from a starting position across space over time, continuously decreasing in density. You will find maps showing the spatial diffusion of phenomena such as water pollution from a point source, oil slicks from a leaking oil tanker, or forest fire from its point of ignition. Sometimes spatial diffusion is considered a good thing. For example, the spread of information, health care, or education into or within a culture are positive examples of diffusion. Other times, spatial diffusion can signal something gone awry in an ecosystem, such as the spread of invasive plant or animal species or the spread of disease. The diffusion of a set of point features is easiest to analyze from maps showing their positions at different time intervals.

Your task when analyzing any sort of **diffusion map** (a map of change in position over time due to a diffusion process) is to understand the dispersal or growth mechanism that lies behind the pattern you see. Professionals use a variety of numerical modeling strategies in diffusion studies, but the high level of statistical expertise required for you to work with these models falls beyond the scope of this book. However, there are simple methods that you can use to analyze the magnitude and direction of diffusion. Let's look at a hypothetical example of point-feature spatial diffusion—the movement of markers outward from an initial point due to the combined influence of coastal currents and winds.

In our example, 30 floating markers equipped with electronic location transmission devices were released from Sandy Point on Waldron Island, one of the San Juan Islands in Washington State. The latitude and longitude of each marker was recorded hourly over a four-hour period, and the location of each marker was plotted on a nautical chart covering the area (figure 18.18). The 120 point symbols on the chart show the positions of the 30 markers at one, two, three, and four hours after release from the initial point at the western tip of Waldron Island. The colors and shapes represent their positions over time.

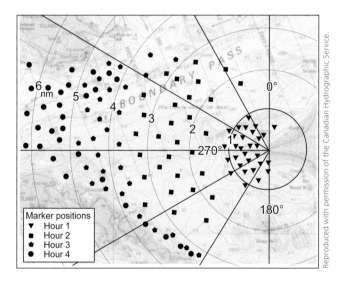

Figure 18.18   Positions of 30 markers at one–hour intervals after their initial release from the western tip of Waldron Island, Washington.

Looking at the hourly positions of the markers plotted on the chart, notice the following:

1. The markers appear to have slowly diffused outward from the release point during the first hour, followed by more rapid diffusion during the second hour, then slowing again.

2. Diffusion is not the same in all directions. We can hypothesize (based on ancillary information) that a combination of coastal current and surface wind has restricted diffusion toward the north and south.

3. Marker movement in the west-southwest direction is essentially stopped when markers reach the surf zone of several small islands around three hours after release.

4. Four hours after release, many of the markers have entered a narrow channel to the west.

These observations constitute a **qualitative analysis** (one that does not involve measurements and numbers) of marker diffusion that gives us several insights into the diffusion process. As we saw in chapter 17, mathematical measures can lessen disagreements arising from different visual impressions of the same distribution on the map; so, it is equally important to conduct a **quantitative analysis** of spatial diffusion. There are several measures of diffusion that will give you further insight into the spatial nature of the movement of the markers. These measures are also evidence for or against conclusions drawn from your qualitative observations. The first set of quantitative measures is based on making point counts within data collection units.

Notice on figure 18.18 the graticule for a polar aspect azimuthal equidistant map projection (see chapter 3 for more on map projections). What you are seeing is actually a polar coordinate grid centered at the initial point of release. The circles are what we call **range rings.** Range rings serve both as data collection unit boundaries and as a circular distance scale, since the rings are spaced at 1-nautical-mile intervals outward from the initial point. The straight lines are true azimuth lines radiating outward from the initial point at 30-degree angular increments. These lines divide the circle into 12 equal **circle sectors** (portions of a circle, like pieces of a pie, enclosed by two radii and an arc; also called **circular sectors**).

Counting the number of markers within range rings and sectors at each hour is straightforward. You can gain further insight into the diffusion process by using a spreadsheet or statistical analysis program to create frequency diagrams of the data. The frequency diagram at the top of figure 18.19 shows the number of markers within each range ring. The four rows of raised columns show the number of markers counted in each range ring

at the four time intervals. The left column in the first row shows 25 markers to be within 1 nautical mile of the initial release point, which reinforces our visual observation of slow diffusion during the first hour.

The other three rows in this diagram are interesting because the shape formed by the columns roughly resembles a "wave" of diffusion moving outward from the hour-one marker positions during the next three hours. The three waves peak at 2–3 nautical miles, 4–5 nautical miles, and 5–6 nautical miles, which indicates a steady diffusion rate for two hours, followed by a slowdown in the movement of markers during the fourth hour.

The frequency diagram at the bottom of figure 18.19 shows the number of markers in each sector during each time period. What stands out on this diagram is the reduction in the **angular range** (the difference between the maximum and minimum true azimuth angles) of the markers over the four hours. Notice the increasing

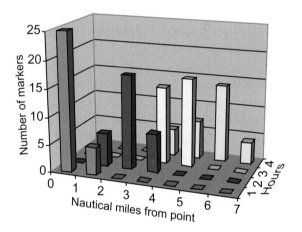

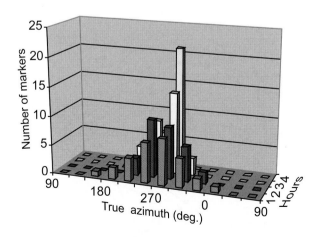

Figure 18.19 Frequency diagrams showing the number of markers within each 1-nautical mile range ring (top) and within each 30-degree circle sector (bottom) at the end of each of the first four hours.

number of markers in the western sector over time. The data on this graph correspond well with the visual observation of an increased concentration of markers in the channel to the west during hours 3 and 4.

Maps showing the movement of each point feature give you additional details about the diffusion process. The map in figure 18.20, for example, shows the movement of each marker over the four-hour time period. The shift lines connecting the marker positions at each hour provide rough approximations of the actual path followed, since shorter time interval positional information was not collected.

The map looks like a maze of intersecting lines, but you can see the overall westerly drift of the markers from the release point. Notice that there is little angular shift among lines after the first hour. This tells you that the westerly currents must have been fairly constant and winds responsible for the motion must have been fairly steady in intensity. Also apparent is the variable amount of movement among markers, particularly the very slow movement of the markers that drifted close to the shore of nearby islands as opposed to the faster movement of the markers in open water.

Measuring the length of each shift line gives you more detailed information about individual marker movement. These measurements allow you to compute the average spatial shift for marker movement for each hour. We converted the length of the 30 shift lines for each hour into nautical miles, then summed the lengths and divided each sum by 30 to obtain average spatial shifts of 0.73, 2.00, 1.65, and 1.20 nautical miles for hours 1

to 4. These values are the speeds of movement for each hour in knots, since a **knot** is 1 nautical mile per hour. The four average spatial shift speeds we calculated support the visual observation of slower diffusion during the first hour followed by more rapid diffusion for hour two, then slowing again.

You can learn more about the details of the hour-to-hour movement of the markers by making a **composite frequency diagram** for the shift line data. The frequency diagram in figure 18.21 shows a composite of the number of markers and the distance they moved in 0.2-nautical-mile increments for each of the first four hours. The first row shows that a roughly equal number of markers moved from 0.2 to 1.2 nautical miles during the first hour. Rows for hours two through four show the longer distances to be close to normally distributed about the average value, particularly in the last two hours. The exceptions are the markers that moved about 1 nautical mile, which are likely those markers in the surf zone of the small islands. You can use the standard deviation that we described earlier and other statistics based on a normal distribution of values to describe these data.

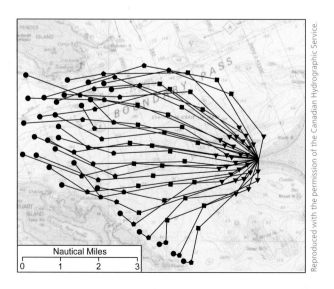

Figure 18.20 This map shows you the approximate paths followed by each of the 30 markers during the first four hours after they were released.

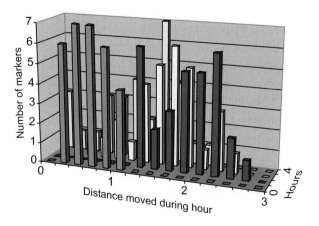

Figure 18.21 This composite frequency diagram shows the number of path lines falling within 0.2-nautical-mile-distance increments at the end of each hour.

A CAUTIONARY TALE

Disease and mortality always lurks in the human landscape. Since the mid-1800s, maps have played a fundamental role in **epidemiology,** the discipline studying the incidence, distribution, and control of disease in a population. Dr. John Snow's map of the terrible cholera outbreak in central London during the summer of 1854 is one of the first epidemiological maps (the center section of the map is shown in figure 18.22). The popular story is that Snow discovered the source of the epidemic to be a particular water pump by plotting the locations of deaths and pumps and seeing which pump was in the center of the deaths. Snow's personal account of the outbreak, published in 1855, paints a more accurate picture of how the map was used:

> "As soon as I became acquainted with the situation and extent of this irruption of cholera, I suspected some contamination of the water of the much-frequented street-pump in Broad Street, near the end of Cambridge Street; but on examining the water, on the evening of the 3rd September, I found so little impurity in it of an organic nature, that I hesitated to come to a conclusion. Further inquiry, however, showed me that there was no other circumstance or agent common to the circumscribed locality in which this sudden increase of cholera occurred, and not extending beyond it, except the water of the above mentioned pump.
>
> …
>
> On proceeding to the spot, I found that nearly all the deaths had taken place within a short distance of the pump. There were only ten deaths in houses situated decidedly nearer to another street pump. In five of these cases the families of the deceased persons informed me that they always went to the pump in Broad Street, as they preferred the water to that of the pump which was nearer. In three other cases, the deceased were children who went to school near the pump in Broad Street. Two of them were known to drink the water; and the parents of the third think it probable that it did so. The other two deaths, beyond the district which this pump supplies, represent only the amount of mortality from cholera that was occurring before the irruption took place.
>
> With regard to the deaths occurring in the locality belonging to the pump, there were sixty-one instances in which I was informed that the deceased persons used to drink the pump-water from Broad Street, either constantly ,or occasionally. In six instances I could get no information, owing to the death or departure of every one connected with the deceased individuals; and in six cases I was informed that the deceased persons did not drink the pump-water before their illness.
>
> The result of the inquiry then was, that there had been no particular outbreak or increase of cholera, in this part of London, except among the persons who were in the habit of drinking the water of the above-mentioned pump-well.
>
> The deaths which occurred during this fatal outbreak of cholera are indicated in the accompanying map, (figure 18.22) as far as I could ascertain them. ? The pump in Broad Street is indicated on the map, as well as all the surrounding pumps to which the public had access at the time.
>
> It requires to be stated that the water of the pump in Marlborough Street, at the end of Carnaby Street, was so impure that many people avoided using it. And I found that the persons who died near this pump in the beginning of September, had water from the Broad Street pump. With regard to the pump in Rupert Street, it will be noticed that some streets which are near to it on the map, are in fact a good way removed, on account of the circuitous road to it. These circumstances being taken into account, it will be observed that the deaths either very much diminished, or ceased altogether, at every point where it becomes decidedly nearer to send to another pump than to the one in Broad Street. It may also be noticed that the deaths are most numerous near to the pump where the water could be more readily obtained. The wide open street in which the pump is situated suffered most, and next the streets branching from it, and especially those parts of them which are nearest to Broad Street. If there have been fewer deaths in the south half of Poland Street than in some other streets leading from Broad Street, it is no doubt because this street is less densely inhabited" (Snow 1885).

Snow's account clearly shows that he did not use his map to discover the likely source of the cholera epidemic, since from the beginning he suspected water contamination from the Broad Street pump. This suspicion came from his previously held belief that cholera is a gastrointestinal disease transmitted by polluted drinking water. Rather, his map is a record of his investigation into the locations of the cholera deaths and water pumps. The map helped him explain to the public that the polluted water had to be from the Broad Street pump because it was the closest source for the victims.

The real story behind Snow's cholera map is valuable lesson for those who believe that cause and effect relationships can be determined by comparing maps alone. Visual and statistical correlations among mapped information have helped epidemiologists test biologically plausible hypothesis about the causes of disease and death. But without a sound hypothesis and determined detective work the associations interpreted from maps may seduce us into thinking that we have learned something about the cause. The resulting claims that you may hear about disease and mortality "hot spots" often create needless worry and distract researchers from the actual causes of disease and death.

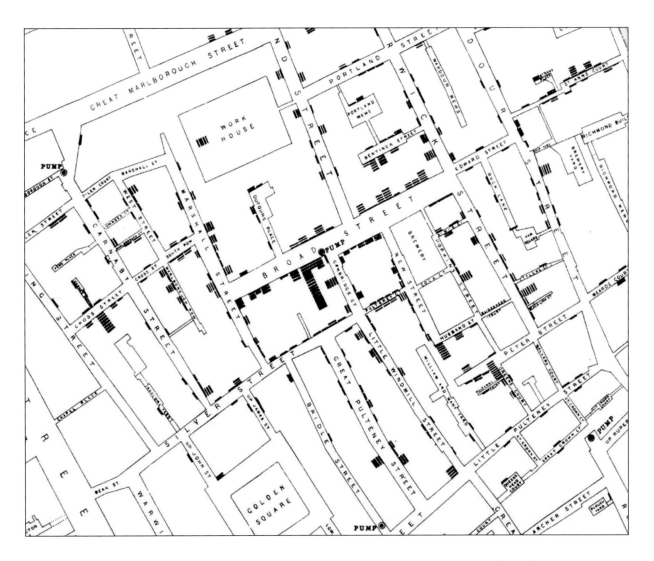

Figure 18.22   Center section of Dr. John Snow's 1855 map of the 1854 cholera epidemic in central London.

SELECTED READINGS

Bailey, T. C., and A. C. Gatrell. 1995. *Interactive spatial data analysis.* New York: John Wiley & Sons, Inc.

Clark, W. A. V., and P. C. Hosking. 1986. *Statistical methods for geographers.* New York: John Wiley & Sons, Inc.

Dunteman, G. H. 1984. *Introduction to multivariate analysis.* Beverly Hills, Calif.: Sage Publications.

Griffith, D. A., and C. G. Armheim. 1997. *Multivariate statistical analysis for geographers.* Upper Saddle River, N.J.: Prentice-Hall.

Griffith, D. A., C. G. Armheim, and J. R. Desloges. 1991. *Statistical analysis for geographers.* Englewood Cliffs, N.J.: Prentice-Hall.

Haining, R. 1990. *Spatial data analysis in the social and environmental sciences.* New York: Cambridge University Press.

Johnson, S. 2007. *The ghost map: The story of London's most terrifying epidemic—and how it changed science, cities and the modern world.* New York: Riverhead Books.

Johnston, R. J. 1980. *Multivariate statistical analysis in geography.* New York: Longman, Inc.

Mitchell, A. 2005. *The ESRI guide to GIS analysis, volume 2: Spatial measurements & statistics.* Redlands, Calif.: ESRI Press.

Morrill, R., G. L. Gaile, and G. I. Thrall. 1988. *Spatial diffusion.* Newbury Park, Calif.: Sage Publications.

Shaw, G., and D. Wheeler. 1985. *Statistical techniques in geographical analysis.* New York: John Wiley & Sons, Inc.

Snow, J. 1885. *On the mode of communication of cholera.* 2nd ed. London: John Churchill

Walford, N. 1995. *Geographical data analysis.* New York: John Wiley & Sons, Inc.

appendix A DIGITAL CARTOGRAPHIC DATA

RASTER DATA
    Scanned maps
    Digital elevation models
    Remote sensing imagery
    Raster thematic maps
VECTOR DATA
    Basemap data
    Operational layers
DATA SOURCES

Digital data are the foundation of computer mapping systems. Each year, more geographical information is gathered in digital form directly in the field. Numerical information suitable for computer mapping is also the product of converting existing printed maps into digital files. Because different data formats have been used in developing digital cartographic databases, mapping software must be matched to these different ways of organizing the data.

There are two fundamentally different data formats. Satellite imagery, air photos, digital elevation models (DEMs), digital scanning, and printing technology use a matrix or grid of cells called **pixels.** In this raster format, data are recorded, stored, and processed on a pixel-by-pixel basis.

The alternative to the raster format is vector format. Think of the environment as a collection of features of varying geometrical character. These features may be represented in a database as two-dimensional points, lines, or areas defined by x- and y-coordinates, or as a three-dimensional surface with an additional z-coordinate.

When discussing digital data products, there are three main characteristics to understand: coverage area (worldwide, continent, country, and so on); resolution or minimum mapping unit (4 kilometer pixels, 1 kilometer, 90 meter, 1 acre, and so on), which relates to the amount of detail you can see; and data size (in terms of both disk storage and also number of datasets contained in the product). For example, knowing that one worldwide DEM product is 500 gigabytes versus another that is 5 gigabytes helps you determine whether your computer is able to handle a particular DEM. Conversely, if you get a smaller dataset, it will be at a coarser resolution, and you may find that it does not provide the detail you need. In these cases, using tiles (pieces) of the higher-resolution data is an option.

It is useful if you can find information on the main benefit or characteristic of a product that you want to use. This helps you choose one product over another. You also need to consider if the data format is something your software can read. A DEM, for example, is directly readable in ArcGIS products. Be aware that for some DEM products the data can come in multiple formats. For example, the U.S. Geological Survey (USGS) seamless data site (http://seamless.usgs.gov) provides DEM data that can be downloaded from in ArcGIS GRID format so that no additional data conversion is required.

When using data that span ocean areas, it might be useful to know whether ocean areas are coded with an attribute value that relates to some measure like bathymetry, or if they are coded as NoData or zero. Codes like these make a difference in whether or not you will have to post-process the data.

RASTER DATA

Below are descriptions of a variety of raster products. We try to provide you with the information you need to help you choose your data wisely.

Scanned maps

The simplest way mapmakers convert existing maps to digital form is to scan the map as a whole. Many raster databases like this are commonly available. The advantage of these raster databases is that they can be created relatively quickly and inexpensively from existing maps simply by scanning them. Many maps have already been scanned, and a good portion of them have been geometrically corrected through rectification so that the scale of the photograph is uniform and planimetrically correct (see chapter 9 for more on rectification). With these data, you can simply overlay your other data layers and they should line up.

While these digital maps are useful for some applications as a background reference, a disadvantage is that only the information on the original map is included in the image. The original map design dominates the display of the information even when other layers are overlaid. Furthermore, the map is represented in the database as a single raster layer. You can't turn off the roads or turn on the rivers in the display as you could if the data were in layers. Since individual features on the map can't be manipulated separately, data flexibility is extremely limited.

However, the advantages of using these digital maps sometimes outweigh the limitations. Using them, you do not have to collect and symbolize all the map layers. They allow you the flexibility of overlaying your own information if they are rectified. Many are readily available and some are also free of cost.

Digital elevation models

Other raster data comes from sources not originally found on a map. For example, elevation data can be gathered using remote sensors to provide wide coverage at varying resolutions. These data, called **digital elevation models (DEMs)** or sometimes **digital terrain models (DTMs)** can then be used to generate terrain and other continuous surfaces. Originally, DEMs were created by digitizing contours on maps and interpolating to a regular grid of elevations. Now they are more often generated using remotely sensed data. DEMs are available at several spatial resolutions, from very high-resolution datasets compiled from lidar, to worldwide datasets at much coarser resolutions.

Remote sensing imagery

Air photos and satellite imagery from remote sensors provide us with invaluable data for many applications. They are often acquired and distributed in digital raster format. The ability to classify and computationally analyze these images allows us to see beyond what might appear to us if we were flying overhead. Sensors that can collect data outside the visible portion of the electromagnetic spectrum provide information about temperature, moisture, and other characteristics of the landscape not visible to the human eye.

Raster thematic maps

Raster thematic data focus on narrow, closely defined topics. Thematic raster data consist of a matrix of cells (or pixels) organized into rows and columns, and each cell contains a value representing thematic information, such as temperature, soil type, or land use. These data are available for a wide range of themes, coverage extents, and resolutions. The Internet will provide you with an amazing source of this type of data, much of which can be downloaded for free.

VECTOR DATA

Cartographers also create and use vector databases. You can generally think of vector data as either basemap data or operational overlays.

Both types of vector data are compiled by either converting existing maps to digital form on a feature-by-feature basis or by **digitizing** (converting to digital format) the information from primary sources such as air photos, satellite imagery, Global Positioning Systems, surveying, census taking, and more. The data layers consist of x-, y-, and sometimes z-coordinates that are separated into different data files based on theme (solids, roads, lakes, etc.) The resulting database is complex in structure but more flexible to use than raster databases.

The problem with vector databases is that data gathering is time-consuming and expensive. And once the data have been collected, they must be conflated. Conflation is a procedure of reconciling the positions of corresponding features in different data layers. The extra time, cost, and effort is often justified, however, because individual features as well as feature layers in the database can be manipulated and tailored to a wide variety of user needs.

Basemap data

Basemap data are used for many purposes by many people. We often think of transportation, boundaries, hydrography (water), hypsography (landforms), and cultural features as basemap data. These types of data make up the majority of what is shown on a reference map. They can also be used more minimally to provide base location information on thematic maps. Often basemap information is obtained from reputable agencies whose primary task is to compile wide coverage, multiple resolution, accurate (to some stated standard), and timely data. The operational layers can then be placed upon these base data to create thematic maps.

Operational layers

Operational overlays relate to specific uses, so they are often collected to fit a narrowly defined application. Soils, land use, parcels, zoning, and habitat units are examples. These data are often compiled with respect to the requirements for a specific project or application. They are available in countless resolution-extent-theme combinations. Again, the Internet is a great source for this type of data, offering reputable sources, such as government mapping agencies, and not-so-reputable sources, such as certain (but not all) personal databases.

Today, people can upload their data to be displayed on virtual globes, such as ArcGlobe or ArcGIS Explorer, so that others can see their data on a standard basemap. These "mash ups" sometimes also provide the opportunity to download the data. Again, be wary of the source if it is not one that you already know to be trustworthy.

DATA SOURCES

Because of the large number and variety of sources from which both vector and raster data can be obtained, we will not attempt to provide an exhaustive list of data providers or datasets. Instead we will narrow our discussion of data sources to those from the USGS that are free of charge, cover large geographic extents, and are widely used in computer mapping.

For the United States, the USGS offers a collection of geospatial data compiled from a variety of sources, including maps, air photos, and remote sensors. Below is a list of public-domain software developed by the USGS and its partners to support a wide variety of natural science research and mapping activities. These can be downloaded from the Internet at http://www.usgs.gov/pubprod/data.html#data.

TABLE A.1   USGS datasets

DLG	Digital line graph	DLGs are digital versions of paper topographic maps. Large-scale (7.5-minute) DLGs are digital versions of USGS 1:20,000-, 1:24,000-, and 1:25,000-scale topographic maps. Intermediate-scale (1:100,000-scale) DLGs are created from USGS 30-minute-by-60-minute topographic maps and are distributed in 30-minute-by-30-minute tiles corresponding to the east or west half of the topographic map. Small-scale (1:2,000,000-scale) DLGs are derived from sectional maps printed in the U.S. National Atlas produced by the USGS. Data files are organized on a state-by-state basis, except that Alaska is not available.
DOQ	Digital orthophoto quadrangle	DOQs are air photos that have been digitally processed into a geometrically corrected image in which distortion from the camera angle and topography has been removed so that they can be used like a map. DOQs are black-and-white, true-color, or color-infrared (CIR) images with 1-meter ground resolution. The USGS produces three types of DOQs for the United States: * Quarter-quad DOQs covering one quarter of a 1:24,000-scale topographic map (3.75 minutes latitude by 3.75 minutes longitude). * Full-quad DOQs for 1:24,000-scale topographic maps covering 7.5 minutes latitude by 7.5 minutes longitude. * Seamless DOQs available for free download from The National Map Seamless Server. DOQs on this site are the most current version and are available for the conterminous United States.
DRG	Digital Raster Graphic	DRGs are scanned digital images of USGS topographic quadrangles. DRGs have been made for all quadrangles in the following USGS topographic map series for the United States and its territories: * 7.5-minute maps of the conterminous United States, Hawaii, and limited areas of Alaska at 1:24,000 and 1:25,000 scale. * 7.5-minute-by-15-minute maps in limited areas of the conterminous United States at 1:25,000 scale. * Pacific Island maps at 1:20,000, 1:24,000, and 1:25,000 scales. * Puerto Rico and the Virgin Islands at 1:20,000 scale. * Alaska at 1:63,360 scale. * 30-minute-by-60-minute maps of the conterminous United States at 1:100,000 scale. * 1-minute-by-2-degree maps of the United States at 1:250,000 scale.
GLCC	Global Land Cover Characterization	GLCC is a series of global land-cover classification datasets based primarily on the classification of 1-km AVHRR (Advanced Very High Resolution Radiometer) satellite image composites from April 1992 through March 1993. Ancillary data sources included digital elevation data, ecoregions interpretation, and country- or regional-level vegetation and land-cover maps.
GTOPO30	Global 30 Arc-Second Elevation Dataset	GTOPO30 is a global raster DEM with a latitude–longitude grid cell spacing of 30 arc-seconds (approximately 1 kilometer by 1 kilometer at the equator). Although the grid is global, it contains only land elevations in meters. Ocean areas are given a no data value of -9999. Data are available in 33 tiles covering the entire earth.

Table A.1 *continued*

HYDRO1k		HYDRO1k is a geographic database developed to provide comprehensive and consistent global coverage of hydrographically related datasets, including streams, drainage basins, and other data layers derived from the USGS GTOPO30 global digital elevation model. HYDRO1k provides georeferenced datasets, both raster and vector, for users who need to analyze and map hydrologic information on a global or continental scale.
LUDA	Land-use and land-cover data	Land-use and land-cover data contain polygons for areas of different land cover and use.
NLCD	National Land Cover Dataset	A land-cover dataset for the conterminous United States on the basis of 1992 Landsat Thematic Mapper imagery and supplemental data. The seamless dataset contains 21 categories of land cover suitable for a variety of applications, including landscape analysis, land management, and environmental modeling. The data are distributed by state as 30-meter resolution raster images in an Albers equal area map projection.
NAPP	National Aerial Photography Program	NAPP provides cloud-free aerial photographs covering the conterminous United States over five-to-seven year cycles. The program began in 1987 and continues to be our most recent and consistent source of high-quality aerial photography. Black-and-white or color-infrared aerial photographs, depending on location and date, are taken from an altitude of 20,000 feet. Each photo is centered on a quarter of a 1:24,000-scale USGS topographic map.
NED	National Elevation Dataset	The NED contains elevation data for the United States. NED data have a resolution of 1 arc-second (approximately 30 meters or 100 feet) for the conterminous United States, Hawaii, and Puerto Rico and a resolution of 2 arc-seconds (approximately 60 meters or 200 feet) for Alaska.
NHD	National Hydrography Dataset	The NHD is a comprehensive set of digital spatial data that contains information about surface water features such as lakes, ponds, streams, rivers, springs, and wells. The NHD is based on USGS 1:100,000-scale digital line graph (DLG) hydrography data integrated with stream reach information from the United States Environmental Protection Agency (EPA).
SRTM	Shuttle Radar Topography Mission	SRTM data at 1 arc-second (30 meter or 100 foot) ground resolution are available for the United States, its territories, and possessions. SRTM data at 3 arc-second (90 meter or 300 foot) spatial resolution are available for the entire earth between 60°N and 56°S latitude. The product consists of seamless raster data available from The National Map Seamless Server.

appendix B **SELECTED NAVIGATION AND GPS ABBREVIATIONS AND ACRONYMS**

2D fix   Two-dimensional fix (latitude-longitude).

2D NAV   Two-dimensional navigation.

3D fix   Three-dimensional fix (latitude-longitude-elevation).

3D NAV   Three-dimensional navigation.

ALT   Altitude (also EL and ELEV).

BRG   Bearing; the direction measured clockwise in degrees from north. This is technically the "azimuth," but most GPS receivers use the term "bearing" instead of azimuth. GPS receivers use either true north or magnetic north, whichever you select in your setup menu.

C/A   Course acquisition code; of the two types of signals sent out by GPS satellites, this one is for unprotected civilian use.

CDI   Course deviation indicator; a graphic way of showing the amount and direction of cross track error (XTE or XTK) in your course.

CMG   Course made good; the compass heading from your starting point to your present position.

COG   Course over the ground; the direction of travel achieved from your starting point to your present location.

CTS   Course to steer; the recommended heading to reach your destination most efficiently.

DMG   Distance made good; distance from your last position to your present position.

DOP   Dilution of precision; same as geometric quality (GQ) (see glossary).

DST   Distance.

DTG   Distance to go; the distance you have yet to travel to reach your destination.

DTK   Desired track; the course you want to travel as given by your "from" and "to" waypoints.

E   East.

EL   Elevation (also ELEV and ALT).

ELEV   Elevation (also EL and ALT).

EPE   Estimated position error; estimate of the error of a fix, given in feet.

ETA   Estimated time of arrival; the time you should reach your destination, based on your current speed.

ETE   Estimated time en route; the amount of time remaining to arrive at your destination based on your current speed.

FT   Feet.

GMT   Greenwich mean time.

GQ   Geometric quality; same as DOP (see glossary).

GRI   Grid.

HDG   Heading; the direction in which you're moving with respect to either true or magnetic north.

K   Kilometers or knots.

LFX   Last fix.

LMK   Landmark.

M   Meters or magnetic.

MAG   Magnetic (north).

MGRS   Military Grid Reference System (see glossary).

MOB   Man overboard; a simple keystroke combination you can use to mark a spot quickly so that you can return to it.

MPH   Miles per hour.

N   North.

NAV   Navigation.

ODOM   Odometer.

P Code   Precision code; of the two types of signals sent out by GPS satellites, this one may be used only by the military.

PPS   Precise positioning service; the accuracy provided by precise (P) codes for military use.

S   South.

SA   Selective availability (see glossary).

SOA   Speed of advance or speed of approach; same as ground speed (SPD) and velocity made good (VMG); your speed in the direction of your destination. If you're heading directly toward your destination, your SOA is the same as your ground speed. If you're not on course, your SOA is less than your ground speed. The GPS receiver shows a negative SOA as a blank line.

SOG   Speed over ground; same as ground speed (SPD), speed of advance or speed of approach (SOA) and velocity made good (VMG); the speed you're traveling over the ground (as opposed to speed through the water).

SQ   Signal quality; the strength of the signals from a satellite.

SPD   Speed or ground speed; same as speed of advance or speed of approach (SOA) and velocity made good (VMG); the speed you're traveling over the ground (as opposed to speed through the water).

SPS   Standard Positioning Service; the accuracy provided by Coarse Acquisition Codes for civilian use.

STM   Statute miles.

TRK   Track; your direction of movement relative to a ground position.

TTF   Time to fix; same as time to first fix (TTFF); the time it takes a GPS receiver to make its first position fix after you turn it on. This is the time it takes the receiver to collect position information from every satellite.

TTG   Time to go; estimate of the time from your current location to your next waypoint, based on your current speed.

TTFF   Time to first fix; same as time to fix (TTF); the time it takes a GPS receiver to make its first position fix after you turn it on. This is the time it takes the receiver to collect position information from every satellite.

VMG   Velocity made good; same as speed of advance or speed of approach (SOA) and speed over ground (SPD); your speed in the direction of your destination. If you're heading directly toward your destination, your SOA is the same as your ground speed. If you're not on course, your SOA is less than your ground speed. The GPS receiver shows a negative SOA as a blank line.

VOG   Velocity over the ground; same as ground speed (SPD) or speed of approach (SOA); the speed you're traveling over the ground (as opposed to speed through the water).

W   West.

WGS84   World Geodetic System 1984 (see glossary).

WPT   Waypoint (see glossary).

XTE   Cross track error (also XTK); the distance from your current location to your desired track (DTK).

XTK   Cross track error (also XTE); the distance from your current location to your desired track (DTK).

appendix C **TABLES**

Table C.1 Metric and English equivalents

Converting from English units		
Multiply	by	to obtain
acres	0.4046856	hectares
acres	43,560.0*	square feet
acres	4,046.856	square meters
acres	0.0015625*	square miles
acres	4,840.0*	square yards
feet	30.48*	centimeters
feet	0.0003048*	kilometers
feet	0.3048*	meters
feet	0.00018939394	miles
square feet	0.000022956	acres
square feet	929.0304*	square centimeters
square feet	0.09290304*	square meters
square feet	0.00000003587	square miles
inches	2.54*	centimeters
inches	0.0254*	meters
inches	0.000015782	miles
inches	0.027777778	yards
square inches	6.4516*	square centimeters
square inches	0.00064516*	square meters
square inches	645.16*	square millimeters
miles	160,934.4*	centimeters
miles	5,280.0*	feet
miles	63,360.0*	inches
miles	1.609344*	kilometers
miles	1,609.344*	meters
miles	1,760.0*	yards
miles	1.15077945	nautical miles
square miles	640.0*	acres
square miles	27,878,400.0*	square feet
square miles	2.589988110647	square kilometers
yards	91.44*	centimeters
yards	0.0009144*	kilometers
yards	0.9144*	meters
yards	0.000568182	miles
square yards	0.000206611	acres
square yards	0.83612736*	square meters
square yards	0.0000003228305	square miles
Converting from metric units		
Multiply	by	to obtain
centimeters	0.03280839895	feet
centimeters	0.3937007874	inches
centimeters	0.00001*	kilometers
centimeters	0.01*	meters

* Constants are exact.

Table C.1   *continued*

centimeters	0.000006213711922	miles
centimeters	0.01093613298	yards
square centimeters	0.001076391042	square feet
square centimeters	0.15500031	square inches
hectares	2.471054073	acres
hectares	107,639.1042	square feet
kilometers	100,000.0*	centimeters
kilometers	1,000.0*	meters
kilometers	1,093.613298	yards
kilometers	3,280.839895	feet
kilometers	39,370.07874	inches
kilometers	0.6213711922	miles
square kilometers	247.1054	acres
square kilometers	10,763,910.42	square feet
square kilometers	0.386102158496	square miles
meters	100.0*	centimeters
meters	3.280839895	feet
meters	39.37007874	inches
meters	0.001*	kilometers
meters	0.0006213711922	miles
meters	1.093613298	yards
square meters	0.0002471054	acres
square meters	10.76391042	square feet
square meters	0.0000003861003	square miles
square millimeters	0.00001076391042	square feet

* Constants are exact.

Converting geographic (angular) units		
Multiply	by	to obtain
degrees	0.017453292	radians
radians	57.2958	degrees

Table C.2   Variation in the length of a degree of latitude

Measured along the meridian on the WGS 84 ellipsoid.

Lat.	Meters	Statute miles	Lat.	Meters	Statute miles	Lat.	Meters	Statute miles
0-1°	110 567.3	68.703	30-31°	110 857.0	68.883	60-61°	111 423.1	69.235
1-2	110 568.0	68.704	31-32	110 874.4	68.894	61-62	111 439.9	69.246
2-3	110 569.4	68.705	32-33	110 892.1	68.905	62-63	111 456.4	69.256
3-4	110 571.4	68.706	33-34	110 910.1	68.916	63-64	111 472.4	69.266
4-5	110 574.1	68.708	34-35	110 928.3	68.928	64-65	111 488.1	69.275
5-6	110 577.6	68.710	35-36	110 946.9	68.939	65-66	111 503.3	69.285
6-7	110 581.6	68.712	36-37	110 965.6	68.951	66-67	111 518.0	69.294
7-8	110 586.4	68.715	37-38	110 984.5	68.962	67-68	111 532.3	69.303
8-9	110 591.8	68.718	38-39	111 003.7	68.974	68-69	111 546.2	69.311
9-10	110 597.8	68.722	39-40	111 023.0	68.986	69-70	111 559.5	69.320
10-11	110 604.5	68.726	40-41	111 042.4	68.998	70-71	111 572.2	69.328
11-12	110 611.9	68.731	41-42	111 061.9	69.011	71-72	111 584.5	69.335
12-13	110 619.8	68.736	42-43	111 081.6	69.023	72-73	111 596.2	69.343
13-14	110 628.4	68.741	43-44	111 101.3	69.035	73-74	111 607.3	69.349
14-15	110 637.6	68.747	44-45	111 121.0	69.047	74-75	111 617.9	69.356
15-16	110 647.5	68.753	45-46	111 140.8	69.060	75-76	111 627.8	69.362
16-17	110 657.8	68.759	46-47	111 160.5	69.072	76-77	111 637.1	69.368
17-18	110 668.8	68.766	47-48	111 180.2	69.084	77-78	111 645.9	69.373
18-19	110 680.4	68.773	48-49	111 199.9	69.096	78-79	111 653.9	69.378
19-20	110 692.4	68.781	49-50	111 219.5	69.108	79-80	111 661.4	69.383
20-21	110 705.1	68.789	50-51	111 239.0	69.121	80-81	111 668.2	69.387
21-22	110 718.2	68.797	51-52	111 258.3	69.133	81-82	111 674.4	69.391
22-23	110 731.8	68.805	52-53	111 277.6	69.145	82-83	111 679.9	69.395
23-24	110 746.0	68.814	53-54	111 296.6	69.156	83-84	111 684.7	69.398
24-25	110 760.6	68.823	54-55	111 315.4	69.168	84-85	111 688.9	69.400
25-26	110 775.6	68.833	55-56	111 334.0	69.180	85-86	111 692.3	69.402
26-27	110 791.1	68.842	56-57	111 352.4	69.191	86-87	111 695.1	69.404
27-28	110 807.0	68.852	57-58	111 370.5	69.202	87-88	111 697.2	69.405
28-29	111 823.3	68.862	58-59	111 388.4	69.213	88-89	111 698.6	69.406
29-30	111 840.0	68.873	59-60	111 405.9	69.224	89-90	111 699.3	69.407

Table C.3 Variation in the length of a degree of longitude

Measured along the parallel on the WGS 84 ellipsoid.

Lat.	Meters	Statute miles	Lat.	Meters	Statute miles	Lat.	Meters	Statute miles
0°	111 321	69.172	30°	96 488	59.956	60°	55 802	34.674
1	111 304	69.162	31	95 506	59.345	61	54 110	33.623
2	111 253	69.130	32	94 495	58.716	62	52 400	32.560
3	111 169	69.078	33	93 455	58.071	63	50 675	31.488
4	111 051	69.005	34	92 387	57.407	64	48 934	30.406
5	110 900	68.911	35	91 290	56.725	65	47 177	29.315
6	110 715	68.795	36	90 166	56.027	66	45 407	28.218
7	110 497	68.660	37	89 014	55.311	67	43 622	27.106
8	110 245	68.504	38	87 835	54.579	68	41 823	25.988
9	109 959	68.326	39	86 629	53.829	69	40.012	24.862
10	109 641	68.129	40	85 396	53.063	70	38 188	23.729
11	109 289	67.910	41	84 137	52.281	71	36 353	22.589
12	108 904	67.670	42	82 853	51.483	72	34 506	21.441
13	108 486	67.410	43	81 543	50.669	73	32 648	20.287
14	108 036	67.131	44	80 208	49.840	74	30 781	19.127
15	107 553	66.830	45	78 849	48.995	75	28 903	17.960
16	107 036	66.510	46	77 466	48.136	76	27 017	16.788
17	106 487	66.169	47	76 058	47.261	77	25 123	15.611
18	105 906	65.808	48	74 628	46.372	78	23 220	14.428
19	105 294	65.427	49	73 174	45.469	79	21 311	13.242
20	104 649	65.026	50	71 698	44.552	80	19 394	12.051
21	103 972	64.606	51	70 200	43.621	81	17 472	10.857
22	103 264	64.166	52	68 680	42.676	82	15 545	9.659
23	102 524	63.706	53	67 140	41.719	83	13 612	8.458
24	101 754	63.228	54	65 578	40.749	84	11 675	7.255
25	100 952	62.729	55	63 996	39.766	85	9 735	6.049
26	100 119	62.212	56	62 395	38.771	86	7 792	4.842
27	99 257	61.676	57	60 774	37.764	87	5 846	3.632
28	98 364	61.122	58	59 135	36.745	88	3 898	2.422
29	97 441	60.548	59	57 478	35.716	89	1 949	1.211
						90	0	0

Table C.4   Latitude-longitude values for the largest city in each state

City, state	Latitude		Longitude	
Birmingham, Alabama	33°	25'N	86°	52'W
Anchorage, Alaska	61	20	149	55
Phoenix, Arizona	33	22	112	05
Little Rock, Arkansas	34	45	92	15
Los Angeles, California	34	15	118	15
Denver, Colorado	39	46	104	59
Hartford, Connecticut	41	53	72	45
Wilmington, Delaware	39	45	75	23
Miami, Florida	25	45	80	11
Atlanta, Georgia	33	45	84	21
Honolulu, Hawaii	21	20	157	50
Boise, Idaho	43	37	116	10
Chicago, Illinois	41	49	87	37
Indianapolis, Indiana	39	50	86	10
Des Moines, Iowa	41	37	93	30
Wichita, Kansas	37	35	97	20
Louisville, Kentucky	38	16	85	30
New Orleans, Louisiana	30	00	90	04
Portland, Maine	43	31	70	20
Baltimore, Maryland	39	22	76	30
Boston, Massachusetts	42	15	71	07
Detroit, Michigan	42	23	83	05
Minneapolis, Minnesota	45	00	93	10
Jackson, Mississippi	32	20	90	10
St. Louis, Missouri	38	40	90	10
Billings, Montana	45	47	108	29
Omaha, Nebraska	41	15	96	05
Las Vegas, Nevada	36	10	115	10
Manchester, New Hampshire	43	00	71	30
Newark, New Jersey	40	40	74	05
Albuquerque, New Mexico	35	06	106	40
New York, New York	40	40	73	58
Charlotte, North Carolina	35	14	81	53
Fargo, North Dakota	46	52	97	00
Cleveland, Ohio	41	30	81	45
Oklahoma City, Oklahoma	35	30	97	30
Portland, Oregon	45	29	122	48

Table C.4   *continued*

Philadelphia, Pennsylvania	44	00	75	05
Providence, Rhode Island	41	52	71	30
Columbia, South Carolina	34	00	81	00

Sioux Falls, South Dakota	43	30	96	50
Memphis, Tennessee	35	08	90	03
Houston, Texas	29	46	95	21
Salt Lake City, Utah	40	46	111	57
Burlington, Vermont	44	30	73	15

Norfolk, Virginia	36	49	76	15
Seattle, Washington	47	36	122	20
Charleston, West Virginia	38	23	81	30
Milwaukee, Wisconsin	43	05	88	00
Cheyenne, Wyoming	41	08	104	47

Table C.5   Areas of quadrilaterals of 1° extent

Lower latitude	Area in square kilometers	Area in square miles	Lower latitude	Area in square kilometers	Area in square miles
0	12,308.09	4,752.16	45	8,686.89	3,354.01
1	12,304.44	4,750.75	46	8,533.30	3,294.71
2	12,297.14	4,747.93	47	8,377.07	3,234.39
3	12,286.21	4,743.71	48	8,218.17	3,173.04
4	12,271.63	4,738.08	49	8,056.69	3,110.69
5	12,253.39	4,731.04	50	7,892.69	3,047.37
6	12,231.56	4,722.61	51	7,726.18	2,983.08
7	12,206.05	4,712.76	52	7,557.23	2,917.85
8	12,176.94	4,701.52	53	7,385.85	2,851.68
9	12,144.23	4,688.89	54	7,212.17	2,784.62
10	12,107.89	4,674.86	55	7,036.18	2,716.67
11	12,067.92	4,659.43	56	6,857.93	2,647.85
12	12,024.41	4,642.63	57	6,677.51	2,578.19
13	11,977.30	4,624.44	58	6,494.94	2,507.70
14	11,926.61	4,604.87	59	6,310.33	2,436.42
15	11,872.35	4,583.92	60	6,123.64	2,364.34
16	11,814.57	4,561.61	61	5,935.01	2,291.51
17	11,753.24	4,537.93	62	5,744.46	2,217.94
18	11,688.41	4,512.90	63	5,552.08	2,143.66
19	11,620.06	4,486.51	64	5,357.88	2,068.68
20	11,548.24	4,458.78	65	5,161.97	1,993.04
21	11,472.95	4,429.71	66	4,964.38	1,916.75
22	11,394.19	4,399.30	67	4,765.19	1,839.84
23	11,312.01	4,367.57	68	4,564.44	1,762.33
24	11,278.21	4,334.52	69	4,362.18	1,684.24
25	11,137.44	4,300.17	70	4,158.56	1,605.62
26	11,045.08	4,264.51	71	3,953.53	1,526.46
27	10,949.38	4,227.56	72	3,747.24	1,446.81
28	10,850.36	4,189.33	73	3,539.73	1,366.69
29	10,748.06	4,149.83	74	3,331.05	1,286.12
30	10,642.47	4,109.06	75	3,121.29	1,205.13
31	10,533.66	4,067.05	76	2,910.51	1,123.75
32	10,421.62	4,023.79	77	2,698.75	1,041.99
33	10,306.39	3,979.30	78	2,486.14	959.90
34	10,188.00	3,933.59	79	2,272.70	877.49
35	10,066.48	3,886.67	80	2,058.51	794.79
36	9,941.87	3,838.56	81	1,843.64	711.83
37	9,814.18	3,789.26	82	1,628.17	628.64
38	9,683.49	3,738.80	83	1,412.17	545.24
39	9,549.80	3,687.18	84	1,195.70	461.66
40	9,413.15	3,634.42	85	978.84	377.93
41	9,273.60	3,580.54	86	761.67	294.08
42	9,131.15	3,525.54	87	544.21	210.12
43	8,985.85	3,469.44	88	326.60	126.10
44	8,837.75	3,412.26	89	108.88	42.04

Glossary

absolute position. A specific geographic position given in coordinates.

absolute relief. The elevation values at locations in the landscape given in reference to a specified datum.

absolute relief mapping method. Mapping method for showing numerical elevation information.

acceleration. The rate of change of velocity.

accessibility. A measure of how reachable locations are from other locations.

accuracy. How well the measured map coordinates conform to the true coordinates of a position.

acre. A 208.7-foot-by-208.7-foot square, or 43,560 square feet.

actual scale. The scale that you measure in the vicinity of any point on the map; actual scale will vary from one location to another.

adjustable-length polar planimeter. A polar planimeter that is more accurate than a fixed length one because the arm length can be set to the shortest distance for each application. See also **polar planimeter.**

admiralty mile. Originally related to the length of a nautical mile measured on the surface of the earth just south of Great Britain, chosen to be 800 feet longer than a statute mile.

aerial mapping camera. A camera specially designed to expose large 9-inch-by-9-inch (23-centimeter-by-23-centimeter) frames on photographic film.

aerial photograph. A vertical photograph of the environment taken from an aircraft, satellite, or other remote platform; the photo is not made on a map projection surface and varies in scale with elevation differences on the ground; commonly called an *air photo.*

aeronautical chart. A map created specifically for air navigation.

agonic line. The line of zero magnetic declination along which the true and magnetic north poles are aligned.

agrimensore. A Roman field surveyor.

Airy 1830 ellipsoid. An ellipsoid defined in the nineteenth century to best fit the regions of Great Britain and Ireland.

Aitoff projection. A modification of the equatorial aspect azimuthal equidistant projection created by David Aitoff that doubles the horizontal scale, resulting in an elliptical projection with the same two-to-one width-to-height ratio as the Mollweide projection; parallels are not straight horizontal lines, and the map is neither equal area nor equidistant.

Aitoff, David. The Russian cartographer who, in 1889, published a modification of the equatorial aspect azimuthal equidistant projection to create the Aitoff projection. See also **Aitoff projection.**

Albers equal area conic projection. A secant case equal area conic projection mathematically devised in 1805 by Heinrich C. Albers.

Albers, Heinrich C. The German mathematician who mathematically devised the Albers equal area conic projection in 1805. See also **Albers equal area conic projection.**

altimeter. An instrument used to measure the height above a vertical datum, usually mean sea level.

altimeter page. A screen on a handheld GPS receiver that shows your current elevation, the maximum and minimum elevation along the route you have followed, an elevation profile for your trip, and your rate of ascent or descent.

amount. The quantity associated with a feature; also called a *measurement.*

anaglyph. A special form of stereopair produced by printing the maps constructed from two vantage points in red and blue and then superimposing one upon the other. When viewed through special glasses equipped with red and blue lenses, the map is seen stereoscopically and the image appears three-dimensional.

angle scale. The system of marks that show angle measurements along the edge of a protractor.

angular range. The difference between the maximum and minimum azimuth angles.

angular shift. The shift in position that would occur if the center of a shift line were plotted relative to an x,y axis relating to compass directions. See also **shift line.**

animated map. A map that gives you a sense that things are changing either over time, through space, or both.

annotated orthophotomap. An orthophoto on which conventional map symbols have been overlaid. See also **orthophoto.**

annotated route line. A line tracing the path along which a feature moves with explanatory labels, point symbols, or other indicators of the changes in location.

annual increase or decrease. The annual change in compass variation (declination) and direction of declination that is shown on charts in minutes of a degree per year. See also **compass variation.**

antipodal meridian. The meridian at the opposite side of the earth from the one being considered.

approximate contour. A contour between the standard contours used to indicate the approximate location of contours where information isn't reliable; usually shown in areas where the vegetative surface cover precludes economically contouring the ground so that the contours will meet National Map Accuracy Standards; also called an *indefinite contour.*

arbitrary grid cell system. A grid cell location system composed of numbered columns and lettered rows, or conversely lettered columns and numbered rows.

arc-degree. The spherical distance along the earth's surface measured in angular units of latitude or longitude. An arc-degree is divided into 60 arc-minutes, and an arc-minute is divided into 60 arc-seconds.

ArcGlobe. An animated virtual globe program developed by ESRI. See also **virtual globe.**

arc-minute and arc-second. See **arc-degree.**

arctan. The angle that has a tangent equal to a given number, calculated as tan^{-1}.

area cartogram. A map made by distorting the geographical size of data collection areas in proportion to their magnitudes; also called a *cartogram.*

area correspondence. A shape index used for describing the degree of similarity between a standard two-dimensional shape and a feature.

area count. A tally of the total area occupied by a certain type of feature within a data collection unit.

area feature. A two-dimensional object or region.

area feature map. A map with symbols showing the existence of two-dimensional area features that are homogeneous within their boundaries.

area pattern. A shape repeated across an area to create structure within an area graphic mark. See also **graphic mark.**

area proportion. A value obtained by dividing an area count by the data collection unit area. See also **count** and **data collection unit.**

Aristotle. A fourth century B.C. Greek philosopher who believed that the earth's sphericity could be proven by careful visual observation.

arithmetic mean. The sum of the values $(\sum x)$ divided by the number of values *(n)*, written mathematically as

$$\bar{x} = \frac{\sum_{i=1}^{n} x_i}{n}$$

Also called the *arithmetic average* or *average.*

arrangement. The ways things are placed on the ground or on a map.

aspect. 1. [When discussing slope] The downslope direction of the maximum vertical change in the surface determined over a given horizontal distance; the same as gradient direction. See also **gradient.** 2. [When discussing map projections] The location of the point or line(s) of tangency on a projection surface.

aspect map. A map constructed from contours or digital elevation model data to show specific slope aspect categories. See also **digital elevation model.**

aspect ratio. Width–height ratio of any quadrilateral. See also **quadrilateral.**

aspect-slope map. A map on which aspect and slope are shown together.

asymmetric link. A one-way link in a network. See also **network.**

attribute. The component of a geographic information system that describes some characteristic or characteristics of a feature. See also **geographic information system.**

attribute accuracy. Fidelity in the description of characteristics of geographic features.

attribute error. Misreporting of the characteristics of a feature.

authalic. Area-preserving.

authalic sphere. A sphere with the same surface area as a specified reference ellipsoid.

autonomous. Independent.

average covariance. The sum of the covariance for all data collection units divided by the number of data collection units (n), written mathematically as

$$\text{cov}_{avg} = \frac{\sum_{i=1}^{n}(x_i - \bar{x})(y_i - \bar{y})}{n}$$

average diversity index. The average diversity for all cells on a map calculated using the equation

$$H_{avg} = \frac{-\sum_{j=1}^{m}\sum_{i=1}^{c} p_i \, ln \, p_i}{m}$$

where p is the proportion in the category, c is the kernel size, and n is the number of categories found in the kernel at map position i for the mapped area containing m grid cells.

average fragmentation index. The average fragmentation for all cells on a map calculated using the equation

$$F_{avg} = \frac{\sum_{i=1}^{m}(n-1)}{m \times (c-1)}$$

where c is the kernel size, and n is the number of categories found in the kernel at map position i for the mapped area containing m grid cells.

average spacing. The average distance between point features or centroids and their neighbors.

average spatial shift. The difference in location for all features between two time periods, calculated by measuring the length of each shift line, adding these lengths together, and dividing the sum of these lengths by the number of features. See also **shift line.**

azimuth. The horizontal angle measured in degrees clockwise from a north reference line to a direction line. See also **reference line** and **direction line.**

azimuthal equidistant projection. A planar projection with straight-line meridians (radiating outward from the pole) and equally spaced parallels; this arrangement of parallels and meridians results in all straight lines drawn from the point of tangency being great circle routes.

azimuthal projection. A projection that preserves global directions; sometimes called a *true direction projection.*

Babylonian sexagesimal system. A system for describing degrees (°), minutes ('), and seconds (") where there are 60 minutes in a degree and 60 seconds in a minute; commonly referred to as *degrees, minutes, seconds (DMS).*

back azimuth. The direction line drawn on a map from the known position of a distant feature back to your position; the opposite direction from a given azimuth calculated by adding 180 degrees to azimuths less than 180 degrees, or subtracting 180 degrees from azimuths greater than 180 degrees; also called a *back sight.*

back bearing. The opposite direction of a bearing; indicating a back bearing is simply a matter of changing letters indicating direction (N, S, E, W for north, south, east, west, respectively).

back sight. A line from your current position backward along your route; also called a *back azimuth.*

backslope. A hillslope element that is the linear portion of the hillslope below the shoulder descending to the concave footslope portion of the hillslope.

backtrack function. A function on GPS receivers that helps you retrace a route back to its starting point.

baseline. 1. In direction finding, the line between you and a reference point, whether it is some object or some known position, along which direction is measured; commonly called a reference line. 2. [Geographer's line] In the U.S. Public Land Survey System, the parallel determined by government land surveyors that intersects the principal meridian to form an initial survey point.

base station. A GPS receiver at a stationary position on a precisely known point, typically a surveyed benchmark.

bathymetric contour. A line connecting points of equal depth below the hydrographic datum; also called an *isobath.*

bathymetric map. A map, such as a nautical chart, that shows water depths.

bearing. Horizontal angles given in degrees ranging only from 0 to 90; in strict nautical terminology, the azimuth of a track.

bearing card. The graphical face of a compass, graduated to show bearings rather than azimuths. See also **surveyor's compass.**

Beidou. The Chinese regional navigation system being developed into a GPS called Compass.

benchmark. A permanently fixed brass plate that has been installed in the ground at a location where the elevation has been determined by precise leveling methods.

benchmarking. See **mark recovery.**

bias. A systematic distortion of the representation, that is, a systematic error—as opposed to a random error.

bifurcation ratio *(Rb).* The ratio of links of one order to the number of links of the next highest order, written symbolically as

$$R_b = \frac{N_x}{N_{(x+1)}}$$

where N_x is the number of links of any order x, and $N_{(x+1)}$ is the number of links of the next highest order.

bivariate mean. The average of the horizontal and vertical coordinates for n features, written as $(\bar{x}, \bar{y})$ and calculated as

$$(\bar{x}, \bar{y}) = \left(\frac{\sum_{i=1}^{n} x_i}{n}, \frac{\sum_{i=1}^{n} y_i}{n} \right)$$

block diagram. A map that portrays a piece of terrain as if it were cut out of the surface of the earth; the vertical sides of the block allow the subsurface geologic information to be shown.

Bluetooth. Short-range wireless networking technology.

branching network. A network that has only one possible path between pairs of places.

British National Grid coordinate system. Great Britain's national grid coordinate system developed by the OSGB and heavily used by land surveyors, as well as for maps based on surveys made by the OSGB or commercial map firms.

British National Grid (BNG) reference system. A grid cell location system commonly used in England, Scotland, and Wales; devised to overlay the British National Grid coordinate system.

buffer. A zone of a specified distance around a feature or features.

building setback line. The distance from a lot line beyond which building or improvements may not extend without permission from an authority.

bump mapping. A method used to give a general indication of the surface texture or land cover, achieved through modification of the original digital elevation model by adding elevation values around randomly scattered points that represent vegetation in the landscape.

cadastre. The written records kept on land parcels.

camera tilt. Tilting of the camera from vertical when an aerial photograph is taken.

camouflage detection film. A special film that is sensitive to near-infrared wavelengths as well as to visible light; commonly called *color infrared (CIR) film.*

cardinal direction. North, south, east, or west.

cardo maximus. A north–south boundary line that began in the middle of the area that was used to survey centuria under the Roman Centuriation system.

carriage. The part of a polar planimeter that is fixed to or slides along the tracer arm and contains the measuring wheel.

carrier phase enhancement. Improved accuracy of GPS positioning using higher-frequency carrier phase GPS signals; also called *real time kinematic (RTK) navigation.*

carrier phase GPS. A method used to increase the accuracy of GPS that involves higher-cost GPS units able to use higher-frequency signals.

carrying contour. The location at which two or more contours merge into a single contour; used to represent vertical or near-vertical topographic features such as cliffs and escarpments.

Cartesian coordinate system. System that allows you to pinpoint any location on a map precisely and objectively by giving its two coordinates (x, y).

cartogram. A map made by distorting the geographical size of data collection areas in proportion to their magnitudes; also called an *area cartogram.*

cartographic abstraction. The process of transforming data that have been collected about our environment into a graphical representation of features and attributes relevant to the purpose of the map.

cartographic generalization. The process of reducing the amount of information on a map by changing the geometric representation of features.

cartographic map. What most people think of as a traditional map drawn on paper or nowadays displayed on computer screens: a graphic representation of the environment (by graphic, we mean that a cartographic map is something that you can see or touch).

cartographic modeling map. A map on which several data variables are combined into a single numerical index; also called a *composite index map* or *composite variable map.*

cartographic selection. The process of deciding which classes of features to show on the map.

cartographic symbolization. The use of signs and graphic symbolism on maps.

case. Used for projections to describe whether the projection surface touches or intersects the generating globe.

cased line symbol. A symbol in which the interior line is bounded by a casing that is shown in a different color.

categorical data. Data that consist of categories used to distinguish different types of features within a map theme; also called *nominal-level data.*

categorical map. A map that has polygons enclosing areas assumed to be uniform or areas to which a single description can apply.

census. A survey that collects data from all the members of a population, whether people, animals, businesses or other entities, within a defined space at a specified time.

Centaurus. The constellation that includes the five stars of the Southern Cross.

central conic projection. A true-perspective conic projection based on a light source at the center of the generating globe.

central cylindrical projection. A true-perspective cylindrical projection based on a light source at the center of the generating globe.

central meridian. The origin of the longitudinal x-coordinates for a map grid; also called the *longitude of origin* or less commonly the *longitude of center.*

central parallel. The origin of the y-coordinates for a map grid; also called the *latitude of origin* or, less commonly, the *latitude of center.*

centrifugal force. The outward force caused by a rotating body's inertia that draws the mass away from its center of rotation.

centroid. Center of area; balance point or center of gravity.

centuria. Subdivisions of land under the Centuriation system into squares about 132 acres (53.5 hectares) in area.

Centuriation system. A system of land surveying used in Roman times in which the land was subdivided into a grid of square parcels; used for the agricultural land in the Po River valley in northern Italy and elsewhere throughout the Roman empire.

change map. A map used to show areas in which attributes have changed over a certain time period or where features have changed location over time.

chart cartridge. A storage device, available from several companies in different formats, that contains electronic charts used for navigation.

Chernoff face. A way to display multiple variables in one glyph by assigning different attributes to different facial characteristics, including head eccentricity, eye size, eye spacing, eye eccentricity, pupil size, eyebrow slant, nose size, mouth shape, mouth size, and mouth opening.

choropleth mapping. A mapping method in which each data collection area is assumed to be homogeneous throughout and is given a particular color lightness, color saturation, or pattern texture depending on the magnitude of the attribute; from the Greek terms *choro* meaning place, space, or land, and *pleth* meaning full.

Chromadepth glasses. Glasses that create a stereoscopic effect using holographic film lenses that combine refraction and diffraction of light to make each lens act like a thick glass prism; these glasses, patented by Chromatek Inc., create the visual impression of colors in the visible spectrum being closer or farther from your eyes.

Chromadepth map. A map that uses the colors in the rainbow (red, orange, yellow, green, blue, indigo, and violet) to create the impression of different heights within a display when viewed with Chromadepth glasses.

chronometer. A portable clock accurate enough for longitude finding, invented in 1762 by Englishman John Harrison.

circuit network. A network that has more than one possible path between some pairs of places.

circular error probable (CEP). A method to describe GPS accuracy by giving the radius of a circle containing a defined percentage of all possible fixes.

circular sector. A portion of a circle, like a pie piece, enclosed by two radii and an arc; also called a *circle sector*.

circumradius. Radius of the circumscribing circle.

circumscribing circle. The smallest circle that contains a given triangle.

civil aeronautical chart. A chart published in the United States, specially designed for use in air navigation, that emphasizes features of greatest aeronautical importance.

Clarke 1866 ellipsoid. The ellipsoid with the best fit for North America in the nineteenth century; used as the basis for latitude and longitude on topographic and other maps produced in Canada, Mexico, and the United States from the late 1800s to about the late 1970s.

classification. Ordering, scaling, or grouping into classes that simplify features and their attributes.

classing method. The procedure used to assign class intervals to numerical distributions.

class interval. The numerical range for a data class.

closed traverse. A survey of a ground path that ends at the point of beginning (POB) or at a previously surveyed position. This type of traverse provides a check against errors in the distances and bearings because they must ultimately lead back to the original starting point.

clustered spatial arrangement. The arrangement that occurs when point features tend to be grouped into one or a few small areas.

coastal relief model. A 3 arc-second digital elevation model (DEM) grid of the seafloor; distributed by the U.S. National Geophysical Data Center. This is the first DEM of the U.S. Coastal Zone, extending from the coastal state boundaries to as far offshore as National Ocean Service hydrographic data will support.

coefficient of areal correspondence (CAC). A commonly used measure of the amount of overlap between two categories of area features.

coefficient of determination. The square of the correlation coefficient in a regression that provides a measure of the strength of the correlation.

cognitive map. A map held solely in the mind's eye; also called a *mental map*.

colatitude. The angular distance from the pole to a boundary point along a meridian.

color hue. What we think of as the "color," such as red, green, or blue.

color infrared (CIR) film. A special film that is sensitive to near-infrared wavelengths as well as to visible light; also called *camouflage detection film*.

color ramp. A continuous succession of colors between a specified beginning and ending color; used in continuous hypsometric tinting.

combined mapping method. A way to show multiple quantitative themes on maps by combining two or more different mapping methods.

community grant. A land grant made to groups of settlers. Individuals in the group were given small tracts to settle and cultivate, but most of the grant was held in common for grazing, timber, and other purposes.

compact shape. A shape in which all points on the boundary are as close as possible to the center. The circle is the most compact two-dimensional shape because its boundary is everywhere equidistant from its center point. The length of its boundary (perimeter) relative to its area is also minimal.

compactness index. A measure of the compactness of a shape using the circle in some way as a standard reference shape.

compass. See **magnetic compass.**

Compass. The global GPS that China is developing by improving its regional Beidou navigation system.

compass card. The graphical face (dial) of a compass marked with directions.

compass housing. The case that encloses all the compass parts.

compass method. 1. [Of distance estimation] Using a magnetic compass and a map for distance estimation in order to determine your position. 2. [Of intersection] A method used to determine the location of a distant feature by locating your map position, sighting on the distant feature, and recording the foresight azimuth as noted on a compass. 3. [Of map orientation] Using a compass and a map that shows the direction of magnetic north to orient a map. 4. [Of resection] Using a compass, rather than a straightedge or visual sighting, to determine the back azimuths in the resection method of locating a position.

compass page. A screen on a handheld GPS unit that indicates cardinal directions on a compass card as well as in written format.

compass point. A direction indicated by arrows or prongs on a compass card.

compass rose. A circular direction indicator printed in one or more places on charts. The outer circle of the compass rose is oriented to true north, while the inner ring is oriented to magnetic north.

compass variation. The angular difference between true and magnetic north shown on a navigational chart; also called *declination*.

complementary formats display. A mapping method in which maps are combined with graphs, plots, tables, text, images, photographs, and other formats for the display of multivariate data.

completeness. The ability of a map projection to show the entire earth.

composite frequency diagram. A diagram that shows the shift in position over time.

composite index map. A map on which several data variables are combined into a single numerical index; also called a *cartographic modeling map* or *composite variable map*. This type of map is made possible by the multiple criteria evaluation (MCE) capability of GIS software.

compositing. The process of combining multiple images as overlapping layers producing a single image.

composition. The variety and quantity of patch types without consideration of the spatial arrangement of patches within the landscape mosaic.

concave slope. An inwardly curving slope.

conceptual accuracy. Using the correct data to represent the real-world feature or phenomena on the map.

conformal. Correct form or shape. A map is conformal when angles on the globe are preserved, thus preserving shape.

conic projection. A map projection made by projecting the earth's surface onto a cone with the line of tangency on the generating globe along any small circle or along two lines of tangency if the cone slices through the generating globe.

connectivity. The functional connections among features, often associated with accessibility.

connectivity measure. A numerical description to objectively compare networks with one another and to relate a network's connectivity to its flow. A measure that tells you the degree to which line features are interconnected.

connectivity of a network. A measure devised by forming a ratio between the actual number of links between nodes in a network and the maximum possible number of links for that network.

connectivity of a place. A measure devised by counting the linkages to or from that place.

constant format display. A map comparison technique whereby a series of maps with the same graphic design are used to depict changes in magnitude or feature type from map to map; also called *small multiples.*

constant isoline interval. An isoline interval that is the same for the entire map.

constant slope path. A path that maintains a constant steepness.

construction line. A line drawn for the purpose of creating or "constructing" a profile.

contagion. The tendency of patch types to be spatially clustered.

contiguity. Boundary connectedness.

contiguous cartogram. A cartogram on which the proximity and contiguity of neighboring areas are maintained, although this is accomplished at the expense of shape distortion.

continuity. Preservation of spatial proximity at all locations; since it isn't possible to preserve continuity at all locations when projecting a spherical surface onto a plane, all world map projections exhibit some degree of discontinuity.

continuous hypsometric tinting. A hypsometric tinting method in which the abrupt change between hypsometric tints is minimized by gradually merging one tint into the next, giving a smooth appearance to the tonal gradation.

continuous phenomena. Geographic phenomena that have continuously changing values over space.

continuous surface. A surface representing a geographic phenomenon that has continuously changing values; a surface that changes smoothly in numerical value across the landscape.

continuous surface map. A map representing a geographic phenomenon that changes smoothly in numerical value across the landscape.

contour. A line of equal elevation above a datum.

contour interval. The vertical distance between contours.

contour line. An isoline of equal elevation that portrays the shape and elevation of the land.

contour line enhancement. A technique used to add relative relief cues to contours while maintaining the absolute portrayal of relief; used to remedy the problem that many map readers have in relating contours to terrain features.

contrast. The relative differences among patch types with respect to an attribute of the patches.

control points. Very accurate points of known accuracy that surveyors use to georeference other map data.

control segment (CS). The segment of the GPS that consists of the set of ground monitoring stations that track the location and health of each satellite.

convex slope. An outwardly curving slope.

core area. The internal area of a patch after a user-specified edge buffer is eliminated.

correction lines. Township lines established every 24 miles north and south of the baseline as part of the U.S. Public Land Survey System and Canada's Dominion Land Survey. Surveyors created correction lines where they readjusted range lines to compensate for the convergence of meridians.

correlation coefficient. A measure of the similarity between two different datasets for the same data collection units; short for the spatial cross-correlation coefficient. A measure of spatial association based on computing the covariance for counts of different point features within data collection units.

correlation coefficient (r). A measure of correlation found by dividing the average covariance by the product of the standard deviations for each dataset, using the equation

$$r = \frac{\sum_{i=1}^{n}(x_i - \bar{x})(y_i - \bar{y})}{n\sigma_x \sigma_y}$$

correspondence. The agreement between points on the globe and points on the projected surface.

corridors of movement. Broad paths of movement compiled by generalizing the route data for individual features.

Cossin, Jean. Originator of the sinusoidal projection, which he used to create a world map in 1570. See also **sinusoidal projection.**

count. The total number of point, line, or area features within a data collection unit.

counting dial. A dial on a polar planimeter that records the number of whole and partial revolutions the measuring wheel makes; also called a *recording dial*.

course. A navigational route.

course deviation indicator (CDI). A graphic display on a GPS unit that shows how far you've strayed left or right of your desired path.

course heading. Direction.

covariance. A measure of the degree to which the values for each data collection unit deviate from the arithmetic mean for all units.

covariance table. A table similar to a distance table on a state highway map or road atlas that has the covariance values (rounded to the nearest whole number) for all combinations of the data collection units.

critical distance. A threshold distance used in fixed distance weighting.

critical values. Values that have special relevance to a map's theme and are used to set class interval limits.

cross fix. A location determined by the resection methods in which lines cross or "resect" at your location.

cross section. A diagram of the vertical section of the ground surface below a profile line.

cross track error (XTE or XTR). How far you've strayed left or right of your desired path when using a GPS unit.

cross-staff. A rudimentary Roman surveying instrument that measures the angle between the North Star and the horizon to lay out the vertical and horizontal boundaries of centuria under the Centuriation system.

Crux Australis. Constellation of five stars used by navigators in the southern hemisphere to find south; also known as the *Southern Cross*.

cultural features. Features produced by humans; examples include roads, boundaries, transmission lines, and buildings.

currency. How up to date a map is.

curvature. The amount that a surface deviates from being linear.

cut contour. A special type of depression contour used to identify the place where a roadway has been dug through the landscape, drastically lowering (cutting) the terrain.

cyclical phenomena. Phenomena that recur at regular intervals.

cylindrical equal area projection. A true-perspective cylindrical projection based on a linear light source akin to a fluorescent light with parallel rays along the polar axis.

cylindrical projection. A map projection made by projecting the earth's surface onto a cylinder that touches the generating globe along any large circle or slices through the generating globe along two small circles.

cylindrical projection family. A projection family based on the use of a cylinder as the developable surface.

dashboard compass. A type of rotating card compass with a floating dial.

dasymetric mapping. A mapping method in which the mapped areas aren't political data collection units such as counties, but rather areas of inherent homogeneity in the data. The boundaries of the original data collection units are modified using related ancillary data, and the attribute values are redistributed within the newly drawn units.

data collection unit. A natural or human-defined unit within which information is gathered for inventory or analysis.

data quality report. Part of the metadata files for the digital dataset that includes information about such aspects as lineage, positional accuracy, and attribute accuracy.

datum contour. The contour that corresponds to mean sea level; also called the *zero contour*.

dead reckoning. Estimating your present position relative to your last accurately determined location by using direction, speed, distance, and time information.

dead reckoning plot. A map of your route shown as a series of intended dead reckoning tracks between points.

decimal degrees (DD). *Decimal degrees (DD) = dd + mm / 60 + ss / 3600*, where *dd* is the number of whole degrees, *mm* is the number of minutes, and *ss* is the number of seconds.

declination. See **compass variation.**

declination diagram. A diagram that shows the angular relationship among grid north, magnetic north, and true north.

decumanus maximus. An east–west boundary line that began in the middle of the area used to survey centuria under the Roman Centuriation system.

degrees, minutes, seconds (DMS) A system for describing for degrees (°), minutes ('), and seconds (") where there are 60 minutes in a degree and 60 seconds in a minute. See also **Babylonian sexagesimal system.**

Delaunay triangulation. A way of partitioning space into triangles by using point features as triangle vertices.

DEM. See **digital elevation model.**

dendritic. Branching.

density. Count per unit area.

density value. A value obtained by dividing the count value by the area of its data collection unit.

Department of Defense (DOD). A U.S. government agency whose mission is to provide national security; three services, Army, Navy, and Air Force, are under the control of the DOD.

Department of Homeland Security (DHS). A department of the federal government charged with protecting the United States from terrorist attacks and responding to natural disasters.

depression contour. A contour used to identify closed depressions or basin-like features; shown on maps as a line with small right-angle ticks added to the downslope (or inside) of the contour.

depth contour (or depth curve). A line of equal water depth below the mean sea level or mean lower low water datum; also called an *isobath*.

Descartes, René. French philosopher and mathematician who invented the Cartesian coordinate system in 1637. See also **Cartesian coordinate system.**

descending orbit. An orbit with a daylight pass over the earth from north to south.

developable surface. A surface that can be flattened onto a plane without geometrical distortion; in the case of map projections, a developable surface is the flat surface onto which the earth's features are projected. There are three developable surfaces: planes, cones, and cylinders.

difference TIN. The triangulated irregular network (TIN) that contains differences in height for all triangle corner points from the two TINs used in TIN volume computation.

differential correction. Correction factor computed by comparing the known geographic coordinates of a fixed, surveyed location with the GPS coordinates of the same location.

differential GPS (DGPS). A form of GPS augmentation in which a receiver fixed at a known ground location corrects the locations that a roving receiver records, thus assuring much higher accuracy.

diffusion. The spread or scatter of some geographical phenomenon over space.

diffusion map. A map of the change in position of features over time due to a diffusion process.

digital elevation model (DEM). A sample of elevations or depths taken on a regular grid; sometimes called a *digital terrain model (DTM)*.

digital image. An image composed of pixels with numbers representing gray scale or color shades.

digital mapping camera. A camera that records aerial photos digitally rather than on film.

digital orthophotoquad (DOQ). An orthophotomap product developed by the U.S. Geological Survey (USGS). DOQs are created from 1:40,000-scale black-and-white aerial photography scanned with pixels of approximately 1-meter spatial resolution to create an orthophoto. This image is used to produce an orthophotomap corresponding to a 1:12,000-scale quarter quadrangle of a 1:24,000-scale U.S. Geological Survey topographic map.

digital polar planimeter. A more accurate, computerized version of a mechanical polar planimeter.

Digital Raster Graphic (DRG). A scanned 1:24,000-scale U.S. Geological Survey topographic map.

digital terrain model (DTM). A sample of elevations or depths taken on a regular grid; commonly called a *digital elevation model (DEM)*.

digitize. To convert paper maps, such as engineering plans or cadastral maps, to digital format.

dilution of precision (DOP). The effect of satellite geometry above the horizon on the accuracy of a GPS position; also referred to as *geometric dilution of precision (GDOP)*.

dimensional instability error. Error caused by the stretching and shrinking of the medium on which the map is produced, caused by temperature and humidity changes.

dip. The downward vertical angle relative to the horizontal plane; commonly used to describe the orientation of geologic beds at the earth's surface.

direct association. A one-to-one relationship in which the occurrence of one thing correlates with the occurrence of another.

direction. The difference between the reference line and the direction line given in angular units.

direction line. The line from you to a distant object.

directionality. Trend for a set of points in which features are farther from a specified point in one direction than in another direction.

discrete feature. A point, line, or area feature, such as an individual house, tree, railroad, or parking lot.

discrete hypsometric tint. A distinct gray tone or color given to the space between contours; also called *discrete layer tint.*

discrete layer tint. A distinct gray tone or color given to the space between contours; also called *discrete hypsometric tint.*

discrete ordinate method. A method to determine the volume of a feature by multiplying the average height or depth of a feature by the feature's area.

dispersion. The process by which features move outward from a starting position across space over time, continuously decreasing in density. See also **diffusion.**

displacement. The adding of space between features on a map so that the features can be visually distinguished and a correct interpretation of their ground relationships can be portrayed.

disproportionate symbol error. Error that occurs when the size of the symbol used to map the feature is not proportional to its ground size.

distance band. A "circle of influence" used in fixed distance weighting.

distance database. A computer database that holds information about the distance between cities and other significant geographical places, such as state and national parks, recreation areas, and historical sites.

distance estimation. A way of determining your ground position using the distance and direction to other objects; also called *range estimation.*

distance inset. A map notation indicating the distance between significant features, such as cities.

distance line. A line to or from your position to a distant feature.

distance segment. Ground distance annotation often added between neighboring places such as highway intersections or towns on highway maps; also known as *first order route distance segment* or *first order distance segment.*

distance table. A table that identifies measured ground distances between a select set of locations.

distance weight. A numeric value associated with a feature; the higher the value, the greater the weight for that feature. A feature with a larger weight will have more influence on the pattern than something with a smaller weight.

diversity. A measure that describes spatial variation in various patch types (such as land-cover classes or ecosystems) within the landscape.

diversity index. Measure of pattern arrangement that considers the relative abundance of each category for the kernel at each cell location.

diversity measure. A measure used in ecological analyses to describe the spatial arrangement of habitat units.

diversity range. The range between the minimum and maximum diversity index values possible for the collection of categories on a map.

DLS. See **Dominion Land Survey.**

DMS. See **degrees, minutes, seconds.**

Dominion Land Survey (DLS). The public land survey system used to divide most of Canada's western provinces into townships and sections for agricultural and other purposes.

donation land claim. A 160-acre (about 65-hectare) parcel granted under the Donation Land Claim (DLC) Act of 1850 that had to be surveyed with north–south and east–west boundaries and whose boundaries had to conform to the Public Land Survey (if the survey had already been made in the region). See also **Donation Land Claim (DLC) Act of 1850.**

Donation Land Claim (DLC) Act of 1850. An act granting 320 acres (almost 130 hectares) of federal land to any qualified settler who had resided on public lands for four years or more in the Oregon Territory (Idaho, Oregon, Washington, and Montana). See also **donation land claim.**

DOP. See **dilution of precision.**

dot density map. A map on which point symbols are used to show variations in density across a surface. This is sometimes called *dot mapping,* but be careful not to confuse this with the point symbol mapping method in which each dot represents one feature.

dot grid counting. A counting method for area estimation in which a square grid of alternating solid and open dots is used to measure an area. You count all the dots that fall entirely within the feature's boundary, then you count the number of either open or solid dots falling on the feature's boundary and add this number to the total of those dots falling entirely within the boundary.

dot size. The diameter of the dots on a dot density map.

dot unit value. The amount that each dot represents on a dot density map; that value must be greater than 1.

DOQ. See **digital orthophotoquad.**

draft. Depth from the water line to the bottom of a boat's keel.

DRG. See **Digital Raster Graphic.**

Dufour, Guillaume H. The Swiss cartographer who developed the partial hachuring method in the mid-nineteenth century that largely replaced the Lehmann hachuring system.

dynamic image map. A display that includes animated sequences of maps shown in a predefined order. The display often allows the map reader to interact with map symbols, so that symbols are linked to text, pictures, or video clips.

dynamic qualitative thematic map. A map that focuses on changes in feature locations and qualitative attributes over time.

earth resources satellite. A satellite launched with the primary mission of providing systematic, repetitive environmental system measurements, such as surface reflectance, wind speed and direction, wave height, surface temperature, surface altitude, cloud cover, and atmospheric water vapor level.

easement. A right held by one person to make limited use of another person's real property (land).

easterly declination. When the compass needle points to the east of true north.

easting. The x-coordinate distance measured east from the origin in a grid coordinate system.

elapsed time. The length of time between a map's date of completion and its date of use.

electromagnetic spectrum. The range of wavelengths from long (radio waves) to short (gamma rays), with visible light wavelengths falling in between long and short.

electronic compass. A compass that measures the relative strengths of magnetic fields passing through two wire coils.

electronic depth measuring instrument. An instrument used to obtain soundings by determining the amount of time that acoustic pulses take to reach the bottom and return to the instrument; this travel time is then converted to distance above the bottom because the velocity of acoustic waves is a known value.

electronic piloting device. A device that links maps to radio and GPS signals so that you can navigate your boat by observing your progress along a path on a map and then making decisions en route.

Elektro. A Russian weather satellite positioned over the equator south of Moscow; also known as *Geostationary Operational Meteorological Satellite (GOMS).*

ellipse. A flattened circle whose longest (major) and shortest (minor) axes are at right angles.

ellipsoid. A three-dimensional figure obtained by rotating an ellipse 180 degrees about its polar (minor) axis.

ellipsoidal height. The distance above or below the surface of the WGS 84 ellipsoid along a line from the surface to the center of the earth.

engineering plan. A very accurate, large-scale reference map.

English units. A system of measurement sometimes referred to as "foot-pound-second," after the base units of length, mass, and time. Used in the United States, United Kingdom, Canada, Australia, India, Malaysia, New Zealand, Republic of Ireland, and other countries; also known as *imperial units.*

Enhanced Thematic Mapper Plus (ETM+). A sensor on Landsats 5 and 7 that provides 30-meter resolution images from the blue (band 1), green (band 2), red (band 3), near-IR (band 4), and mid-IR (bands 5 and 7) portions of the electromagnetic spectrum, as well as a 60-meter resolution thermal infrared band (band 6) and a 15-meter resolution panchromatic band (band 8).

en route aeronautical chart. A digital chart used for aeronautical navigation.

Environmental Systems Research Institute. See **ESRI.**

ephemeral stream. A stream that flows only during and immediately after periods of heavy rainfall or snowmelt.

ephemeris. A table that charts the movement of celestial bodies (planets, moons, etc.) and predicts their positions at a given time; a table of each satellite's exact position at any time.

epidemiology. The discipline studying the incidence, distribution, spread, and control of disease in a population.

Epsilon. The dimmest of the five stars in the Southern Cross; it is offset below the center of the cross.

equal area. The property of preserving the relative size of regions everywhere on the map projection; this property is also called *equivalence.*

equal frequency intervals. A procedure used to assign class intervals to a numerical distribution so that the numbers of features in each class are as close as possible to equal.

equal range intervals. A procedure used to assign class intervals to a numerical distribution so that the range of data values is divided by the desired number of classes to obtain equal range intervals.

equator. The imaginary circle around the earth that is equidistant from the north and south poles.

equatorial aspect. The aspect that occurs when a tangent case projection point or line of tangency is at or along the equator.

equatorial axis. The axis from the center of the earth to the equator.

equatorial plane. The same plane as the earth's rotation; the orbital plane for geostationary satellites.

equidistance. The preservation of spherical great circle distance on a map projection.

equidistant. Truly distance-preserving, as in a map projection.

equidistant cylindrical projection. See **equirectangular projection.**

equirectangular projection. A cylindrical map projection on which parallels and meridians are mapped as a grid of equally spaced horizontal and vertical lines twice as wide as high; also called the *equidistant cylindrical projection* or *geographic projection.*

equivalence. A map projection property that truly represents the areas of different regions on the earth. See also **equal area.**

Eratosthenes. Greek scholar, head of the then-famous library and museum in Alexandria, Egypt, who around 250 BC made the first scientifically based estimate of the earth's circumference.

error. The degree of uncertainty in a measurement.

error budget. An expression that describes the contribution of each source of error to the overall error of a satellite range measurement.

error triangle. A small triangle created using the resection method; you can assume your position is within this triangle.

ESRI. Developer and marketer of a full line of GIS products and services.

estimated position error (EPE). An estimate of a horizontal GPS position's accuracy.

estimated time of arrival (ETA). The time that you estimate you will arrive at your intended destination (used in GPS navigation).

ETOPO2. A global equal-angle DEM grid with a spatial resolution of 2 arc-minutes; this is currently the dataset most used for global terrain mapping.

ETOPO5. An equal-angle DEM grid at a spatial resolution of 5 arc-minutes, meaning that each cell is given the average elevation of its 5-minute-by-5-minute quadrilateral on the ground.

Euclidean distance. The straight-line (shortest) distance between two points on a flat surface.

Euclidian geometry. The geometry of planes and solids based on Euclid's axioms, such as the premise that parallel lines never cross, that the shortest distance is a straight line, and that space is three-dimensional.

European Datum of 1950 (ED 50). Datum created after World War II to be a consistent reference datum for most of western Europe; however, Belgium, France, Great Britain, Ireland, Sweden, Switzerland, and the Netherlands continue to retain and use their own national datums. Latitudes and longitudes in this datum are based on the International Ellipsoid of 1924.

European Terrestrial Reference System 1989 (ETRS 89). A reference system, valid across Europe, that is based on the WGS 84 datum.

factual error. Error on a map due to a feature being left off by mistake, a feature existing on the map but not in the environment, a symbol misplaced on the map, or a symbol that remains on the map after the feature has disappeared from the environment.

false accuracy. The reporting of map use results at a higher level of accuracy than it is legitimately possible to get from the map.

false easting. A value added to all x-coordinates so that there are no negative eastings in the grid coordinate system zone.

false northing. A value added to all y-coordinates so that there are no negative northings in the grid coordinate system zone.

false precision. The reporting of results at a higher level of precision than it is legitimately possible to obtain from the map.

fathom. The equivalent of 6 feet; used as the unit of measurement for depth.

feathering. A technique used to drop contours in areas of high relief where coalescence would cause the lines to bleed together; also known as *feathering-out treatment.*

feature. A spatial component of a geographic information system.

Federal Aviation Administration (FAA). A U.S. government agency responsible for regulating and overseeing all aspects of civil aviation in the United States.

Feng-Yun. A Chinese weather satellite.

fiducial mark. A small registration mark exposed on the film at the edge of an air photo to facilitate geometrical calculations.

field conditions. Conditions in the region in which a GPS receiver is being used; these conditions include location characteristics, length of time at the location, and satellite geometry (arrangement of the satellites in the sky).

field observation. The observation of features *in situ*—at the place they are found in the environment.

fill contour. A contour that shows where the terrain was raised to support a road or railway grade.

first order polygon contiguity. A distance weight that uses polygons that share a boundary with a target polygon. The contiguous polygons are considered to influence the target polygon and are included in computations for the target polygon; also called *polygon contiguity weight.*

first order route distance segment. Ground distance between major cities, often added to highway maps.

fiscal cadastre. The written record of property valuation kept on land parcels.

fishnet map. A map that resembles a net or wire mesh draped over the terrain. It is produced by combining closely spaced, parallel terrain profiles across the landform at right angles to each other; also called a *wireframe map.*

fix. A position determined using a navigational method for positional determination, such as GPS.

fixed distance weight. A distance weight in which you impose a distance band on the analysis.

flight line. The path that an aircraft follows, typically north–south or east–west, to acquire air photos.

floating disk. A free-floating part of a compass that indicates magnetic north because it has been magnetized so that it will align itself with the earth's magnetic field.

flow line. A proportional line symbol used on flow maps. See **flow map.**

flow map. A map that shows ratio-level changes in magnitude along a line feature.

fly-through. View of a relief portrayal from an animated sequence of vantage points, giving the impression that you're flying and that the terrain is passing under and around you.

footslope. A hillslope element at the bottom of the backslope that either has a linear slope or is slightly concave.

foresight. A line from your current position forward to a feature along your route.

forester compass. A reversed card compass designed for use by foresters to plan timber cutting lines.

forwardlap. Overlap between photos along a flight line.

fragmentation. A metric often used in environmental and landscape analysis to describe the disintegration of larger landscape patches into areas that are smaller and more detached.

fragmentation index. A quantitative description of the degree to which land in different land-cover categories are clustered or scattered on the landscape.

fragmentation index range. The range between the minimum and maximum fragmentation index values possible for the collection of categories on a map.

fragmentation map. A map that shows the total number of land-cover (or other) categories found in the neighborhood of each cell.

fragmentation measure. A measure used in ecological analyses to describe the spatial arrangement of habitat.

frame. An individual picture produced by a camera.

frequency diagram. A graphical display showing the proportion of features that fall into each category of data; also called a *histogram*.

Galileo. A 30-satellite GPS the European Union is developing that is intended to provide greater precision than is currently available and improve coverage at higher latitudes.

Gall-Peters projection. A secant case of the cylindrical equal area projection that lessens shape distortion in higher latitudes by placing lines of tangency at 45°N and 45°S.

Gauss, Carl. A mathematician who, along with Johann Krüger, worked out formulas describing the geometric distortion for the transverse Mercator projection and the equations for making it on the ellipsoid.

GDOP. Geometric dilution of precision. See also **dilution of precision (DOP).**

general triangle. A triangle that has no 90° angle (unlike a right triangle).

generating globe. A globe reduced to the scale of the desired flat map.

geocaching. Outdoor treasure hunting using GPS and other navigational aids, such as maps.

geocentric latitude. Latitude on a sphere; the angle between the line from a point on the surface of the earth toward the earth's center and then to the equator.

geocentric view. A viewpoint in which, rather than relating everything to your own location, you learn to mentally orient yourself with respect to the external environment.

geocentrically. Relative to a geographical direction system.

geodetic latitude. Latitude on an ellipsoid; the angle between the line perpendicular to the surface of the earth toward the earth's center and then to the equator.

Geodetic Reference System of 1980 (GRS 80). Ellipsoid which is essentially identical to the WGS 84 ellipsoid.

geographer's line. See **baseline.**

geographical mile. 6,087.1 feet; one minute of longitude along the equator.

geographical north. The north pole of the axis of earth rotation; also called *true north*.

geographic coordinate. Latitude and longitude coordinate that pinpoints a place on the earth's surface.

geographic coordinate system. A coordinate system based on geodetic latitude and longitude.

geographic information system (GIS). Automated, spatially referenced system for the capture, storage, retrieval, analysis, and display of data about the earth.

geographic projection. A cylindrical map projection on which parallels and meridians are mapped as a grid of equally spaced horizontal and vertical lines twice as wide as high; also called the *equidistant cylindrical projection* or *equirectangular projection*.

geoid. Equigravitational surface (mean sea level) used as the reference for elevations.

geoid model. A lookup table stored in a GPS receiver that contains information used to interpolate height above the ellipsoid (referred to as orthometric height).

geoidal undulation. The difference between the geoid and the ellipsoid.

geometric dilution of precision. See **dilution of precision.**

geometric point symbol. A simple symbol such as a circle, square, or triangle.

geometric symbol. A simple shape, such as a square, circle, or triangle, used to represent a feature.

geometrical reference framework. A framework you can use to easily describe and determine locations, distances, directions, and other geographic relationships.

geomorphology. The study of landforms and terrain-forming processes.

georeferencing. The procedure used to bring data layers into alignment via known ground location control points or, alternatively, the procedure of bringing a map or data layers into alignment with the earth's surface via a common coordinate system.

Geostationary Operational Environmental Satellite (GOES) system. A U.S. weather satellite system.

Geostationary Operational Meteorological Satellite (GOMS). A Russian weather satellite positioned over the equator south of Moscow; also known as *Elektro*.

geostationary orbit. An orbit in which the satellite is always in the same position with respect to the rotating earth. By orbiting in the same direction and at the same rate as earth, the satellite appears stationary (synchronous) with respect to the rotation of the earth; also called *Clarke's orbit* because Arthur C. Clarke first proposed the idea of using a geostationary orbit for communications satellites.

GIS. See **geographic information system.**

glacial contour. A contour that represents the surface of an ice mass or permanent snow field at the date of the photography used to compile the feature.

global geoid. Slightly undulating, nearly ellipsoidal surface that best fits mean sea level for all the earth's oceans.

Global Navigation Satellite System (GNSS). The generic term for a worldwide radio navigation system that provides autonomous geographic positioning; more commonly known as a *Global Positioning System (GPS)*.

Global Positioning System (GPS). A worldwide satellite system that provides autonomous navigation and positioning; a "constellation" of earth-orbiting satellites makes it possible to pinpoint geographic location and height above the ellipsoid with a high degree of accuracy.

GLOBE. A 30 arc-second DEM grid for the world's land areas distributed by the National Geophysical Data Center.

globe. A round model of the world; a spherical representation of the earth.

GLONASS. The Russian GPS, originally completed in 1995 with 24 satellites; Global Navigation Satellite System.

glyph. A pictographic point symbol based on an abstract human face. See also **Chernoff face.**

gnomonic projection. A planar projection based on a light source at the center of the generating globe.

GOES East. A geostationary weather satellite that obtains images of the eastern United States.

GOES West. A geostationary weather satellite that obtains images of the western United States.

Goode interrupted homolosine projection. A pseudocylindrical equal area map projection created by compositing twelve segments that form six interrupted lobes; constructed in 1923 by Goode from the uninterrupted homolosine projection.

Goode, J. Paul. An American geography professor who constructed the uninterrupted and interrupted homolosine projections.

gore. A north–south strip of the earth mapped with the transverse Mercator projection that is used in the construction of globes.

government lot. A designation used by the USPLSS for small, oddly shaped parcels, particularly along water bodies. See also **U.S. Public Land Survey System.**

GPS. See **Global Positioning System.**

GPS augmentation. Improvement of GPS positioning using external information, often integrated into the calculation process, to improve the accuracy, availability, or reliability of the satellite signal.

GPS aviation unit. A device that links GPS to digital aeronautical charts.

GPS chartplotter. An electronic map display in which positional data from a built-in GPS receiver are used to superimpose a boat's current position on a nautical chart display screen.

GPS grade. A categorization of GPS receivers into three groups based on the level of accuracy and special functions they offer.

GPS satellite page. A screen on a handheld GPS unit that shows the positions of satellites within range of the receiver and the strength of signals received from each satellite.

GPS III. The next generation of the NAVSTAR-GPS being developed by the U.S. Department of Defense.

GPS time. A definition of time that begins at midnight Saturday, January 5th, 1980, and is counted in weeks and seconds of a week from this instant.

GPS week. A week that begins at the Saturday–Sunday midnight transition.

gradient. A vector that describes both magnitude and direction; used to determine the maximum amount of vertical change on a surface and the direction in which that change occurs.

gradient magnitude. The maximum slope ratio.

gradient path. The steepest route down a hill or the steepest grade up a mountain.

gradient vector map. A map that shows with vector lines the maximum slope magnitude and downward direction.

graduated symbol. A symbol that is scaled proportionately to a range of data values; each data value is symbolized to show its location in the progression of smaller to larger data values.

graphic mark. A point, line, area, or pixel on a map that is used to represent a geographic feature.

graticule. The arrangement of parallels and meridians draped over a spherical or ellipsoidal approximation of the earth.

great circle. The largest possible circle that can be drawn on the surface of the spherical earth. Its circumference is that of the sphere and its center is the center of the earth. All great circles divide the earth into halves.

great circle route. The shortest path between two points on the globe.

Greenwich mean time (GMT). The time at the prime meridian in Greenwich, England.

Greenwich meridian. International prime meridian defined by the north–south optical axis of a telescope at the Royal Observatory in Greenwich, a suburb of London; selected in 1884 at the International Meridian Conference.

grid azimuth. The horizontal angle from a grid north reference line to a direction line.

grid bearing notation. The system of using the map grid to express direction.

grid cell counting. A counting method for area estimation using a grid of small, evenly spaced squares superimposed on the map.

grid convergence. The slight rotation of grid lines from horizontal and vertical at grid coordinate system zone edges.

grid coordinate system. A rectangular coordinate system superimposed mathematically onto a map projection.

grid north. The northerly direction indicated by the north–south grid lines or ticks used on a map.

ground distance multiplier. A coefficient used to convert map distance to ground distance when taking slope error into account.

GTOPO30. Similar to the GLOBE grid, this 30-arc-second digital elevation model grid for the world's land areas is distributed by the U.S. Geological Survey.

gyrostabilized mount. A mount that uses gyroscopes to stabilize side-to-side camera motion.

hachure. A thin, short line arranged so that it faces downhill in the direction of the steepest slope.

half section. One half of one section (320 acres) in the U.S. Public Land Survey System and Canada's Dominion Land Survey.

half-point. One of the eight compass points indicating NNE, ENE, ESE, SSE, SSW, WSW, WNW, and NNW.

hand bearing compass. A type of rotating card compass with a floating dial commonly used by mariners.

handheld GPS receiver. A portable, low-cost GPS unit for recreational use.

Harrison, John. Englishman who, in 1762, invented the chronometer, a clock accurate and portable enough for longitude finding.

heading. Direction.

height above the ellipsoid (HAE). Elevation relative to the surface of the ellipsoid based on the distance from the earth's center to the earth's surface at a given point; this is the elevation most often found by GPS receivers.

hemisphere. Half of the earth.

heredia. A subdivision of land allotted to a family under the Roman Centuriation system; 1/100th of a centuria.

Heron of Alexandria. A scholar who lived in approximately 60 AD who is attributed with Heron's formula, which is the formula for the area of a triangle. See also **Heron's formula.**

Heron's formula. A formula for finding the area of a triangle given the lengths of its sides *a, b,* and *c*

$$area = \sqrt{s(s-a)(s-b)(s-c)}$$

where *s* is the semiperimeter of the triangle *(a+b+c)/2*; attributed to Heron of Alexandria, ca. 60 AD.

hillshading. A technique used to enhance the three-dimensional appearance of terrain features that creates the patterns of light and shadow on hills that would be produced by illumination from an imaginary light source positioned at the upper left corner of the map; also called *relief shading* or *shaded relief.*

hillsign. A crude line drawing of highly stylized hills and mountains, used on both large- and small-scale maps.

hillslope element. A division of a hillside defined by its position and slope, which can have considerable influence on geomorphic and hydrologic processes.

Hipparchus. (190–125 BC) A Greek astronomer and mathematician who proposed that a set of equally spaced east–west lines called parallels be drawn on maps; also credited with inventing the stereographic map projection.

histogram. A graphical display showing the proportion of features that fall into each data category; also called a *frequency diagram.*

homogeneous. Uniform in structure or composition throughout.

homolosine projection. An equal area map projection that is a composite of two pseudocylindrical projections—the sinusoidal projection (used for the area from 40°N to 40°S latitude) and the Mollweide projection (used to cover the area from 40°N to 40°S to the respective poles). See also **Goode, J. Paul.**

horizontal dilution of precision (HDOP). The effect of satellite geometry above the horizon on the accuracy of a horizontal GPS position.

hot spot analysis. An analysis used to identify clustering in sets of points.

hydrographic chart. A nautical chart published for an inland water area.

hydrographic organization or agency. An oceanic mapping agency that provides nautical charts.

hypsometric tinting. A method of "coloring between contour lines" that enhances the relative relief cues for contours while maintaining the absolute portrayal of relief; also called *layer tinting* or *hypsometric coloring.*

identification key. An aid to an image interpreter that shows the typical appearance of a feature on an aerial photograph.

IKONOS. From the Greek word for "image," a commercial satellite system that provides images from space at the same resolution as aerial photography.

illumination. The amount of light that hits a surface.

illumination map. Map that takes into account the location of the sun and any objects that impede the sun's illumination, such as hills, trees, or buildings.

image. A graphic representation of the geographical environment collected by electromagnetic sensors.

image interpretation. Using photographic and digital images to identify objects and their attributes.

image map. A map made by superimposing traditional map symbols over an image base.

impedance. Distance decay.

imperial units. A system of measurement sometimes referred to as "foot-pound-second," after the base units of length, mass and time; used in the United States, United Kingdom, Canada, Australia, India, Malaysia, New Zealand, Republic of Ireland and other countries; also known as *English units.*

inclination. The angle between the equatorial plane and the satellite orbital plane.

indefinite contour. A contour between the standard contour lines used to indicate the approximate location of contours where the elevation information isn't reliable; also called an *approximate contour.*

independence. The spatial condition that exists when two features or events have no influence on each other.

index contour. A labeled contour usually drawn with a thicker line than the intermediate contours.

index mark. The mark at the center of the protractor circle from which angles are measured.

Indian Regional Navigational Satellite System (IRNSS). The GPS that India is developing.

indirect association. A correspondence that is two or more steps removed from a direct relationship; if A is related to B, and B is related to C, then A has an indirect association with C.

inertial navigation system (INS). A system used to fill gaps in GPS coverage. An INS uses gyroscope-like devices that measure up-down and sideways changes in a vehicle's angular velocity.

inference. The process of deriving conclusions based on what you already know.

initialize. To be provided with position and velocity data from a remote source, such as another GPS user or a GPS receiver.

initial point. A surveyed starting point used in establishing land partitioning in the U.S. Public Land Survey System. See also **U.S. Public Land Survey System.**

inland chart cartridge. A storage device that contains a collection of electronic charts covering all or part of a state.

inradius. Radius of the inscribed circle.

INS. See **inertial navigation system.**

inscribed circle. The largest circle contained in a given triangle.

inspection method 1. [Of distance estimation] Using one or more linear features or prominent objects in your vicinity for distance estimation in order to determine your position. 2. [Of intersection] A method used to find the position of something to plot on the map using a straightedge and visual sightlines. 3. [Of map orientation] Using one or more linear features or prominent objects in your vicinity to orient a map.

instrument flight rules, high-altitude chart. A chart designed for navigation at or above 18,000 feet above mean sea level.

instrument flight rules, low-altitude chart. A chart that provides aeronautical information for navigation below 18,000 feet above mean sea level.

intended dead reckoning track. The route you intend to follow between selected points.

interactive map. A map that gives you control over such things as which images are shown on the screen, sequencing of images, and movement from one vantage point to another.

intermediate contour. A contour line that is not labeled and is located between index contours.

intermittent stream. A stream that flows a large portion of the year but ceases to flow occasionally or seasonally.

International Atomic Time or Temps Atomique International (TAI). A time scale defined as the weighted average of the time kept by about 200 atomic clocks in over 50 national laboratories worldwide.

International Earth Rotation Service (IERS). The agency that provides the correction data to relate Universal Time Coordinated to International Atomic Time.

International Ellipsoid of 1924. The ellipsoid used as the basis for the latitudes and longitudes of the European Datum of 1950. See also **European Datum of 1950.**

International Meridian Conference. Held in Washington, D.C., in 1884. This was where the Greenwich meridian was selected as the international standard for the prime meridian.

international nautical mile. 6,076.1 feet or 1,852 meters; equals one minute of latitude on a perfect sphere whose surface area is equal to the surface area of the ellipsoidal earth

interrupted projection. A projection in which the generating globe is segmented in order to minimize the distortion within any lobe of the projection.

interrupted homolosine projection. See **Goode interrupted homolosine projection.**

intersection. Area of overlap between features.

interspersion. The intermixing of patches of different types.

interval-level data. Data consisting of numerical values on a magnitude scale that has an arbitrary zero point (such as temperature in degrees Fahrenheit).

intervisibility map. A map of the visible or hidden areas on the ground when viewed from a given vantage point; also called a *viewshed map.*

inverse distance squared weight. A distance weight in which the influence of distance drops off according to the square of the distance; influence decreases at an increasing rate with distance.

inverse distance weight. A distance weight based on the concept of impedance. All features impact or influence all other features, but the farther away something is, the smaller the impact it has. The decrease of influence with distance is linear.

inverse Euclidean distance weight. A straight-line distance weight.

inverse Manhattan distance weight. A distance weight that measures distance along line segments that are arranged at right angles to each other.

Irish National Grid reference system. Grid cell location system used in Ireland, including Northern Ireland.

irregular surface area. Computation of the area of a feature that takes the slope of the landscape into account.

isarithm. A line of equal value used to map continuous data surfaces by connecting points of a selected value; also known as an *isoline*.

island polygon. An isolated polygon that does not share any boundaries with other polygons.

isobar. An isoline connecting points of equal atmospheric pressure.

isobath. An isoline connecting points of equal depth below the hydrographic datum; also called a *bathymetric contour*.

isochrone. An isoline connecting points of equal time.

isogonic line. An isoline connecting points of constant angular difference between true and magnetic north.

isohyet. An isoline connecting points of equal precipitation.

isolation. The distance between patches across space.

isoline. A line of equal value used to map continuous data surfaces by connecting points of a selected value; also known as an *isarithm*.

isopleth. An isoline drawn across a density surface created from statistical data. Isopleths look identical to standard isolines, but they differ in that they show a conceptual density or surface where the values can't physically exist at points.

isotherm. An isoline connecting points of equal temperature.

jewel pivot. The part of the compass that allows the needle to float freely.

kernel. A moving matrix of a neighborhood of pixels (commonly 3 by 3) used in spatial analyses of raster data.

king's case. A case in which contiguity is defined by the four adjacent quadrats sharing a row or column edge with the center quadrat as well as diagonal neighbors, expanding the neighborhood to eight quadrats.

knot. A speed of one nautical mile per hour; used in computing the speed of ships, planes, and wind.

Krüger, Johann. A mathematician who, along with Carl Gauss, worked out formulas describing the geometric distortion for the transverse Mercator projection and the equations for making it on the ellipsoid.

laddered contour label. A label placed in-line with the label on the adjacent index contours so that you can easily read the elevation values from one index contour to the next.

lag distance. The systematic difference in distance between two spatial phenomena.

lagged spatial association. This occurs when there is a positive or negative association between two sets of features, but one set is spatially shifted systematically relative to the other.

Lambert azimuthal equal area projection. A planar equal area projection usually restricted to a hemisphere, with polar and equatorial aspects used most often; invented by Johann Heinrich Lambert in 1772.

Lambert conformal conic projection. A secant case normal-aspect conic projection with its two standard parallels placed so as to minimize the map's overall scale distortion; invented by Johann Heinrich Lambert in 1772.

Lambert, Johann Heinrich. The mathematician and cartographer who in 1772 published equations for the Lambert azimuthal equal area and Lambert conformal conic map projections. See also **Lambert azimuthal equal area projection** and **Lambert conformal conic projection.**

land conveyance. The transfer of the title to land by one or more persons to another or others.

land grant. An area of land whose title was given to its owner before the territory was part of the United States. After the territory was acquired by the United States, the title was officially confirmed.

land information system (LIS). A geographic information system based on the cadastre.

land mile. 1,760 yards (about 1,609 meters) or 5,280 feet; the commonly used distance of a mile on the ground, commonly referred to as a *statute mile*.

land navigation. Navigation over the terrain, often restricted to specific routes such as trails, roads, subways, or railroads.

Land Ordinance of 1785. The ordinance that established the U.S. Public Land Survey System.

land parcel. A division of property with some implication for landownership or land use; the smallest unit of ownership or, as in the case of a farm field, a unit of uniform use.

land partitioning. The dividing of property into parcels.

land record. A publicly owned and managed system, defined by state or provincial standards, that records real estate ownership, transfers, taxation, and development.

Landsat. A NASA project to monitor the earth from space with the following satellites: Landsat 1 (placed in orbit in 1972), Landsat 2 (1975), Landsat 3 (1978), Landsat 4 (1982), Landsat 5 (1984), and Landsat 7 (1999); Landsat 5 and 7 are currently in operation; Landsat 6, launched in 1993, failed to reach orbit.

landscape drawing. An artistically rendered, oblique-perspective map that provides a realistic terrain portrayal.

landscape metric. A quantitative description of patterns in categories on maps that relate to the spatial characteristics of patches.

large scale. In a two-way grouping of map scales, when the numerical value of the representative fraction 1/x is larger than 1:500,000, the map is considered large scale; in a three-way grouping of map scales, when the numerical value of the representative fraction 1/x is larger than 1:250,000, the map is considered large scale.

laser range finder. A "point and read" instrument with an infrared laser beam built into binoculars or a monocular (one eyepiece) instrument, which measures distance up to 1,000 yards (300 meters).

latitude. The angular distance of a parallel from the equator; the numbering system for parallels; the north-south angular distance from the equator to the place of interest.

latitude of origin. The geographical origin of the y-coordinates; also called the *central parallel* or, less commonly, the *latitude of center.*

Law of Cosines. A law from spherical trigonometry that can be used to calculate one side of a general triangle when the angle opposite and the other two sides are known; the law states that: $c^2 = a^2 + b^2 - 2ab\,cos(c)$ where c is the side opposite angle c, and that a and b are the two sides enclosing c.

Law of Sines. A law from spherical trigonometry that states that if the sides of a triangle are a, b and c and the angles opposite those sides are A, B and C, then $a/sin(A) = b/sin(B) = c/sin(C)$.

layer tinting. A method of coloring between contour lines that enhances the relative relief cues for contours while maintaining the absolute portrayal of relief; also called *hypsometric tinting* or *hypsometric coloring*, and used to refer to coloring between isolines.

lead line. A line with a lead weight tied to one end that is lowered into the water to determine depth by markings on the line.

least squares criterion. The criterion most commonly used for "best fit" in a regression; it minimizes the squared distances between the y values of the points and the y values on the regression line.

legal cadastre. The written records concerning proprietary interests in land parcels.

legal subdivision (LSD). A fraction of a section in Canada's Dominion Land Survey.

legend disclaimer. A note in the map legend that indicates something about the map's accuracy.

legible. Able to be seen and recognized.

Lehmann system. A hachuring system named after its founder, Johann G. Lehmann.

Lehmann, Johann G. A Saxon military officer who introduced the Lehmann method of hachuring to Europe in 1799. See also **Lehmann system.**

lensatic compass. A type of rotating card compass.

leveling. Surveyor's method to determine elevations relative to the point where mean sea level is defined, using horizontally aligned telescopes and vertically aligned leveling rods.

light detection and ranging (lidar). A remote sensing system that uses rapid pulses of laser light striking the surface of the earth to determine elevations.

limnologist. A scientist who studies lakes and streams.

line feature. A one-dimensional feature (such as a boundary line) or a two-dimensional feature conceived of as a line for mapping purposes (such as roads and streams, which have width in reality but are mapped as lines).

line feature map. A map that has symbols showing line features.

line generalization. The smoothing of linear features and edges of areal features on a map so that there is a desired loss of detail.

line of tangency. The line at which the projection surface is in contact with the spherical or ellipsoidal generating globe. This line on the map is true in scale to the equivalent line on the spherical or ellipsoidal earth surface.

line pattern. A repetition of lines to create an areal pattern.

lineage. The history of source material used to create the digital datasets for mapping, as well as any mathematical transformations of the source material.

linear connections. Roads, rivers, bus lines, airline routes, and a host of other connections that link one place with another.

linear interpolation. A method used to predict an unknown value between two known values under the assumption that the rate of change is constant.

linear slope. A slope that increases or decreases in elevation at a constant or uniform rate.

linear surface. A surface assumed to increase or decrease in height at a constant or uniform rate between two points.

link. The linear connection between two nodes.

LIS. See **land information system.**

lobe. A section of an interrupted projection. See also **Goode interrupted homolosine projection.**

local anomaly. A deviation from the normal compass display caused by local magnetism, such as that produced by magnetic ore bodies.

local disturbance. A compass deviation caused by such anomalies as power lines, thunderstorms, and iron objects.

local geoid. Surface slightly above or below (usually within 2 meters) the global geoid elevations. This difference is due to the vertical datum being used for the region is mean sea level at one or more locations rather than average sea level for all oceans.

locational reference system. A way of pinpointing the position of things in space.

locked on. Condition in which a GPS receiver is receiving the satellite signals.

logical consistency. The internal consistency of a representation; such consistency helps to ensure that the proper relationships are conveyed.

long lot. Narrow ribbon parcel used during the French settlement of North America. A long lot's boundaries ran back from the waterfront as roughly parallel lines, allowing a settler to have a dock on the river, a home on the natural levee formed by the river, and a narrow strip of farmland that often ended at the edge of a marsh or swamp.

longitude. The angle, measured along the equator, between the intersection of the prime meridian and the point where the meridian for the feature of interest intersects the equator.

longitude of origin. The geographical origin of the longitudinal x-coordinates; also called the *central meridian* or, less commonly, the *longitude of center.*

longitudinal distance. The distance between meridians.

LORAN. Abbreviation for Long Range Navigation. A navigation system based on radio signals from multiple transmitters used to determine the location and speed of the receiver; used all over the world for marine navigation.

lot. A special type of parcel within a legal subdivision that is recorded on a map.

low-altitude air photo. An air photo taken anywhere from just above the ground to around 15,000 feet above the surface.

loxodromic curve. The rhumb-line path on the earth that crosses each meridian at the same acute angle, thus forming a spiral that converges on the north or south pole.

lubber's line. A line parallel to the centerline of a ship, marked on the edge of the bowl of a mariner's compass. This line allows directions to be read relative to the direction of travel.

magnetic azimuth. The horizontal angle measured from a magnetic north reference line to a direction line.

magnetic compass. An instrument for indicating geographical directions magnetically, as indicated by the alignment of a magnetic needle or floating disk to the earth's magnetic field.

magnetic declination. The angular difference between true and magnetic north; called *compass variation* on charts.

magnetic north. The direction of your compass needle when it is aligned with the earth's magnetic field.

Maidenhead grid system. A proprietary grid used by amateur (ham) radio operators; based on partitions of the earth into progressively smaller quadrilaterals of latitude and longitude.

majority filter function. A function that eliminates isolated single cells that are likely to be misclassified by assigning them the value that appears most often in their immediate neighborhood.

Manhattan distance. Distance measured along a route that has segments oriented only vertically or horizontally; its name is drawn from the rectangular street network found in New York's downtown Manhattan district.

map. A spatial representation of the environment that is presented graphically; a collection of graphic symbols used to represent a portion of the earth.

map accuracy. The fidelity with which a map represents geographic phenomena at a given scale.

map analysis. In this stage of map use, you make counts and measurements, and look for evidence of spatial structure and correspondence.

map comparison. The process of visually or quantitatively examining spatial associations between features on different maps.

map divider. An instrument that allows you to measure distance incrementally as you "walk" it along a path.

map interpretation. An effort to understand why geographical features are found where they are and the spatial relations that they exhibit. When you interpret maps, you normally draw on some combination of your personal knowledge, fieldwork, written documents, interviews with experts, or other maps and images.

map measurer. A device consisting of a wheel and one or more circular distance dials that is ideal for measuring the length of curved lines on maps.

map overlay analysis tool. A GIS function that allows you to superimpose maps to better understand the spatial relationships among features using quantitative measures such as intersection and union.

map page. A screen on a handheld GPS unit that shows a crude basemap of your position with features such as political boundaries, streets and highways, cities, railway lines, lakes and rivers, and navigational aids, as well as your current heading and speed of travel.

map projection families. A way to organize the wide variety of projections into a limited number of groups on the basis of shared properties or attributes.

map reading. When you determine what the mapmakers have depicted, how they've gone about it, and what artifacts of the cartographic method deserve special attention.

map scale. The relationship between distances on the map and their corresponding ground distances.

map transformation process. The process that includes getting environmental information onto a map through the transforming steps of data collection, manipulation, and display, along with the extraction of information from the map through map reading, analysis, and interpretation.

map use. The process of obtaining useful information from one or more maps to help you understand the environment and improve your mental map. Map use consists of three main activities: reading, analysis, and interpretation.

mapping grade GPS. A GPS unit that offers an enhanced level of accuracy and is designed to facilitate GIS data collection; also called *resource grade GPS*.

mapping method. A commonly used technique for making maps such as choropleth mapping, dot density mapping, dasymetric mapping, and contouring.

mapping period. The time between initial data collection and final map printing.

mariner's compass. A rotating card compass in which the magnetized compass card is floating in a bowl of clear liquid and is centered and stabilized by a vertical pin in the middle of the bowl, allowing it to turn freely to magnetic north with changes in boat direction.

Marinus of Tyre. Person thought to have constructed the first equirectangular map projection in about 100 AD.

mark recovery. Identification of the location of old survey points or marks in the field; sometimes improperly called *benchmarking*.

matrix. A row and column grid of cells or pixels. See also **raster data.**

maximum elevation figure. A number on an aeronautical chart that indicates what altitude above mean sea level should be maintained to clear all terrain features and obstructions such as towers and antennae.

maximum slope path. A path that does not exceed a specified slope angle.

mean center. The x–y coordinate calculated as the average of all x-coordinates and all y-coordinates; also called the *bivariate mean.*

mean center of population. Described by the U.S. Census Bureau as "the point at which an imaginary, flat, weightless, and rigid map of the United States would balance perfectly if weights of identical value were placed on it so that each weight represented the location of one person on the date of the census."

mean high water (MHW). The datum used to define high water to support harbor and river navigation.

mean low water (MLW). Average of all the low tide levels recorded over the 19-year metonic cycle; one of two datums used in North America to define low water and used on Canadian nautical charts as the datum for sounding values.

mean lower low water (MLLW). The arithmetic average of the lower of the two daily low tides recorded over the 19-year metonic cycle; one of two datums used in North America to define low water as the datum for sounding values; the official U.S. National Ocean Service nautical chart datum.

mean sea level (MSL). The average of all low and high tides at a particular starting location over the metonic cycle; the datum used to give the numerical elevation at individual survey points on aeronautical charts, topographic maps, engineering plans, and other large-scale maps.

mean sun. A hypothetical sun that moves along the equator at the average speed with which the real sun appears to move across the sky.

measurement. The process of determining the quantity associated with features; reported as an amount, intensity, or magnitude.

measurement level. A way to characterize the nature of numerical information about features. There are four basic measurement levels: nominal, ordinal, interval, and ratio.

medium earth orbit. An orbit at an altitude of about 20,000 kilometers (12,000 miles) above the earth's surface.

medium scale. In a common three-way grouping of map scales, when the numerical value of the representative fraction 1/x is between 1:250,000 and 1:1,000,000, the map is considered medium scale.

mental map. A map held solely in the mind's eye; also known as a *cognitive map.*

Mercator projection. A tangent-case cylindrical conformal projection; the only projection on which all rhumb lines are straight lines on the map.

Mercator, Gerhardus. Constructed the Mercator projection in 1569.

meridian. A north–south line on the earth that extends between the north and south poles.

meridional arc. An arc of a great circle between two parallels measured along a meridian.

metadata. Information about the types of information shown on a map, such as the different sets of data used to create the map.

Meteosat system. A European weather satellite system with satellites identified as Meteosat-8 and Meteosat-9 over the Atlantic Ocean and Meteosat-6 and Meteosat-7 over the Indian Ocean.

meter. 39.37 inches, slightly longer than a yard. First defined as one ten-millionth of the distance from the equator to north pole along a meridian. Today the meter is defined relative to the speed of light and is 1/299,792,458th of the distance light travels in a vacuum in one second.

metes and bounds. A form of land description that requires minimal surveying skill to delineate a property boundary; the legal property description is tied to landmarks.

method of intersection. A method using foresights to find the position of something in the field or on a map.

metonic cycle. The 19-year cycle of the lunar phases and days of the year.

metric system. A system of measurement initially based on a fraction of the earth's polar circumference. See also **meter.**

mileage map. A distance cartogram on which routes between major locations are shown by straight lines labeled with the distance along each route.

Military Grid Reference System (MGRS). A grid cell location system used by North American Treaty Organization (NATO) nations; the confusion of using long numerical coordinates and numerical grid zone specifications is alleviated by substituting single letters for several numerals.

military time. Hours are numbered from 00 to 23, so that midnight is 00, 1 AM is 01, 1 PM is 13, and so on; minutes and seconds are expressed the same as in regular time.

mimetic symbol. A symbol that "mimes" (is a miniature caricature of) the thing it represents; often created as a combination of geometric shapes such as a square with a triangle on top to represent a house.

Minard, Charles. A French pioneer in the use of graphics in statistics and engineering; famous for his dramatic flow map of the disastrous losses suffered by Napoleon's army in the Russian campaign of 1812.

minimum distance tolerance. A threshold size dimension below which a feature is not counted.

minimum mapping unit (MMU). The smallest allowable size for a group of cells of a given class in a classified raster image.

minutes of angle (MOA). The measurement used for a subtended angle; one minute of angle (1/60th of a degree) will subtend approximately 1 inch at 300 feet, 2 inches at 600 feet, and so forth.

mission planning. Planning done prior to going out to the field in order to identify the optimal times for recording GPS data.

MLS. See **mean sea level.**

Mollweide projection. An elliptical equal area projection that most commonly uses the equator as the standard parallel and the prime meridian as the central meridian; invented in 1805 by Carl B. Mollweide.

Mollweide, Carl B. A German mathematician who invented the Mollweide projection in 1805.

monument. A fixed object established by surveyors when they determine the exact location of a point, often placed at lot corners and street intersections during the original survey.

Moran's *I*. A spatial autocorrelation measure that indicates if features are arranged in a clustered, random, or regular manner.

mosaic. Refers to the entire set of landscape patches that cover a region.

movement. The process whereby something changes location.

multipath error. Error in a GPS position due to reflection of the satellite signal off objects in the environment before the signal reaches the GPS receiver.

multiple display map. A mapping method for multivariate data in which a number of individual map displays are used; these displays can be either of constant format (small multiples) or of complementary formats.

multipurpose cadastre. The body of land records containing information about land parcel features and their attributes, including such things as land slope and aspect, soil characteristics, drainage, vegetation cover, number of residents, building construction, access road width and surface material, utility service, and zoning restrictions.

multiscale maps. A series of maps at varying scales, each depicting an amount and type of information appropriate to the particular scale.

multisensor receiver. A receiver that integrates information from multiple sensors, such as GPS receiver, a wireless signal detector, and an altimeter.

multispectral. An imaging system that is sensitive to two or more spectral bands; imagery that shows information gathered in different spectral bands.

multivariate map. A map that shows more than one theme; it may show a composite of two or more related themes, or several attributes of a feature may be incorporated into the same symbol.

NAD. North American Datum.

nadir. The point on the ground directly below the camera when an air photo is taken.

Napier's analogies. A set of trigonometric formulas useful in the solution of oblique spherical triangles.

National Aerial Photography Program (NAPP). A U.S. government program established in 1987 to develop a cloud-free aerial photography database of consistent scale, orientation, and high-image quality.

National Aeronautics and Space Administration (NASA). A U.S. government agency whose mission is to pioneer the future in space exploration, science, and aeronautics.

National Elevation Dataset (NED). A digital elevation model of 1-arc-second resolution distributed by the U.S. Geological Survey covering the continental United States, Hawaii, and the U.S. island territories; the resolution for Alaska's grid is 2 arc-seconds.

National Geodetic Vertical Datum of 1929 (NGVD 29). Datum defined by the observed heights of mean sea level at 26 North American tide gauges, as well as the elevations of benchmarks resulting from over 60,000 miles (96,560 kilometers) of leveling across the continent, totaling over 500,000 vertical control points.

National Geospatial-Intelligence Agency (NGA). A U.S. government agency whose mission is to provide timely, relevant, and accurate geospatial intelligence in support of national security; formerly called the National Imagery and Mapping Agency (NIMA).

National High Altitude Photography (NHAP) program. A program operated by the U.S. Geological Survey (USGS) from 1980 to 1989, in which black-and-white and color-infrared photography of the conterminous United States was taken from aircraft at an altitude of 40,000 feet (13,000 meters).

National Oceanic and Atmospheric Administration (NOAA). A U.S. government agency whose mission is to understand and predict changes in the earth's environment and manage coastal and marine resources to meet the nation's economic, social, and environmental needs.

natural breaks. Gaps in a distribution of data that can be used to define natural data groupings for purposes of class interval selection.

natural feature. A physical feature in the landscape, including mountains, valleys, plains, lakes, rivers, and vegetation.

nautical chart. A map created specifically for water navigation.

nautical mile. 1,852 meters exactly (about 1.15 statute miles); used around the world for maritime and aviation purposes. The original nautical mile was defined as 1 minute of latitude measured along a meridian.

navigation. A two-part activity that involves planning your route to get from one place to another and then following that route in the field.

navigation chart. A map designed to be used to navigate an airplane or boat between locations.

NAVSTAR Global Positioning System. The global positioning system created and managed by the U.S. Department of Defense; acronym derived from Navigation System with Time and Ranging.

nearest neighbor statistic (NNS). A mathematical measure that indicates if point features are arranged in a clustered, random, or regular manner; compares the average spacing with the expected mean nearest neighbor distance associated with a random distribution of point features.

near-infrared (near-IR). Wavelengths from 0.7 to 0.9 μm that are slightly longer than visible red in the electromagnetic spectrum.

near-polar orbit. A satellite orbit inclined about 10 degrees from the polar axis.

near-ultraviolet (near-UV). 0.3 to 0.4 μm wavelengths, slightly shorter than visible light in the electromagnetic spectrum.

network. A set of places (nodes) and their connections.

network hierarchy. A network in which nodes can contain more than one level of connectivity; a single place may, for example, be connected by roads with different surface or traffic flow characteristics.

Newton, Sir Isaac. Scientist who developed the theory of gravity in the 1660s.

node. A point at which a line ends or where two or more lines come together or intersect.

nominal level. A measurement level that tells you only to which category (class) a feature belongs.

nominal-level data. Data that consist of assignment to categories; also called *categorical data.*

noncontiguous cartogram. A cartogram on which the shape of geographical areas is maintained at the expense of proximity and contiguity with neighboring geographical units.

nonparametric. Statistical methods often referred to as "distribution-free" because they do not rely on an assumption that the data fit a particular probability distribution, such as a normal or bell curve.

normal aspect. The most commonly used aspect for each of the three projection families based on a developable surface. For planar projections, normal aspect is polar; for conic projections, it is oblique (somewhere between the equator and a pole); for cylindrical projections, the normal aspect is equatorial.

normalization. A process in which data attributes are adjusted to facilitate comparison between data collection units; an example is to convert the absolute number of people found in data collection units of varying size to the number of people per square mile.

North American Datum of 1927 (NAD 27). A system of control points surveyed from the 1920s to the early 1980s relative to the surface of the Clarke 1866 ellipsoid.

North American Datum of 1983 (NAD 83). System of NAD 27 control points corrected for surveying errors where required. Also added were the geodetic latitudes and longitudes of thousands of more recently acquired points determined relative to the Geodetic Reference System of 1980 (GRS 80).

North American Vertical Datum of 1988 (NAVD 88). Late 1980s' adjustment of the National Geodetic Vertical Datum of 1929 (NGVD 29) using new data; mean sea level for the continent was defined at one tidal station on the St. Lawrence River at Rimouski, Quebec, Canada.

north celestial pole. The spot in the sky directly above the true north pole.

North Star. The star positioned in the northern hemisphere sky less than one degree away from the north celestial pole; used by navigators to find true north because the star is so far away from the earth that the angle from the horizon to the North Star is the same as the latitude; also called *Polaris.*

northing. The y-coordinate distance measured north from the origin in a grid coordinate system.

numerical error. Error introduced when processing numerical data; occurs, for example, when numbers are rounded or when faulty computer processing compromises the data.

oblate ellipsoid. A sphere slightly flattened at the poles that is used as an approximation to the earth for mapping purposes; obtained by rotating an ellipse 180 degrees about its polar axis.

oblique aspect. Any alignment of a map projection's point or line(s) of tangency other than the equatorial, polar, or transverse aspects.

oblique perspective map. A map that gives us an oblique bird's-eye view of the landscape; these maps have a more three-dimensional look to them than planimetric maps.

oblique stepped surface map. A map made by dividing the region into data collection areas and showing magnitude data by proportionately varying the heights of the areas; also called *prism map.*

observation noise. Random noise in the GPS signal partially associated with ionospheric influences.

offshore chart cartridge. A storage device that contains a collection of electronic charts that include sounding, navigational aid, bottom condition, and harbor information.

open traverse. A survey of a ground path that does not end at the point of beginning (POB) or at a previously surveyed position.

ordering. The process of assigning a numeric connectivity value to the links in a network.

orders of hierarchy. The levels of links used to describe the hierarchical nature of the network.

ordinal-level data. Data that are ranked according to a less-than/greater-than rule; how much more or less one class is than another isn't specified.

Ordnance Survey Great Britain 1936 (OSGB 36). Datum for geodetic latitude and longitude coordinates based on the Airy 1830 ellipsoid.

Ordnance Survey of Great Britain (OSGB). Great Britain's national mapping agency.

Oregon Lambert system. A single grid coordinate system that replaced Oregon's two state plane coordinate zones with the same central meridian but a different parallel of origin.

orient (a map). To position a map so that features and directions on the map are aligned with corresponding features and directions on the ground.

orienteering. Using a map and compass (or GPS) to find your way across unfamiliar terrain.

orienteering compass. A compass consisting of a compass card, north arrow, and rotating magnetic needle on which the housing is mounted on a transparent rectangular base that facilitates use with a map.

origin. The point of intersection of the x- and y-axes and designated (0,0) in a two-dimensional Cartesian coordinate system.

orthographic projection. A planar projection with the light source at an infinite distance from the generating globe so that all rays are parallel; how the earth would appear if viewed from a distant planet.

orthoimage. A satellite image that has been corrected planimetrically. See also **orthophoto.**

orthometric height. Height above the mean sea level datum or geoid.

orthophanic. A projection that appears to be a natural and correct representation of the earth.

orthophoto (orthophotograph). An aerial photo that has been geometrically corrected to remove image distortions caused by camera optics, camera tilt, and elevation differences so that its scale is uniform and its features can be overlain on a planimetric map.

orthophotomap. A conventional map such as a topographic quadrangle that uses an orthophoto as the base.

Oswald Winkel. A German cartographer who in 1921 constructed the Winkel Tripel projection.

overlap. Portion of the ground duplicated on successive air photos along a flight line.

overlay. The process of superimposing two or more maps to better understand the relationships among geographic features and their attributes.

Page and Grid reference system. A proprietary grid reference system developed by Thomas Brothers Maps that assigns a unique page number to each geographical area; map grids are half-mile squares, ranging from grid A1 through grid J7, and the grids always appear at the same location on each page.

panchromatic film. Film sensitive to the full wavelength range of visible light.

parallel rule. A device used to determine directions; consists of two straightedges connected by metal bars that allow the straightedges to remain parallel as they are separated.

parallel. An imaginary east–west line on the earth that can be used to define locations north or south of the equator. See also **latitude.**

parametric measure. A statistical calculation in which certain assumptions are made about the parameters of the population from which a sample is taken; the assumption that data are normally distributed is most common.

parcel. An area with some implication for landownership or land use; also called a *tract.*

partial hachuring system. A hachuring method on north-oriented maps in which hachures are eliminated on the northwest sides of hills which greatly improves the three-dimensional impression of relief and makes terrain features far easier to identify. This system, developed by the Swiss cartographer Guillaume H. Dufour in the mid-nineteenth century, largely replaced the Lehmann system.

partially connected network. A network in which links between some pairs of nodes pass through at least one additional node and are thus indirect.

patch. An ecological data collection unit; an individual landscape area characterized by a single land-cover class.

patch shape complexity. The geometry of patches, ranging from simple and compact, to irregular and convoluted.

patch size. The area of a patch.

pattern orientation. The orientation of graphic marks that are replicated to create areal patterns within map symbols.

pedometer. A device that estimates distance traveled on foot by counting the number of steps.

percent rise. The slope ratio expressed as a percent (multiplied by 100); also called *percent slope.*

percentage. A proportional measure obtained by dividing the number of features in a subcategory by the total number of features in a data collection unit, and then multiplying this value by 100.

percentage connection. A measure devised by forming a ratio between the actual number of links between nodes in a network and the maximum possible number of links for the network, and then multiplying this value by 100.

perennial stream. A stream that flows all year long.

perimeter-area fractal dimension. A shape complexity metric based on the relative amount of perimeter length per unit area.

perspective view regional map. A map created by draping imagery or relief shading over an underlying fishnet surface in oblique view.

Peters, Arno. A German historian and journalist who in 1967 devised a map based on Gall's equal area projection and presented it in 1973 as a "new invention" superior to the Mercator world projection.

Peutinger table. A medieval copy of a simplified road map made by the Romans.

photo rectification. The process of removing any scale distortion in an aerial photo due to camera tilt by physically altering the image geometry.

photogrammetrist. A person trained to compile reliable data from aerial photographs.

physical distance. Distance measured in standard units such as feet or meters.

physiological error. Error introduced by the mapmaker due to physiological conditions, such as tiredness, poor eyesight, or a shaky hand.

pictographic symbol. A symbol designed to look like a miniature version or caricature of the feature it represents.

piloting. Planning your intended route and then determining your position, direction of movement, or distance traveled along that route in the field by paying attention to landmarks, aids to navigation, and water depth.

pixel. Picture element; a cell at the intersection of rows and columns in a digital image. See also **raster data.**

plan curvature. Curvature that is perpendicular to the direction of the maximum slope; also called *planform curvature.*

planar projection. A map projection made by projecting the earth's surface onto a flat plane that touches the generating globe at a point or slices through the generating globe along a small circle; also called an *azimuthal projection.*

planar projection family. A projection family based on the use of a plane as the developable surface.

planform curvature. See **plan curvature.**

planimetric coordinate method. A method of area measurement based on finding the areas of trapezoids formed by drawing horizontal lines from the boundary points to the vertical y-axis.

planimetric map. A map that accurately represents only the horizontal positions of features and not the vertical positions that a topographic or bathymetric map shows.

planimetric perspective map. A map that gives an oblique view of the mapped area with no elevation information.

planimetric surface area. The values found in most area calculation reports and GIS databases that are based on the assumption that the earth's surface is flat.

plat. A map drawn to scale showing the lots into which an area has been divided.

platform. The vessel, craft, or instrument from which remote sensing is undertaken.

platted lot. A parcel identified by lot number in a platted subdivision.

platted subdivision. A subdivision mapped for public record to show legal descriptions of the subdivided area.

PLSS. See **U. S. Public Land Survey System.**

pocket stereoscope. The portable version of a stereoscope used to view the overlapping area on adjacent aerial photos in three dimensions.

point count. A value that gives you information about the abundance of point features in each data collection unit.

point feature. A truly zero-dimensional feature (such as a boundary marker) or a two-dimensional feature (such as a city) that has area but is represented as a point for mapping purposes at smaller scales.

point feature map. A map that has symbols showing the existence of point features at specific locations; a point can be a one-, two-, or three-dimensional feature conceived as a point for mapping purposes.

point of tangency. The point where a planar projection surface touches the spherical or ellipsoidal generating globe approximating the earth's shape.

point sample. A sampling method in which data are collected at point locations.

POIs. Points of interest.

point-to-point correspondence. Condition that exists when every point on the earth is transformed to a corresponding point on the map projection.

polar aspect. A planar map projection is in polar aspect when its point of tangency is at the north or south pole or when its circle of tangency encircles the north or south pole.

polar axis. The axis from the center of the earth to the poles.

polar planimeter. A mechanical instrument used for area measurement on maps and aerial photographs.

Polaris. The star positioned in the northern hemisphere sky less than one degree away from the north celestial pole. Polaris is used by navigators to find true north; also called the *North Star.*

polygon contiguity weight. A distance weight that uses polygons that share a boundary with a target polygon. The contiguous polygons are considered to influence the target polygon and are included in computations for the target polygon; also called *first order polygon contiguity.*

population count. Data collected for every element of a theme.

population density. Population per unit area such as a square mile or kilometer.

posigrade orbit. The orbit of an object such as a satellite that moves in the same direction as the earth's rotation.

position dilution of precision (PDOP). The effect of satellite location geometry above the horizon on the accuracy of a three-dimensional GPS position.

positional accuracy. A measure of accuracy determined by comparing the horizontal and vertical position provided by a GPS receiver to the actual ground coordinates of that position on the earth's surface.

positional error. The distance between the coordinates of a feature on the map and the feature's actual location on earth.

post-processing. Correction of GPS position data using information that is downloaded to a computer from a ground station and processed using special software.

precision. The amount of detail used to report a measurement.

prime meridian. The zero meridian (0°) used as the reference from which longitude east and west is measured.

principal meridian. A meridian determined by government land surveyors that intersects a surveyed baseline to establish an initial point in the U.S. Public Land Survey System and Canada's Dominion Land Survey. See also **baseline.**

principal point. The point at the center of an aerial photograph where straight lines that connect opposite pairs of fiducial marks intersect.

principal scale. The scale of the generating globe used to make a map projection.

prism map. A map made by dividing a region into data collection areas and showing magnitude by proportionally varying the heights of areas; also called an *oblique stepped surface map.*

private grant. A land grant issued by the government to an individual as a reward for service to the government; such parcels could subsequently be sold by the owner.

profile. A graphic diagram that shows the change in elevation of a surface along a path on the ground; commonly called a *terrain profile.*

profile curvature. Curvature that is parallel to the direction of the maximum slope.

profile line. A line on a map along which a terrain profile is constructed.

projection surface. The mapping surface onto which the sphere or ellipsoid representing the earth is projected. See also **developable surface.**

propaganda map. A map of persuasion designed to distort or misrepresent the truth.

proportional symbol. A symbol used to represent an exact data value by scaling the symbol's visual variable to be directly proportional to the value it represents.

proprietary grid. A map-related grid-cell location system developed by a private company.

protractor. A measuring device that divides a circle into equal angular intervals, usually degrees.

proximity. Nearness of features or locations in space.

pseudocontiguous cartogram. A cartogram on which the transformed data-collection areas share common boundaries, but the boundaries aren't the same as those found on the earth or represented on a conventional map.

pseudocylindrical map projection. Similar to a cylindrical projection in that parallels are horizontal lines and meridians are equally spaced; the difference is that all meridians except the vertical-line central meridian are curved instead of straight.

Public Land Survey System (PLSS). See **U.S. Public Land Survey System.**

pueblo grant. A land grant issued by the U.S. government to communities of Native Americans.

Pythagorean theorem. A calculation used to measure the hypotenuse of a right triangle. This theorem tells us that if a, b, and c represent the sides of a right triangle, and c is the hypotenuse, then $c^2 = a^2 + b^2$.

quadrants. Four divisions of the Cartesian coordinate system, formed by the intersection of the two axes, labeled counter-clockwise starting from the upper right quadrant by the Roman numerals I, II, III, and IV; because the signs for both coordinates are positive in quadrant I, it is the quadrant used for map grids.

quadrat. A square or rectangular data collection or sampling unit.

quadrilateral. The area bounded by equal increments of latitude and longitude.

qualitative analysis. An analysis that involves only nominal or categorical data.

qualitative information. Data that have classes varying in type but not quantity; these data tell you what things exist—lakes, rivers, roads, cities, farms, and so on—but not how much of each thing exists. See also **nominal-level data.**

quantitative change map. A map that shows the increase or decrease in an attribute value over a specified time period.

quantitative information. Numerical data that represent an amount, magnitude, or intensity. See also **measurement level.**

quantitative thematic map. A map that shows one or more themes of quantitative information.

quarter section. One half of one half of a square-mile section (160 acres) in the U.S. Public Land Survey System and Canada's Dominion Land Survey.

quarter-point. One of the 16 compass points indicating N by E, NE by N, NE by N, and so on.

quarter-quarter section. One quarter of one quarter of a section or one-sixteenth of a section (40 acres) in the U.S. Public Land Survey System and Canada's Dominion Land Survey.

quartile. One class in a four-class quantitative data grouping.

Quasi-Zenith Satellite System (QZSS). The proposed Japanese regional GPS that will provide better coverage for the Japanese Islands.

QuickBird. A near-polar, sun-synchronous satellite launched in 2001 by the DigitalGlobe corporation to obtain vertical images of the same area every 3.5 days in a 13.5-kilometer-wide (8-mile-wide) swath.

quintile. One class in a five-class quantitative data grouping.

radius of curvature. The measure of how curved a surface is. On an oblate ellipsoid, the radius of curvature is greatest at the pole and smallest at the equator. On a sphere it is equal everywhere.

railroad map. A map of a railroad network, common in the late 1800s' settlement period. Railroad companies sometimes structured such maps to create a favorable impression of their service in a region.

raised relief globe. A globe on which the terrain surface is raised up from the base elevation and exaggerated in order to give a more realistic impression of relief.

raised relief topographic map. A flat topographic map that has been printed on a sheet of plastic and then, using heat, vacuum-formed into a three-dimensional model of the landscape.

random sample. A sampling method in which all locations or all features in population are given an equal probability of being selected for the sample.

random spatial arrangement. Arrangement that occurs when there is no apparent order or pattern in the spacing between features.

range. A north–south column of townships 6 miles wide bounded on the east and west by surveyed range lines in the U.S. Public Land Survey System and Canada's Dominion Land Survey.

range estimation A way of finding your ground (or map) position determining your distance and direction to other positions; also called *distance estimation*.

range grading. Map portrayal of a distribution of data using a limited number of quantitative map symbols, each representing a different range of data values.

range line. Meridian surveyed at 6-mile intervals east and west of the principal meridian in the U.S. Public Land Survey System and Canada's Dominion Land Survey.

range noise. Noise in the GPS distance measurement primarily attributed to observation noise plus random parts of the multipath error.

range ring. A zone of a specified distance around a feature that serves as both a data collection unit boundary and a circular distance scale.

ranging. Finding the distance from one location to another.

raster data. Data collected by grid cells or pixels in a row and column grid or matrix.

raster slope map. A slope zone map generated using a GIS and raster elevation data.

ratio-level data. Data that consist of numerical values on a magnitude scale that has a zero point that denotes absence of the phenomena.

real-time correction. Correction of the GPS position using information from a ground station that is instantaneously sent to the portable receiver using a radio link.

real time kinematic (RTK) navigation. Improved accuracy of GPS positioning using higher frequency carrier phase GPS signals; also called *carrier phase enhancement*.

recoverable survey mark. A survey mark at a position in the field that is positively identifiable and can be easily found again.

recreation grade GPS. An inexpensive, commonly available, and user-friendly GPS unit with a relatively low level of accuracy; used primarily for outdoor sports or recreational activities.

rectangular cartogram. A pseudocontiguous cartogram made by transforming data collection regions into rectangles proportional in size to the magnitude being shown.

rectangular distance table. A table in which locations are listed alphabetically along the top and side to create a row–column matrix of distances.

rectified air photo. An air photo that has had its geometry corrected to remove systematic scale distortions.

rectifying sphere. An oblate ellipsoid used to define the circumference of a spherical earth where the length of meridians from equator to pole on the ellipsoid equals one-quarter of the spherical circumference.

reference line. The line between you and a reference point (some object or known position), along which direction is measured; sometimes called a *baseline*.

reference map. A map that serves as a reference library of geographic information; such a map provides information about the locations of features, allowing you to estimate directions and distances between these features.

reference point (control point). A previously surveyed point of known integrity, such as a PLSS section corner.

reflexive distance. A distance relationship that occurs when point *A* is the nearest neighbor of point *B,* and point *B* is also the nearest neighbor of point *A*.

register. Part of the cadastre that provides information concerning landownership.

regression. A statistical analysis that provides an indication of the degree of linearity in relations between the datasets.

regression line. The straight line of "best fit" drawn through a scatterplot of two variables.

regular spatial arrangement. The arrangement that exists when there is equal spacing between environmental features.

regular time. The numbers used to identify the 24 hours in a day; the numbers from 1 to 12 are used to specify hours from both midnight to noon (AM) and from noon to midnight (PM).

relative abundance. A value determined by first finding the average number of features in each map unit, then identifying the above- or below-average units on the map.

relative relief. The local range between high and low elevations.

relative relief mapping method. A mapping method designed to give a general impression of relative heights of different landform features.

reliability diagram. A simple outline map showing variation in the nature of source data used to produce the map.

relief. The vertical differences on the earth's surface.

relief displacement. The radial displacement of higher objects from the center of a vertical aerial photo; such displacement is evident in the apparent "leaning out" of the tops of higher trees, towers, and buildings on the photo.

relief model. A physical model that is in effect a chunk of a giant globe, often constructed to show the curvature of the earth.

relief portrayal. A mapping technique used to portray the terrain surface; a deliberate attempt on the part of the mapmaker to give you a sense of the three-dimensional nature of the surface.

relief reversal. The undesirable effect where hills look like valleys and valley bottoms look like ridgetops; this effect is most pronounced when illumination by the sun or an imaginary light source comes from the lower right corner of the map.

relief shading. A technique used to mimic natural patterns of light and shadow on the landscape so as to enhance the three-dimensional appearance of terrain features; the desirable effect is created by illumination from an imaginary light source at the upper left corner of the map; also called *hillshading* or *shaded relief.*

remote sensing. The collection of images of the earth (and other planetary bodies) from energy-sensing devices.

representation. A portrayal that stands for the environment; it is both a likeness and a simplified model.

representative fraction (RF). A common way to describe map scale; the ratio between map and ground distance.

rescale. To enlarge or reduce a map.

resection method. Technique for locating position that involves plotting lines that cross, or "resect" at your position.

resource grade GPS. A GPS unit that offers a higher level of accuracy than inexpensive handheld units used by the average person. Resource grade GPS is designed to facilitate GIS data collection; also called *mapping grade GPS.*

reversed card compass. A compass that directly displays the angle between true north and the object on which you are sighting; direction notation on the compass card reads counter-clockwise.

RF. See **representative fraction.**

rhumb line. A line of constant compass direction; a direction line extended so that it crosses each meridian at the same angle.

right circular cone. A cone whose volume is *1/3πr²h,* where *h* is the cone height and *πr²* is the area of its circular base.

right-of-way. Real property (land) designated for a specific purpose such as access.

right triangle. A triangle that has a 90-degree angle.

rise. Elevation difference between two locations.

road and recreation atlas. A book for travel and trip planning that contains a series of maps covering large regions.

road map. A map designed to facilitate road navigation but used for many general purposes.

roamer. A clear plastic device that has calibrated rulers etched into its surface that match standard topographic map scales; used to make simple measurements using grid coordinates on maps.

Robinson projection. A pseudocylindrical projection that is neither equal area nor conformal.

Robinson, Arthur H. An American academic cartographer who in 1963 constructed the Robinson projection.

rook's case. Contiguity defined by the four adjacent quadrats sharing a row or column edge with the center quadrat.

root mean square error (RMSE). The square root of the average of the squared discrepancies in position *(d)* of well-defined points *(n)* determined from the map and compared with higher accuracy surveyed locations of each point:

$$RMSE = \sqrt{\sum_i^n d_i^{\,2} / n}$$

rotating card compass. A compass on which the magnetic needle and compass card are joined and work as a single floating unit.

rotating needle compass. A compass that has the direction system (compass card) inscribed on the housing and uses a magnetized needle balanced on a center pin that rotates independently of the compass card.

route segment. See **distance segment.**

route sequence. Coordinates for landmarks or destinations stored as waypoints in a GPS receiver's memory and organized into a path to follow.

rover. See **roving receiver.**

roving receiver. A GPS receiver used in the field that is not located at a fixed position; also called a *rover.*

run. Horizontal ground distance component of a "rise over run" slope determination.

sample. Data collected for only a portion of the region or population used to make inferences about the characteristics of the population from which the sample was drawn.

sampling method. The means of obtaining a sample from the spatial setting or the total population.

sampling site. A location where counts or measurements are taken.

Sanson-Flamsteed projection. Another term for the sinusoidal map projection. The projection was named for Nicholas Sanson, who in the seventeenth century used the sinusoidal projection for continents and atlas maps of the world; and John Flamsteed, who used the sinusoidal projection for star maps.

satellite geometry. Overhead alignment of the satellite constellation at a given moment.

satellite health. The condition of a satellite in terms of clock error and technical malfunctions.

satellite image map. An orthoimage on which conventional map symbols have been overlaid.

scale. See **map scale.**

scale bar (bar scale). A way to express map scale using what looks like a small ruler printed on the map.

scale conversion. Conversion of map scale between a verbal scale, representative fraction, and scale bar.

scale factor (SF). The relation between the denominators of the representative fractions for the actual and principal scales at particular locations on the map; defined as

$$SF = \frac{Actual\ Scale}{Principal\ Scale} .$$

scale variation. Variation in scale from place to place on a map as a result of the projection process; no map projection maintains correct scale throughout.

scatterplot. A graph on which the relation between two variables is shown with increasing values for one variable scaled along the x-axis and increasing values for a second variable scaled along the y-axis.

secant case. The case that occurs when the projection surface slices through the generating globe at either a circle of tangency (for planar projections) or along two lines of tangency (for conic and cylindrical projections).

second order route distance segment. A segment that is longer than a first order segment distance and has as its endpoints major cities and intersections.

second of day (SOD). A definition of time noted as a number between 0 and 86,400, the number of seconds in a day (used in GPS navigation).

second of week (SOW). A definition of time that starts on a given day and denotes the seconds passed in that week as a number between 0 and 604,800 (60 seconds x 60 minutes x 24 hours x 7 days).

section. One square mile (640 acres) in the U.S. Public Land Survey System and the equivalent 2.59 square kilometers in the Canada's Dominion Land Survey; some sections may actually be less than 640 acres due to convergence of meridians.

sectional aeronautical chart. A chart at 1:500,000 scale designed for visual navigation in slow-moving to medium-speed aircraft.

segment. See **distance segment.**

segmented symbol. A symbol in which the parts show the relative magnitudes of subcategories of attributes; pie charts are an excellent example.

segmenting. Breaking features into parts at locations where some attribute changes.

seigneuries. Land grants given to soldiers and other elite citizens under the French feudal system.

selective availability (SA). Intentional degradation of GPS signals by the U.S. Department of Defense intended to prevent hostile forces from fully utilizing the capabilities of the system in times of military conflict.

semimajor axis. The slightly longer equatorial radius of the ellipsoid.

semiminor axis. The slightly shorter polar radius of the ellipsoid.

semiperimeter. Half of the perimeter of a triangle.

sensor. An instrument that records electromagnetic or acoustical energy.

sextant. A handheld instrument historically used to measure the angle (or altitude) of a celestial body above the earth's horizon.

SF. See **scale factor.**

shaded relief. See **relief shading.**

Shannon diversity index *(H)*. A population diversity measure; this index is calculated by finding the proportion of each category (p_i) present for the kernel (spatial neighborhood) at each cell location; each proportion is then multiplied by its natural logarithm $(ln\ p_i)$, and the products are summed and multiplied by -1, using the equation:

$$H = -\sum_{i=1}^{c} p_i\ ln\ p_i$$

shape. The perimeter form of an area feature.

shape index. A method for describing the shape of an area feature in numerical terms.

shift line. A line between the location of individual features at two different times. See also **angular shift**.

shoulder. A convex hillslope element that is below the summit and above the linear backslope.

Shreve order. A method of stream ordering in which the uppermost stream links have an order of one, two first order links combine to form an order two stream, and so on.

Shuttle Radar Topography Mission (SRTM) data. A digital elevation model of 1 arc-second resolution for the United States and 3 arc-second resolution for the rest of the world created from SRTM data.

sidelap. Overlap between photos on adjacent flight lines.

sidereal day. The period it takes for the earth to complete one rotation about its axis with respect to the stars.

sight. Part of a compass used to help you determine angles more accurately.

sight line. A visual direction line.

single theme map. A map that features one theme or variable, shown within a context of base reference information.

sinusoidal projection. An equal area pseudocylindrical projection, equidistant in the east–west direction and in the north–south direction along the central meridian, but only in these directions.

slab summation method. A method to determine the volume of a feature outlined on a topographic map by summing the volumes of horizontal slabs whose thickness is the vertical difference between contours.

slope. The vertical change in the elevation of the land surface, determined over a given horizontal distance; the amount of rise or fall of the ground surface.

slope angle. The inverse tangent *(tan⁻¹)* of the slope ratio (the angle whose tangent is the slope ratio); slope angle values range from 0 to 90 degrees.

slope curvature. The curvature of a surface, defined as being linear, convex, or concave.

slope error. The difference between the true ground and measured map distance due to landform slope.

slope map. A map that shows the steepness of the terrain.

slope measurement error. Error due to deviation of the landform from a linear surface within the distance over which the slope is computed.

slope percentage. The slope ratio expressed as a percent (multiplied by 100); also called *percent rise* and *percent slope*.

slope ratio. The ratio of the elevation difference (rise) and the ground distance (run) between two points.

slope template. A scaled template of selected contour spacings, chosen so that slope categories can be easily identified; also called a *tick sheet*.

slope zone map. A map of specific slope categories.

small circle. Any circle on the earth's surface whose plane intersects the interior of the sphere at any location other than the center; its circumference is smaller than a great circle.

small multiples. A series of map displays with the same graphic design but with different data depicted; used to depict changes in magnitude or type from multiple to multiple (map to map); also called *constant format displays*.

small scale. In a two-way grouping of map scales, a map is considered small scale when the numerical value of the representative fraction 1/x is smaller than 1:500,000; in a three-way grouping of map scales, a map is considered small scale when the numerical value of the representative fraction 1/x is smaller than 1:1,000,000.

smoothing error. Reduction in the length of a linear feature such as a road or river due to the generalization process (the process through which mapmakers abstract lines by straightening curves and smoothing irregularities).

solar altitude. The angle of the sun above the horizon.

solar azimuth. The angular direction of the sun, measured clockwise from north in degrees from 0 to 360.

solar day. 24 hours.

solar ephemeris. A table that gives information about the sun's position.

sounding. A water depth reading.

sounding pole. A pole marked with water depth values.

source scale. The scale at which information was originally derived from its source.

Southern Cross. Constellation of five stars used by navigators in the southern hemisphere to find south; also called *Crux Australis*.

space segment (SS). The segment of the GPS that consists of a constellation of GPS satellites.

space trilateration. A method of determining absolute ground positions using satellite signal velocity and travel time to measure the distance between the known locations of two or more satellites and the GPS receiver.

spacing. The locational arrangement of objects with respect to each other rather than relative to data collection units.

spatial association. The degree to which two categories overlap on the map; the association is positive when two sets of values increase or decrease similarly in space; if values of one dataset increase while those of the other decrease, the association is negative.

spatial autocorrelation. A concept that deals with the degree to which features on the earth's surface are both spatially and numerically similar to other features located nearby.

spatial configuration. The spatial character and arrangement, position, or orientation of patches within the patch class or landscape mosaic.

spatial cross-correlation. The similarity between two different datasets for the same spatial data collection units.

spatial cross-correlation coefficient. A measure of the similarity between two different datasets for the same spatial data collection units; also called *correlation coefficient*.

Spatial Data Transfer Standard (SDTS). Protocol developed under the leadership of the U.S. Geological Survey; this standard dictates that datasets used in mapping and GIS will have an attached metadata file that contains a data quality report.

spatial dependency. This relation is said to exist and be positive when the category ranks of two distributions increase or decrease similarly in space, as calculated in Spearman's rank correlation measure; if the category ranks in one distribution increase while those of the other decrease, negative spatial dependency is indicated.

spatial diffusion. The transfer or movement of things, ideas, or people from place to place; the process by which features move outward from a starting position across space over time, continuously decreasing in density.

spatial pattern. The repetitive or structured arrangement of features on the ground or map.

spatial shift. The difference in location for individual features between two time periods.

SPC. See **state plane coordinate system.**

Spearman's rank correlation coefficient (ρ). A nonparametric measure of correlation that shows the degree to which two datasets are spatially independent, without having to assume that each dataset has a normal distribution of values.

spectral band. A narrow region of the electromagnetic spectrum.

speed over the ground (SOG). The speed you are traveling.

sphere. A three-dimensional solid having all points on its surface the same distance from its center; used to describe the general shape of the earth.

spherical coordinate method. A method of area measurement based on the use of geographic coordinates (latitude and longitude) in calculating the area of spherical triangles.

spot elevation. A location where the elevation of the land surface has been determined (relative to mean sea level) and is depicted numerically on the map.

squared rank difference. The square of the difference between the ranks for two categories.

stacked scale bar. Two scale bars using different distance units that are stacked one on top of the other.

stadia. A built-in pair of horizontal wires spaced to subtend (bracket) a specific small angle in an optical sighting device.

stadia principle. A principle of distance determination based on using a constant angle to determine a distance between a marker and a sighting device.

stadia rod. A graduated rod used with stadia-equipped telescopes; the number of ruling units on the rod that the horizontal stadia wires span is used to compute the intervening distance by using simple trigonometry.

stadion. In Greek times, a stadion varied from 200 to 210 modern yards (182 to 192 meters).

standard deviation. A statistic used as a measure of the dispersion or variation in a distribution; equal to the square root of the variance.

standard deviational ellipse. An ellipse formed when the standard deviation of the x- and y-coordinates from their average values is used to define the major and minor axes of the figure.

standard symbol. A special type of mimetic symbol used as a standard in certain mapping practices or for certain map products.

standardized data. Data that have been adjusted to conform to a common predefined format.

star chart. A chart used to plot stellar and planetary positions throughout the year.

state plane coordinate (SPC) system. A grid coordinate system that was created in the 1930s by the land surveying profession in the United States as a way to define property boundaries that would simplify computation of land parcel perimeters and areas.

statute mile. 1,760 yards (about 1,609 meters) or 5,280 feet; the commonly used measure of land distance in the United States; sometimes referred to as a *land mile.*

stereographic projection. A planar conformal projection based on a light source on the surface of the generating globe opposite the point of tangency.

stereopair. A set of two air photos taken from slightly different vantage points; when viewed together, these photos give the impression of a three-dimensional surface.

stereoscope. The optical stereovision aid used to view photographic stereopairs.

stereoscopic polarization or projection. A technically complex way to display the terrain stereoscopically using a special computer monitor that alternates two maps at least 30 times a second. The first map is displayed with horizontally polarized light, the second with vertically polarized light. Map viewers wear special goggles with polarizing filters that allow the right eye to see only the horizontally polarized map and the left eye the vertically polarized map, so that the terrain is seen stereoscopically in the mind.

stereoscopy. Any technique capable of creating the illusion of depth in a photograph, map, or other two-dimensional image.

Strahler order. A method of stream ordering in which the uppermost stream links are assigned an order of one, and stream order increases by one when streams of the same order intersect.

stratified sample. A sampling method in which the region or population is divided into a number of distinct categories based on prior knowedge so that samples can be collected from each "strata."

street segment. The portion of a street between two adjacent intersections, or between the intersection of a dead-end street and its endpoint.

strength of association. An indication of whether spatial associations are strong or weak.

strip method. A measurement method for area estimation using a series of equally spaced parallel lines that define narrow strips spanning the feature.

subdivision. The degree to which a patch type is divided up into separate constituent patches.

subdivision plat. A map that shows subdivided lots.

submerged contour. A contour that depicts the former river channel in an area that was inundated due to damming; also called an *underwater contour.*

sun-synchronous orbit. A special case of the near-polar orbit in which a satellite passes over the same part of the earth at roughly the same local time each day.

superimposing maps. A technique that shows change by overlaying a map of one date on that of another date.

supplementary contour. A contour placed in an area where elevation change is minimal; supplementary contours are added to help delineate small features that otherwise would be missed by the contour interval used for the map as a whole.

surficial geology map. A map showing bands of differing surface rock types at the earth's surface.

survey. To use linear and angular measurements and apply principles of geometry and trigonometry to determine the exact boundaries, position, or extent of a tract of land.

survey (congressional) township. See **township.**

surveying grade GPS. GPS unit that provides the highest accuracy and is used for precision surveying applications.

surveyor's compass. A compass with quadrants from 0 to 90 degrees, with the north and south directions identified as zero degrees and the east and west directions labeled as 90 degrees.

Swiss coordinate system. The grid coordinate system used in Switzerland for land surveying and topographic mapping by the Swiss Federal Office of Topography (swisstopo), the Swiss national mapping agency.

Swiss Federal Office of Topography (swisstopo). The Swiss national mapping agency.

symbolization. The use of visual variables to represent data attributes.

symmetric link. A link in which movement occurs in both directions.

system of land rights transfer. System to convey a parcel of land without ambiguity from one owner to the next. See also **land conveyance.**

system of measurement. A set of units used to specify anything that can be measured.

systematic sample. A sample arranged so as to collect every nth item from the population; in space, a systematic sample is often set up as a rectangular or triangular grid of sample points or quadrats, or as equally spaced parallel transect lines.

Systeme Probatoire d'Observation de la Terre (SPOT). A system of sun-synchronous near-polar orbiting land resources satellites operated by the SPOT Image Corporation that includes SPOT 1 (launched in 1986), SPOT 2 (1990), SPOT 3 (1993), SPOT 4 (1998), and SPOT 5 (2002); SPOT 3, 4, and 5 are still in operation.

tangent case. The case that occurs when the projection surface touches the generating globe at either a point of tangency for planar projections or along a line of tangency for conic and cylindrical projections.

temperature compensated altimeter. An altimeter designed to compensate for the effect of temperature changes.

temperature gradient. The difference in air temperature between different locations.

temporal accuracy. A measure of accuracy that tells us how well a mapped representation reflects the temporal nature of the mapped features.

temporal change study. An analysis for a continuous surface of the variation in the same attribute at different times.

temporal stability. A measure of how long something stays the same over time.

terminal area chart. A chart at 1:250,000 scale that depicts the airspace around major airports.

terrain profile. See **profile.**

terrain surface. A three-dimensional representation of elevation data.

Thales of Miletus. A Greek scholar who used the gnomonic projection in the sixth century BC for star charts.

thematic map. A map that focuses on a specific subject and is organized so that the subject stands out above the geographical setting; examples include maps of zoning classes, land cover, vegetation zones, species ranges, and population density.

Thomas Brothers Maps. A well-known map publisher.

Thomas Guides. A grid-related product supplied through special arrangement with Trimble Navigation and Thomas Brothers Maps.

tick sheet. A scaled template of selected contour lines drawn so that slope categories can be easily identified; also called a *slope template.*

time composite map. A map that shows change over time by superimposing data for several dates.

time dilution of precision (TDOP). The effect of satellite geometry above the horizon on the accuracy of a GPS position that relates to timing of the satellite signals.

time series maps. A series of choropleth, prism, or other maps used to show changes over time.

TIN. See **triangulated irregular network.**

Tobler's First Law of Geography. All things being equal, everything is related to everything else, but closer things are more related than distant things.

topographic map. A map showing the three-dimensional nature of the terrain surface (elevations and landforms) as well as other ground features.

topographic map series. A set of topographic reference maps produced in a national mapping series, such as U.S. Geological Survey topographic maps.

topological error. Error that occurs when multiple layers of maps on which features are not in perfect registry are overlaid, resulting in small lines or polygons that are unique features when they should be part of larger adjacent features.

total covariance. The sum of the covariance for all data collection units.

total rank difference squared *(D²).* The square of the difference between the ranks for two categories summed to give a total.

totally connected network. A network in which all pairs of nodes are connected by the most direct link.

township. A 6 × 6 mile block of land arranged in rows and columns, bound by township lines on the north and south and range lines on the east and west; used in the U.S. Public Land Survey System and Canada's Dominion Land Survey; also called *survey (congressional) townships* to distinguish them from political townships.

Township and Range System. See **U.S. Public Land Survey System.**

township line. Parallel surveyed at 6-mile intervals north and south of the base line in the U.S. Public Land Survey System and Canada's Dominion Land Survey.

tract. An area with some implication for landownership or land use; also called a *parcel.*

transect line. A line segment along which statistical samples are collected.

transit map. A map used to navigate a transit system such as a subway or bus system; routes are commonly simplified and stretched to show the sequence of stops clearly.

transverse aspect. The aspect that occurs, in cylindrical projections, when the line of tangency for the projection is shifted 90 degrees from the equator so that it follows a pair of meridians (any selected meridian and its antipodal meridian).

transverse Mercator projection. A conformal cylindrical projection invented by Johann Heinrich Lambert in 1772 by rotating the Mercator projection by 90 degrees so that the line of tangency is a pair of meridians; also called the *Gauss-Krüger projection* in Europe.

travel time. The time it takes to travel by standard means between locations.

travel-time map. A cartogram that shows the travel time between locations by straight lines labeled with the time it takes to travel along each route.

traverse. See **open traverse** and **closed traverse.**

triangular distance table. A distance table in which place names appear only once; the table's upper right half is the same as the lower left half to avoid duplication.

triangulated irregular network (TIN). A representation of the land surface made up of irregularly distributed points with three-dimensional coordinates (x, y, and z) and lines that connect them to create a network of nonoverlapping triangles.

trilateration. A method for finding the location of a distant feature using only the distances from known locations to the feature.

Trimble Atlas. Supplied by Thomas Brothers Maps, a coordinate grid designed to increase the locational precision of Page and Grid references.

Trimble Navigation. A Global Positioning System (GPS) receiver vendor.

trip computer page. A screen on a handheld GPS unit that gives you detailed information about the course you are following; you can include on the page an odometer showing how far you have traveled, the length of time you were moving and stopped, your maximum speed, and your current elevation.

tripel. A German word meaning a combination of three elements; used to indicate that the Winkel tripel projection minimizes the geometric distortion of area, shapes, and distances.

TripTik. A general purpose map published by the American Automobile Association on which a planned route between a departure point and arrival destination has been marked.

TripTik Travel Planner. An online mapping application that highlights on a basemap the best route to follow between two locations.

Tropic of Cancer. The parallel of latitude where the sun is directly overhead on the summer solstice.

Tropic of Capricorn. The parallel of latitude where the sun is directly overhead on the winter solstice.

true azimuth. The angle clockwise from a true north reference line to a direction line.

true direction projection. A projection that preserves global directions; also referred to as *azimuthal projections*.

true north. The north pole of the axis of earth rotation; also called *geographical north*.

true north reference line. A line from any point on earth to the north pole.

true perspective projection. A projection created with a light source and generating globe.

true perspective conic projection. A map projection made by projecting the earth's surface onto a cone tangent to the generating globe along a small circle (usually in the mid-latitude meridian).

true perspective cylindrical projection. A map projection made by projecting the earth's surface onto a cylinder tangent to the generating globe along a great circle (usually the equator).

true perspective planar projection. A map projection made by projecting the earth's surface onto a plane tangent to the generating globe at a point.

typify. To provide a typical example of the essential characteristics of a geographic feature.

ubiquitous positioning. The next generation GPS, which is expected to work with high-level precision globally, be available at all times, and provide this enhanced service at an acceptable cost.

ultraviolet (UV). Wavelength of energy in the electromagnetic spectrum just shorter than visible light. See also **electromagnetic spectrum.**

uncertainty. The determination of what is missing in the map representation; a relative measure of the discrepancy between the environment and its representation on a map.

unclassed choropleth map. A choropleth map on which each data collection area is given a lightness, saturation, or texture proportional to its magnitude.

underwater contour. A contour that depicts a former river channel in an area inundated due to damming; also called a *submerged contour.*

uninterrupted homolosine projection. See **homolosine projection.**

union. Area covered by one or the other of two features.

unit area value. The ground area per grid cell or dot in computing the area of a region.

United States National Grid (USNG). Officially known as the United States National Grid for Spatial Addressing, this is a grid cell location system based on the UTM coordinate system and the alphanumeric grid square referencing system used in the Military Grid Reference System (MGRS).

unity. A scale factor of 1.

universal polar stereographic (UPS) system. A grid coordinate system that covers a circular region over each pole; consists of a north zone and a south zone, each superimposed upon a secant-case polar stereographic projection centered on the respective pole.

Universal Time (UT). A time scale tied to the rotation of the earth based on the apparent motion of the mean sun with respect to the meridian that runs through the Greenwich Observatory at 0° longitude; the same as *Greenwich mean time (GMT).*

Universal Time Coordinated (UTC). A time frame based on atomic clocks and aligned with the 0° longitude time zone; estimated and disseminated by the United States Naval Observatory.

universal transverse Mercator (UTM) system. A grid coordinate system extending around the world from 84°N to 80°S; the UTM system has 60 north–south zones, each 6 degrees in longitude; each of the 60 zones has its own central meridian and uses a secant-case transverse Mercator projection centered on the zone's central meridian.

urban canyon. A location such as a street in an urban landscape surrounded by very tall buildings.

USB. A plug-and-play interface between a computer and add-on devices.

U.S. Geological Survey (USGS). A U.S. government agency that is a multidisciplinary science organization focused on biology, geography, geology, geospatial information, and water. It provides timely, relevant, and impartial study of the landscape and natural resources. It is also the U.S. agency responsible for the national topographic map series.

U.S. National Geophysical Data Center (NGDC). A U.S. government agency whose mission is to provide long-term scientific stewardship for the nation's geophysical data.

USPLSS. See **U.S. Public Land Survey System.**

U.S. Public Land Survey System (USPLSS or PLSS). A regular, systematic partitioning of land into easily understood parcels prior to settlement, established through the Land Ordinance of 1785; otherwise known as the *Township and Range System.*

user segment (US). The segment of a Global Positioning System that consists of the receiver, which has an antenna tuned to the radio frequencies transmitted by the satellites.

UT1. The same as the uncorrected Universal Time from meridian observations or from more modern methods involving Global Positioning System satellites.

UV. See **ultraviolet.**

UV haze filter. A filter that corrects for UV effects on photographic film (UV effects cause images to look bluish and other colors to be modified); this filter also eliminates haze (caused by dust particles in the air) from aerial photos.

value added. A term denoting that a map or chart has been enhanced with extra features to increase its utility.

vanishing point. The point on the horizon where parallel lines appear to converge.

variable contour line density. A mapping technique in which different contour intervals are used on different parts of the same map; most commonly used when the terrain changes markedly from one region on the map to another.

variable isoline interval. An isoline interval that is not constant throughout the data range but is larger or smaller as the values increase.

variable scale bar. A scale bar stretched to match the local map scale; a variable scale bar can be used on a map on which scale varies symmetrically in both the north–south and east–west directions, as on the Mercator projection.

variable smoothing. Different levels of line generalization applied to various features or parts of features on the map.

variance. The average of the squared deviations of individual values in a distribution from their means, written mathematically as

$$var. = \frac{\sum_{i=1}^{n}(x_i - \bar{x})^2}{n}$$

variance–mean ratio (VMR). A mathematical measure that indicates if features are arranged spatially in a clustered, random, or regular manner; this measure is used only for features, not their attributes.

vehicle navigation system. A GPS- (or radio-) enabled system used in passenger cars, trucks, buses, delivery vehicles, and rental cars that provides route information.

velocity. Speed of movement.

verbal scale. A way to express map scale in commonly used map and ground units using a descriptive expression such as so many "inches to 1 mile" or so many "centimeters to 1 kilometer."

Vernier scale. A counting dial on a polar planimeter that has a graduated scale calibrated to indicate fractional parts of the subdivisions of the larger scale; allows the partial revolutions to be determined with greater precision.

vertical dilution of precision (VDOP). The effect of satellite geometry above the horizon on the accuracy of the elevation component of a GPS position.

vertical exaggeration (VE). Exaggeration that occurs when the ratio between the map's vertical scale and its horizontal scale is greater than 1. VE exists on a map when the vertical scale exceeds the horizontal scale.

vertical photograph. An air photo taken straight down when the aircraft is flying parallel to the ground; the principal point on the photo is the nadir point on the ground.

vertical reference datum. An arbitrary surface with an elevation of zero.

viewing mirror. Part of a compass used to help you determine angles in the field more accurately.

viewshed map. A map of the visible or hidden areas on the ground when viewed from a given vantage point; also called an *intervisibility map.*

virtual globe. A 3D software model representing the earth.

visibility analysis. An analysis that provides a way to see what environmental features are visible from a given vantage point. See also **viewshed map.**

visual field effect. The effect that occurs when a map symbol's appearance is modified by nearby symbols.

visual hierarchy. The visual impression that categories of features are ordered or prioritized.

visual variable. A property of a symbol that carries either a qualitative or quantitative message to the map reader. Visual variables include color hue, color value, color saturation, size, shape, orientation, pattern arrangement, pattern orientation, and pattern texture.

volume. The amount of space occupied by a three-dimensional object, calculated as area times height or depth.

VOR (VHF Omnidirectional Range) navigational beacon. A VHF (very high frequency) beacon used for aeronautical navigation. VOR beacons are usually located at or near airports. The compass roses on aeronautical charts are centered on these beacons.

WAAS. See **Wide Area Augmentation System.**

waypoint. A specified geographical location, spot, or destination defined by longitude and latitude that is used with a Global Positioning System for navigational purposes.

weight. The process of assigning a value to a feature according to its importance; a measure of the relative contribution or influence an individual makes to the whole.

weighted standard deviational ellipse. The standard deviational ellipse calculated using the locations influenced by a weight that indicates which points are more important than others; calculations are based on either Euclidean or Manhattan distance.

westerly declination. Declination that occurs when the compass needle points to the west of true north.

Wide Area Augmentation System (WAAS). A GPS augmentation system created by the U.S. Federal Aviation Administration; this system uses additional signals to increase the reliability, integrity, accuracy, and availability of GPS.

wide-area master station (WMS). A ground-based facility that calculates corrections for WAAS using data received from wide-area reference stations. A WMS sends information about satellite positions and satellite health status to WAAS satellites.

wide-area reference station (WRS). A ground-based facility that gathers GPS data used for WAAS corrections.

Winkel tripel projection. A compromise projection that is not equal area, conformal, or equidistant, but rather minimizes all three forms of geometric distortion; this map projection was constructed as a mathematical combination of two existing projections.

wireframe map. A map that resembles a net or wire mesh draped over the terrain. It is produced by combining terrain profiles drawn across the landform at right angles to each other; also called a *fishnet map.*

Wisconsin transverse Mercator (WTM) system. A special state coordinate grid for Wisconsin formed by shifting the central meridian of a UTM zone to the center of the state.

WMS. See **wide-area master station.**

world aeronautical chart. A chart at 1:1,000,000 scale that covers land areas at a standard size and scale for navigation by moderate-speed aircraft and aircraft operating at high altitudes.

World Geodetic System of 1984 (WGS 84). Ellipsoid that provides an excellent average fit for the entire earth. It is updated with vastly superior surveying equipment coupled with millions of observations of satellite orbits. Accuracy is achieved because the elliptical shape of each satellite orbit monitored at ground receiving stations mirrors the earth's shape.

WRS. See **wide-area reference station.**

WTM. See **Wisconsin transverse Mercator system.**

x-axis. The horizontal axis of a two-dimensional Cartesian coordinate system.

x-slope. The slope in the *x* (easting) direction when determining the gradient.

y-axis. The vertical axis of a two-dimensional Cartesian coordinate system.

yard. Three feet or 36 inches; decreed by King Henry I to be the distance from the tip of his nose to the end of his thumb with arm outstretched.

y-slope. The slope in the *y* (northing) direction when determining the gradient.

zero contour. The contour that corresponds to mean sea level; also called the *datum contour.*

zone. The set of cells or pixels with the same value or classification in a raster image.

zone-of-indifference weight. A composite distance weighting that combines a fixed distance component with an inverse distance component. Features within a critical distance of a feature are given a constant weight; but once the critical distance is exceeded, the level of impact declines.

Index

Related titles from ESRI Press

Designed Maps: A Sourcebook for GIS Users
ISBN 978-1-58948-160-2

Designing Better Maps: A Guide for GIS Users
ISBN 978-1-58948-089-6

Cartographic Relief Presentation
ISBN 978-1-58948-026-1

Cartographica Extraordinaire: The Historical Map Transformed
ISBN 978-1-58948-044-5

ESRI Map Book, Volume 23
ISBN 978-1-58948-193-0

ESRI Press publishes books about the science, application, and technology of GIS. Ask for these titles at your local bookstore or order by calling 1-800-447-9778. You can also read book descriptions, read reviews, and shop online at www.esri.com/esripress. Outside the United States, contact your local ESRI distributor.

To the owner of this book,

Thank you for purchasing and using *Map Use,* 6th edition. Your comments will help to ensure that future editions of *Map Use* provide the best possible text. Please take a minute to answer the following questions and return this page.

1. Did you purchase the book for a course or for personal use?

Course _____ Which course? _____

Personal use _____

2. Did the book meet your expectations? Please explain.

_____

_____

_____

_____

3. What did you like most about *Map Use,* 6th edition?

_____

_____

_____

_____

4. What would you change to make the book better?

_____

_____

_____

_____

5. Are you interested in reviewing books for ESRI Press?

Yes _____   No _____

6. Are you considering writing a book yourself? If yes, please describe briefly:

_____

_____

_____

_____

Name:

Affiliation:

E-mail:

ESRI
Attn: ESRI Press Manager
380 New York Street
Redlands, CA 92373